Water Quality Management

Water Quality Management

PETER A. KRENKEL
Water Resources Center
Desert Research Institute
University of Nevada System
Reno, Nevada

VLADIMIR NOVOTNY
Department of Civil Engineering
Marquette University
Milwaukee, Wisconsin

ACADEMIC PRESS

A Subsidiary of Harcourt Brace Jovanovich, Publishers
New York   London   Toronto   Sydney   San Francisco   1980

ACADEMIC PRESS, INC.
111 Fifth Avenue, New York, New York 10003

United Kingdom Edition published by
ACADEMIC PRESS, INC. (LONDON) LTD.
24/28 Oval Road, London NW1   7DX

Library of Congress Cataloging in Publication Data

Krenkel, Peter A.
    Water quality management.

    Includes bibliographies and index.
    1. Water quality management. I. Novotny, Vladimir,
1938-   joint author. II. Title.
TD365.K73   628.1'68   80-516
ISBN 0-12-426150-7

PRINTED IN THE UNITED STATES OF AMERICA

80 81 82 83   9 8 7 6 5 4 3 2 1

Contents

Preface

Our environment is polluted. The air we breathe is contaminated by organic chemicals and toxic metals, precipitation in the form of acid rain is falling on large portions of our planet, and heavy pollution loads reach our surface waters from urban areas and agricultural lands. Yet we are living in an environmentally conscious society, and for the last 100 years, man has attempted to eliminate or at least to control this pollution. These efforts have sometimes been successful, but they have often failed because of insufficient economic resources, lack of knowledge about the affected ecosystems, or deficient legislation.

In the late 1960s and 1970s we encountered an era of great concern for the environment. Although this widespread movement was many times poorly informed and even naive, it did produce public awareness of many environmental problems in the United States. This resulted in the most comprehensive water pollution control legislation ever passed: the Water Pollution Control Amendments of 1972 (PL 92–500) and the Clean Water Act of 1977 (PL 95–217). This legislation has affected and will affect all United States pollution abatement efforts for decades to come. This era also produced, however, the doctrinaire philosophy of "zero discharge," which is practically nonattainable.

In the early 1980s we are entering another era of environmental and social awareness that is more sober and realistic. In addition, the costs of pollution control, when combined with rising energy costs, have made commensurate social benefits mandatory. We have realized that our resources are not unlimited, and that it is impossible to eliminate all pollution reaching our surface waters. However, we can reduce pollution loads to levels where the receiving land and water resources can accept them without deleterious effects. The primary purpose of this book is to identify such levels, to recognize the consequences and impacts of pollution on receiving waters, and to propose remedial measures to correct unfavorable water quality conditions.

This treatise should become a handy reference to those working in the protection of water resources, as well as to those who want to use these resources for beneficial purposes, including waste assimilation.

A question is posed in some environmental circles as to whether we need to analyze surface water systems when point source pollution can be controlled only by imposing strict and uniform effluent limitations. The answer is clearly affirmative. For example, in 1980 a large municipality located in the midwestern United States designed and seriously presented a more than $1.5 billion pollution abatement program in which they proposed waste-sludge digestion instead of existing sludge-drying and disposal. The facility would have complied with effluent standards existing at that time. However, it would have quadrupled the ammonia load to the already overloaded receiving water. This would have led to serious water quality problems. In some other situations, strict effluent standards are not necessary and lead to waste and overuse of financial resources.

On the other hand, if pollution abatement measures are designed to comply with the nature and limits of receiving waters, remarkable success can be achieved. The River Thames in London is alive again, and recently fish have been caught after almost a century of septic summer conditions. The Stockholm harbor now has good water quality, suitable for recreation and fish, and fish propagation and significant water quality improvements have been observed in the Delaware River and elsewhere. The most outstanding water quality management system can be found in the Ruhr area of the Federal Republic of Germany, a highly industrialized area of about 10 million inhabitants. This industrial district relies on water resources of a relatively few small streams for its water supply and wastewater disposal. Yet these streams and reservoirs are managed to that level that provides satisfactory water supply, ample water recreation, and safe disposal of wastewaters.

The materials contained in this volume were gathered by the authors during many years of teaching, research, and consulting on water quality management. The framework of this treatise is based on the lecture notes of the authors for a water quality management course taught during many years at Vanderbilt University, the University of Tennessee at Knoxville and at Chattanooga, the University of Nevada at Reno, and Marquette University. Numerous students were exposed to the various topics on water quality management, and the resulting experience is contained in the text.

However, the materials presented in this book go beyond a common graduate level course. The senior author has served as the principal responsible for environmental quality at the Tennessee Valley Authority and as a consultant to many industries and to the World Health Organization. The junior author was involved for several years as a principal research engineer in water quality research in central Europe. This combined knowledge and experience was incorporated in the text, which has made the book somewhat unique, since both United States and European water quality management philosophy and knowledge have been unified and found to be quite compatible.

The book coverage is broad. The topics include the fundamentals of water quality; legal aspects; physical, chemical, and biological dimensions of water quality; water quality requirements; pollution inputs from both point and nonpoint sources; stream and reservoir water quality modeling; eutrophication; thermal pollution; and groundwater quality. The reader may also find detailed discussions on water quality parameters and characteristics; hydrologic and hydraulic aspects of water quality; mixing; simple and complex water quality models; waste assimilative capacity determination; effluent outfall design; and many other practical engineering topics.

The book is oriented primarily for practicing environmental engineers and professionals involved in pollution abatement programs including planning, management, and enforcement. It also can be used by graduate environmental students undertaking studies in water quality management and/or as a fundamental reference. In addition, many other professionals involved in water quality management or water resources problems will find the materials in this book quite useful.

Basically the book is divided into three sections. The first section deals with the fundamentals of water quality; the second section is devoted primarily to water quality modeling and systems analysis of streams, reservoirs, and estuaries. The third section discusses practical water quality topics and problems such as eutrophication, thermal pollution, outfall design, and groundwater pollution. By omitting the mathematical sections, the book may be used by the novice.

It has taken many years to write and complete this book. Therefore, it is appropriate to acknowledge those who contributed to this endeavor.

They include Gracie Krenkel and Lynn Novotny for patience and understanding; Marquette University for a general creative environment; the Desert Research Institute for logistic support; and the many students undertaking the course material, whose successes have made the effort worthwhile.

1 | *Introduction*

"Everything living is created from water" is an ancient quotation which closely describes the importance of water. Man and living nature can neither develop nor survive without water. Man has been concerned with water from the very beginning of his existence. In addition to water's being essential to his diet, it is also the means by which he can banish hunger, develop energy, drive industry, promote trade and transport, enjoy recreation, and, finally, remove and dispose of the impurities and by-products produced by his cultural activities.

The quantity of water on earth appears to be enormous when it is considered that more than two-thirds of its surface is covered by water. However, 97.2% is seawater, 2.05% is frozen water, and only 0.65% is fresh water on land and/or in the atmosphere (Anon., 1977). Also, the availability of water is a stochastic (random) phenomenon. Water may be available at the wrong place, at the wrong time, and possess undesirable quality. Many ancient civilizations flourished because of the presence of water and diminished when water became scarce or of poor quality.

Water is inequitably distributed among peoples, nations, and/or states. Although the quantity of fresh water can be slightly increased by such unconventional means as desalination or transportation of arctic glaciers, the costs of these processes are beyond the reach of many of those areas

1

that need them the most. Thus, the amount and locale of fresh water are essentially constant.

Until recently, man was concerned more with water quantity than water quality, with his interest in quality being mostly passive. During prehistoric times, man sought locations with an abundance of good quality water and avoided places with water shortages and/or poor water quality. With the establishment of permanent settlements and increasing population, the opportunities for avoiding poor water quality became limited. However, because most larger cities were located on or near a large source of water, and because most of the human waste materials were not directly discharged into the surface waters, the streams and rivers were usually able to assimilate the residual pollution.

A growing concentration of industry and an increased population density resulted in greater amounts of impurities and waste materials that reached surface waters and groundwater aquifers. The first noticeable deterioration of water quality occurred when household sewage and industrial wastes were connected to existing storm water conveyance systems in large urban areas. By the end of the last century, the Thames River near London and some other rivers near major European cities were so polluted that certain local sections of the rivers were devoid of dissolved oxygen. This condition resulted in septic conditions which were detectable by offensive odors for large distances. By the 1950s, an anaerobic reach of the Thames River, some 20 km long, persisted during several months in most years, and the smell of hydrogen sulfide could sometimes be detected several kilometers away (Gameson *et al.*, 1972). By this time, many rivers and other water bodies had reached a state where they could no longer satisfactorily assimilate the discharged wastewaters. The original freshwater biota had been eliminated and some of the receiving waters had become open sewers.

Several major European cities and industrial areas, e.g., London, Paris, Moscow, Rome, and the Ruhr industrial district in Germany, are located on relatively small streams, which brought about an earlier concern with water quality there than had occurred in the United States. The success of concentrated efforts to abate pollution in the Thames Estuary was reported in the early 1970s, when the entire estuary remained aerobic and fishing was once again a viable endeavor, even during low-flow summer periods. The dramatic improvement in the Thames Estuary is shown in Fig. 1.1 (Gameson *et al.*, 1972).

In Germany, the first efforts to control and effectively manage surface water quality date back to the end of the last century. In the early 1900s, an agency responsible for basin-wide water quality management was created for the Emscher River Basin. This was followed with the estab-

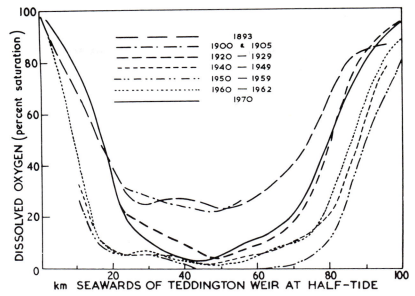

Fig. 1.1. Oxygen sag curves for the Thames Estuary (Gameson *et al.*, 1972).

lishment of six additional river basin authorities in the Ruhr area, a highly industrialized and densely populated area of the German Federal Republic.

Although the pioneering work on quantifying the waste assimilative capacity of rivers began on the Ohio River in the 1920s, large scale efforts to control wastewater discharges and receiving water quality did not start until after World War II. Even before the passage of the present comprehensive water quality legislation, the Federal Water Pollution Control Amendments of 1972 (PL-92-500), areawide water quality management was attempted and intrabasin agencies responsible for water quality management were established. The Delaware River Basin Commission is one of the best examples of a coordinated effort of four states to control the water quality of this important area. The Tennessee Valley Authority (TVA) is a federal agency responsible for water resources development, power production, and, to a minor extent, water quality control. However, its environmental efforts in recent years have been minimal because of disinterest on the part of the General Manager's office.

There are also significant differences in water quality considerations between developed and underdeveloped countries. In developed and highly industrialized countries, the present concern is with the conservation of water resources. Although pollution is increasing, the developed countries often rely both on sophisticated wastewater treatment technology and on the assimilative capacity of receiving waters to satisfactorily accept some pollution. Water used for domestic water supplies is reasonably safe from contaminants, since adequate sanitary control practices have long been established.

In most developing countries, the situation is quite different. Many of them are situated in areas where water is scarce, and others are located in areas with unregulated flow where devastating floods are followed by periods of little or no flow. The vast majority of people living in the undeveloped countries still rely on surface waters as their primary source of water and, simultaneously, their means of waste disposal. A majority of these populations must use sources that are not protected from pollution and/or contamination. Therefore, epidemic outbreaks of waterborne disease are common. It has been estimated that in the developing countries, some 5 million people die each year from enteric diseases, and 500 million people per year suffer adverse effects from such diseases.

Water quality near overpopulated large cities of the undeveloped nations has reached catastrophic proportions similar to those existing in large industrialized areas of the United States and Europe during the first half of this century. Only 7% of the 2430 large Indian cities are served by treated water, and wastewater treatment facilities are adequate in only a few instances (the population of India exceeded 547 million according to the 1971 census). For example, the Kalu and Ulnas rivers near Bombay receive practically untreated wastewaters from the city and a variety of chemical and industrial wastes (Pavanello and Mohanrao, 1972). The resulting disastrous effects can be seen not only in the rivers, which are practically septic, but also in the destruction of marine life in the estuary of Ulnas in the Arabian Sea. The river Chillwong flowing through Djakarta in Indonesia is now so polluted that the septic conditions which have developed result in a continuous obnoxious hydrogen sulfide odor.

Since 1960, serious problems with river pollution by heavy metals and other pollutants have been continuously reported in Japan. Although Japan is a highly industrialized country, population density and industrial production are such that they magnify water quality problems to levels yet unobserved in other parts of the world.

Minamata disease, Itai-Itai disease, and Yusho disease are relatively new medical terms, all originating from Japan and all the result of water pollution in a highly industrialized society.

Minamata disease is caused by methylmercury. The name results from incidents where the inhabitants of Minamata Bay were stricken by the symptoms after ingesting fish contaminated by methylmercury discharged from an industrial plant. Later incidents occurred in Niigata with similar debilitating effects. A total of 168 people were seriously affected in these areas, among which 52 deaths occurred (Krenkel, 1973). The Itai-Itai disease resulted from people's ingesting fish contaminated with cadmium, the cadmium replacing calcium deposits in the bone (Kobayashi, 1970). Itai-Itai means "Ouch-Ouch," the name being associated with the pain inflicted by the condition. The Yusho disease (oil disease) originated from the ingestion of rice oil contaminated with polychlorinated biphenyls (PCBs), a material now found in many fish in the United States.

DEFINITION OF WATER QUALITY MANAGEMENT

Water quality is not synonymous with water pollution and, similarly, water quality management should not be equated only with water pollution control. Water quality management deals with all aspects of water quality problems relating to the many beneficial uses of water, while water pollution control usually connotes adequate treatment and disposal of wastewater.

Water uses consist of intake, on site, and instream flow uses. Intake uses include water for domestic, agricultural, and industrial purposes, or uses that actually remove water from the source. On site uses primarily consist of water consumed by swamps, wetlands, evaporation from water bodies, natural vegetation, and unirrigated crops and wildlife. Finally, flow uses include water for estuaries, navigation, wastewater dilution, hydroelectric power production, and fish, wildlife, and recreation purposes (Natl. Water Commis., 1973).

Intake water may be measured in two ways: by the amount withdrawn and the amount consumed. Projected uses for the United States and the world are presented in Figs. 1.2 and 1.3. Water withdrawn is water diverted from its natural course for a beneficial use. Water consumed or consumptive use is water that is incorporated into a product or lost to the atmosphere by evapotranspiration and thus not reusable. Consumptive use is the real indicator of water demand, inasmuch as most of the nonconsumptive water can be, or is, reused. It should be noted that water reuse is thus not a new concept.

Examination of Fig. 1.2 might lead to the conclusion that there is no water shortage in the United States. However, experience in recent years has demonstrated that flow averages are misleading. For example, precipitation in the Pacific Northwest may exceed 100 inches per year while precipitation in the Pacific Southwest may be less than 1 inch per year. Proper water resource management implies that the supply and demand of

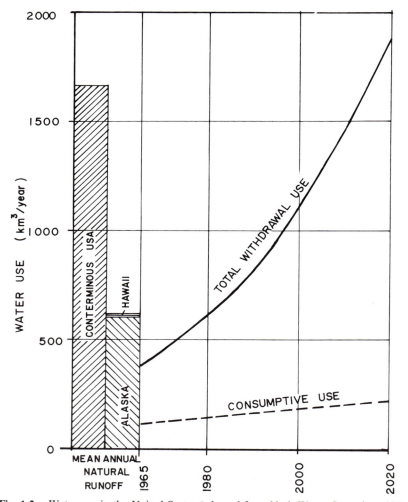

Fig. 1.2.   Water use in the United States (adapted from Natl. Water Commis., 1973).

water are equivalent. Although water supply can be regulated by storage, reservoir capacity is not unlimited, as shown by the severe shortages in California during the drought of 1976–1977 when many reservoirs were depleted.

As previously stated, the amount of water on earth is constant. It can neither be increased nor diminished. While the supply of water remains constant, the demand for water is significantly increasing. It is expected that by the year 2000, the world's population will reach more than 6 bil-

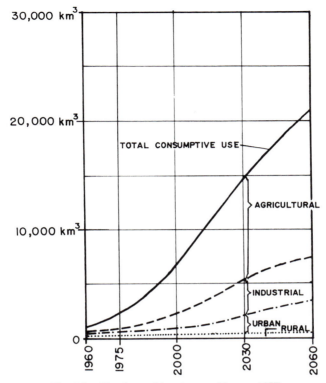

Fig. 1.3. Yearly world water use (Anon., 1977).

lion, or a doubling during one generation. This increase, along with increasingly higher living standards, will place an enormous strain on the earth's water resources. The demand for water has accelerated and will continue to do so.

Although major portions of the chemical, biological, and physical composition of surface waters still can be related to natural processes, the influence of man's activities has become significant and prevailing in many areas. Without proper control, man's cultural and economic activities can and will adversely affect all components of the environment including air, water, and land resources.

At the 1977 United Nations Conference on Water, it was noted that "We cannot rely solely on new water resources to meet our greatly increased needs" and, furthermore, that "Overuse or mining of underground sources must be avoided." Fortunately, there is an immense potential in water economy to be served by more efficient delivery and use and through the reclaiming of water bodies that once produced clean water but have since become polluted (Anon., 1977).

Water quality management thus brings a new dimension to the common water resource problem, i.e., the optimization of water quality for all beneficial uses. It is implied that water should be managed so that no use at any one location will be detrimental to its use at another location. It is thereby distinguished from water quantity management which is the engineering of water resource systems so that enough water will be provided to all potential users within the region of interest.

The water resource systems for which quantity and quality are to be managed include the following components:

Hydrologic Systems deal primarily with water quantity and quality of precipitation, melted snow, and glacier water.

Overland Flow Management includes the estimation and control of quantity and quality from nonpoint sources and their delivery to the nearest receiving water. The problem areas include storm water runoff, agricultural runoff, and land–water management systems.

Streams are the primary conveyance systems and where the most profound water quality modifications take place.

Lakes and Reservoirs, which may or not be stratified, represent significant sources of water, pollution sinks, and locations where large amounts of organic material can be produced by photosynthesis.

Estuaries are the transition zones between rivers and oceans or large lakes. Because the flow in estuaries is significantly diminished, they may represent a final disposal site for sediments and concomitant sediment adsorbed contaminants.

Groundwater Systems are an integral part of the overall hydrologic transport system. The flow of many rivers, especially during low-flow periods, is closely related to groundwater seepage and storage. Similarly, the water quality of shallow groundwater aquifers can be closely correlated with that of the surface water at the recharge zone.

Oceans represent the alpha and omega of the hydrologic cycle. Some near-shore zones of oceans near industrial and large urban centers have become highly polluted, primarily because of sludge disposal practices. In addition, several large accidental spills from oil tankers and oil wells have had a deleterious effect on marine life and shoreline aesthetics.

In managing water quality, the factors and inputs that must be considered are not restricted to man-made sources. Natural causes of water quality changes such as geologic formations, vegetation, geographic factors, and natural eutrophication also are of import.

The investigation and management of water resources systems for

[handwritten margin note: Most water qual. modifications here!]

water quality must include consideration and evaluation of (a) the physical, chemical, and biological composition of headwaters and significant groundwater discharges; (b) water quantity and quality requirements for all existing and potential water uses; (c) the means of water withdrawal and their effect on water quality and quantity; (d) the existing and future water and wastewater treatment technology used to alter water quality; (e) the wastewater outfall configuration and effluent mixing; (f) the eutrophication status of the receiving waters; (g) the waste assimilative capacity of the receiving waters; (h) the ecological changes that might be caused by wastewater discharges; and (i) the potential effects of the discharge of heated waters.

POLLUTION AND WATER QUALITY MANAGEMENT

The question immediately arises as to what constitutes pollution and whether the term pollution is synonymous with water quality. As previously stated, water quality is a reflection or response of water composition to all possible inputs and processes, whether natural or cultural. The Latin word *pollutus* means to soil or defile, while Webster's dictionary defines pollution as the state of being "physically impure or unclean, befoul, dirty, or taint."

Pollution can be defined as the addition of something to the water which changes its natural quality so that the downstream riparian proprietor does not obtain the natural water of the stream transported to him.

The state of California (McKee and Wolf, 1963) differentiates between pollution, contamination, and nuisance as follows: Pollution adversely and unreasonably impairs the beneficial use of water even though no actual health hazard is involved. Contamination causes an actual hazard to public health, and immediate action can be taken by the public health agency. A nuisance is the damage that results from odors or unsightliness caused by what the law terms "unreasonable practices" in the disposal of sewage or industrial wastes.

The World Health Organization (WHO) considers waters to be polluted *Definition* when they are altered in composition or condition, directly or indirectly, as a result of man's activities so that they become unsuitable, or less suitable, for any or all of the functions or purposes for which they would be suitable in their natural state.

Examples of pollution include the addition of a substance that by itself or in combination with other substances can cause toxicity, the addition of organic substances that can cause dissolved oxygen depletion and subsequent modification of the aquatic biota, the raising of water temperature, and the excessive addition of fresh water to seawater which may cause a sudden change in the osmotic pressure within marine organisms' cells and lead to their death. Significant shellfish kills have been reported when flood waters were diverted into brackish bays or estuaries. Figure 1.4 shows reported cases of fish kills caused by pollution, although the numbers are somewhat misleading. For example, in 1961, little interest was demonstrated in fish kills and only several states submitted reports, while later data reflect an avid interest in fish kills and reports from all 50 states.

Pollution should not always be related to wastewater discharges, inasmuch as, in some circumstances, a wastewater can be discharged into a water course without measurably deteriorating water quality and without damaging water use (Kneese and Bower, 1971). The use of the term pollution usually connotates damages associated with water quality alterations. On the other hand, activities which are not associated with

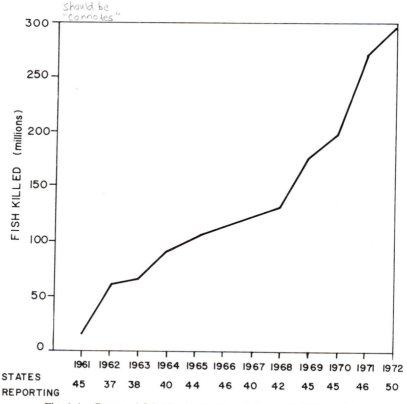

Fig. 1.4.   Reported fish kills in the United States (USEPA, 1975).

wastewater discharges can cause significant and sometimes catastrophic conditions.

The diversion of water and a reduction in flow could be considered pollution because the streams' capability to assimilate wastes is reduced. Cutting down a forest significantly increases the sediment and nutrient loadings to streams and lakes and may result in deterioration of water quality. Stream channelization and the removal of trees along the river banks may have significant ecological effects, possibly causing destruction of the indigenous biota in the stream. In addition, removal of stream bank trees and vegetation may cause a significant increase in water temperature. An example of damages to a stream or a lake by stream channelization has been reported in Florida (Patrick, 1975). Prior to channelization, the Kissimmee and Taylor creeks entering Lake Okeechobee were able to satisfactorily assimilate all of the organic wastes discharged into them. Subsequent to channelization, the organic wastes discharged were quickly deposited into Lake Okeechobee which resulted in a severe increase in the nutrient level and concomitant anaerobic conditions in the lake.

Another way in which man can cause a deterioration of water quality is by overintensive agricultural activities and farming. Excessive fertilization causes nutrient loadings in excess of what would occur under natural conditions. For example, livestock feedlots represent a most significant source of pollution which can be equated in magnitude to the pollution generated from developing urban areas.

Point-source pollution can be controlled; however, problems exist with the control of so-called "nonpoint-source" pollution. It should be noted that the control of pollution implies the regulation of wastewater discharges originating from individual outfalls and nonpoint sources. The purposes of all pollution control endeavors should be (1) to protect the assimilative capacity of surface waters, (2) to protect shell- and finfish and wildlife, (3) to preserve or restore the aesthetic and recreational value of surface waters, and (4) to protect humans from adverse water quality conditions (Patrick, 1975).

THE ECONOMIC IMPACT OF WATER QUALITY

As previously indicated, water is a resource necessary for the sustenance of life. Not only would life be nonexistent without water, but also economic development would cease. In addition, the water must be of adequate and/or acceptable quality.

The economic value of water cannot be directly measured using traditional economic approaches because it does not have a direct market value, and users usually do not have a choice of where the water will be delivered or at what price. Under normal circumstances, the price of water is determined by the cost of alterations necessary to make it suitable for the intended beneficial use. Only in times and places of shortage does a disproportion between the supply and demand cause an increase in the price of water, with the purpose being to limit demand. For example, the price of water rose significantly during the 1976–1977 drought period in portions of California.

Deteriorating and unacceptable water quality has the same impact on the economy as water shortage. Again, the economic value of quality cannot be defined as a free market value based on the demand and supply. It is difficult to imagine that a water user would have the option to buy water at different levels of quality, and the demand for a given water quality would then determine its value. It must be concluded that, fortunately or unfortunately, the interactions of market forces do not perform satisfactorily in the cases of water quality control or wastewater disposal. The doctrinaire approach of economists may dictate the rationale for water pollution control. However, at the present time, an upstream polluter is not directly affected by the deterioration of water quality caused by his discharge unless his intake is located downstream from his outfall. Economists refer to water pollution as "technological external diseconomies" (Kneese and Bower, 1971). This may be defined as a situation where a particular action, e.g., a wastewater discharge, produces economic changes such as higher cost or less valuable products, and the cost is transferred from the decision unit-polluter to a managerially independent unit downstream user. The transfer is by a technical or physical linkage between the production processes and not by a market transaction.

In the absence of market forces to determine the value of water quality, the value must be determined by institutional and/or legislative means. The value of water quality determined in this manner is based on an overall comparison of alternatives with and without pollution or adverse water quality. Knowing what the overall economic impact of "good" water quality conditions would be as compared to "bad" or existing water quality conditions, the delegated authorities and/or public must evaluate the benefits, both monetarily and socially, associated with improved water quality conditions. However, the decision as to what level of water quality is acceptable is not based only on benefits.

The economic impact of water quality or pollution depends on such factors as water availability, water use, and the public perception of water quality. Even minimal contamination may have a devastating impact on a water supply source for a community. On the other hand, water used for navigation can tolerate much higher levels of pollution with only marginal economic impact. Similarly, biodegradable organic pollution can have a significant effect on fish and shellfish production but only a marginal or even beneficial effect on water used for irrigation.

The goal of "zero" pollution is still a distant utopia, even in the most advanced countries, let alone the less developed ones. One must always consider the enormous cost of improving and maintaining water quality, especially when extremely high removal efficiencies are required.

In a classical economic analysis, the efficiency of a project or an action is evaluated by comparing the profit with the cost. As previously mentioned, this approach may not be feasible with water quality. Although the costs associated with water quality control can usually be quantified, the evaluation of benefits attributed to water quality improvement is quite difficult. Many times, the benefits cannot be expressed in monetary terms or even be given a value. This is particularly true when high degrees of waste load reduction are required.

For example, changing the quality of a water body from septic to aerobic conditions may require a relatively low-capital expenditure but demonstrate dramatic economic improvement by establishing a viable fishery. Obviously, because of the low cost and easily demonstrable benefits, the economic efficiency of this low-degree treatment is high. On the other hand, the removal of an exotic, allegedly carcinogenic organic material, occurring in trace quantities, may have a long-term favorable public health effect, but require very expensive treatment technology. In this case, an apparent low-economic efficiency indicator will be obtained.

The economic value of water quality improvement also depends on the economic state of the area being considered. Inasmuch as the costs of pollution control are high, it is obvious that basic needs such as food and housing will have a higher priority in developing nations than in highly developed ones. Thus, the value of water quality enhancement will increase with an increase in the standard of living. When society is able to spend more time in the pursuit of recreation and nature's wonders, the economic value of improving water quality and its concomitant benefits increase dramatically. In addition, this advanced society will be more aware of water quality and the potential adverse public health effects of poor water quality. Water quality enhancement in advanced societies may have the following economic impacts: (a) increased income by providing jobs associated with an improved recreation industry, (b) reduced cost of treatment for various uses of water, (c) increased value of property located near the water, (d) expanded economic development of the region, (e) decreased damages to health and property associated with adverse water quality conditions, (f) increased income to commercial and sport fisheries because of increased productivity, and (g) more jobs and income to industries associated with pollution control.

Obviously, society must dictate the level of water quality that will yield desirable economic impact and the level of pollution that is acceptable. It is interesting to note that the economic value of good water quality can usually be correlated with the economic standard of the society and, to a lesser degree, to such factors as population density, water availability, and the natural beauty of the region.

LAND USE AND WATER QUALITY

The concept of correlating water quality with upstream land use is not a new one. However, its importance was recognized only after it was realized that water pollution could not be remedied by removing only portions of settleable solids and biodegradable organics from wastewater outfalls. In fact, the beneficial effects of conventional treatment technology applied to these effluents may only be marginal when they are discharged into lakes and reservoirs. In many cases, the nutrients causing proliferation of algae and acceleration of eutrophication originate from nonpoint sources.

The problem of land use and its effect on water quality is associated with urban and agricultural developments. Spreading and uncontrolled urban and watershore developments can easily result in a lowering of water quality. Bare soils in areas under development give rise to sediment washout into receiving waters and the subsequent formation of sediment deposits. Over 100 tons of sediment per year can be lost from 1 hectare of developing urban land. In addition, such sediments may carry large amounts of organics, nutrients, heavy metals, and pesticides. The imperviousness of urban areas increases runoff; thus even small rains may wash significant amounts of pollutants into surface waters.

Septic tank systems, although primitive and inefficient, are still the only disposal systems available to some 30% of the population in the United States. This practice will continue in the foreseeable future since the sewering of widely dispersed populations is simply not economical (Lienesch, 1977). Although many such systems are improperly designed, located on unsuitable soils or in areas with a high groundwater table, and are failing after only a few years of operation, the principle of subsurface household disposal systems should not be eliminated. Adequately designed subsurface disposal systems keep pollutants out of surface waters.

However, it should be noted that subsurface disposal systems in dense lake shore or river bank developments are often the primary source of nitrates, ammonia, coliform bacteria, and other pollutants in the adjacent surface waters because of the higher mobility of these materials in some soils and shallow groundwater aquifers.

The uncontrolled "urban sprawl" may have other adverse but indirect impacts on water quality such as the loss of valuable farm and forest lands possessing lower pollution potential, overtaxation and general deterioration of urban centers, overloading of treatment facilities, and a lack of services in the suburbs. Nonurban land uses such as farm barnyards and animal feedlots, as well as farming on impermeable soils without adequate erosion control, can also produce excessive amounts of pollution.

The control of pollution related to land use can result from short-term managerial practices or long-term planning. Simple zoning without evaluation of the consequences is not adequate. A planning process which will reduce pollution from lands must include factors such as the quantification and qualification of water resources in the area, the determination of the extent of pollution originating from nonpoint sources as compared to that from effluent outfalls, soil characterization, and the adequacy of existing and planned wastewater treatment facilities. Such a planning process must be conducted by an authoritative planning agency or commission. This entity not only must have adequate technical, economic, and manpower resources to conduct the planning process, but also it must have sufficient authority to convey the results to the local, state, or federal agencies responsible for implementation.

There are five general approaches to land use control related to water quality (Lienesch, 1977): zoning control, critical area protection, environmental impact review, property acquisition, and taxation and charges.

Since 1926, zoning has been the principal means of land use control in the United States. The basic principle of zoning is to divide the land into districts in which certain uses are permitted. The rules used in zoning to protect water quality include specifying minimum lot sizes where septic tanks are to be used, reducing storm water runoff by restricting population density, and limiting developments on areas with poor soils, sloping, and hilly contours and flood plains. In addition, developments in areas of natural beauty may be limited and open spaces may be required as buffer zones between urban developments and water courses.

Some states have authority to acquire land to protect prime agricultural areas and to maintain open space and areas of natural beauty. These programs can also be used to protect water quality. Taxes and charges have been applied to the use of land in ways that are applicable to water quality control. For example, many states use preferential taxation schemes to help preserve farmland. Such systems can be used to encourage nonde-velopment for water quality enhancement purposes, such as buffers along rivers or lakes.

In the past, the legal basis on which designated planning agencies con-trolled land use for water quality improvement relied on local and state laws and zoning ordinances. However, with the passage of Public Law 92-500, a new dimension to land use planning has evolved through Section 208, the Act's areawide waste treatment management planning provi-sions. Under this section, the governor of each state designates his state's "Waste Treatment Management Areas" and appoints an organization for each area that can develop effective areawide waste treatment manage-ment plans for that area. The implications of Section 208 are such that water quality management practices will dictate land use planning. As stated by Billings (1976), the major architect of the Act, "What Congress intended was that development in all areas of every state should be re-viewed in the context of water quality."

REFERENCES

Anon. (1977). *Aqua, United Nations Water Conference* **1**(1), 7.

Billings, L. G. (1976). The evolution of 208 water quality planning. *Civ. Eng.* 54–55 (November).

Gameson, A. L. H., Barrett, M. J., and Shewbridge, J. S. (1972). The aerobic Thames Estuary. *In* "Advances in Water Pollution Research" (S. H. Jenkins, ed.). Pergamon, Oxford.

Kneese, A. V., and Bower, B. T. (1971). "Managing Water Quality: Economics, Technology, Institutions," 2nd ed. Johns Hopkins Press, Baltimore, Maryland.

Kobayashi, J. (1970). Relation between the Itai-Itai disease and the pollution of river water by cadmium from a mine. *In* "Advances in Water Pollution Research," Vol. I, Pergamon, Oxford.

Krenkel, P. A. (1973). Mercury in the environment. *Crit. Rev. Environ. Control* **3**(3), 303–373.

Lienesch, W. C. (1977). Legal and institutional approaches to water quality management planning and implementation. Contract No. 68-01-3564, U.S. Environ. Prot. Agency, Washington, D.C.

McKee, J. E., and Wolf, H. W. (1963). "Water Quality Criteria," 2nd ed. State Water Pollut. Control Board, Publ. No. N3A, Sacramento, California.

National Water Commission (1973). Water policies for the future. Rep. to Pres. U.S. Gov. Printing Office, Washington, D.C.

Patrick, R. (1975). Some thoughts concerning correct management of water quality control. *In* "Urbanization and Water Quality Control" (W. Whipple, Jr., ed.). Am. Water Resour. Assoc., Minneapolis, Minnesota.

Pavanello, R., and Mohanrao, G. J. (1972). Consideration of water pollution control problems in developing countries. *Proc. Int. Conf. Int. Assoc. Water Pollut. Res., 6th, Israel.*

U.S. Environmental Protection Agency (1975). Fish kills caused by pollution in 1975. Office Planning and Standards, Monitoring and Data Support Div., Washington, D.C.

2 | *Legislation*

INTRODUCTION

Before proceeding with the technical aspects of water quality management, it is appropriate to discuss the rules and regulations governing water quality control and the climate under which they were developed. The passage of the Federal Water Pollution Control Amendments of 1972 (FWPCA, PL-92-500) marked the culmination of efforts, spanning almost eighty years, to control water pollution. It is the most comprehensive environmental water quality control program ever formulated by the United States Government and knowledge of its development and promulgation is essential.

Proper application of the law requires knowledge of several controversial issues, as well as the philosophies of their proponents and opponents. For example, the current statutory requirements for wastewater treatment are based on what can be accomplished technologically rather than what water quality requires, as dictated by its intended use. Opponents of this philosophy claim that in many stream segments, overprotection and excessive costs will result, while proponents insist that technology-based effluent standards are the only enforceable ones.

As will be subsequently demonstrated, the emphasis on water pollution

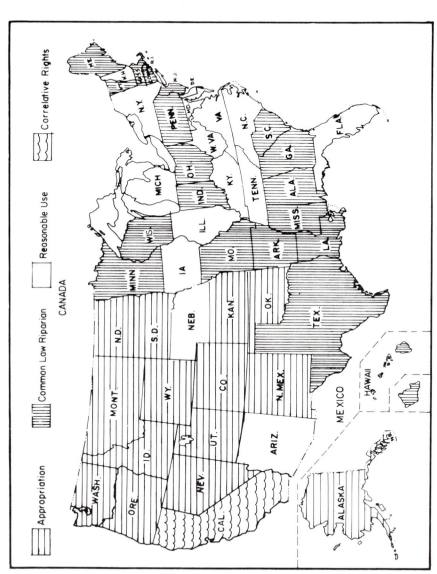

Fig. 2.2.   Groundwater summary of water laws (Beuscher, 1967).

Because rivers are easier to quantify than other water categories, laws concerning them have evolved more rapidly. It should be noted that all underground water is considered to be percolating until the existence of a stream is proven. Then, the same laws applying to a surface stream are applicable to subsurface flow (Ohio Legis. Serv. Commis., 1955).

Riparian Water Laws

Nearly all of the states east of the Mississippi River follow the Riparian Doctrine. The key features of this law of water rights are summarized here.

The owner of land adjacent to a stream is entitled to receive the full natural flow of the stream undiminished in quantity and unimpaired in quality. The riparian landowner has a legal privilege to use the water at any time, subject only to the limitation that the use be reasonable. The right is a natural right that can be transferred, sold, or granted to another person as property (Ohio Legis. Serv. Commis., 1955).

During times of water shortage, all riparian owners have equal rights to the reasonable use of water and the supply is shared, although domestic use may have preference over "lower" uses.

The Riparian Doctrine is based on Code Napoleon, which was taken from Roman Civil Law. It was part of the early law of the original states and not a part of English Common Law (Ohio Legis. Serv. Commis., 1955). It basically presumes that there is an abundance of water and, as a consequence, users under the Riparian rule lack an adequate record-keeping system. Thus, public planners and private water resource consultants and investors are usually confronted with uncertainties in water resource development, such as a lack of information on existing and potential demand or supply. These uncertainties have not yet caused major problems in the East where water supplies have been abundant. However, if demand increases as the projections indicate, greater stability will be required, e.g., by enactment of permit legislation. In the permit system, the prospective riparian owners and users are given permits to use water in accord with specified conditions of water use. Permit legislation has been adopted in many midwestern states.

It should be noted that riparian rights do not allow transfer of water from one basin to another. In addition, natural lakes have the same status as streams. Also, riparian rights are transferred to the new owner when property is sold and are not affected by use or nonuse.

In summary, riparian landowners and users are protected from withdrawals or uses of water that unreasonably diminish its quality or quantity. Where diversions or uses have been shown to be unreasonable, the riparian owners adversely affected have been compensated, or the uses have been enjoined.

Appropriation Water Laws

The system of water law adopted by most western states is known as the law of appropriation, which is best stated as: ''First in time is first in right.'' The basic tenets of the system are (1) a water right can be acquired only by the acquiring party diverting the water from the water course and applying it to a beneficial use, and (2) in accordance with the date of acquisition, an earlier acquired water right shall have priority over other later acquired water rights. Water in excess of that needed to satisfy existing uses is viewed as unappropriated water, available for appropriation and application of the water to a beneficial use. The process of appropriation can continue until all of the water in a stream is subject to rights of use through withdrawals from the stream. In times of shortages, the earliest claimants take full share, and others may do without water. If the right is not used, it is lost. In addition, the right is not identified by ownership of riparian lands.

Several problems are connected with this doctrine. The appropriation doctrine of the West made it virtually impossible to preserve instream values or to acquire a water right pursuant to a diversion if the intended use was not for an economic purpose. Hence, neither instream values nor out-of-stream noneconomic values could be protected. The following examples may assist in understanding the dilemma (Natl. Water Commis., 1973).

Since water rights only can be acquired by the diversion of water from a stream, a resort could not protect a beautiful waterfall which attracted visitors and guests. Such natural waterfalls are instream values. Or, if the flow in some streams was augmented to improve water quality, the released water could be diverted for use by others, thus frustrating the purpose of the release by impairing water quality. Since state statutes generally appeared to equate beneficial use with economic use, even diversion of water for such purposes as to protect wildlife or to develop a waterfowl marsh would not qualify, and water for such purposes could not be acquired.

Under the appropriation doctrine, there is no natural flow motion (Beuscher, 1967). The appropriators can take as much water as they are entitled to, even though it exhausts the water course. Some western states, however, permit the states to file for, and ultimately acquire, a right to the unappropriated flow and thus preserve such flow, if desired. It should be noted that an appropriator can store water in reservoirs in accordance with a storage appropriation; however, appropriations for direct use and storage may be kept separate.

Groundwater

As previously noted, groundwaters are usually classified as either subterranean streams or percolating water. Since proven underground streams are subject to the rules of surface streams, it remains only to discuss so-called percolating water.

Percolating waters have been governed by the legal maxim *Cujus est solem ejus ad collum et ad inferno,* or "He who owns the soil owns from the heavens to the depths" (Busby, 1954), which is known as the English rule. Thus, unless a groundwater can be shown to be a stream, the landowner has no limitations on its use. However, some states have applied the reasonable use rule: *Sic utere tuo et alienum non Laedas,* or "So use your own as not to injure another." In either case, the user is usually charged with a mandate not to pollute percolating waters.

California established the principle of Correlative Rights that holds that an owner can only have as much water as will satisfy his "beneficial use" needs. Furthermore, in times of shortage, the water is allocated on the basis of each owner's equitable portion.

For an excellent discussion of water use laws, the reader is referred to Ohio Legis. Serv. Commis. (1955). Figure 2.3, which was taken from that work, presents a schematic summary of the various principles presented. Because each state has its own modifications, individual state statutes should be consulted.

AMBIENT STANDARDS VERSUS EFFLUENT STANDARDS

Probably no other issue has caused more controversy in water pollution control than whether to base standards on what exists in the river or what is emitted from the pipe. It is interesting to note that while effluent standards are promulgated by PL 92-500, in so-called water quality limiting cases, ambient or stream standards are utilized.

Stream standards are based on the establishment of threshold values for a particular contaminant which, in turn, are founded on the intended use of the water. On the other hand, effluent standards dictate the amount of a

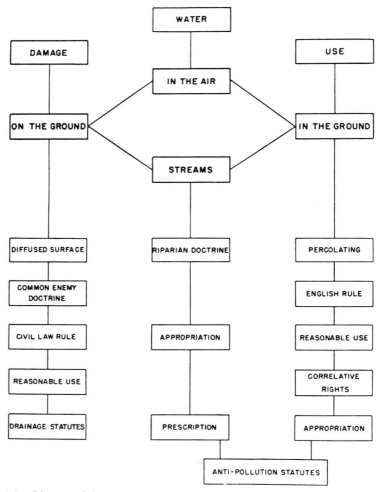

Fig. 2.3.   Diagram of classes of water and water laws (Ohio Legis. Serv. Commis., 1955).

contaminant that can be discharged from a pipe regardless of the size of a river or the intended use of its water.

Both standards have two additional classifications. Stream or ambient standards can be based on dilution requirements to maintain a certain concentration of contaminant or standards measured in the river which depend on a particular value of a contaminant based on the intended use of the water. The former is no longer popular and is not used in the United States except for "rule of thumb" purposes. Effluent standards can be based on either concentration and/or total mass discharged or on the degree of treatment required in order to maintain a specific waste load allocation.

The opponents of stream standards claim that the minimum permissible quality of water tends to become its maximum quality because dischargers will attempt to use the least possible treatment that will satisfy the standard. In addition, stream standards tend to become fixed values rather than judgment guides and may result in arbitrary decisions. If stream classification is involved, the future use of the stream may be fixed in advance. However, probably the most valid criticism of stream standards is that they are difficult to enforce. Two questions immediately arise. How are the rivers' waste assimilative capacity assets to be divided among several dischargers? How is the violating waste discharger found when the stream standards are exceeded?

On the other hand, a major advantage of stream standards is that they take into account the receiving water's natural ability to assimilate some waste materials without causing deleterious effects. Thus, a discharger located on a large stream is not unnecessarily penalized as is the case with effluent standards. It follows that the degree of treatment imposed on a small industry with a small waste volume is less than would be required for a large installation of the same type and in the same vicinity. Provision can also be made for regulation revision as conditions require and with less administrative difficulty as is the case with effluent standards. The prime argument for stream or ambient standards probably is the fact that "treatment for treatment sake" is avoided; thus the consumer and/or taxpayer has minimized unnecessary costs.

CRITERIA VERSUS STANDARDS

The differentiation between these terms has caused continuous controversy, and the establishment of the new water quality criteria and the toxic substances criteria under Section 307(a) of PL 92-500 has exacerbated the controversy. The prime source of contention is that criteria, once announced and published, tend to be "engraved in stone" and are accepted as standards by the regulatory agencies.

McKee (1960) aptly presented the differences between standards and criteria as follows:

> The term standard applies to any definite rule, principle, or measure established by authority. The fact that it has been established by authority makes a standard somewhat rigid, official, or quasi-legal; but this fact does not necessarily mean that

the standard is fair, equitable, or based on sound scientific knowledge, for it may have been established somewhat arbitrarily on the basis of inadequate technical data tempered by a cautious factor of safety. Where health is involved and where scientific data are sparse, such arbitrary standards may be justified.

A criterion designates a means by which anything is tried in forming a correct judgment concerning it. Unlike a standard, it carries no connotation of authority other than that of fairness and equity; nor does it imply an ideal condition. When scientific data are being accumulated to serve as yardsticks of water quality, without regard for legal authority, the term criterion is most applicable.

McKee also presented a case for a water quality objective, which "represents an aim or a goal toward which to strive, and it may designate an ideal condition". It is interesting that several "objectives" are included in PL 92-500.

THE DEVELOPMENT OF WATER POLLUTION CONTROL LEGISLATION IN THE UNITED STATES

A substantial body of administrative rules and regulations govern all environmentally related projects including those dealing with or affecting water quality. Since adverse water quality or polluted waters are at least a public nuisance, if not a public health hazard, various classical court litigations and rulings have evolved between the polluters as defendants and downstream water users as plaintiffs. Most of the rulings in these cases treated pollution as a nuisance, trespassing, or negligence.

In recent decades, the interest in protection of the environment has created a new interest in water quality among lawyers, judges, and legislators. Special courses on environmental law are now taught in most law schools, and a new era of attorneys study recent landmark decisions dealing with water quality and the protection of the environment. Many law firms now specialize in handling environmental problems.

The term environmental law primarily refers to that body of law which seeks to regulate individual, corporate, and governmental behavior in order to prevent adverse environmental consequences (Wengert, 1975). It is a form of social control having as its objective the preservation of environmental values such as ecological balance, natural beauty, public health, and resources productivity, in order to maintain a stable and satisfactory level of living and quality of life for present and future generations. It is based on legislative enactments as well as judicial law resulting from specific litigations (Wengert, 1975). The primary objects of environmental law are the individual, firm, or government agency whose actions may be alleged or construed to pose a threat of damage to the environment.

Events Prior to 1948

The control of water pollution in the United States has been a most pervasive problem in which probably the most difficult problem has been the traditional battle between "states rights" and federal power. Even today, most states are still jealous over their roles in the control of water pollution and believe that if they were left alone, they would ultimately solve the nation's water quality problems. In this regard, it is instructive to note that until 1948, legal authority for water pollution control was almost exclusively vested in the states and localities, and most states had a responsible agency, although the legal powers of such agencies varied greatly. Until the beginning of federal input into the problem in 1948, a major obstacle had been funding for Publically Owned Treatment Works (POTW). It is not surprising that people are not willing to devote funds to something they cannot see. The average person has no comprehension of the underground network alone and the concomitant expense that is required for an adequate sewerage works.

The genesis of the federal water pollution control program was in the passage of the Public Health Service Act in 1912, by which the Public Health Service (PHS), then a part of the United States Treasury Department, was authorized by the Congress to investigate pollution in navigable waters. As previously mentioned, this interest was instilled by the spread of waterborne disease rather than an interest in improving water quality. The only other major federal water pollution control legislation passed until 1948 was the Oil Pollution Act of 1924, which was designed to prevent the discharge of oil into navigable waters and coastal areas. Enforcement of the Act was vested in the Department of the Army and the United States Coast Guard. It was not too effective, however, because of the necessity to prove "gross negligence" on the part of the offender, which is most difficult.

The Water Pollution Control Act of 1948 (PL 80-845)

The first major federal water pollution control legislation, passed in 1948, was effective for five years, but was extended for three more years in 1953. The Act established some authority for the federal government to have a role in abating interstate water pollution control and had a provision for POTW construction grants. However, the loan program was never funded (Davies, 1970). A significant accomplishment of the Act was the establishment of the Robert A. Taft Sanitary Engineering Center in Cincinnati, Ohio. Also, the PHS established a Division of Water Pollution Control, whose philosophy was that the initiative for water pollution control should rest with the states and not with the federal government. A somewhat decentralized organization resulted because of the need for field offices that were to work closely with the state health departments. It is significant to note that all subsequent water pollution control legislation has been amendments to this Act.

The Water Pollution Control Act Amendments of 1956 (PL 84-660)

In 1956, the first permanent water pollution control legislation was passed by the Congress. This Act still promoted jurisdiction by the states and limited action to interstate waters, which were defined as waters which flow across, or form a part of, the boundaries between two or more states. The Act provided for a conference among interested parties, a public hearing if no action was taken after six months, and then court action after waiting another six months. A major shortcoming was that state consent or a Governor's request had to be obtained prior to the commencement of any enforcement activities. Other provisions of the act included the establishment of research fellowships, construction grants for POTWs, a water pollution control advisory board, and administration of the Act by the Surgeon General of the PHS.

The Water Pollution Control Act Amendments of 1961 (PL 87-88)

The 1961 amendments reflected unhappiness on the part of the Congress with the stature of the program within the Department of Health, Education, and Welfare (HEW) and transferred administration of the Act to the Secretary of HEW who also replaced the Surgeon General as the Board Chairman of the Water Pollution Control Advisory Board. The Act also provided for the establishment of seven additional laboratories for water pollution control and an increase in the authorization for the construction grants program. Grants to states for water pollution control programs were increased, as were monies for fellowship grants and demonstration and research grants. Probably the most significant change was that enforcement provisions were extended to navigable waters. In addition, under certain conditions, the Attorney General of the United

States was authorized to file suit against polluters on behalf of the United States.

The Water Quality Act of 1965 (PL 89-234)

Evidently, the Congress was still not happy with the progress being made in water pollution control, and the discontent was promoted by Senator Edmund Muskie, Chairman of the newly formed Senate Subcommittee on Air and Water Pollution. Commencing in 1963, Muskie's efforts to introduce drastic changes in the law resulted in the 1965 Water Quality Act.

Increases were authorized in construction grants, authorization was given to seek remedies for stormwater overflow pollution, regional POTWs were encouraged, and additional monies for research and demonstration were authorized. In addition, water pollution control was given a new status by its transfer out of the PHS and directly under an assistant secretary of HEW. Its new name was the Federal Water Pollution Control Administration (FWPCA).

Probably the most important part of the Act was the requirement that each state must submit, for approval by the Secretary of HEW, mandatory water quality standards applicable to interstate waters or portions thereof. If a state failed in this endeavor, the Secretary could establish the standards, subject to review by a hearing board.

Eight months later, in accord with a new reorganization plan submitted to the Congress by President Lyndon Johnson, the newly created FWPCA was transferred out of HEW and placed in the Department of the Interior.

The Clean Water Restoration Act of 1966 (PL 89-753)

This Act removed the previously existing dollar limit on construction grants, provided for grants to states for planning agencies involved in comprehensive water quality control and abatement plans for a basin, increased grants for research and development, and emphasized demonstration grants.

In addition, specific areas of research were delineated for support, including estuaries, eutrophication, and thermal pollution. Provision for the training of state personnel was also made. A significant section transferred responsibility for the Oil Pollution Act of 1924 from the Secretary of the Army to the Secretary of the Interior.

The Water Quality Improvement Act of 1970 (PL 91-224)

Considerable debate between the House and the Senate over proposed revisions concerned with oil pollution, thermal pollution, and enforcement methodology delayed the passage of this legislation and resulted in many deficiencies in the Act. It did not impose a clearly defined level of treatment, utilized only water quality standards for enforcement, and still required the consent of the governor of the state where pollution originated before enforcement procedures on interstate waters could be instigated. Thus, another mechanism for improving the quality of the nation's waters was sought. It should be noted that the title of the FWPCA was changed under this Act to the Federal Water Quality Administration (FWQA), but still remained in the Department of the Interior.

The Refuse Act of 1899 (33 USC 407)

Because of the inadequacies of water pollution control enforcement proceedings under existing legislation, President Richard Nixon, in December of 1970, announced a new program to control water pollution from industrial sources. This program evolved around the permit authority contained in the Refuse Act. While the Refuse Act was an archaic statute designed to protect navigation and not water quality, it has several provisions that enhance enforcement procedures.

In essence, the Refuse Act states that it is unlawful to place any material, except sewage and runoff, into a navigable waterway or tributary thereof without a permit from the Department of the Army. Furthermore, penalties were attached that were enforceable. An interesting proviso, known as a *qui tam* action, gave impetus to the private citizen to seek out violations. The action's name originates from the Latin, *Qui tam pro domino vege quam pro se ipso sequitar,* or "He who brings the action as well for the king as for himself." Thus, one-half of the fines were to be paid to the person(s) giving information leading to conviction.

The United States Supreme Court in 1966 upheld the applicability of this law to pollution control, and thus the Corps of Engineers permit program was viable. The act was used to obtain an injunction against the Florida Power and Light Company to abate the discharge of heated waters into Biscayne Bay. It was also used to control the discharge of mercury from chlor-alkali plants during the so-called "mercury crisis." Before issuing permits, however, tacit approval was to be obtained through review by the Environmental Protection Agency (EPA), which was formed in 1970.

While these procedures appeared to be helping to improve water quality, they did not apply to municipal sewage nor runoff, and a court decision in 1971 stopped the issuance of permits and invalidated the program. Thus, it was clear that new legislation was needed.

At this point, it should be noted that yet another change, previously alluded to, had been made in the administration of water pollution control activities. On July 9, 1970, President Richard Nixon presented to the Congress Reorganization Plan Number 3 of 1970, which transferred the FWQA, along with other environmental agencies, into a newly created Environmental Protection Agency. This move finally gave water pollution control the status that it deserved, inasmuch as the administrator of EPA is a cabinet level appointee and reports directly to the President.

The Water Pollution Control Act Amendments of 1972 (PL 92-500)

PL 92-500 was enacted by Congress over Presidential veto and became effective October 18, 1972. The 1972 Amendments have dominated recent literature concerning federal efforts in the field of water quality control. A dominant theme is whether expeditious long-term progress is possible and whether the enhancement of water quality is commensurate with the large sum of public funds being committed. There is no doubt that PL 92-500 constitutes one of the most powerful pieces of legislation in the history of the environmental quality movement in the United States.

PL 92-500 accomplishes three basic tasks: (1) regulation of discharges from point sources (primarily industrial plants, municipal sewage treatment plants, and agricultural feedlots); (2) regulation of spills of oil and hazardous substances; and (3) financial assistance for wastewater treatment plant construction. Other provisions regulate vessel wastewater and disposal of dredged material. In addition, provision is made for federal support of research and other demonstration projects, as well as support for state water pollution control programs.

Provisions of the Act

Two features of the act are particularly significant. First, Congress adopted a goal of no discharge of pollutants whatsoever by 1985. This is in contrast with prior legislation which simply required sufficient control of discharges to maintain the ambient quality of receiving waters at a specified level. Second, the 1972 Amendments contain a tough enforcement scheme.

Effluent Limitations

The Act specifies that effluent limitations, based on application of "the best practicable control technology currently available (BPCTCA) as defined by the EPA administrator," shall be achieved by July 1, 1977. By July 1, 1983, discharges from point sources must achieve that degree of effluent reduction which can be achieved through "application of the best available technology economically achievable (BATEA) for such category or class, which will result in reasonable further progress toward the national goal of eliminating the discharge of all pollutants." The degree of effluent reduction required under these standards is imposed even though it results in a degree of effluent control greater than that required to meet established water quality standards. In short, the point of reference is what technology can achieve rather than what acceptable water quality requires. However, as a minimum, in cases in which the 1977 standard or the 1983 standard technology will not result in achieving water quality standards, even more stringent controls may be required. These cases are called water quality limiting.

In addition to the preceding standards, the Act established the concept of a technologically based "standard of performance" for certain specified categories of new sources. The established standard of performance must reflect "the greatest degree of effluent reduction which the Administrator determines to be achievable through application of the best available control technology, processes, operating methods, or other alternatives including, where practicable, a standard permitting no discharge of pollutants."

Effluent limitations are imposed on POTWs. The standard to be achieved by July 1, 1977, is "based upon secondary treatment as defined by the Administrator." The standard to be achieved by July 1, 1983, is based upon the "application of the best practicable waste treatment technology over the life of the works." Effluent limitations must be established for the discharge of toxic pollutants, and pretreatment standards must be established for the discharge of industrial wastes into POTWs.

Enforcement Scheme

The enforcement scheme is built around a permit system referred to as the National Pollutant Discharge Elimination System (NPDES). This system serves as the basic mechanism for enforcing the effluent and water quality standards applicable to direct dischargers, and a point source discharge requires an NPDES permit. The permit, among other things, establishes specific effluent limitations on a facility-by-facility basis which implement the technology-based standards. In addition, it specifies a compliance schedule which must be met by the discharger and requires compliance with other relevant state and local pollution control laws, if more stringent.

The Act establishes a system under which permits are initially to be issued by EPA, with the states subsequently taking over the permit-issuing function (except for federal facilities), subject to federal procedural requirements and to authority in EPA to veto any permit which it considers not to be in compliance with the Act. At the present time approximately one-half of the states have assumed responsibility for the NPDES system in their states.

Planning Studies

Section 208 of the Act is probably the most far-reaching pollution abatement ever enacted inasmuch as it is really a land use planning scheme. For the first time, it was realized that the control of point sources would not solve all of the pollution problems in the United States. Instead of a nonpoint source regulatory program which would require federal manpower beyond the most grandiose expectations (Billings, 1976), the Congress gave an incentive to the development of state- and areawide management programs to implement abatement measures for all point and nonpoint pollutant sources. All states were to develop a continuous planning process and elaborate state- and areawide water quality management plans.

The planning efforts are not only to produce a zoning plan but also provide the "best management practices" within a basin or designated area to limit pollution. The governor of each state also is to designate the planning areas and management agencies that will implement the plan (USEPA, 1976).

In many urban areas, insufficient wastewater treatment plant capacity has hindered residential and commercial development. For a 20-year period, the planning documents must therefore delineate the needs for wastewater treatment works including storm water collection and treatment.

Once the areawide plans are approved, EPA will provide funds and construction grants only to those works included in the plan. It is also significant to note that the plans may affect the issuance of NPDES permits.

The planning, which is a continuous process, must ensure that the wastewater discharges, present and future, are not in excess of the natural waste assimilative capacity of the receiving waters. In cases where receiving waters cannot assimilate the discharges, the plan must allocate the maximum waste discharge loads according to the river waste assimilative capacity.

While the planning concept is excellent, the architects of the plan were not well versed in water quality modeling. For example, the state of Tennessee was given $50,000 to model all of the streams in Tennessee.

Other Provisions

A special provision in the Act that is important to the power industry permits the administrator, in certain cases, to establish a less stringent effluent standard for the thermal component of discharges than required by application of the best practical (1977), best available (1983), or new source technology-based standards. Unlike the basic effluent limitations, this special provision permits the administrator to consider the effect of the discharge upon water quality and establish an effluent limitation which will "assure the protection and propagation of a balanced, indigenous population of shellfish, fish, and wildlife in and on the body of water into which the discharge is to be made." The burden is upon the discharger to demonstrate to the satisfaction of the administrator that a less stringent effluent limitation will accomplish the foregoing objective, and the granting of the less stringent effluent limitation is discretionary with the administrator. It should be noted that this provision is not an exemption or variance in the true sense of these words. It is interesting to note that as of this writing, all of the Section 316(a) demonstrations have been successful with the exception of a few special cases with unusual hydraulic regimes. Inasmuch as each of these "demonstrations" has cost between $.05 and 1 million, it would appear that the only beneficiary of the provision has been the biologist and, furthermore, that Congress was correct in assuming that thermal pollution may not be as harmful as it first appeared.

Section 316(b) of the Act requires that "the location, design, construction, and capacity of cooling water intake structures reflect the best technology available for minimizing environmental impact." At existing power plants, this demonstration is based upon an assessment of the significance of the losses of primarily fish eggs, larvae, and juveniles via entrainment of the organisms through the plant with the cooling water or via impingement upon the intake screens. If the losses are considered to be insignificant, no modifications to the plant intake are required.

Another section of the Act, the effects of which have yet to be observed, is Section 307(a), which deals with toxic pollutants. This section requires EPA to establish a list of toxic pollutants and define the best available treatment for them to be required by 1983. Because EPA had not complied with the law in publishing a list of toxic substances and the best available treatment for them, several environmental groups sued EPA, and the result was a consent decree (dated June 7, 1976), which forced

EPA to agree to a list of toxic materials and dates by which they would promulgate the best available treatment for them. It should be noted that these treatment requirements, when published, could have an effect on existing NPDES permits whose discharges contain them.

The Act also established reporting and monitoring requirements, a requirement for every discharge to develop a spill-prevention control and countermeasure plan for oil and hazardous materials, increased efforts in research, and demonstration and public participation. In addition to provisions for citizens' suits, the Act provided for fines for violators and even imprisonment of corporate officials.

An interesting provision of the Act was the establishment of a National Water Quality Commission, which was to evaluate the accomplishments of the Act and its commensurate public benefits and report to the Congress in 1975. The Commission was chaired by Vice President Nelson Rockefeller and submitted an excellent report. In essence, the report stated that while advances had been made in water quality improvement, the costs associated with these advances and, in particular, the costs associated with compliance with the 1983 requirements did not demonstrate commensurate social benefits.

The Clean Water Act Amendments of 1977 (PL 95-217)

The new amendments to the Act, which were published in 1978, contained these significant changes. Municipalities and certain industries were given an extension of the July 1977 deadline to comply with the best practical treatment standards if failure to do so was the result of circumstances beyond their control, a provision was made to extend the deadline to July 1, 1983, for POTWs and to January 1, 1979, for industries. Controversy between the Senate and the House on the 1983 BATEA requirements resulted in a compromise where all industry must meet what is called "Best Conventional Technology" (BCT) for "conventional" pollutants by July 1, 1984. In defining BCT, EPA must consider the age of the plant, energy requirements, and the cost and benefits of requiring a technology more advanced than BPCTCA.

A major revision in the law allows EPA to add or remove toxic materials without the previously required formal hearings. In addition, all industries must meet BATEA for defined toxics by July 1, 1984. While no waiver is allowed for toxic pollutants, BATEA for conventional pollutants, nontoxic, and nonconventional pollutants may be waived, depending on the applicant's satisfying the EPA administrator that specific criteria are met as dictated in the Act. Also, a provision was made to allow the use of long outfalls in coastal areas, without secondary treatment, if approved by the administrator, again using criteria delineated in the Act. This provision will resolve the controversy over requiring secondary treatment when outfalls as long as 25,600 feet (Passaic Valley Sanitary District, New Jersey) have been constructed in lieu of secondary treatment and satisfy water quality standards. A prime consideration in any waiver is that national water quality standards must be maintained.

Other provisions of the Act include (1) relaxing pretreatment requirements for industries discharging into POTWs; (2) exemption of industries discharging less than 25,000 gallons per day from the industrial cost recovery provision of PL 92-500; (3) the use of *ad valorem* taxes by municipalities; (4) incentives for ''innovative'' technology; (5) a requirement that alternative treatment methods be considered before awarding grants; (6) continuing emphasis on ''208'' planning studies; and (7) a change in the dredging permit program, where permits will be required from the Corps of Engineers for projects with significant potential adverse environmental effects. This also resolves problems occurring because of a

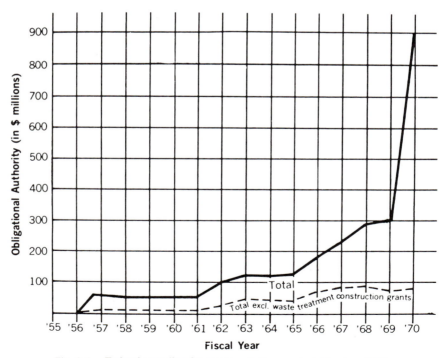

Fig. 2.4.   Federal spending for water pollution control (Davies, 1970).

court decision that apparently extended the need for dredging permits to include even farm ponds. Increasing state participation in dredging is also provided for.

The Act authorizes \$25 billion for construction grants over the next five years and mandates that 25% of the individual state's funds will be allocated for collection systems. A significant provision of the Act specifically states that federal agencies, such as TVA, who claimed that they were not subject to state requirements because they were not a person, must now comply with all substantive and procedural federal, state, interstate, and local requirements. It is interesting to compare the present construction grants authorization with Fig. 2.4, taken from Davies (1970). Obviously, the water pollution control effort by the PHS could have been much greater had they had the present monies given to EPA.

Inasmuch as the interpretation and implementation of the 1977 Amendments are not yet delineated in full, the reader is referred to the Act itself for qualifications and details of the provisions mentioned.

OTHER LAWS AFFECTING WATER QUALITY MANAGEMENT

Among the complicating factors in water quality management today are the plethora of environmental laws affecting the decision-making process and the number of government agencies having a role in their implementation. In addition to local, state, and interstate authorities, there are at least 40 federal agencies having some responsibility for water resources (DeWeerdt and Glick, 1973).

Recent legislation that may affect water quality includes:

1. The National Environmental Policy Act of 1969 (PL 91-190)
2. The Clean Air Amendments of 1970 (PL 91-604) and 1977 (PL 95-95)
3. The Federal Environmental Pesticide Control Act of 1972 (PL 92-516)
4. The Rare and Endangered Species Act of 1973 (PL 93-205)
5. The Safe Drinking Water Act of 1974 (PL 93-523)
6. The Resource Conservation and Recovery Act of 1976 (PL 94-580)
7. The Toxic Substances Control Act of 1976 (PL 94-469)
8. The Water Resources Planning Act of 1965 (PL 89-80)
9. The Wild and Scenic Rivers Act of 1968 (PL 90-542)

Inasmuch as a discussion of each of these laws would occupy a treatise, the reader should be aware of their content and refer to the specific laws when they may affect a particular project or proposal. The law that probably has the most significant effect on water resources projects is The National Environmental Policy Act of 1969 (NEPA). Therefore, a discussion of this legislation is in order.

The purpose of NEPA was to create and maintain conditions under which man and nature can exist in productive harmony and fulfill the social, economic, and other requirements of present and future Americans. Title I of the Act created the national environmental policy, and Title II established the Council on Environmental Quality, a three-man council appointed by the President of the United States.

Through Executive Order 11514, the President directed federal agencies to develop procedures and initiate measures necessary to meet the national environmental goals. The Council on Environmental Quality (CEQ) was directed to prepare guidelines to assist federal agencies in implementing the Act's requirement for environmental impact statements.

The Act required that for any proposed activity significantly impacting the quality of the human environment, a statement should be prepared that includes (1) a description of the proposed action and its effect on the environment, (2) any adverse effects that cannot be avoided, (3) alternatives to the proposed action, (4) short-term uses of the environment versus long-term productivity, and (5) any irretrievable or irreversible commitment of natural resources.

There are four major stages in the preparation of an environmental impact statement (EIS): (1) the initiating agency must decide whether the proposed action requires an EIS; (2) if the decision is affirmative, then a draft statement is prepared; (3) the draft statement is circulated to all concerned federal, state, and local agencies, the Council on Environmental Quality, and the general public, which requires a minimum of 45 days; and (4) the agency prepares and makes available a final statement incorporating the comments received.

The Tellico Project

An example of how these new laws have caused changes in water resources planning is the Tellico Project, which is a dam that TVA began construction on prior to the passage of NEPA. The TVA had spent $15 million on the concrete structure when an environmentalist group filed for an injunction against the agency, stating that an EIS had not been filed on the project as required by NEPA. The project was stopped, and TVA was instructed to prepare an EIS. Stopping a project such as this one, however, has its own environmental and economic consequences. For example, construction workers were idle, drainage problems were created, and the normal sequence of construction was interrupted. The TVA prepared an EIS for the Little Tennessee River Dam. The plaintiffs then charged that certain portions of the statement were inadequate, specifically those portions which dealt with a description of water quality, the economic considerations, and the change from a free-flowing river into an impoundment. However, the Federal District Court found that, in spite of the newness of the Act and the unfamiliarity with the process, TVA's EIS was adequate. The Court, therefore, decided in favor of TVA, and work resumed on the project.

To once again stop the project, the plaintiffs then utilized the Rare and Endangered Species Act, with the discovery of a previously unknown species of fish, the snail darter, which was subsequently declared a rare and endangered species by the Department of the Interior. This time, the claim was made that closure of the dam would alter the native habitat of the snail darter, which was in clear violation of the Act. By this time, $116 million had been spent on the project. The United States District Court refused to grant an injunction, finding it unreasonable and against the intent of Congress to stop a project so near completion. The case was appealed and subsequently decided against the agency by the 6th Circuit Court of Appeals. The TVA then took the case to the Supreme Court which denied certiorari, which in effect upheld the decision of the 6th Circuit Court. Senator Howard Baker then established a review committee, through legislation, that would have the power to grant exemptions to the Rare and Endangered Species Act. When the cabinet level committee voted unanimously not to exempt Tellico because its costs clearly outweighed the benefits, Baker attempted to overrule the committee and indicated that he would propose the abolition of the committee that he had created because he did not agree with its decision on Tellico. An amendment to the Public Works Appropriations bill to exempt Tellico from the Rare and Endangered Species Act was ultimately passed by the Congress and signed by President Jimmy Carter. It is interesting to note that the Cherokee Indian tribe has now filed suit to stop Tellico, claiming that closure of the dam will inundate their tribal burial grounds, thus depriving them of their religious rights.

THE STATE'S ROLE IN WATER POLLUTION CONTROL

As previously mentioned, until the passage of the Federal Water Pollution Control Acts, primary responsibility for water pollution control was vested in the states. As the federal role has become greater, the states have also increased their interest via new legislation and increasing staff capabilities. For example, the state of Tennessee had probably less than

20 people working on water pollution control prior to the 1960s and now has over 200. Of course, much of the funding for this effort is from federal sources, but state monies also have increased.

It is also significant to note that, as is the case with the federal government, a majority of the states have taken water pollution control out of the traditional health departments. In 1972, only 19 states had kept this function in the respective state health department (Haskell and Price, 1973).

In addition to the permit authority, if given to the states by EPA, the states are responsible for developing the comprehensive State Water Quality Management Plans. The principal components of these plans are (USEPA, 1976):

Water Quality Analysis Program
    Water quality assessments including nonpoint sources
    Inventories and projections of wastewater discharges
    Revisions of water quality standards
    Wasteload allocations
Water Quality Implementation Program
    Municipal and industrial treatment works program
    Urban stormwater management programs
    Residual waste management programs
    Nonpoint source management programs
    Target abatement dates
    Regulatory programs
    Management agency and institutional arrangements

The first set of elements provides technical direction for the state Water Quality Management Plan in the form of water quality goals and evaluation of permissible levels of pollutant loadings in receiving waters. The second set of elements involves the determination of particular abatement measures, regulatory controls, and financial and management arrangements to meet the water quality goals.

THE AMERICAN JUDICIAL SYSTEM AND WATER QUALITY

Water quality practitioners in the United States must realize the profound impact that the United States courts can have on actions involving water quality and the environment in general. No one involved in planning and implementing public and private projects affecting water quality can rely on federal or state statutory laws without considering possible court actions by those affected if water quality is worsened or altered.

There have been many court cases where polluters, either factual or potential, were involved as defendants.

The American legal system, which developed from English Common Law, has been a judge-made law rather than legislated or statutory law, although statutory enactments and administrative rules have become more important in the past 50 years (Wengert, 1975). Most of the water quality litigations are decided by a judge according to the Equity Law and primarily involve injunctions or specific orders from the court restraining certain types of actions or regulating other actions. Damages resulting from pollution are settled according to Common Law, and most involve monetary reparations or restitution of property.

Many court decisions involving water quality are based on the law of nuisance. The judges use experts as witnesses presented by both plaintiffs and defendants in order to formulate their opinions, and it is here that many water quality specialists participate.

The judges have the authority to establish their own standards and limitations where the nuisance or trespassing doctrines can be applied. In the past, these doctrines were usually applied when actual physical damage was proven by the plaintiff. As stated by Judge John Grady, this situation may no longer be true because ''It is the ability of courts of equity to give a more speedy, effectual, and permanent remedy in cases of public nuisance. They can not only prevent the nuisance that is threatened, and before irreparable mischief ensues, but arrest or abate those in progress and, by perpetual injunction, protect the public from them in the future'' (*Illinois v. the City of Milwaukee*, 1977).

In litigations before the Equity Court, plaintiffs must clearly show convincing evidence that there is a hazard to the health, morals, or safety of the community or to an individual whether that hazard eventuated in an injury. A discharge of sewage or other wastewater may constitute such a health hazard.

In a landmark case (*Illinois v. the City of Milwaukee*, 1977), the federal judge set effluent standards for the city of Milwaukee that are far more restrictive than those required by PL 92-500.

Biochemical oxygen demand	5 mg/liter
Suspended solids	5 mg/liter
Fecal coliforms	40/100 ml
Phosphorus	1 mg/liter
Free chlorine residual	0 mg/liter

In addition, no storm water overflows or effluent by-passes will be allowed. The city of Milwaukee has estimated that the total cost of achieving these limitations will ultimately reach $1.5 billion. From the testimony of many expert witnesses, the judge reached the conclusion that the effluent discharges from the city of Milwaukee were causing a nuisance and a potential health hazard to the people of Illinois who use the water for drinking and swimming. In addition, Judge Grady found that the waters of Lake Michigan, within the territorial boundaries of Illinois, were undergoing accelerated eutrophication, some of which was caused by nutrients emanating from the Milwaukee discharges. The 7th Circuit Court of Appeals upheld the legal basis for the action, the federal common law of nuisance, which it said was not preempted by the FWPCA. The court found, however, that the factual evidence did not justify effluent standards more restrictive than those imposed by relevant EPA regulations.

It should be noted that the Water Pollution Control Act recognizes the importance and the role that courts can have in implementing water quality goals. Private legal actions are likely to have a more significant role in the preservation of water quality through enforcement proceedings authorized by legislation. The proceedings authorized by legislative action emphasize control over all pollution sources without regard to possible injuries caused by individual sources. This may avoid a drawback encountered under the common law inasmuch as provisions allowing citizen suits, through the courts, on both the regulators and the polluters provide the advantage of both private and public systems for pollution control (Cox and Walker, 1975).

Despite the overall judicial approval for pollution control measures, courts occasionally have disapproved pollution regulations that are not soundly conceived or that impose unique and special burdens upon private property owners for the common good. These cases have included preventing riparian owners from boating, fishing, and swimming in order to protect water supplies (Lienesch, 1977).

The courts will undoubtedly be called upon in the future to decide whether a particular pollution control regulation is valid or invalid. Based on experience with past pollution control litigation, the courts may be expected to be supportive of most water quality requirements and even may go beyond statutory requirements in cases where public health, the absence of nuisance, or natural beauty is to be protected.

REFERENCES

Beuscher, J. H. (1967). "Water Rights." College Printing and Typing, Madison, Wisconsin.

Billings, L. G. (1976). The evolution of 208 water quality planning. *Civ. Eng.* **46** (11), 54–55.

Busby, C. E. (1954). American water rights law. *South Carolina Law Quart.* **5**, 111.

Cox, W. E., and Walker, W. R. (1975). Water quality rights in an urbanizing environment. *In* "Urbanization and Water Quality Control" (W. Whipple, ed.). Am. Water Resour. Assoc., Minneapolis, Minnesota.

Davies, J. C. III (1970). "The Politics of Pollution." Pegasus, New York.

DeWeerdt, J. L., and Glick, P. M. (1973). A summary–digest of the federal water laws and programs. Natl. Water Commis., Arlington, Virginia (May).

Haskell, E. H., and Price, V. S. (1973). "State Environmental Management." Praeger, New York.

Lienesch, W. C. (1977). Legal and institutional approaches to water quality management and implementation. Contract No.68-01-3564, U.S. Environ. Prot. Agency, Washington D.C.

McKee, J. E. (1960). The need for water quality criteria. *In Proc. Conf. Physiol. Aspects of Water Quality, Div. Water Supply and Pollut. Control.* U.S. Public Health Serv., Washington, D.C.

National Water Commission (1973). Water policies for the future. Final Rep. to Pres. Washington, D.C.

Ohio Legislative Service Commission (1955). Water rights in Ohio: Research report number 1. (January).

U.S. Environmental Protection Agency (1976). Guidelines for state and areawide water quality management program development. Washington, D.C.

Wengert, N. I. (1975). Environmental law. *In* "Environmental Design for Public Projects" (D. W. Hendricks, E. C. Vlachos, L. S. Tucser, and J. C. Kellogg, eds.). Water Resour. Publ., Fort Collins, Colorado.

3 | Physical and Chemical Parameters of Water Quality

INTRODUCTION

The term "water quality" is not a new concept. Ancient British Common Law, from which the riparian water rights were derived, stated that the user of water was not entitled to diminish it in quality (McGauhey, 1968). However, the question as to what constituted quality was neither posed nor answered. Even at the beginning of this century, water quality evaluation relied more or less on subjective observations rather than on scientific measurements. Water quality was expressed in terms such as "fresh," "dirty," "faulty," and "potable." Only water used for a public water supply was analyzed by early chemical and bacteriological methods in order to determine its safety for human consumption. Recent developments in water quality surveillance methods and consequent concern with the protection of human health and the aquatic environment have broadened the qualification and quantification of water quality.

It has been only in the past few decades that water quality engineering and science have begun to develop from an art to a science. New wastewater treatment methods and the discovery and evaluation of new harmful water contaminants and pollutants have stimulated research activities seeking new and improved methods of water quality analysis. Increased emphasis has also been placed on the removal of secondary pollutants such as nutrients or refractory organics and on water reuse.

refractory organics resistant to treatment

There is no way to define water quality precisely. It must be understood that the quantification of water quality may have a different meaning to an aquatic scientist concerned with aquatic life, a farmer concerned with irrigation, or a public health official concerned with the protection of human health. In each case, a different picture of water quality can be developed based on the indicators and criteria used for a particular evaluation. Therefore water quality should be related to the anticipated beneficial use of the water such as fish and wildlife protection, irrigation, or public water supply.

Quant. of water quality depends upon water use.

Numerous indicators can be used for evaluating water quality. These range from very simple observations such as the measurement of turbidity by the Secchi disk to very sophisticated water quality indices and analytical methods. An indicator selected for a particular water quality evaluation using an appropriate commonly accepted analytical method may be called a parameter, provided the indicator is accepted by the scientific and engineering community and/or instituted by legislation. A parameter or indicator can either be descriptive or have a numerical value.

wa. parameter (indicator) can be either descriptive or quantitative.

CLASSIFICATION OF POLLUTANTS

Common water quality parameters and indicators can be categorized into four groups as proposed by Klein (1962), i.e., chemical, physical, physiological and biological. A brief discussion of the general effects of each of these catagories is in order (Krenkel, 1974).

Chemical Pollutants

One may broadly classify chemical materials into organic and inorganic pollutants, where organics are categorized as compounds containing a carbon atom.

Major effects of inorganic materials include changes in the pH of the water and toxicity caused by materials such as heavy metals. In addition, inert insoluble inorganics may result in sludge deposits on the receiving water bottom and inhibit benthic biological activity.

Affects of inorg. - pH change - toxicity (eg heavy metals)

The emphasis on organic materials concerns their potential for depleting dissolved oxygen and the concomitant effects on aquatic life. When a biologically oxidizable material is introduced into a receiving

Prim. concern @ organics Depl. of O_2

water containing appropriate bacteria, nutrients, and dissolved oxygen, the bacteria will oxidize the organic material to carbon dioxide and water, using necessary amounts of oxygen. If sufficient organic material is present, the dissolved oxygen levels thus may be lowered to values inimical to aquatic life.

It should be noted, however, that organic materials may be toxic, and inorganic materials may exhibit an oxygen demand. For example, sulfurous acid may deplete oxygen by being oxidized to sulfuric acid, and the toxic effects of certain organic materials, such as pesticides, are well known.

Other organic materials that may cause problems include oil, with its tendency to form surface films, and taste- and odor-producing substances like phenols and materials used as fungicides, such as the mercurials.

Of relatively new concern are the trihalogenated methanes, which are apparently formed in the water treatment plant by the process of chlorination. In addition, the new interest in water reuse has led to increasing interest in the materials that are refractory to treatment. While conclusive evidence as to their harmful effects has not been found, evidence to the contrary is also lacking.

Physical Pollutants

Physical pollutants include color, temperature, foam, suspended solids, turbidity, and radioactivity. Obviously, these characteristics may be associated with chemical pollutants and may result from the discharge of chemicals into the receiving water.

Color is undesirable in drinking water and for certain industrial uses, although it is not necessarily harmful. Water is never colorless. Even pure tap water may have a pale green-blue tint in large volumes. Natural color exists in water primarily as negatively charged colloidal particles. During floods, stream water may be colored by suspended particles of clay and silt. Color caused by suspended particles is referred to as apparent color and is differentiated from true color which is due to vegetable and organic extracts. Color is measured by comparing the sample to a standard potassium chloroplatinate extract (Sawyer and McCarty, 1967).

Turbidity may be caused by a wide variety of suspended materials ranging from colloidal to coarse dispersions. In rivers under flood conditions, most of the turbidity will be the result of relatively coarse dispersions, primarily clay and silt. In lakes or other quiescent waters, most of the turbidity during summer and fall periods is caused by planktonic microorganisms (zooplankton and phytoplankton). In glacier-fed streams

and lakes, the blue and green color is caused by silt particles produced by
the grinding action of the glaciers. Substantial amounts of suspended particles originate from soil erosion, dust, and dirt accumulation in urban
areas and from wastewater discharges. Turbidity in the laboratory is measured by comparing the water sample to a standard silica solution or by
using the Jackson candle. Under field conditions, turbidity is measured by
the Secchi disk and is expressed as the depth of water at which the visibility of the disk disappears. Turbidity may not be harmful to fish; however,
reductions in sunlight intensity caused by turbidity may cause a reduction
in productivity. It should be noted that turbidity is not synonymous with
suspended solids inasmuch as it is a measure of the interference of the
passage of light through the water, while suspended solids are measured
by evaporation and/or filtration.

[handwritten margin note: Turb. not synon @ s.s.]

Temperature is a relatively new concern, primarily because of the
increasing use of water for cooling purposes. Temperature increases may
affect the potability of water and, according to McKee and Wolf (1963),
temperatures above 15°C are objectionable in drinking water. Temperature effects are treated separately in Chapter 14. It is important to note
that temperature significantly affects physical, biological, and chemical
processes. In general, a 10°C rise in temperature will result in a doubling
of a reaction rate, and vice versa.

Suspended solids may be contributed by erosion processes in addition
to wastewater discharges. The material may be inorganic and/or organic
and may result in inhibition of photosynthesis, a retardation of benthic
activity, a reduction in waste assimilative capacity, and, in general, nuisance conditions. Considerable controversy exists over the measurement
of suspended solids and their potential effects on the aquatic environment.

Foam from various industrial waste contributions and detergents is primarily objectionable from the aesthetic standpoint. However, agents
causing foaming in a waterway may have adverse physiological effects. In
addition, surface active agents may cause a reduction in the rate of oxygen gas transfer, thus resulting in a reduction in waste assimilative
capacity.

With the advancement of the nuclear industry and the uncontrolled
testing of nuclear weapons in some countries, a concern for the protection
of humans and the environment against excessive radiation effects is
increasing. The shortage of fossil fuels for energy production undoubtedly
will lead to increased nuclear energy production which will necessitate all
possible precautions against hazards resulting from the contamination of
the environment by radioactive wastes.

Natural radioactivity is always present in the environment, caused by either radioactive minerals or cosmic radiation. But radiation, when absorbed by living organisms in quantities substantially above that of natural background, is recognized as injurious (FWPCA, 1968). Therefore, it is necessary to prevent the entry of excess radioactivity into the environment.

Upon introduction into an aquatic environment, radioactive wastes can (1) remain in solution or in suspension, (2) precipitate and settle to the bottom, or (3) be taken up by plants and animals. Radioactivity is concentrated biologically by uptake directly from the water and passage through food webs, and chemically and physically by adsorption, ion exchange, coprecipitation, flocculation, and sedimentation. Many silts and clay particles have ion-exchange capacity, and upon settling they can reduce the level of soluble radioactivity.

Radioactive pollutants in the aquatic environment may be cycled through water, sediment, and biota. In addition, radioactive contamination may accumulate in some organisms. Some forms of algae have been known to concentrate radiophosphorus by factors up to 10 (Bevis, 1960).

The 1962 Drinking Water Standards (USDHEW, PHS, 1962) established limits for radium-226 and strontium-90 as measures of radioactivity of drinking water supplies. In addition to these two isotopes, overall radioactivity is measured as gross beta radiation.

The principal source of strontium-90 has been fallout from nuclear weapon testing in the atmosphere. Since the international effort to eliminate nuclear testing resulted in a reduction and control of these tests, substantial decreases in strontium-90 levels have been achieved. The permissible level of 10 pCi/liter is substantially greater than the highest level found in public water supplies to date (Clark *et al.*, 1971).

With the expected increase of nuclear energy production and the increased use of radioactive materials, further expansion of monitoring and surveillance will be necessary.

Physiological Pollutants

Taste and odor causing.

Probably the most difficult water quality problem presented to the environmental engineer is that imposed by public opinion, and taste and odor complaints in a drinking water supply are among the most troublesome. While odor may not necessarily indicate serious pollution problems, the layman associates odors and bad taste with water pollution. Inasmuch as palatability is associated with the taste of water and food, taste and odor problems are particularly objectionable in waters used for drinking and food processing.

Some substances can impart objectionable taste and odor to a water in very small amounts. For example, hydrogen sulfide can be smelled at concentrations on the order of 0.0011 mg/liter, and chlorophenols can be tasted at concentrations as low as 0.001 mg/liter.

In addition, many taste- and odor-producing materials possess the ability to taint fish flesh. Thus, minute quantities of certain substances may make fish inedible and yet not cause adverse physiological effects on either fish or man.

Biological Pollutants

Biological pollutants can be classified in two categories: primary and secondary. Primary biological pollutants are those that may cause disease, while secondary biological pollutants are substances added to the water that may result in increases in biological growths.

[handwritten margin note: Primary (disease causing) bacteria virus protozoans helminths]

As previously mentioned, the impetus for early water pollution control efforts in the United States was motivated by waterborne disease. Until the recent concern over the formation of trihalogenated methanes by chlorination of a water supply, the chlorination process was considered mandatory for the protection of public health.

The microbiological agents involved in waterborne diseases include bacteria, virus, protozoans, and the helminths. Viral-borne diseases include infectious hepatitis and poliomyelitis, while bacterial-borne diseases include typhoid fever, paratyphoid fever, cholera, and bacterial dysentery. Giardiasis and amoebic dysentery exemplify protozoan-borne diseases, and shistosomiasis is associated with the helminths. While these diseases are not considered to be significant in the United States today, sporadic epidemics do occur and are usually traced to inadequate water quality control. It should be noted, however, that waterborne disease is still a major problem in many other countries. To appreciate the potential effects of a laxity in water treatment process control, one only needs to consider the 1892 cholera epidemic in Hamburg, Germany, where 17,000 cases of cholera occurred, resulting in 8,065 deaths in two months because of inadequate water treatment (Maxcy, 1956).

[handwritten margin note: secondary (cause incr. in bio. growth) nutrients]

Excessive growths of phytoplankton, macrophytes, and fungi may result from the addition of certain nutrients to the water. These increases in productivity may be called secondary pollution effects and are discussed in detail in Chapter 13 on eutrophication.

have been reported at concentrations of 11–40 mg/liter of nitrate and
have been the basis for the recommended limits of 45 mg/liter in drinking
water (USEPA, 1976b).

Sulfur

While sulfur per se is not of major concern in water quality management, the role of sulfur in oxidation and reduction processes is important.
Hydrogen sulfide, which is formed under reducing conditions, is a major
cause of corrosion in treatment works and exemplifies the odor associated
with sewage. On the other hand, sulfate ion, which is formed under oxidizing conditions, demonstrates a cathartic effect in humans when present
in excessive amounts.

cathartic →
cleanses the
bowels.

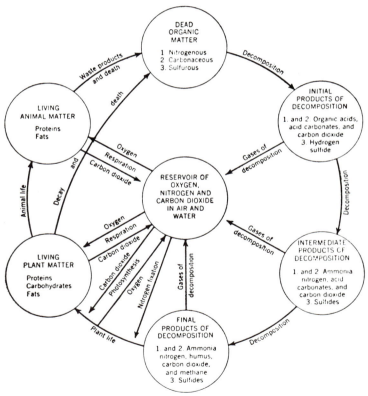

Fig. 3.1.   Cycle of nitrogen, carbon, and sulfur in anaerobic decomposition (Imhoff and
Fair, 1956).

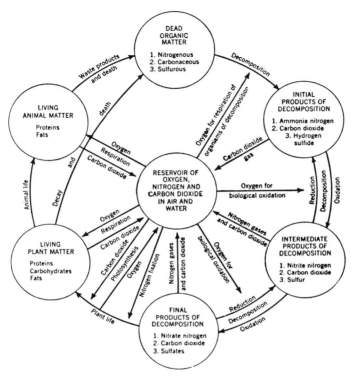

Fig. 3.2. Cycle of nitrogen, carbon, and sulfur in aerobic decomposition (Imhoff and Fair, 1956).

Because of the importance of carbon, nitrogen, and sulfur in water quality management problems, their cycles are presented under both aerobic and anaerobic conditions in Figs. 3.1 and 3.2 (Imhoff and Fair, 1956).

Toxic Organic Chemicals

During the past few decades, controversy has arisen over the benefits and environmental costs of the use of pesticides. The genesis of concern was probably in the book *Silent Spring*, written by Rachel Carson in 1962 (Carson, 1962). While the problems expressed by Carson were overemphasized, they did form a basis for the environmental movement of the 1960s.

The monitoring of DDT and dieldrin in fish species in the Great Lakes added greater evidence to the pesticide problem (Johnson and Ball, 1972).

High salmon mortalities were observed in Lake Michigan where DDT residues in fish from Lake Michigan averaged two to five times higher than those from Lake Superior, and nearly 60 times higher than those from the Pacific coast. Tests for disease conditions among the dying fish were negative, and DDT was considered to be the most probable cause of mortality (Johnson and Pecor, 1969). Many of the organic chemicals, including DDT, can accumulate in fish and biota tissues with resulting concentrations much higher than in the surrounding water. Dieldrin, one of the chlorinated hydrocarbon pesticides, is extremely toxic to animals and fish. Even concentrations of 0.05 μg/liter have been shown to be detrimental to fish and other aquatic animals (FWPCA, 1968). In 1969, the United States Food and Drug Administration stopped the commercial sale of Lake Michigan fish, primarily coho salmon, after finding fish samples which exceeded 0.3 ppm of dieldrin in their tissues.

Severe limitations of commercial fishing and sportfishing have continued. High concentrations of PCBs (polychlorinated biphenyls) were found in fish samples from the Great Lakes. Contamination of coho salmon from Lake Michigan by PCBs has been established, and the combination of relatively high concentrations of PCB and pesticide residues in coho salmon eggs may be particularly significant if these compounds have synergistic or additive toxic effects. Recent work in Sweden (Anon., 1970) has suggested that PCB residues in salmon eggs may be responsible for reduced embryo survival and poor hatching success. More recent PCB contamination has been observed in the Hudson River in New York, and the Coosa River in Georgia.

[handwritten margin note: Synergistic → working together.]

In addition, DDT concentrations have been found in fish in TVA's Wheeler Reservoir at concentrations of 1200 ppm. Even though these are concentration levels dangerous to man, TVA chose to ignore the problem because the management was not interested (Krenkel, 1979).

Chlorinated hydrocarbons are considered as potentially dangerous carcinogenic materials. They are formed by the substitution of a chlorine atom for hydrogen in the hydrocarbon molecule. Besides pesticides and PCBs, recent research indicates that the formation of chlorinated hydrocarbon can be related to the use of chlorine for the disinfection of water supplies and organic wastes. Most surface waters contain some organics, e.g., those elutriated from fallen leaves, which are amenable to chlorination and the subsequent formation of organic hydrocarbons.

[handwritten margin note: elutriate → wash away due to force of raindrops.]

While the effects of these materials are somewhat controversial, there is no doubt that trihalogenated methanes are formed during water treatment, as demonstrated in Table 3.1 (USEPA, 1976a). It is interesting to note that the discovery of trihalogenated methanes in public drinking water gave impetus to the passage of the Safe Drinking Water Act of 1974.

[handwritten margin note: Trihalogenated methanes are formed during water treatment with Cl.]

TABLE 3.1 Trihalogenated methane content of water from water treatment plant

Sample Source	Free Chlorine (ppm)	Concentration (μg/liter)		
		Chloro-form	Bromo-dichloro-methane	Dibromo-chloro-methane
Raw river water	0.0	0.9	*b*	*b*
River water treated with chlorine and alum-chlorine (contact time 80 min)	6	22.1	6.3	0.7
3-day-old settled water	2	60.8	18.0	1.1
Water flowing from settled areas to filters*a*	2.2	127	21.9	2.4
Filter effluent	unknown	83.9	18.0	1.7
Finished water	1.75	94.0	20.8	2.0

a Carbon slurry added at this point.
*b*None detected. If present, the concentration is less than 0.1 μg/liter.

Another example of contemporary water contamination by organics is the appearance of organic mercury in fish and other organisms because of the introduction of mercury into the environment as a result of its use as a fungicide and in industrial processes. While the real hazard of this problem is also controversial, there is no doubt of its adverse effect on American fisheries.

Table 3.2 shows some of the organic chemicals of present concern in the Great Lakes. Continued research is needed to understand the significance of pesticides and other organic chemicals with respect to fish and aquatic biota inasmuch as they can significantly impact humans, commercial fishing, the sale of fish products, and sportfishing.

Some mention should be made of the manner in which trace organics are measured in water supplies even though the methodology has been subjected to criticism. The tests utilized are called carbon chloroform extract (CCE) and carbon alcohol extract (CAE). Both tests utilize activated carbon to collect the adsorbable organics contained in a known volume of water passed through the carbon filter. The CCE adsorbents are extracted by using chloroform and the CAE adsorbents are extracted by using alcohol. While the exact nature of these materials is not known, CCE has been shown to contain pesticides and petrochemicals while CAE consists, at least partially, of naturally occuring substances. The recommended cri-

TABLE 3.2 Organic chemicals of present concern in the Great Lakes [a]

Organic Chemical	Recommended Limits[b]
Phenolics	1 μg/liter (tainting)
PCBs	0.001 μg/liter (aquatic life)
Phthalate	3 μg/liter (aquatic life)
DDT	0.001 μg/liter (aquatic life)
DDD	—
DDE	—
Heptachlor	0.001 μg/liter (aquatic life)
Aldrin	0.003 μg/liter (aquatic life)
Dieldrin	(summation Aldrin + Dieldrin)
Lindane	4 μg/liter drinking water, 0.01 μg/liter aquatic life
Heptachlor epoxide	—
Methoxychlor	100 μg/liter drinking water, 0.03 μg/liter aquatic life
Organic mercury	2 μg/liter drinking water, 0.05 μg/liter aquatic life

[a] Anon. (1974).
[b] EPA recommendation (FWPCA, 1968).

teria for drinking water are CCE = 0.7 mg/liter and CAE = 3 mg/liter
(USEPA, 1973).

Acidity, Alkalinity, and pH

One of the primary indicators used for the evaluation of surface waters
for their suitability for various beneficial uses is pH, which expresses the
molar concentration of the hydrogen ion as its negative logarithm. When
an acid is added to water, it ionizes and the hydrogen-ion concentration
increases. When a base is added the hydroxyl ion will neutralize the disso-
ciated hydrogen ions resulting in a decrease in their concentration, since
pH decreases as the concentration of hydrogen ion increases.

Most natural waters are buffered by a carbon dioxide–bicarbonate
system (Sawyer and McCarty, 1967). Acidity in most surface waters represents the amount of carbonic acid present and for all practical purposes,
if the pH of the water is below 8.5, some acidity is present. From a sanitary or public health viewpoint, acidity has relatively low importance
since no deleterious effects from carbon dioxide have been recognized.
However, excess carbon dioxide may have adverse effects on humans or
aquatic biota. Excess acidity also may be harmful to concrete structures
because of the dissolving and corrosive properties of free carbon dioxide.

The alkalinity of a water is a measure of its buffering capacity or its ability to neutralize acids or bases. While many materials may contribute to
alkalinity, it is usually assumed to be the bicarbonate ion concentration in

natural waters (pH 6–9). It is correctly defined as the sum of the carbonate, hydroxyl, and bicarbonate ion concentration on an equivalent basis, other materials being insignificant.

Solids

Solids may be classified as total solids, dissolved solids, suspended solids, and settleable solids. These categories may be further categorized as organic or volatile solids, and inorganic or nonvolatile solids. Total solids, as the term indicates, represent the total amount of dissolved and particulate matter present in water. This is measured by evaporating the sample and weighing the residue. The difference between the residue after evaporation and the weight of ash after burning the residue under high temperature represents the volatile solids which can be correlated to the organic content of the solids. Dissolved and particulate solids can be separated by filtration of the sample. The residue of the filtered sample after evaporation represents the total dissolved solids, while the dried residue on the filter is referred to as the total suspended solids. Settleable solids, which are important in the design and operation of wastewater treatment plants, are determined by allowing a sample of the solids containing water to settle for a predetermined period of time and measuring the volume of the settled material relative to the volume of the vessel used for the test.

The specific conductance of a water may be related to the dissolved solids content and may be used to correlate with certain chemical constituents. Specific conductance is defined as the reciprocal of the resistance of a column of water 1 cm long with a cross section of 1 cm² and is reported as Mhos/cm or Siemens/cm. Most potable water has a specific conductance of 50 to 1500 μMhos/cm; seawater shows a specific conductance of 500×10^{-4} Mhos/cm, and distilled water, 0.5 to 4 μMhos/cm. An estimate of the dissolved solids content of a water may be obtained by multiplying the specific conductivity by factors ranging from 0.5 to 0.9 (Sawyer and McCarty, 1967). A typical factor for surface waters is 640 mg of TDS $\cong$ 1000 millimhos/liter.

Hardness

Because of the advent of detergents, the major concern with hardness today is its correlation with scaling. Traditionally, hardness was associated with the additional soap required to create cleansing action in a "hard" water. There also has been some controversy over the hardness of a water and the incidence of heart disease, although the beneficial and detrimental effects have not been conclusively demonstrated (Muss, 1962). A recent report from Texas researchers indicated that people in the Upper Rio Grande Valley of Texas had the lowest incidence of cardiovascular disease in the United States. The low morbidity was attributed to the high lithium content of the hard water comprising the potable water supply of the 24 West Texas communities examined (Hansen, 1978).

Hardness is caused by the sum of the alkali earth elements, although the major constituents are usually calcium and magnesium. Expressed in its usual terms of parts per million of $CaCO_3$, water may be classified as

0– 75 mg/liter	Soft
75–150 mg/liter	Moderately hard
150–300 mg/liter	Hard
>300 mg/liter	Extremely hard

It is significant to note that hardness has an antagonistic effect on the toxicity of the heavy metals to aquatic life. While the mechanism is speculative, there is no doubt that an increase in hardness of a water decreases the toxicity of heavy metals.

Chlorides

Chlorides may demonstrate an adverse physiological effect when present in concentrations greater than 250 mg/liter and with people who are not acclimated. However, a local population that is acclimated to the chloride content may not exhibit adverse effects from excessive chloride concentrations. Because of the high chloride content of urine, chlorides have sometimes been used as an indication of pollution.

Iron and Manganese

Both iron and manganese may be present in significant quantities in soils as their oxides and hydroxides. Iron in soils is mostly insoluble and may be dissolved by water containing free CO_2 in the same manner as calcium and magnesium carbonates are dissolved. In this reaction, the iron does not occur if oxygen is present. However, under anaerobic conditions both iron and manganese can be reduced and solution occurs without difficulty.

The fact that high amounts of iron and manganese are associated with anaerobic conditions has several implications. In reservoirs, it may indicate a lack of oxygen in the hypolimnion when the reservoir is stratified.

Groundwaters that contain high iron and manganese content are always devoid of dissolved oxygen and are high in carbon dioxide content. The high carbon dioxide content may be the result of bacterial oxidation of organic material, and the absence of dissolved oxygen shows that an- aerobic conditions have developed. Even a good groundwater aquifer oc- casionally may produce a water high in iron and manganese content if high amounts of organic matter are introduced into the soil near the re- charge area.

Neither iron nor manganese is harmful to humans in reasonable concen- trations. However, when water containing high amounts of Fe^{2+} or Mn^{2+} ions is exposed to air, oxidation of these ions will result in insoluble Fe^{3+} and $Mn(IV)$ compounds, and the water may become turbid because of the resulting precipitate. Both iron and manganese may cause stains on textile and white paper products, impair laundry operations, stain fixtures, and cause the development of growths in pipes and other plumbing fixtures.

Metals

Because of the exposure given to heavy metals by the news media and because of unfortunate poisonings that have occurred, metals are proba- bly of most concern to the layman. The previously mentioned incidents of mercury and cadmium poisoning in Japan have been widely publicized, although the public health significance of these metals in the concentra- tion found elsewhere has been subjected to question. The metals of major concern, along with their alleged effects and recommended concentra- tions, are shown in Table 3.3.

The presence of metals in water is usually related to man's cultural activities including the use of automobiles (lead), industrial production, mining, vector control, and macrophyte and microphyte control. Urban runoff may contain significant concentrations of lead, zinc, and copper and, to a lesser extent, chromium, mercury, and nickel. Corrosion may also contribute metallic materials. It should be noted, however, that natu- rally occurring concentrations of metals in water may be significant, in particular with respect to groundwater. For example, in many instances the concentration of arsenic in groundwater naturally exceeds the recom- mended criteria.

Sometimes, the metals natl. occur- ring in waters can be significant.

Other Materials

Other inorganic materials of concern in water include ammonia, as- bestos, barium, chlorides, cyanides, fluorides, magnesium, nitrates,

TABLE 3.3 Metals that may be found in water

Metal	Sources	Alleged Effects	Recommended Criteria[a]	
			Drinking Water	Aquatic Life/Other
Aluminum	water treatment, soil, cooking utensils	none known		
Arsenic	pesticides, soil, foods, industrial wastes	chronic, cumulative carcinogenic, cardiovascular effects	50 µg/liter	0.01 mg/liter 100 µg/liter irrigation
Cadmium	industrial wastes, soil	arteriosclerosis, carcinogenic Itai-Itai disease, kidney damage	10 µg/liter	0.4–1.2 µg/liter (see reference)
Chromium	industrial wastes	carcinogenic, ulceration (hexavalent form)	50 µg/liter	100 µg/liter
Copper	industrial wastes, piping, algicide	emetic effect, liver damage, taste	1 µg/liter	0.1 LC$_{50}^{96}$
Iron	soil, industrial wastes	primarily aesthetic, stains	0.3 mg/liter	1 mg/liter
Lead	soil, industrial wastes, piping	cumulative, plumbism, nephritis	50 µg/liter	0.01 LC$_{50}^{96}$
Manganese	soil, industrial wastes	primarily aesthetic, stains, toxic at high levels	50 µg/liter	100 µg/liter
Mercury	soil, industrial wastes, fungicides	may be methylated, organic forms may cause minamata disease	2 µg/liter	0.05 µg/liter
Nickel	industrial wastes	dermatitis		0.01 LC$_{50}^{96}$
Silver	industrial wastes, soil	cumulative, discoloration of skin, argyria	0.05 mg/liter	0.05 LC$_{50}^{96}$
Tin	industrial wastes	none known at levels found in water		
Zinc	industrial wastes, soil	Metallic taste to water, milky appearance above 5 mg/liter	5 mg/liter	0.01 LC$_{50}^{96}$

[a]Note: The recommended criteria are from USEPA (1976) and are subject to change.

phenols, phosphorus, potassium, selenium, sodium, and sulfates. Some of these substances previously have been discussed and others, such as phosphorus and nitrogen, will be discussed in the chapter on eutrophication (Chapter 13).

Inasmuch as fluorides in water supplies are controversial, consideration of their significance is in order. Fluorides have been found to occur naturally in some groundwater at very high concentrations. As a result of evidence that people ingesting these waters demonstrated a much lower incidence of dental caries, subsequent epidemiological studies showed that fluoride, in a concentration of approximately 1 mg/liter, was indeed beneficial to young people. While excessive concentrations may result in mottling of teeth, exhaustive studies have found no adverse physiological effects. Thus, there is no doubt that addition of fluorides to a drinking water supply is beneficial to the younger population. It is interesting to note that one would have to drink 1057 gallons of water with a concentration of 1 mg/liter of fluoride to obtain a fatal dose (4 g NaF). The recommendations for fluoride in water are as follows (USEPA, 1976b):

Annual Average Air Temperature	Recommendation
66°F or lower	1.5 mg/liter
66–79°F	1.3 mg/liter
80°F or higher	1.2 mg/liter

The other water quality parameters of significance, along with their alleged effects and recommended water quality criteria, are given in Table 3.4.

WATER QUALITY INDEX

A numerical classification of WQ based on the phys., chem., and bio. water qual. parameters indicated previously.

For many years, there have been efforts to develop an overall water quality index which would use selected physical chemical, biological, and microbiological indicators in order to classify surface waters according to their quality. The first such efforts were connected with attempts to assign chemical values to the Kolkwitz–Marson saprobien system. Based on these studies, some European countries adopted a system of water quality classes.

Research investigations are under way in the United States to develop a numerical index which would estimate the overall water quality based on several selected indicators. A parameter described as a Water Quality

TABLE 3.4 Nonmetallic water quality parameters

Substance	Sources	Alleged Effects	Recommended Criteria[a] Drinking Water	Aquatic Life/Other
Ammonia	industrial wastes, domestic wastes, fertilizer	pH dependent	0.5 mg/liter	0.02 mg/liter NH_3^+
Asbestos	industrial wastes, mining, piping soil, mining	carcinogenic		
Barium	industrial wastes, soil	muscle stimulant	1 mg/liter	
Boron		toxic to vegetation	0.1 LC_{50}^{96}	750 μg/liter irrigation
Carbon–chloroform Extract, carbon-alcohol extract	organics, pesticides petrochemicals	indication of organics	CCE = 0.7 mg/liter CAE = 3.0 mg/liter	
Chlorides	wastewaters, industrial wastes, soil, ocean	laxative, pollution indicator	250 mg/liter	
Cyanide	industrial wastes, pesticides	adverse metabolic effects	0.2 mg/liter	5 μg/liter
Fluorides	industrial wastes, soil	fluorosis	(see text)	
Magnesium	soil, industrial wastes	laxative, hardness	125 mg/liter	
Nitrates	industrial & domestic wastes, soil, fertilizer	methemaglobenemia	45 mg/liter	
Phenols	industrial wastes	taste and odor, fish tainting, formation of chlorophenols	1 μg/liter	1 μg/liter
Selenium	industrial wastes, soil, pesticides	carcinogenic, dental caries, toxicity to animals, antagonistic to mercury, dermatitis	10 μg/liter	0.01 LC_{50}^{96}
Sodium, potassium	soil, water treatment industrial wastes	heart disease, soil permeability	no recomm. (see Agr. Uses, SAR)	
Sulfates	soil, industrial wastes	laxative, $MgSO_4$ and Na_2SO_4	250 mg/liter	

[a]Note: Recommended criteria are from USEPA (1976b) and USEPA (1973) and are subject to change.

Index (WQI) was developed at the National Sanitation Foundation in 1970 (Brown *et al.*, 1970). The WQI is an additive model, expressed as (Brown *et al.*, 1972)

$$WQI = \sum_{i=1}^{n} w_i q_i, \tag{3.1}$$

where WQI is a number between 0 and 100; q_i is the quality of the *i*th parameter, a number between 0 and 100; w_i is the unit weight of the *i*th parameter, a number between 0 and 100; and *n* is the number of parameters. Parameter selections, q_i and w_i, were determined by a panel of experts in water quality management. The list of parameters includes dissolved oxygen, fecal coliforms, pH, BOD_5, nitrate, phosphate, temperature, turbidity, and total solids. The ranking and ranges of parameters are contained in Brown *et al.*, 1970. For comparison, Harkins (1974) proposed another Water Quality Index in 1974.

These overall water quality indices are still in the experimental stage, and further evaluation is necessary before they can be considered as practical.

THE NATURAL QUALITY OF WATER

It should be realized that the basic chemical and biochemical parameters comprising water quality are the result of natural causes, which may be extremely difficult to control. The Nile and Mississippi Rivers carried high suspended-sediment loads thousands of years before man settled in the areas. The sediment-laden Nile River, with nutrients adsorbed on the sediment particles, was actually beneficial to the Nile delta farmers who relied on the natural supply of fertilized water during the monsoon season floods. The construction of the Aswan Dam, which was designed to improve food production in the Nile valley, actually diminished the natural supply of fertilizer to the farmers and resulted in adverse effects to the agricultural economy of Egypt's only fertile agricultural area. In addition, the backwaters created by the dam have caused an increase in the incidence of schistosomiasis.

Similarly, the high salinity of the Colorado River is primarily caused by the arid conditions in the Colorado River drainage area, which result in the high salt content of the soils and subsoils.

Natural and background water quality should be measured rather than estimated. In almost every river basin, headwater reaches exist which are representative of undisturbed drainage areas. Information on the basic mineral composition of streams may be obtained by analyzing the water quality of the major headwater sources.

The background water quality depends on many factors. Surface water is basically a mixture of groundwater (higher mineral content) and surface (low mineral content) runoff. Their composition depends on the character of the drainage area, the permeability of the soils, the elevation and slope, the climatic conditions, the percent of forest cover, the general geology of the area, the erosion rates, the ion concentration in the precipitation, and the natural biota of the surface waters.

Meteorological Water

Meteorological water is precipitation (rain and snow) that is collected directly into containers such as cisterns. It would appear that this water should be quite pure; however, it is often contaminated by the collection device and should be disinfected before use.

A new interest in precipitation has been demonstrated because of the acid rainfall problem and the apparent presence of heavy metals. For example, intensive studies in two Delaware watersheds have shown cadmium, lead, and mercury to be present in measurable quantities (Biggs *et al.*, 1972):

Metal	Maximum Conc. (ppb)	Watershed 1 Average Conc. (ppb)	Watershed 2 Average Conc. (ppb)
Cadmium	80.0	19.5	16.0
Lead	60.0	9.0	12.7
Mercury	1.8	0.5	0.44

In addition, rainfall collected near a lake in Ontario that has become acidic in recent years has demonstrated sulfate values of 11 ppm maximum, and 7 ppm for a 10-month average. The pH ranged from 3.9 to 4.3, and it was stated that "virtually all fish in the lake had been eliminated" (Beamish, 1974). It is interesting to note that the major source of these materials has been attributed to the TVA.

River Waters

River waters serve many uses. Inasmuch as water is known as the "universal solvent," it follows that runoff will contain anything that the water will dissolve as it flows across the ground. Thus turbidity, taste and odor, color and dissolved minerals are expected to be found in a river water. The BOD will usually be from 1 to 2 ppm, and the dissolved oxygen less than saturation. In addition, because of natural flow variations, the quality characteristics are expected to be quite variable.

Impoundments, Lakes, and Ponds

An ideal source of water is one that is constant in quantity and quality. While an impounded water may supply a constant quantity of water, its composition may vary with season and depth. Under stratified conditions, a lake may demonstrate high oxygen levels and high temperatures in the epilimnion, and low oxygen levels and low temperatures in the hypolimnion. In addition, hypolimnetic waters may contain undesirable levels of iron and manganese. These conditions may change with the fall overturn when the lake will demonstrate homogeneous, but different, water quality conditions.

Groundwater

Groundwater is probably the best source of water inasmuch as it may be constant in quantity and is usually constant in quality. It is also usually free from organic odors although it may be deficient in dissolved oxygen concentration. It also may contain excessive quantities of carbon dioxide, hydrogen sulfide, methane, iron, manganese, and hardness. In many areas of the United States, groundwater comprises the major source of water, even though it may contain objectionable substances such as arsenic, selenium, and fluorides.

Spring Waters

Spring waters are not thought to be ideal sources of water, although some areas use them exclusively. They may show high turbidity subsequent to precipitation, and their flow may not be dependable. In addition, spring waters are subject to contamination by solution channels.

Ocean Water

In order to complete this discussion, the composition of sea water should be noted. Table 3.5 shows the composition of seawater found in

TABLE 3.5   Seawater composition[a]

Cation	Ppm	Anion	Ppm
Ca^{2+}	1,000	HCO_3^-	115
Mg^{2+}	5,234	Cl^-	26,790
Na^+	22,950	Br^-	41
K^+	486	SO_4^{2-}	2,757

[a]From Nordell (1961).

most oceans. It should be noted that these values, expressed as equivalent $CaCO_3$, are usually constant throughout the world.

REFERENCES

Anon. (1970). PCB-Indications of Effects on Fish. *Proc. PCB Conf.*, p. 59067, Wenner-Gren Center, Stockholm, Sweden (September 29).

Anon. (1974). Work plan. IJC Menomonee River Pilot Watershed Study, Wisconsin Dept. Nat. Resour., Univ. of Wisconsin, Southeast Wisconsin Region. Plan. Comm., Madison (September).

Beamish, R. J. (1974). Loss of fish populations from unexploited remote lakes in Ontario, Canada as a consequence of atmospheric fallout of acid. *Water Res.* **8**, 85–95, (January).

Bevis, H. A. (1960). Significance of radioactivity in water supply and treatment. *J. Am. Water Works Assoc.*, **52**, (7), 841–846, (July).

Biggs, R. B., Millen, J. C., and Ottley, M. J. (1972). Trace metals in several Delaware Watersheds. Univ. of Delaware Water Resour. Center, Newark, Delaware. (June).

Brown, R. M., McClelland, N. I., Deininger, R. A., and Tozer, R. G. (1972). "A water quality index crashing the psychological barrier. *Proc. Int. Conf. Int. Assoc. Water Pollut. Res.*, 6th, Jerusalem, Israel.

Brown, R. M., McClelland, N. I., Deininger, R. A., and Tozer, R. G. (1970). A water quality index–do we dare? *Water Sewage Works* (October).

Carson, R. (1962). "Silent Spring. Houghton, Boston, Massachusetts.

Clark, J. W., Viessman, W., Jr., and Hammer, M. S. (1971). "Water Supply and Pollution Control." International Textbook, New York.

Federal Water Pollution Control Administration (1968). Water quality criteria. Rep. Natl. Tech. Adv. Comm., Washington, D.C.

Hansen, R. E. (1978). "Update," *J. Am. Water Works Assoc.* (November).

Harkins, R. D. (1974). An Objective Water Quality Index. *J. Water Pollut. Control Fed.*, **46**, 558.

Imhoff, K., and Fair, G. M. (1956). "Sewage Treatment," 2nd ed. Wiley, New York.

Johnson, H. E., and Ball, R. C. (1972). "Organic Pesticide Pollution in an Aquatic Environment" (S. D. Faust, ed.). Amer. Chem. Soc., Washington, D.C.

Johnson, H. E., and Pecor, C. (1969). *Trans. North Am. Wildl. Nat. Resour. Conf.*, 34th, 159–166.

Klein, L. (1962). "River Pollution: Causes and Effects, Vol. II." Butterworth, London.

Krenkel, P. A. (1974). Sources and classification of water pollutants. *In* "Industrial Pollution" (I. Sax, ed.) Van Nostrand-Reinhold, Princeton, New Jersey.

Krenkel, P. A. (1979). "Personal Communication." Tenn. Valley Auth., Chattanooga, Tennessee.

Maxcy, K. F. (1956). "Preventive Medicine and Public Health." Appleton, New York.

McGauhey, P. H. (1968). "Engineering Management of Water Quality." McGraw-Hill, New York.

McKee, J. E., and Wolf, H. (1963). Water quality criteria. State Water Qual. Control Board, Publ. 3-A, Sacramento, California.

Muss, D. L. (1962). Relation between water quality and deaths from cardiovascular disease. *J. Am. Water Works Assoc.* **54**, 1371–1378.

National Academy of Sciences, and National Academy of Engineering (1973). Water quality criteria–1972. EPA-R3-73-033 (March).

Nordell, E. (1961). "Water Treatment for Industrial and Other Uses." Van Nostrand-Reinhold, Princeton, New Jersey.

Sawyer, C. N., and McCarty, P. L. (1967). "Chemistry for Sanitary Engineers." McGraw-Hill, New York.

U.S. Department of Health, Education and Welfare, Public Health Service (1962). Drinking water standards. Washington, D.C.

U.S. Environmental Protection Agency (1973). Drinking water standards. Advisory Comm. Revision Appl. of Drinking Water Standards (September).

U.S. Environmental Protection Agency (1976a). Disinfection of wastewater. Task Force Rep., EPA-430-9-75-012, Washington, D.C. (March).

U.S. Environmental Protection Agency (1976b). Quality criteria for water. EPA, 440-9-76-023, Washington, D.C.

4 | Biological Aspects of Water Quality

INTRODUCTION

Inasmuch as the major impetus to water quality control is the protection and propagation of aquatic life, it is imperative that the water quality engineer have some understanding of the biological sciences. A major problem in the past has been associated with a lack of communication between the engineer and biologist. If we are to attain the goal of maintaining the integrity of our waterways, the biologist must understand what information is needed by the engineer, and the engineer must comprehend the data collection, analysis, and interpretation problems of the biologist.

[margin note: Major impetus of WQ control is prot. + prop. of aquatic life. ∴ Need understanding of biol.]

This chapter is not intended to be a treatise on aquatic biology but, instead, plans to delineate fundamental biological principles as applied to water quality management. Two biological areas are of prime interest to water quality control. One is concerned with the transmission of water-borne disease, and the other deals with the entire food web as it pertains to the protection and propagation of the biota. The trend toward quantitative biology is encouraged by the authors inasmuch as numerical indices, such as a species diversity index, are much more useful than mere taxonomic enumeration.

[margin note: Interest Areas ① Disease prop. ② Biota protection]

71

Ecosystems

Aquatic environments are examples of ecosystems. The science dealing with ecosystems is called ecology, which has been defined as the study of interrelationships between living organisms and their environment. The word is derived from the Greek *oikos,* which means "house" or "dwelling place" and the ending for science or study, "-ology" (Sewel, 1975).

Until recently, the science of ecology was primarily concerned with the taxonomy (identification and classification) of nature's structure and the narrow functions of living species. Subsequent to the environmental movement, the ecologists focused their attention on the dynamics of ecosystems and their behavior and response to various inputs to the systems. The science of ecology is now able to more quantitatively predict the response of ecological systems to new inputs and varying environmental conditions. It is interesting to note, in this regard, that mathematics is becoming increasingly predominant in biological science curricula.

Ecosystems can be divided into two components, the biotic, or living, and the abiotic, or nonliving. Ecosystems are considered to be in a dynamic equilibrium with their inputs and surroundings. When one of the inputs is changed, the ecosystem will readjust to a new equilibrium.

Sun energy reaching water surfaces is the major input to aquatic ecosystems. However, other energy inputs, such as organics in wastewaters, can significantly contribute to the overall composition and behavior of these systems.

The biotic component of aquatic ecosystems can be divided into three groups of organisms.

(1) Producers can accept inorganic materials and energy from the sun to produce new organic matter. This process is called photosynthesis, and the group of participating organisms are known as autotrophic. These include green aquatic plants attached to the bottom or vascular aquatic plants and microorganic planktonic organisms known as phytoplankton. A special group of autotrophic microorganisms utilizing energy obtained from the oxidation of ammonia commonly called nitrifiers are known as chemoautotrophs. Their energy results from the oxidation of reduced inorganic compounds. As with the photoautotrophs, the chemoautotrophs use carbon dioxide as their sole carbon source. However, the photoautotrophs obtain their energy from sunlight.

The producers are the essential building blocks in the construction of basic primary-organic matter and the only organisms which can utilize inorganic materials and convert them to new organic tissues. Typical examples include algae, aquatic weeds, and diatoms. Since most of these organisms rely on radiation from the sun for their source of energy, they will be found in large quantities only in the upper zone of water bodies, the depth of which will be limited by the depth of appreciable solar radiation penetration. This water depth is called the euphotic zone and depends on the turbidity of the water.

(2) Consumers are organisms which utilize organic carbon as their source of energy. They apparently depend on producers as their source of nourishment. The primary consumers can utilize plant and algal organic tissue directly, while secondary consumers must obtain their nourishment from the primary consumers.

(3) Decomposers are the organisms, primarily bacteria and fungi, that can decompose organic matter to basic minerals and organic residues which may then become again available to the producers. The decomposers thus represent the other end of the production of organic matter and its cycle.

The heterotrophic microorganisms, including both primary consumers and decomposers, are important in purifying wastewaters and surface water polluted by organics. Heterotrophs may use a multitude of carbon compounds as a carbon source. These bacteria and fungi are necessary for the process known as stream self-purification.

stream self-purification by producers and primary consumers.

The relative distribution of producers, consumers, and decomposers depends on the type and amount of energy stored or entering the system. Different types of organisms will be emerging, growing, and dying at rates affected by ambient conditions. The overall growth rate for each group of organisms is a summation of their growth rate, death rate, and grazing by other organisms. If a water body initially possesses a low number of consumers or decomposers, producers will develop, sometimes in very large quantities, depending upon optimal energy from the sun and mineral nutrient conditions. As the overall organic matter concentration increases, both consumers and decomposers will develop, and when the system energy inputs diminish, the decomposers will liquidate the organic matter. The process is schematically shown in Fig. 4.1.

The food chain or food web is the method by which energy is transferred through the biosphere as shown in Fig. 4.2. The ingestion of a smaller consumer by a larger one establishes a trophic level, with the producers at the lowest trophic level. On the other hand, in the terrestrial environment, man is at the highest trophic level.

Of most concern to the biologist is the potential concentration increase of toxic materials that can occur via the food chain. The process, called biomagnification, is of particular concern today because of the wide distribution of certain exotic organics, such as pesticides, and the ability of cer-

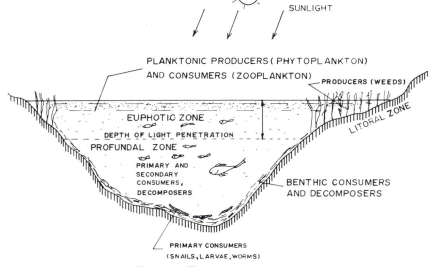

Fig. 4.1.   The aquatic ecosystem.

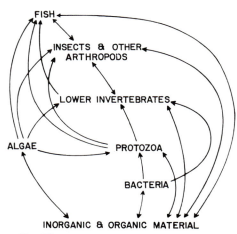

Fig. 4.2.   Simplified aquatic food web. (Parker & Krenkel, 1967)

TABLE 4.1 Concentration of
DDT in a marine environment

Species	Trophic Level	DDT (ppm)
Oar weed	1	0.003
Sea urchin	2	0.05
Lobster	3	0.024
Shag liver	4	2.87
Cormorant liver	5	4.14

tain organisms to concentrate them. Probably the most publicized bio-magnification demonstrated has been that of DDT, as shown in Table 4.1 (Robinson *et al.*, 1967). It is interesting to note that the concentration of DDT was 2.87/0.003, or 957 times through four trophic levels.

Several other definitions should be mentioned. Plankton are organisms of small size that drift in the water subject to the action of the water. They may be defined further as zooplankton, or animal plankton, and phytoplankton, or plant plankton. Nekton are organisms of larger size, such as fish, that swim freely and determine their own distribution in water, while seston are nonliving and living bodies of plants and animals that float or swim in the water.

Periphyton are attached organisms, as are sessile organisms. Productivity is a time-rate unit of the total amount of organisms grown, while primary productivity is the production of photosynthetic organisms. Biomass is the weight of all life in a specified unit of the environment and standing crop is the biota present in an environment at a selected point in time. Organisms living in symbiosis are living together for their mutual benefit. Finally, synergism is the combination of two materials which, by themselves, may be innocuous, but add to each other's toxicity. Antagonism is the antithesis of synergism.

ECOSYSTEM RESPONSE TO STRESS

When a wastewater is introduced into a receiving water, it is obvious that the biota will be stressed and respond accordingly. An outstanding pictorial representation of the effects of an organic wastewater discharge (a sewage) on a river was presented by Bartsch and Ingram (1959) and will be used as a basis for this discussion.

Figure 4.3 shows a hypothetical river along the horizontal axis, with an organic waste (sewage) being discharged into the river at day and/or mile

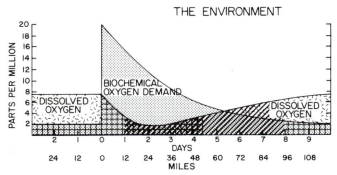

Fig. 4.3. The Effects of a wastewater discharge on the dissolved oxygen content of a waterway (Bartsch and Ingram, 1959).

zero. Both BOD and DO are plotted before and after the discharge. Prior to the discharge, the DO is near 7 ppm and the BOD near 2 ppm, values that might be expected in a relatively unpolluted river during the summer. The wastewater discharge contributes some 20 ppm of BOD, which is subsequently oxidized, and approaches the original river BOD value at day 9. On the other hand, the DO is seen to fall at the point of discharge to some minimum value near day 2, and then approach normal conditions near day 9.

The decrease in BOD and the decline and rise in DO depict the process known as stream self-purification, which is highly dependent on the rate at which oxygen is absorbed back into the stream (reaeration), which, in turn, is dependent on the turbulence intensity and the water depth.

Figure 4.4 shows the same situation with the potential effects of photosynthesis superimposed. Note that while oxygen is produced during the daytime and may even approach supersaturation, the photosynthetic organisms behave as bacteria during the nighttime hours, utilizing oxygen and producing carbon dioxide. Under certain conditions, more oxygen may be used by respiration than produced by photosynthesis. It is obvious that when the upper curve is integrated with the lower curve, cyclic variations in DO content will occur and must be accounted for.

Again referring to the same conditions depicted in Fig. 4.3, several other stream phenomena are illustrated in Fig. 4.5, including the nitrification process and anaerobiosis. In addition, the growth of bacteria is shown as a typical logarithmic growth curve subsequent to the addition of an organic wastewater. It should be noted that while anaerobic conditions are unlikely to occur with today's regulations, the processes illustrated may very well occur at the mud–water interface, which may be anoxic. It is also pertinent to observe that methane may be produced under anoxic

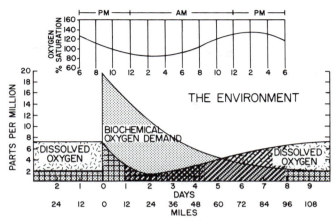

Fig. 4.4. The effects of photosynthesis on the situation depicted in Fig. 4.3 (Bartsch and Ingram, 1959).

conditions in addition to odor producing reduced sulfur materials. The nitrification process is important because the conversion of organic nitrogen to NH_4^+ nitrogen to NO_2^- nitrogen to NO_3^- utilizes oxygen and must be accounted for in describing the oxygen balance of a receiving water.

Figure 4.6 demonstrates the food chain and shows how one species proliferates at the expense of others. Note that the bacteria are eaten by the

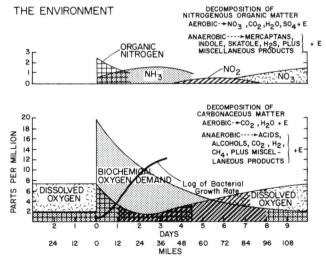

Fig. 4.5. The effects of nitrogenous oxygen demand and anaerobic-decomposition on the situation depicted in Fig. 4.3 (Bartsch and Ingram, 1959).

THE BIOTA

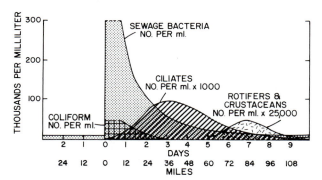

Fig. 4.6. The downstream biota as a result of the discharge depicted in Fig. 4.3 (Bartsch and Ingram, 1959).

protozoans, whose population is minimal prior to the introduction of the sewage and subsequent prolific growth of the bacteria. After a few days, the environment becomes ideal for the ciliated protozoans, and they predominate in the zooplankton population. Further downstream, the rotifers and crustacea predate the protozoans and then become the dominant microscopic zooplankton in the river.

The final curve in this series represents the principle upon which current quantitative biological observations are based. Figure 4.7 illustrates two biological measurements. The upper curve shows the number of

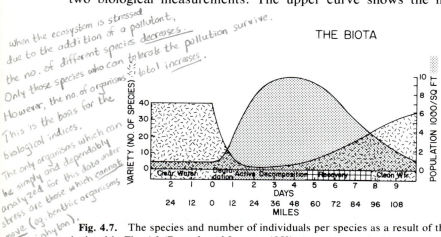

THE BIOTA

When the ecosystem is stressed due to the addition of a pollutant, the no. of different species decreases. Only those species who can tolerate the pollution survive. However, the no. of organisms / total increases.

This is the basis for the biological indices.

The only organisms which can be simply and dependably analyzed for this data under stress are those which cannot move (eg. benthic organisms and periphyton).

Fig. 4.7. The species and number of individuals per species as a result of the discharge depicted in Fig. 4.3 (Bartsch and Ingram, 1959).

species present, and the lower curve depicts the number of individuals per species present. Again, the situation described in Fig. 4.3 is repeated, only now it shows the effects of stress on the biota. Note that prior to the wastewater discharge, there are many species present but with few individuals per species. Subsequent to the induced stress, however, the pollution tolerant organisms proliferate while others disappear. Obviously, the number of individuals per species increases dramatically. This effect of stress on the biological community is the basis for the biological indices which will be discussed subsequently.

While the phenomena described occur with respect to many species, the measurement of each component of the food chain would be prohibitively expensive. Furthermore, it has been the authors' experience that the only organisms that can be simply and dependably analyzed for species and individuals per species under stress are the benthic organisms and the periphyton because they cannot move and avoid the stress. The plankton move with the current, and the nekton move at will. Thus, measured numbers and species of these organisms may not be indicative of the true composition of the biota. It is interesting to note in this regard that fish may be attracted toward effluents because of their food content; thus, fish populations in or near an effluent plume may not reflect the true census. On the other hand, fish have the ability to detect and avoid harmful materials at very low concentrations, again resulting in biased observations in a survey.

Many times the surfaces on which periphyton can grow are either sparse or nonuniform. Thus, an artificial substrate is recommended for quantitative analysis. The most commonly used artificial substrate is the standard 25×75 mm glass microscopic slide (APHA/AWWA/WPCF, 1975). The slides are mounted on a rack constructed of plexiglass and styrofoam, submerged below the water surface and anchored to the bottom for a period of about two weeks, depending on the season. The slides can then be taken back to the laboratory and subjected to analysis, as described by Patrick *et al.* (1954).

BIOLOGICAL INDICES OF WATER QUALITY

As shown in the previous section, when sewage or other wastewater is introduced into a receiving water, the aquatic ecosystem must readjust to different environmental conditions and achieve a new equilibrium. Surface waters have a capability to accept and reduce pollution by various processes including dilution, sedimentation, oxidation, and biological degradation. Wastewaters containing organics and nutrients serve as food to various microorganisms and other aquatic species. The species feeding on the pollution will develop in large numbers and may become predominant, and other less tolerant species may not survive. Patrick (1953) defined pollution as "anything which brings about a reduction in the diversity of aquatic life and eventually destroys the balance of life in a stream."

Organisms are known to respond to various food levels and ambient stresses. Some are more tolerant to pollution and even use waste organics and minerals as their food, while others will experience a reduction in their development and/or die if their ambient conditions are disturbed by the presence of a contaminant. Typical "clean water" organisms include diatoms, many benthic invertebrates, insect larvae, shellfish, and fin fish such as salmon, bass, and trout. Organisms which prefer high concentrations of organics and can tolerate higher pollution levels include heterotrophic bacteria, fungi, snails, worms, and fish such as carp, catfish, and bream. A skilled limnologist can evaluate the state of water quality according to the number and distribution of planktonic and benthic organisms and microorganisms.

Several methods of water quality classification have been developed using biological indices. One of the first was developed by Kolkwitz and Marsson (1908) and is widely used throughout Europe. This system is based on the tolerance of aquatic organisms to various levels of pollution. Other systems which followed are based on aquatic species diversity as affected by pollution.

Biological indices offer some advantages when compared to chemical indicators. For example, chemical surveys indicate stream conditions only at the time of sampling, and, sometimes, the true picture of water quality may not be detected because of lack of data or the limited number of parameters surveyed. Biological methods, on the other hand, are more comprehensive. They reflect the overall water quality conditions over a period of time preceding the survey. However, the lack of unification among biologists and the approximate nature of biological methods precludes an accurate quantification of water quality using only biological indicators. Obviously, close collaboration among water quality specialists in the environmental engineering, chemical, and biological fields is necessary in order to properly and adequately quantify water quality conditions during a survey.

Saprobic System

Kolkwitz and Marsson (1908) attempted to classify aquatic organisms and plants according to the ranges of conditions which occur in surface waters subjected to different levels of pollution caused by sewage discharges. Such classification requires that the respective organisms are uniquely dependent, within relatively narrow ranges, on the chemical composition of the water for their distribution and development. These organisms were classified as being polysaprobic, α- or β-mesosaprobic, or oligosaprobic.

Kolkwitz and Marsson (1908) defined the zones of pollution as:

The polysaprobic zone is characterized by a high concentration of decomposable complex components, a decrease of oxygen, the formation of hydrogen sulfide in the mud, and an increase in carbon dioxide. Organisms generally occur in large numbers but with certain monotony. The bacterial content as estimated by the standard gelatin plate, may exceed 1 million per milliliter of water. Organisms with high oxygen requirements are absent, and fishes usually avoid remaining in this zone for any length of time. The mud is black and odorous because of the presence of sulfides.

The zone of mesoprobia is divided into an α- and/or a β-saprobic section, and it generally follows the polysaprobic zone. In the α-section, the oxygen content may be high but is usually less than 50% of saturation. Chlorophyll containing organisms may develop, and the oxygen content will thus vary between day and night. Odor conditions caused by hydrogen sulfide are ameliorated by the oxygen, and the black muds disappear. Bacterial counts in the α-mesosaprobic zone are usually less than 100,000 per milliliter. The protein components contained in the water in this section are probably already decomposed to amino acids.

In the β-mesosaprobien zone, the decomposition products already approach mineralization, and the oxygen content is greater than 50% of saturation. In addition, bacterial counts are always less than 100,000 per milliliter.

All organisms of the mesoprobic section usually are resistant to minor action by sewage and its decomposition products. Diatomaceae, Schizophyceae, many Chlorophyceae, and some higher plant organisms are present. Higher and lower animal organisms are also found in a great number of individuals and varieties, although the macrofauna are still restricted in species in the α-saprobic zone, while a greater diversity of plants and animals exists in the β-mesosaprobic zone.

The oligosaprobic zone is the domain of (practically) pure water. If it is preceded by a self-purification process locally or chronologically, it succeeds the mesosaprobic zone and then represents the end of mineralization. The oxygen content of the water is near the saturation limit and occasionally may even exceed saturation. The content of organic nitrogen usually does not exceed 1 mg/liter. The water is generally transparent to a considerable depth, except at times of abundant plant growth. The number of bacteria developed on standard nutrient gelatin is generally low. Many species of animals and plants, including fish, are found in this zone.

Some investigators found the Kolkwitz–Marsson division into four quality classes insufficient on both ends of the quality scale. Kathsaprobity and xenosaprobity were added on the clean water side to represent water with purer quality than in the oligosaprobic zone. These "ultra-clean" zones may be typical for some mountain lakes and streams, especially those fed by glaciers. Four additional classes were added to represent water quality worse than that occurring in the polysaprobic zone and were primarily designed for sewage and wastewater effluents (Sládeček, 1973).

Several indices have been developed to specify the degree of saprobity at a sampling station. Pantle and Buck (1955) calculated a saprobein index (S) for each sampling station by the following formula, which uses biological analysis of organisms found in the water and benthos samples, as

$$S = \Sigma s \cdot \frac{h}{\Sigma_h}, \tag{4.1}$$

where s is the degree of saprobity for a species and h is the relative frequency of each species. The value of S ranges from 1 to 4. Lower values are indicators of clean water, and higher ones indicate pollution. Indicator values of various species have been listed by Liebman (1962). The value of h is 1 if the species is found only by chance, 3 if the species occurs frequently, and 5 if the species is abundant. The degree of pollution of the sampling site is then determined from the following relationship (Pantle and Buck, 1955):

S	Degree of Pollution	Saprobity
1.0–1.5	very slight	oligosaprobic
1.5–2.5	moderate	β-mesosaprobic
2.5–3.5	strong	α-mesosaprobic
3.5–4.0	very strong	polysaprobic

A classification enabling extension of the saprobien system into "ultra-clean" and "sewage" zones was proposed by Zelinka and Marvan (1961). A saprobien valency and indicator value are obtained for each species. The saprobien valency depends upon the relative frequency of the species at different levels of pollution and varies from 1 to 10. The sum of the valencies in the different saprobic zones is 10 for each species. A numerical indicator value of each species is also given by the authors with a value of 1 assigned if the species has a small indicator value and 5 if the species has a high indicator value. The values characterizing the degree of

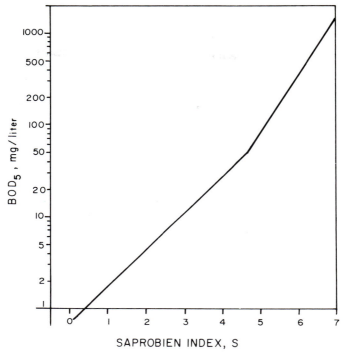

Fig. 4.8. The relationship of average BOD to the saprobian index (Sládeček and Tucek, 1974).

pollution are then determined by a mathematical equation

$$S = \frac{\Sigma_i g_i h_i S_i}{\Sigma_i g_i h_i}, \qquad (4.2)$$

where g_i is a weight factor assigned to the organism. Table 4.2 shows some typical European benthic invertebrates and their saprobic values.

A close correlation of the saprobien index to BOD_5 levels has been observed as shown in Fig. 4.8. The saprobien index was also correlated with less success to percent oxygen saturation and ammonia concentration (Von Tümpling, 1969).

Biota Composition

Based on the fact that pollution will effect a reduction in the number of species present, Patrick (1949) proposed a system that purportedly describes the biological condition of a receiving water based on arbitrary groupings of organisms that would respond similarly to given environ-

Pollution reduces the no. of species present.

TABLE 4.2 Examples of benthic invertebrates evaluated by the saprobien valency and the saprobien index S^a

Taxon	x	o	β	α	p	G	S
Planaria alpina Dana	10	—	—	—	—	5	0.0
Planaria gonocephala (Duges)	7	3	+	—	—	4	0.3
Limnodrilus hoffmeisteri Clap.	—	—	—	4	6	3	3.6
Tubifex tubifex (O.F. Muller)	—	—	+	2	8	4	3.8
Glossosiphonia complanata (L.)	—	+	6	4	—	3	2.4
Haemopis sanguisuga L.	—	3	7	+	—	4	1.7
Ancylus fluviatilis (O.F. Muller)	1	4	3	2	—	1	1.35
Asellus aquaticus (L.) Racov	—	—	2	8	+	4	2.8
Gammarus pulex (L.) fossarum Koch	4	3	3	—	—	2	0.65
Baetis alpinus (Pict.)	8	2	—	—	—	4	0.2
Baetis bioculatus (L.)	—	1	6	3	—	3	2.25
Baetis gemellus Etn.	7	3	—	—	—	4	0.3
Baetis pumilus (Burm.)	1	4	4	1	—	1	1.55
Baetis rhodani (Pict.)	3	3	3	1	—	1	1.05
Baetis scambus Etn.	—	5	5	—	—	3	1.5
Baetis tenax Etn.	1	5	4	—	—	2	1.35
Baetis vernus Curt.	—	2	5	3	—	2	2.15
Amphinemura sulcicollis Steph.	3	5	2	—	—	2	0.85
Amphinemura triangularis Ris.	1	6	3	—	—	3	1.25
Capnia bifrons (Newm.)	1	6	3	—	—	3	1.25
Chloroperla torrentium (Pict.)	4	6	+	—	—	3	0.4
Isoperla diformis Klp.	—	5	5	—	—	3	1.5
Isoperla grammatica Desp.	—	3	6	1	—	3	1.75
Isoperla rivulorum (Pict.)	7	3	—	—	—	4	0.3
Leuctra aurita Navas.	+	5	5	—	—	3	1.5
Leuctra albida Kny.	+	5	5	—	—	3	1.5
Leuctra braueri Kny.	2	5	3	—	—	3	1.15
Leuctra fusca (L.)	—	2	5	3	—	2	2.15
Leuctra hippopus Kny.	7	3	—	—	—	4	0.3
Leuctra inermis Kny.	4	4	2	—	—	2	0.8
Leuctra major Brk.	—	4	5	1	—	2	1.65
Leuctra nigra (Oliv.) Kny.	1	5	4	—	—	3	1.35
Leuctra prima Kny.	6	4	—	—	—	3	0.4
Leuctra rosinae Kny.	9	1	—	—	—	5	0.1
Nemoura avicularis Mort.	2	5	3	—	—	2	1.15
Nemoura cabrica (Steph.)	5	5	—	—	—	3	0.5
Nemoura cinerea Retius	—	4	4	2	—	2	1.8
Nemoura marginata Ris.	2	5	3	—	—	2	1.15
Perla burmeisteriana Claas	—	5	5	+	—	3	0.5
Perla marginata (Panz.)	4	5	1	—	—	2	0.65
Taeniopteryx hubaulti Aubert	7	3	—	—	—	4	0.3
Taeniopterix nebulosa (L.)	—	5	5	—	—	3	1.5
Rhyacophila nubila Zett.	—	5	5	—	—	3	1.5
Rhyacophila septentrionalis McLach.	1	5	4	—	—	2	1.35
Rhyacophila vulgaris Pict.	3	5	2	—	—	2	0.85
Simuliidae g. sp.	3	3	2	2	—	1	1.15
Simulium hirtipes Fries.	7	3	—	—	—	4	0.3

^{a}From Sládeček (1973). x, Xenosaprobity; o, oligosaprobity; β, β-mesosaprobity; α, α-mesosaprobity; p, polysaprobity; G, indicative weight of the species; S, extended saprobic index.

TABLE 4.3 Taxonomic categories of organisms[a]

Group	Organisms
1	Blue-green algae, certain green algae, and certain rotifers
2	Oligochaetes, leeches, and pulmonate snails
3	Protozoa
4	Diatoms, red algae, and most green algae
5	All rotifers not in Group 1, clams, prosobranch snails, and tricladid worms
6	All insects and crustacea
7	All fish

[a]From Patrick (1950).

mental conditions. The groups are shown in Table 4.3, and the groups included in various stream conditions are shown in Table 4.4. The system compares the number of species in different groups found at a particular station with those found at "healthy" stations.

Figure 4.9 shows the application of this method to a river in Pennsylvania subjected to an organic waste discharge. The ordinate shows the number of species present as a percent of the number of species in that group present at "healthy" stations. The numbers above the bars are the numbers of species in groups at a particular sampling location, while the double width of a bar indicates that a large number of individuals of a species are present.

Wurtz (1955) classified organisms into tolerant and intolerant groups that were burrowing, sessile, foraging, and Pelagic. As shown in Fig. 4.10,

Pelagic (def) - Living in the open sea.

Sessile (def) - permanently attached or attached directly by the base.

TABLE 4.4 Stream condition[a]

Stream Condition	Results
Healthy	Groups 4, 6, and 7 each contain more than 50% of number of species found in that group at 9 typical "healthy" stations.
Semihealthy	(a) Either or both groups 6 and 7 less than 50%, and group 1 or 2 less than 100%, or
	(b) Either group 6 or 7 less than 50%, and groups 1, 2, and 4 100% or more, or group 4 contains exceptionally large number of individuals.
Polluted	(a) Either or both groups 6 and 7 are absent, and groups 1 and 2 50% or more, or
	(b) Groups 6 and 7 both present but less than 50% and groups 1 and 2 100% or more.
Very polluted	(a) Groups 6 and 7 both absent and group 4 less than 50%, or
	(b) Either Group 6 or 7 is present and group 1 or 2 less than 50%.

[a]From Patrick (1950).

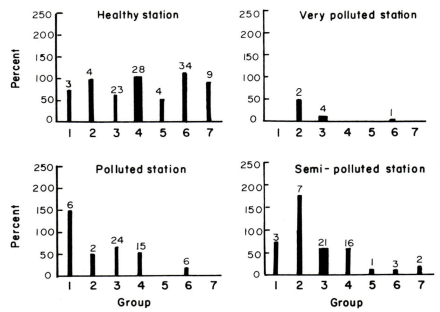

Fig. 4.9.   Histograms showing various degrees of pollution (Patrick, 1950).

each category is plotted above and below a baseline denoting species that are tolerant and nontolerant, with the height of each column representing the percentage of total number of species found at the station. The number above or below the bar represents the number of species in each category. Wurtz considered a stream to be "clean" if the nontolerant species represented more than 50% of the community.

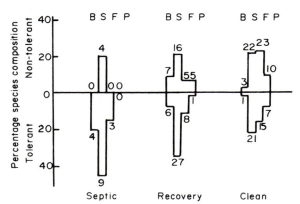

Fig. 4.10.   Histograms showing various degrees of pollution (Wurtz, 1955).

Beck (1955), also using the tolerance or nontolerance of organisms to organic pollution and the concomitant reduction in species, proposed a biotic index which was based on categorizing the macroinvertebrates into tolerant and nontolerant species. Assuming that the presence or absence of tolerant (class I) organisms is more important than the nontolerant (class II) forms, the former are multiplied by 2, with the resulting formulation as

$$\text{biotic index} = 2(n\text{class I}) + (n\text{class II}), \qquad (4.3)$$

where n is the number of species present. As the biotic index approached zero, Beck found that the river conditions tended to be anoxic, and clean streams were found to result in values greater than 10.

More recent approaches to biotic indices have been proposed by Chutter (1972) and Hilsenhoff (1972), both of whom used tolerance levels arbitrarily assigned to benthic organisms.

Hilsenhoff (1972) has proposed a biotic index based on tolerance levels of resident benthic organisms for Wisconsin surface waters. The data for this index are obtained from samples of the stream substrate taken in an appropriate riffle area using a simple D-frame aquatic net. The organisms collected are then identified (preferably to species level, but generic names are sufficient) and counted. The quality values assigned to each species by Hilsenhoff are then multiplied by the quantity of each species collected, with a maximum number of 25 allowed for each. This number is then divided by the total number of organisms in the sample. The mathematical formulation of the Hilsenhoff biotic index (BI) is

$$\text{BI} = \sum_{i=1}^{n} [(q)(s)]/t, \qquad (4.4)$$

where q is the assigned quality value of the organism, s is the number of species i, t is the total number of organisms, and n is the number of species. For example, assume that the following organisms have been collected:

Species	Number of Organisms, s	Quality Value, q
Gamarus	11	2
Tipula	2	2
Hydropsyche	25	3
Stenelimis	7	3
Brachycentrus	20	1
Baetis	25	3
Helichus	2	1
Dicranota	1	0
	$t = 93$	

The resulting biotic index for these data is $\frac{219}{93}$ or 2.35. This index appears to be similar to the European saprobien index but deficient in the extensive scientific knowledge gathered in the development and application of the saprobien system.

Gabriel (1946) used the relationship between the number of producers (P), decomposers (reducers) (R), and consumers (C) in the following biological index, I, of organic pollution:

$$I = \frac{2\ P}{(R + C)}. \qquad (4.5)$$

This index is based on the assumption that reducers and consumers are abundant in polluted waters, resulting in a low index. Further downstream, below the source of pollution, the number of autotrophic (producing) organisms increases with a consequent reduction of consumers and reducers, resulting in an increase of the index value.

It should be noted that while these attempts to quantify biological data are useful and commendable, they all suffer from several shortcomings. For example, organisms are not all tolerant or nontolerant. Thus, the results depend on the investigator's groupings for which category is appropriate for a given organism. In addition, the use of such indices may result in a loss of information because of the method of data reduction. Finally, the relative abundance of individuals of different species may not be taken into account, nor the predominant species in the community. From the biologist's standpoint, it would appear desirable to use the usual tabulation of data in conjunction with one of the described indices.

Species Diversity Relationships

Margalef (1951) proposed the following index to describe species diversity:

$$d = (s - 1)/\ln n, \qquad (4.6)$$

where d is the species diversity index, s is the number of species, and n is the total number of individuals. Wilhm (1972) reported that values obtained with this measure varied from 0.08 at an enriched station to 1.43 at a clean water station. In an earlier paper, Wilhm and Dorris (1968) pointed out that it was possible to obtain the same diversity index with communities having the same number of species and individuals, even though the distributions of individuals among species significantly differed. Their hypothetical situation is shown below (Wilhm and Dorris, 1968):

Individuals in Species i (n_i)

Community	n_1	n_2	n_3	n_4	n_5	Total Individuals	Total Species
A	20	20	20	20	20	100	5
B	40	30	15	10	5	100	5
C	96	1	1	1	1	100	5

In 1957, an index of diversity derived from information theory was proposed by Margalef (1957) which related the concept of species diversity to the concept of information content as developed by Shannon and Weaver (1963). Diversity is equated with the uncertainty which exists concerning the species of an individual selected at random from the population. The more species present in a community and the more equal their abundance, the greater their diversity. This index of diversity, I, is determined by the following relationship:

$$I = - \sum_i p_i \log_e p_i, \qquad (4.7)$$

where $p_i = n_i/N$, n_i is the number of individuals in the ith species and N is the total number of individuals. If only one species is present in the population, the index is equal to zero, and as the number of species increases,

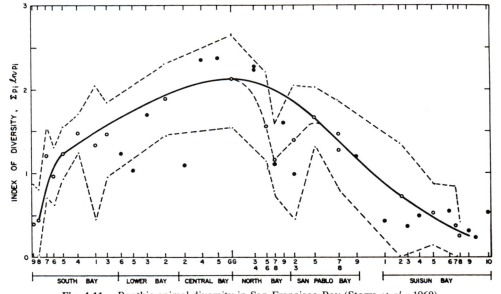

Fig. 4.11. Benthic animal diversity in San Francisco Bay (Storrs *et al.*, 1969).

the index increases. Equation (4.6), known as Shannon's index, is probably one of the most satisfactory expressions of species diversity and has been widely used to ascertain the water quality of surface waters.

The species diversity index ascertained by Shannon's formula was used for evaluation of the water quality of San Francisco Bay as shown in Fig. 4.11. In that study, the diversity index, based on benthic animal diversity, was closely related to the salinity of the overlying water and the percentage of sand in the bottom layer.

Wilhm (1972) suggested a scale for evaluating pollution effects based on benthic macroinvertebrate community diversity. If Shannon's formula is used, values less than 1 indicate heavy pollution, values from 1 to 3 represent intermediate conditions, and values of more than 3 characterize clean waters. Staub *et al.* (1970) used Shannon's formula to evaluate the effects of industrial wastes on primary planktonic producers. They proposed a scale in which a diversity value of 0 to 1.0 indicated heavy pollution, 1.0 to 2.0 moderate pollution, 2.0 to 3.0 light pollution, and 3.0 to 4.5 slight pollution.

Sequential Comparison Index

Finally, the sequential comparison index, which is a diversity index designed for nonbiologists, was proposed and developed by Cairns *et al.* (1968). In this method, a sample of microorganisms collected in a stream is placed on a slide and examined sequentially with a microscope. Observations proceed from left to right and organisms are compared one by one to the preceding organism. If the organism appears similar to the previous one, it is part of the same "run"; if not, it is part of a new run. The more runs for a given number of specimens, the greater the biological diversity. The diversity index is then determined as

$$DI = \frac{\text{number of runs}}{\text{number of specimens}}.$$

Pollution effects will result in low values of the sequential diversity index.

Since this method was designed for nonbiologists and is lacking any scientific identification and qualification of the specimens, it is obviously not as adequate as methods designed for qualified biologists and only should be used if no other estimation is possible.

TOXICITY BIOASSAYS

The presence of toxic substances in surface waters is highly undesirable. As previously stated, they can damage the aquatic ecosystem, cause fish kills, and/or contaminate fish tissues so much that the fish become unfit for commercial fishing or sportfishing. In addition, toxic materials can affect the biological processes occurring in the aquatic environment and reduce the waste assimilative capacity. Furthermore, toxic substances cause many fish kills in the United States each year. Thus, toxicity is one of the primary biological indicators of the quality of surface waters and wastewater effluents.

Theoretically, every compound can be toxic (even kitchen salt), depending only on its concentration and the tolerance of the biota in the aquatic environment to a particular toxicant. Two kinds of toxicity can be distinguished: acute toxicity, in which the toxic effect (death or visible damage to the organisms) can be identified in a short period of time, and chronic toxicity, in which the effects of the toxic materials can be detected only after long periods of time, usually months or years. Chronic toxicity may sometimes involve the accumulation of a toxic compound in the tissue or vital parts of the organisms. In addition, chronic toxicity is important in establishing maximum allowable concentrations of various components in drinking water supplies and in fish designated for human consumption.

Toxicity is usually considered a man-caused cultural phenomenon resulting from the discharge of contaminated wastewaters. However, sometimes toxicity is caused by natural biological processes taking place in surface waters. Hydrogen sulfide or hydroxyamine, both known by-products of natural decay processes, can be toxic to aquatic organisms.

A major problem with toxic substances is the lack of quantitative data delineating "safe" concentrations of the many new compounds that are discharged into receiving waters. Even materials that are relatively well known are not well documented in terms of their toxicity to aquatic life. The consent decree reached on June 7, 1976 (Chapter 2) recognized the importance of toxic materials and resulted in the delineation of 65 toxic pollutants requiring priority treatment. This list has been expanded to 129 compounds and classes of compounds and could result in thousands of priority pollutants if all of the compounds in each of the classes were included. The problem of chemical analysis of these materials is exacerbated by the paucity of data on biological variability, synergism and antagonism, and long-term effects of the mixtures of toxic materials that may result in a waterbody.

Because of the importance of quantifying the effects of toxic materials on the biota, the biologist (and the engineer) has struggled for many years with a means for assessing toxic effects on aquatic life. As will be discussed subsequently, existing techniques leave much to be desired; however, the standard bioassay procedures do form a basis for the development of water quality criteria for the protection of aquatic life. Furthermore, mixtures of unknown materials can be used to determine their toxicity, even though the precise content of the mixture may not be known.

The potential applications of toxicity bioassays include determination of the need for dilution, treatment, or elimination of potentially hazardous substances present in some wastewater effluents, the potential effectiveness of a proposed treatment scheme, and the monitoring of treatment efficiency. Although the toxicity bioassay does not necessarily involve chemical knowledge of the toxicant, chemical analysis and toxicant identification should always be attempted for proper interpretation of the tests.

Toxicity established by a toxicity bioassay is more or less limited to the locality in which the sample was taken inasmuch as there are many additional factors which can affect the level of toxicity. These may include temperature, pH (especially for ammonia and amine toxicants), hardness, organic content, dissolved oxygen content, adsorptivity and/or immobilization of toxic components by suspended particles, and other factors (Sládeček *et al.*, 1975).

The presence of toxic components can be qualitatively identified by field observations or even by some simple laboratory BOD analyses. An experienced biologist can detect toxicity if the benthic or free-flowing water organisms typical for the stream are missing or are in limited numbers, and a microbiologist can detect toxicity if proper growth of bacteria on the plates does not occur. BOD tests performed either by the standard bottle technique or by a respirometer may show an increase in oxygen demand with increased dilution if toxic materials are present. However, a quantitative numerical identification of toxicity should be performed under standard laboratory conditions with specified selected test organisms in order to verify the toxicity.

Acute Toxicity

The purpose of the laboratory toxicity bioassay is to demonstrate maximum concentrations of a substance which will not cause adverse effects on the test organism. Since the test usually involves observations of the death of test organisms which may differ slightly in their tolerance, the "lethal" concentration is a statistical quantity based on the death of 50% of the test organisms during a specified time period. This quantity or concentration has been called the median tolerance limit or TL_m, and is usually estimated at 24, 48, 72, or 96 hours. The trend today is to use the symbol LC_{50}, which denotes the lethal concentration for 50% of the individuals. The term EC_{50} is used to describe adverse effects other than death in 50% of the test organisms within the prescribed test period.

Ideally, the test organisms should include representatives from four *Test*
groups, i.e., microorganisms, plants, invertebrates, and fish. The test *organisms*
organisms should be amenable to captivity, accurately identified, relatively uniform in size and healthy, and acclimated to laboratory conditions. The number of organisms in each test group should be at least ten,
and the tests should be conducted at uniform temperatures (20°–25°C for
warm water biota and 18°–21°C for cold water biota). The dead organisms
should be removed immediately.

The test organisms are placed in containers with various dilutions of the
toxicant plus one container with test water only. Feeding of the organisms
should be avoided during the test. The number of organisms surviving
after the specified test time period(s) (preferably 48 hours or more) is then
plotted v. the logarithm of the concentration of the toxicant in the containers. The concentration in which 50% of the test organisms survive for a
given time is then denoted as the LC_{50}. The graphical technique eliminates possible errors caused by one sick organism, different tolerances,
and/or different conditions in one container.

The lack of standardized bioassay methodology has been a major
problem in the interpretation of bioassay results. *Standard Methods*
(APHA/AWWA/WPCF, 1975) has developed a consensus method for
bioassays; however, EPA has developed their own procedures, which are
described in USEPA (1975) and USEPA (1978). The EPA (USEPA, 1978)
describes preliminary, short-term, static, range-finding tests to be used in
defining the recommended long-term (96 hours), flow-through definitive
tests for use in determining acute toxicity. The details of the test procedures are described in all of these documents including recommended test
organisms, water quality, equipment, and analysis.

The graphical procedures for estimating the LC_{50} are still not standard- *what is a*
ized, inasmuch as logarithmic scales, arithmetic scales, and probit scales *probit scale?*
are all used, even though the probit scale is theoretically sounder and
many times will assist in linearizing the relationship. For many of the tests
quoted in the literature, the procedure was to plot the concentration of
waste on a logarithmic or arithmetic scale and the percentage survivors on
an arithmetic scale. The median or LC_{50} concentration was determined by
connecting the two highest points on the graph which are separated by the
50% survival line and to connect them with a straight line, thus ignoring
all of the other data points. The procedure is shown in Fig. 4.12. As
pointed out by Krenkel (1979), this procedure can lead to considerable
error, as opposed to the mathematically correct method of using a
least-squares analysis or, at the very least, a "line of best fit." In the
authors' opinion, the method of Litchfield and Wilcoxon (1949), as
described in the EPA document (USEPA, 1978) is rational and should be
standardized for acute toxicity bioassay tests. This procedure uses

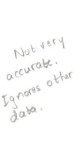

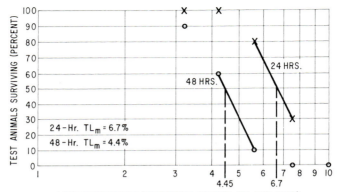

Fig. 4.12. Determination of the LC_{50}.

logarithmic–probit graph paper, a mathematically fitted line of best fit, and statistics to describe the "goodness of fit."

Computer programs are also available for determining the LC_{50} and associated statistical parameters as described by Sprague (1973), although, as noted by Sprague, the results of the computer output should always be checked by a graph in order to determine the reasonableness of the results.

Since the standard acute toxicity bioassay procedure yields the concentration of toxic material at which only 50% of the test organisms will live, it is standard practice to apply an application factor to the LC_{50} value obtained. The usual factors applied to the LC_{50} are 0.01 or 0.1, both of which are questionable. Beak (1958), a noted Canadian fisheries biologist, has stated that the application factors are nothing more than "an intelligent guess."

It should be noted that while the use of application factors is not truly scientific, the assumption is made that the biota will be protected from toxic effects of the material in question. There is no guarantee, however, that the resulting "safe" concentrations will not result in sublethal effects or even be "safe." When the cost of attaining the extremely small quantity of a substance is considered, it would appear to give impetus to research designed to more precisely define toxicity. In this regard, it is important to note that no single application factor is valid for all toxicants, wastes, or species.

If two or more toxic materials are present, it is recommended by the Committee on Water Quality Criteria (NAS/NAE, 1972) that their toxic effects are additive:

$$\frac{C_a}{L_a} + \frac{C_b}{L_b} + \cdots + \frac{C_n}{L_n} \leqq 1.0, \tag{4.8}$$

where C_a, C_b, ..., C_n are the measured or expected concentrations of the toxic materials and L_a, L_b, . . . , L_n are the allowable concentrations as determined by using recommended application factors on bioassays performed under local conditions. If the sum of Eq. (4.8) is greater than 1.0, one or more of the substances must be restricted.

Toxicity Curves   *Give median survival time vs. waste concentration.*

While the acute toxicity bioassay yields useful information for water pollution control endeavors, it does not give information concerning long-term effects, nor does it establish the shape of the toxicity curve from which useful information can be obtained. Long-term effects may include adverse effects on growth, reproductive capacity, disease resistance, predation, and so on.

As pointed out by Warren (1971), many attempts have been made by biologists to mathematically describe relationships between the median survival times of organisms and the concentrations of toxicants. Typical curves derived are shown in Fig. 4.13, taken from Warren (1971). Curve A can be described by

$$(c - a)^n(T - b) = K, \tag{4.9}$$

where c is the toxicant concentration, T is the exposure time, a is the incipient lethal level or threshold concentration, b is the threshold reaction time, and n and K are constants. The incipient lethal level is defined by Warren (1971) as "the level the species could tolerate indefinitely," and the threshold reaction time as the "minimum length of exposure the animal can tolerate before reacting by dying or collapsing, no matter what the level of the lethal agent may be."

Where the relationship follows a straight line, such as in B (Fig. 4.13), it may be described by

$$c^n T = K. \tag{4.10}$$

The problem with these types of relationships is that they may differ depending upon the organism and the toxicant. Furthermore, extrapolation beyond the observations made may lead to erroneous conclusions. Curves C, D, and E demonstrate other relationships found by various investigators. Changes in the shape of such curves may be attributed to changes in the mode of toxicity by the toxicant or interaction with other substances or conditions (Warren, 1971).

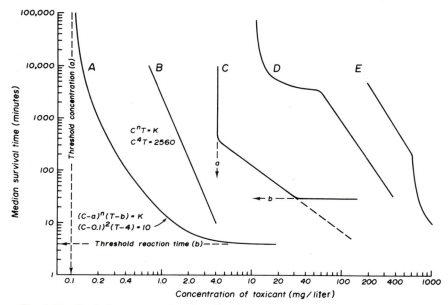

Fig. 4.13. Typical curves used to describe the relationship between the LC_{50} and toxicant concentration (Warren, 1971).

In any case, the development of toxicity curves, as shown in Fig. 4.13, is invaluable in determination of "safe" levels of toxicants, even though they are tedious to develop.

The effects of synergism and antagonism should be noted, especially when attempting to interpret the results of toxicity curves. Hardness is quite antagonistic to heavy metals, as the actual mechanism is not completely understood. The toxicity of many materials is affected by pH, the phenomenon being exemplified by the acute toxicity of ammonia at high pH values. Obviously, increases in temperature and/or decreases in dissolved oxygen will increase toxicity because of the resulting increase in the metabolic rate of the organism.

A most important factor, overlooked by many investigators, is acclimation. Speakman and Krenkel (1972) have shown that effects of temperature change on bluegill (*Lepomis macrochirus*) were dramatically dependent on the history of temperature exposure of the fish.

Finally, the concept of biological availability should be mentioned. Most bioassay tests are performed with laboratory reagent grade chemicals and the results related to total concentrations of the toxicant concentration in an aquatic system. Recent work by Gachter *et al*. (1978) has demonstrated that as much as two-thirds of the copper in lake water is not physiologically available to phytoplankton. The issue also has been raised recently in the Great Lakes over the biological availability of phosphorus and thus the need for control of nonavailable forms of phosphorus.

In summary, it may be stated that short-term acute toxicity bioassays are quite useful for water pollution control endeavors. However, the use of the more difficult to obtain toxicity curves is encouraged inasmuch as the acute toxicity bioassays alone do not yield information that will ensure protection of the aquatic biota. The traditional use of the short-term acute toxicity bioassay is demonstrated by the next example.

*The genera-
tion and use
of toxicity
curves encour-
aged along w/
bioassays.*

A factory producing cosmetic products is discharging wastewaters containing highly toxic compounds; thus, several fish kills have occurred in the river downstream from the outfall. The wastewater discharge is 0.2 m^3/sec. The critical low flow of the river is 15 m^3/sec.

Accurate chemical analysis of the toxic substances in the wastewater is quite complex, and only approximate values can be obtained. The wastewater contains several milligrams per liter of toxic tertiary and quaternary amines and chloramines.

A short-term acute toxicity bioassay has been performed using bluegill fish as the test organisms. The results of the toxicity test were as follows:

Container	Dilution Factor	%Fish Survival (after 96 hr)
1	1:1000	100
2	1:500	100
3	1:200	90
4	1:100	90
5	1:50	80
6	1:20	40
7	1:10	10
8	1:5	0
9	blank	100

The results of the toxicity test are plotted in Fig. 4.14. From the log C (in percent) *v*. % survival plot, the TL_m^{96} concentration of the wastewater is about 4.1%. This means the dilution ratio to obtain 50% survival after 96 hr is 1:25. If the wastewater flow is 0.2 m^3/sec, the stream dilution volume, DV, which would have the same toxic effect (i.e., 50% survival after 96 hr) is

$$DV = \frac{100 - TL_m^{96}}{TL_m} \times (\text{effluent flow})$$

$$= \frac{100 - 4.1}{4.1} \, 0.2$$

$$= 4.68 \text{ m}^3/\text{sec.}$$

[handwritten: $100\,(0.2) = 4.1\,(15 + 0.2)$]
[handwritten: $\dfrac{(100 - 4.1)\,0.2}{4.1} =$]

The dilution volume must be divided by the application factor, F_a, to obtain a "safe" river flow, Q_s.

$$Q_s = \frac{(DV)}{(F_a)} = \frac{4.68}{(0.1)}$$

$$= 46.8 \text{ m}^3/\text{sec.}$$

Since the river flow is only 15 m³/sec, the dilution of the wastewater by the river water is not sufficient and the toxic substances must be removed. The approximate removal can be estimated by comparing Q_s and Q. The required percent removal will be

$$100 \left(\frac{Q_s - Q}{Q_s}\right) = 100 \left(\frac{Q_s - 15}{46.8}\right), \text{ or } 68\%.$$

Since the toxicity will probably change with removal of the toxicant, the final effluent must again be subjected to study.

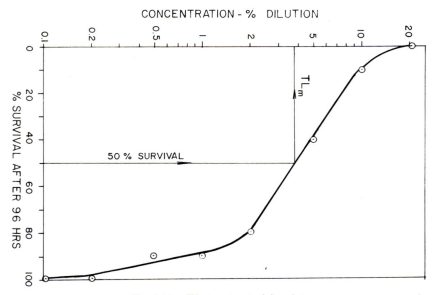

Fig. 4.14. Wastewater toxicity plot.

BACTERIOLOGICAL INDICES OF WATER QUALITY

Probably the most important criteria for adjudicating the safety of a water supply is whether the water is free from pathogenic (disease causing) microorganisms. Microbiological indicators of water quality or pollution are therefore of particular concern because of their relationship to human and animal health. Water contaminated by pathogenic microorganisms may penetrate into private and/or public water supplies either before or after treatment. Furthermore, pathogenic organisms can be ingested by people engaged in swimming and other water contact activities and can cause contamination of shellfish.

As previously stated, people have been concerned with the disease-carrying potential of water since the middle of the last century, when severe outbreaks of cholera were proven to be the result of sewage contamination of public water supplies. Although waterborne disease outbreaks have been minimized in the United States, the potential for waterborne disease epidemics always exists should vigilence and adequate treatment diminish. In fact, the development of modern water treatment processes can be attributed to epidemiological investigations of waterborne diseases. The reader is referred to Maxcy (1956) for fascinating discussions of the classic studies of waterborne diseases.

Waterborne Diseases

As mentioned in Chapter 3, water may be the carrier of pathogens including viruses, bacteria, protozoa, and the helminths. Of most current concern in the United States are outbreaks of infectious hepatitis, a viral-borne disease; giardiasis, caused by a protozoan; and, of course, gastroenteritis, with unknown etiological agents. Table 4.5 presents a list of waterborne disease outbreaks in the United States from 1946 to 1974 as summarized by Pipes (1978), from data taken from Craun and McCabe (1973), and Craun *et al*. (1976).

According to a recent survey from the United States Public Health Service Center for Disease Control (CDC), giardiasis has become the number one United States waterborne disease. More than 15% of 415,000 stool specimens examined throughout the country contained some parasitic infection, with the presence of *Giardia lamblia* predominating. Furthermore, cases of giardiasis were reported in almost every state (Hanson, 1978).

Diseases with etiological agents thought to be transmitted by water are shown in Table 4.6 (Pipes, 1978). The incidence of waterborne disease in

TABLE 4.5 Waterborne disease outbreaks: distribution by etiology, 1946–1974

	Outbreaks for Water Systems			
	Community		Other	
Disease	Number	Percent	Number	Percent
Gastroenteritis (unknown etiology)	71	52.2	153	47.6
Infectious hepatitis	22	16.2	44	13.7
Shigellosis	13	9.6	33	10.3
Chemical poisoning	8	5.9	13	4.0
Giardiasis	7	5.1	8	2.5
Typhoid	6	4.4	51	15.9
Salmonellosis	6	4.4	9	2.8
Amebiasis	1	0.7	4	1.2
Poliomyelitis	1	0.7		
Enteropathogenic *E. coli*			4	1.2
Tularemia			2	0.6
Leptospirosis	1	0.7		
Total outbreaks	136		321	

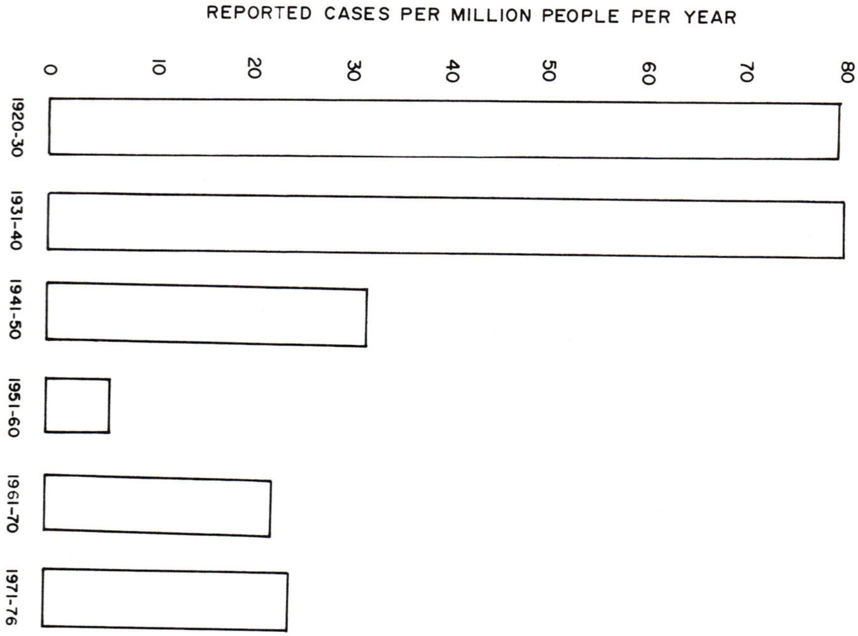

Fig. 4.15. Incidence of waterborne disease, 1920–1976 (Craun, 1977).

TABLE 4.6 Waterborne infectious diseases

	Occurs in United States	Sometimes Reported
Bacterial Diseases		
Bacillary dysentery (*Shigella* spp.)	yes	yes
Cholera (*Vibrio cholerae*)	no	—
Diarrhea Enteropathogenic *E. coli*	yes	yes
Liptospirosis (*Leptospira* spp.)	yes	yes
Salmonellosis (*Salmonella* spp.)	yes	yes
Typhoid Fever (*Salmonella typhi*)	yes	yes
Tularemia (*Francisella tularensis*)	yes	yes
Yersinosis (*Yersinia pseudotuberculosis*)	yes	yes
Unknown Etiology		
Diarrhea, acute, undifferentiated	yes	yes
Gastroenteritis, acute, benign, self-limiting	yes	no
Viral Diseases		
Gastroenteritis, Norwalk type agents	yes	once
Hepatitis A (hepatitis virus)	yes	yes
Parasitic Diseases		
Amebic Dysentery (*Endamoeba histolytica*)	yes	yes
Ascariosis (*Ascaris lumbricoides*)	yes	yes
Balantidial Dysentery (*Balantidium coli*)	yes	yes
Dracontiasis (*Dracunculus medinensis*)	no	—
Enterobiasis (*Enterobias vermicularis*)	yes	yes
Giardiasis (*Giardia lamblia*)	yes	yes
Hepatic capillariasis (*Capillaria hepatica*)	yes	yes
Hydatidosis (*Echinococcus granulosus*)	yes	yes
Schistosomiasis (*Schistosoma* spp.)	no	—
Trichuriosis (*Trichuris trichura*)	yes	yes

the United States from 1920 to 1976 is shown in Fig. 4.15 (Craun, 1977). As pointed out by Pipes (1978), the number of documented outbreaks has significantly diminished since 1920. Furthermore, it is interesting to speculate on the reason(s) for the increases shown subsequent to the 1951–1960 period. It is quite probable that better reporting is responsible, although it is possible that treatment plants have been overloaded with poorer quality water. It should be noted that all of the data are probably low because of the lack of reporting and/or lack of recognition of the clinical syndromes of a particular disease.

Craun *et al*. (1976) indicated that waterborne outbreaks are many times associated with deficiencies in water treatment, primarily associated with inadequate disinfection practices or lack of groundwater treatment. Furthermore, the association of waterborne diseases with the recreational use of water is somewhat nebulous. Pipes (1978) concluded that water contact recreation implies an increased risk of infections and irritations of the skin, ears, nose, and upper respiratory tract, although the risk of contracting the more well-defined enteric diseases by this route is minimal.

One may conclude that the incidence of waterborne disease is increasing in the United States and that the potential for greater increases is significant. Thus, consideration of the potential for waterborne disease is mandatory in water quality management studies.

Indicator Organisms

For practical reasons, it is not realistic to attempt to monitor a water supply for all of the potential pathogens that could be found; however, an emergency or epidemic may require such detailed investigations. Therefore, for routine monitoring and water quality evaluation, an indicator organism is usually used. In addition, the indicator organism may be used to evaluate the efficacy of water and wastewater treatment processes and to give an estimation of the fate of bacteria in receiving waters. The search for an ideal indicator of the presence of pathogenic organisms in water has continued since the discovery that water was a vehicle of transmission for certain diseases.

Ideally, the best indicator of the presence of pathogenic organisms in water or wastewater is the specific pathogens themselves. In practice, however, this could never be a suitable approach because of the variety of pathogens involved and the fact that the laboratory procedures are complex, time consuming, and, for the most part, rather insensitive (Kabler, 1959; Geldreich, 1972). Also, because of the comparatively low numbers of pathogens usually existing, very large water samples are needed for the detection procedure. With the possible exception of some work by Gallagher and Spino (1968), which demonstrated that a quantitative procedure is possible for the isolation of Salmonella sp., the prospect of using specific pathogens as indicators of potential health hazards in surface waters is remote at the present time. Thus, reliance must be placed on an indicator organism.

Total Coliforms

Since the latter part of the last century when the coliform group was first isolated from human feces by Escherich, this group of microorganisms has been the most important indicator of unsanitary or possible disease producing conditions in waters. According to *Standard Methods*

(APHA/AWWA/WPCF, 1975), the coliform group comprises all of the aerobic and facultative anaerobic, gram-negative, nonspore-forming, rod-shaped bacteria which ferment lactose with gas formation within 48 hr at 35°C. This group has been used as an indicator through the years because of the belief that the coliform group compares similarly in many aspects to the common enteric pathogens. It has been observed (Kabler, 1959) that the coliform group and the pathogenic enteric bacteria have survival rates of the same order of magnitude under similar environmental conditions of temperature, pH, disinfection, or extended exposure to soil or to fresh, polluted, or salt waters. Another reason why coliform bacteria indicators have been so popular throughout the years is the extreme abundance of these bacteria in human feces, estimated to be an average of 1.95 billion bacteria per person per day (Geldreich *et al.*, 1962). Thus, the presence of the coliform group in surface waters may indicate contamination by sewage and/or human and animal excreta.

A major problem with using the coliform group as an indicator is that, in addition to being found in the feces of warm-blooded animals and humans, coliforms can be also isolated from the gut of cold-blooded animals, soils, and possibly other sources. High coliform counts have been observed during spring and summer floods when surface waters carry large amounts of sediments and in urban surface runoff. Thus, the total coliform count alone may not be a reliable indicator of the potential contamination by pathogenic microorganisms and, if positive, should be complemented with additional microbiological, chemical, or biological observations.

The coliform group includes organisms that differ in biochemical and serological characteristics and in their origin. *Escherichia coli* is a known inhabitant of the intestinal tracts of humans and warm-blooded animals. *Aerobactor aerogenes, Aerobactor cloacae,* and *Escherichia freundii* are frequently found in other sources. The intermediate-aerogenes cloacae (IAC) subgroups may also be found in fecal discharges, but in much smaller numbers than *E. coli*. With the exception of *E. coli,* all of the other groups of coliforms can be found in soils.

Briefly, the test involves taking appropriate water samples, innoculating a series of fermentation tubes with the "suspect" water, and incubating. If the coliform group is present, gas is produced in a vial contained in the fermentation tubes, and the test is said to be a positive "presumptive" test. The test may be further refined by a competent microbiologist by means of several additional steps including microscopic examination. The EPA has also approved a test using a membrane filter through which a known quantity of water is filtered; the filter is placed in an appropriate medium and incubated, and the coliform colonies formed are directly counted. The multiple-tube fermentation test is still recommended, however.

In order to obtain bacterial densities, the fermentation tube results are subjected to statistical analysis resulting in a most probable number of bacteria (MPN) per unit volume of water, which is a modal value. The direct plate counts, while being statistically more reliable than the MPN procedure, are still not absolute numbers (APHA/AWWA/WPCF, 1975).

In either case, it is important to note that the results of these tests only indicate the "potential" presence or absence of pathogenic organisms. Therefore, positive tests must be subjected to additional analysis before one concludes that a water is "safe" or "unsafe" from the bacteriological standpoint.

Fecal Coliforms

For many years, there has been interest in distinguishing fecal and non-fecal coliforms. In 1904, Eijkman first proposed an elevated temperature test for this purpose (Geldreich *et al.*, 1962). The so-called "IMVIC" procedure described in *Standard Methods* (APHA/AWWA/WPCF, 1975) for differentiating the coliform group into *Escherichia coli, Aerobacter aerogenes*, and *Escherichia freundii* has been in various stages of development for many years. The primary reason for the intense interest in the fecal group is the conviction held by most scientists that these organisms represent a more sensitive measure of health hazards because of their definite origin in the feces of humans and warm-blooded animals (Kabler and Clark, 1960, and Geldreich, 1970). As stated by Geldreich (1970), the fecal coliform test is the most accurate bacteriological measurement presently available for detecting warm-blooded animal feces in polluted waters. Other work by Geldreich *et al.* (1961) showed that very few coliforms traceable to warm-blooded animals are associated with soils, plants, or insects. This lends further support to the usefulness and continuation of this indicator group for surface water quality evaluations.

Fecal coliform bacteria behave similarly to common enteric pathogens. According to Van Donsel *et al.* (1967), a close relationship exists between the growth and survival of fecal coliforms and the pathogens *Salmonella* and *Shigella*. This is one of the most important features of a good indicator system, especially when it is not possible to monitor the pathogens themselves. The fact that the fecal coliforms can be found in the feces of both humans and warm-blooded animals may be an advantage rather than a deficiency, inasmuch as some of the pathogenic microorganisms dangerous to humans can be carried by animals.

Fecal Streptococci

Fecal streptococci live and multiply in the intestines of humans and other warm-blooded animals. Unlike the coliform microorganisms, they do not multiply in surface waters; therefore, the quantity of fecal streptococci found in surface waters is always less than that in the pertinent upstream sewage discharges. There is considerable variation in types and numbers of these bacteria, and, therefore, the streptococcal group is not recognized as an official water quality indicator in the United States. Some of the members of the group can also be found in soils and the feces of freshwater fish (Geldreich and Kenner, 1969).

Other Microbiological Indicators

Several other indicators have been suggested for the evaluation of the quality of surface waters because of the interest in diseases caused by waterborne enteric viruses. Viruses are also more resistant to chlorination and common treatment methods than bacteria. However, the problem with using viruses as indicators of pathogenic contamination is that the methods are not as simple and accurate as those for bacteria, and, in addition, the density of enteric viruses in surface waters is about 4 to 5 orders of magnitude less than that for bacteria. Typical ratios of coliform bacteria to viruses are approximately 92,000:1 in the case of raw sewage and about 50,000:1 for polluted waters. Thus, the coliform bacteria indicator is far superior to a virus indicator.

A two-year survey of viruses in the sewage from Haifa, Israel (Buras, 1976), revealed that enteric viruses are always present in Haifa's wastewater. The virus count fluctuated and was usually low in the winter, the highest numbers of viruses being observed in the April to September period. The values ranged from 600 PFU/100 ml to 17,600 PFU/100 ml (PFU = Plaque Forming Units). The most prevalent virus groups were Polio F and $ECHO_4$.

Fecal Coliforms – Fecal Streptococci Ratio

Interest has been expressed in the ratio of fecal coliform (FC) to fecal streptococci (FS) because fecal coliform counts are much higher than fecal streptococci if the water is contaminated by human excreta. The opposite is true for animal feces. The FC/FS ratios reported for a number of different animal feces are (Geldreich *et al.*, 1968): rat, −0.04; rabbit, −0.0004; cat, −0.29; dog, −0.02, as compared to man, −4.4. Thus, the FC/FS for water contaminated by domestic sewage should be considerably higher than that of surface runoff. Ratios of FC/FS for some stormwater runoff events have been reported within the range of 0.04 to 0.26. On the other hand, sewage showed FC/FS ratios of 4.0 to 27.9 (Geldreich *et al.*, 1968).

Modeling of Bacterial Decay

Assuming that the disappearance of the coliform group of bacteria in a receiving water indicates a similar decline in pathogenic microorganisms, the inclusion of bacterial decay phenomena in water quality modeling is important.

The decay of bacterial count may be expressed as

$$\frac{N_t}{N_0} = 10^{-k_b t} \tag{4.11}$$

where N_t is the count after t days, N_0 is the initial count, and k_b is the bacterial decay rate constant, which is dependent on the temperature of the water, as shown in Fig. 4.16 (Kehr and Butterfield, 1943).

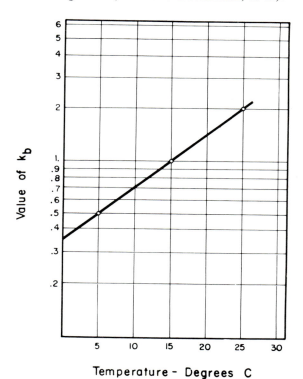

Fig. 4.16.   Effect of temperature on bacterial death rate (Kehr and Butterfield, 1943).

As pointed out by Camp (1963), many investigators express the rate of decline of bacteria in terms of the time required for a 90% reduction in count, which may be computed from Eq. (4.11) as

$$t_{g0} = \frac{1}{k_b}. \tag{4.12}$$

It should be noted that the viability of bacteria in seawater is much less than in fresh water, the mortality rate being some 6 to 8 times greater in seawater as noted by Camp (1963).

With regard to stream surveys, it is important to recognize that the total bacterial plate count used for estimating the total microbiological population is useful as an overall indicator of the microbiological state of a stream reach or a lake. However, the total count is by no means an indicator of pollution or contamination by pathogenic microorganisms, since many of the bacteria are indigenous to natural waters.

EPILOGUE

Should the reader tend to dismiss the importance of waterborne diseases in the United States, he should note that the problem is not exclusively one of foreign countries. The cholera epidemics that occurred in the United States during the 1800s are quite illustrative. Nashville, Tennessee, alone, was subjected to numerous outbreaks of cholera. During the four years prior to 1873, five epidemics occurred, hundreds of people dying with each epidemic. On June 20, 1872, 72 Nashvillians died from cholera and 750 people died in Nashville during the month of June alone. Six major epidemics took 3000 lives in Nashville, and many Tennessee cities were almost abandoned because of fear (Davis, 1964).

REFERENCES

American Public Health Association, American Water Works Association, and Water Pollution Control Federation (1975). "Standard Methods for the Examination of Water and Wastewater," 14th ed. Washington, D.C.

Bartsch, A. F., and Ingram, W. M. (1959). Streamlife and the pollution environment. *Public Works,* **90,** 104–110.

Beak, T. W. (1958). Toleration of fish to toxic pollution. *J. Fish. Res. Board Can.* **15,** 559–572.

Beck, W. M. (1955). Suggested method for reporting biotic data. *Sewage Ind. Wastes* **27,** 1193.

Buras, N. (1976). Concentration of enteric viruses in wastewater and effluent: A two year survey. *Water Res.* **10,** 295–298.

Cairns, J., Albaugh, D. W., Bursey, F., and Chanay, M. D. (1968). The sequential comparison index—a simplified method for nonbiologists to estimate relative differences in biological diversity in stream pollution studies. *J. Water Pollut. Control Fed.* **40,** 1607–1613.

Camp, T. R. (1963). "Water and Its Impurities." Van Nostrand-Reinhold, Princeton, New Jersey.

Chutter, F. M. (1972). Empirical biotic index of the quality of water in South African streams and rivers. *Water Res.* **6,** 19–30.

Craun, G. F. (1977). Waterborne outbreaks. *J. Water Pollut. Control Fed.* **49,** 1268.

Craun, G. F., and McCabe, L. J. (1973). Review of the causes of waterborne disease outbreaks. *J. Am. Water Works Assoc.* **65,** 74.

Craun, G. F., McCabe, L. J., and Hughes, J. M. (1976). Waterborne disease outbreaks in the U.S.—1971–1974. *J. Am. Water Works Assoc.* (August).

Davis, L. (1964). When cholera laid our cities low. *Nashville Tennessean Mag.* (July 12).

Gabriel, J. (1946). *Cas. Lek. Cesk.* **85,** 1425, Prague, Czechoslovakia.

Gachter, R., Davis, J. S., and Mares, A. (1978). Regulation of copper availability to phytoplankton by macromolecules in lake water. *Environ. Sci. Technol.* (December).

Gallagher, T. P., and Spino, D. G. (1968). The significance of numbers of coliform bacteria as an indicator of enteric pathogens. *Water Res.* **2,** 169.

Geldreich, E. F. (1970). Applying bacteriological parameters to recreational water quality. *J. Am. Water Works Assoc.* **61,** 113.

Geldreich, E. E. (1972). Buffalo Lake recreational water quality: A study in bacteriological data interpretation. *Water Res.* **6,** 913.

Geldreich, E. E., and Kenner, B. A. (1969). Concepts of fecal streptococci in stream pollution. *J. Water Pollut. Control Fed.* **41,** 2336.

Geldreich, E. E., Kenner, B. A., and Kabler, P. W. (1961). Occurrence of coliform, fecal coliform and streptococci on vegetation and insects. *Appl. Microbiol.* **12,** 63.

Geldreich, E. E., Bordner, R. H., Huff, C. B., Clark, H. F., and Kabler, P. W. (1962). Type distribution of coliform bacteria in the feces of warm-blooded animals. *J. Water Pollut. Control Fed.* **34,** 295.

Geldreich, E. E., Best, L. D., Kenner, B. A., and Van Donsel, D. J. (1968). The bacteriological aspects of stormwater pollution. *J. Water Pollut. Control Fed.* **40,** 1861.

Hanson, R. E. (1978). "Update," *J. Am. Water Works Assoc.* (December).

Hilsenhoff, W. L. (1972). Stream classification and sampling data for Wisconsin. Unpublished. Univ. of Wisconsin, Madison.

Kabler, P. (1959). Removal of pathogenic microorganisms by sewage treatment process. *Sewage Ind. Wastes* **31,** 1373.

Kabler, P. W., and Clark, H. F. (1960). Coliform group and fecal coliform organisms as indicators of pollution of drinking water. *J. Am. Water Works Assoc.* **52,** 1578.

Kehr, R. W., and Butterfield, C. T. (1943). Notes on the relation between coliforms and enteric pathogens. *Public Health Rep.* 589 (April).

Kolkwitz, R., and Marsson, M. (1908). "Ökologie der pflanzlichen saprobien (Ecology of plant saprobia). *Ber. Dtsch. Bot. Ges.* **26,** 505–519.

Krenkel, P. A. (1979). Dichotomies in the establishment of water quality criteria. *J. Water Pollut. Control Fed.* (September).

Liebman, H. (1962). "Handboch der Frischwasser-und Abwasserbiologie" (Handbook of Fresh water and Wastewater Biology), Vol. 1. Oldenbourg Vlg. Munich, Germany.

Litchfield, J. T., and Wilcoxon, F. (1949). A simple method of evaluating dose-effect experiments. *J. Pharmacol. Exp. Ther.* **96,** 99.

Margalef, R. (1951). Diversity of species in natural communities. *Publ. Inst. Biol. Apl., Barcelona* **9,** 5–28.

Margalef, R. (1957). La theoria de la informacion en ecologia. *Mem. R. Acad. Cienc. Artes Barcelona.* **32,** 373–449.

Maxcy, K. F. (1956). "Preventive Medicine and Public Health." Appleton, New York.

National Academy of Sciences, and National Academy of Engineering (1972). Water quality criteria 1972. Rep. Comm. Water Qual. Criteria, Washington, D.C.

Pantle, R., and Buck, H. (1955). Die biologische überwachung der gewässer und die ergebnisse (Biological observations of surface waters and their results). *Gas Wasserfach* **96,** 604.

Parker, F. L., and Krenkel, P. A. (1967). Thermal pollution—state-of-the-art., Tech Report No. 3, *Environ. and Water Resour. Eng.* iii–4. Vanderbilt Univ., Nashville, Tennessee.

Patrick, R. (1949). A proposed biological measure of stream conditions based on a survey of the Conestoga Basin, Lancaster County, Pennsylvania. *Proc. Acad. Nat. Sci. Philadelphia* **101**, 277–342.

Patrick, R. (1950). Biological measure of stream conditions. *Sewage Ind. Wastes* **22**, 926–938.

Patrick, R. (1953). Biological phases of stream conditions. *Proc. P. Acad. Sci.* **27**, 33–36.

Patrick, R., Hohn, M. H., and Wallace, J. H. (1954). A new method for determining the pattern of the diatom flora. *Bull. Philadelphia Acad. Nat. Sci.* **295**, 1.

Pipes, W. O. (1978). Water quality and health significance of bacterial indicators of pollution. Workshop Proc., Drexel Univ. (April).

Robinson, J., Richardson, A. N., Crabtree, A. N., Coulson, J. C., and Potts, G. R. (1967). "Organo-chlorine residues in marine organisms," *Nature* **214**, 1307–1311.

Sewel, G. H. (1975). "Environmental Quality Management." Prentice-Hall, Englewood Cliffs, New Jersey.

Shannon, C. E., and Weaver, W. (1963). "The Mathematical Theory of Communication." Univ. of Illinois Press, Urbana.

Sládeček, V. (1973). The reality of three British biotic indices. *Water Res.* **7**, 995–1002.

Sládeček, V., and Tuček, F. (1974). Vztah saprobního indexu k BSK$_5$ (A relationship of the saprobien index to BOD$_5$). *Vodní Hospod.* **12**, 322.

Sládeček, V., Zelinka, M., Marvan, P., and Rothschein, J. (1975). Ceskoslovenska vodohospodarska toxicologie (Czechoslovak water resources toxicology). *Vodní Hospod.* **5**, 137–140.

Speakman, J. N., and Krenkel, P. A. (1972). Quantification of the effects of rate of temperature change on aquatic biota, *Water Res.* **6**, 1283–1290.

Sprague, J. B. (1973). The ABC's of pollutant bioassay using fish. *In* "Biological Methods for the Assessment of Water Quality" (J. Cairns Jr. and K. L. Dickson, eds.). Am. Soc. for Testing and Mater., Philadelphia, Pennsylvania.

Staub, R., Appling, J. W., Hofstetter, A. M., and Haas, I. J. (1970). The effect of industrial wastes of Memphis and Shelby County on primary planktonic producers. *BioScience* **20**, 905–912.

Storrs, P. N., Pearson, E. A., Ludwig, H. F., Walsh, R., and Stann, E. J. (1969). "Estuarine water quality and biologic population indices. *Proc. Int. Conf. Int. Assoc. Water Pollut. Res.*, 4th, Prague, Czechoslovakia.

U.S. Environmental Protection Agency (1975). Methods for acute toxicity tests with fish, macroinvertebrates and amphibians. EPA-660/3-75-009, Natl. Environ. Res. Ctr., Corvallis, Oregon (April).

U. S. Environmental Protection Agency (1978). Methods for measuring the acute toxicity of effluents to aquatic organisms. EPA-600/4-78-012, Environ. Mon. & Support Lab., Cincinnati, Ohio (January).

Van Donsel, D. J., Geldreich, E. E., and Clarke, N. A. (1967). Seasonal variations in survival of indicator bacteria in soil and their contribution to storm water runoff. *Appl. Microbiol.* **15**, 1362.

Von Tümpling, W. (1969). Suggested classification of water quality based on biological characteristics. *Proc. Int. Conf. Int. Assoc. Water Pollut. Res.*, 4th, Prague, Czechoslovakia.

Warren, C. E. (1971). "Biology and Water Pollution Control." W. B. Sanders, Philadelphia, Pennsylvania.

Wilhm, J. (1972). Graphic and mathematical analyses of biotic communities in polluted streams. *Annu. Rev. Entomol.* **17**, 223–252.

Wilhm, J. L., and Dorris, T. C. (1968). Biological parameters for water quality criteria. *J. Biol. Sci.* **18** (6), 477–480.

Wurtz, C. B. (1955). Stream biota and stream pollution. *Sewage Ind. Wastes* **27**, 1270–1278.

Zelinka, M., and Marvan, P. (1961). Zur präzisierung der biologischen klassifikation der reinheit fliessender gewässer (Contribution to the accuracy of the biological classification of purity of flowing waters). *Arch. Hydrobiol.* **57**, 389–407.

5 | Quality Requirements for the Beneficial Uses of Water

INTRODUCTION

As previously mentioned, the need and desire for good quality water is as old as civilization itself. Nile River water was purified before the pharoahs used it, and the Romans brought good quality water from large distances by means of the famous aquaducts and had elaborate sewer and pollution control systems.

Although many large epidemic outbreaks can be related to poor water quality, as discussed in Chapter 4, it was not until the middle part of the last century that the link between poor water quality and such outbreaks was clearly established. For example, cholera appeared in London during the summers of 1848 and 1854, when more than 25,000 people died as a result of the absence of effective sewerage and subsequent contamination of water supply sources.

The early water quality requirements and standards were concerned with the safety for drinking purposes and the prevention of epidemic outbreaks. Thus, bacteriological examination and safety were of prime importance. Efforts to establish standards of purity for effluents can be dated as long ago as 1868, when the Rivers Pollution Commission in Great Britain proposed limits for a wide variety of substances. Because of the difficulty of developing and enforcing the criteria for many substances, only a few substances were accepted (Lovett, 1958; Porges, 1972).

Later, the health and pollution control authorities became concerned with the protection of fish and the prevention of septic conditions of surface waters. Therefore, the dissolved oxygen concentration was included as a primary parameter for water quality evaluations. Various dissolved oxygen standards and criteria were proposed for fish and water quality protection. The minimal accepted dissolved oxygen values ranged from 3 mg/liter for polluted rivers to 5–6 mg/liter for streams designed for trout and salmon protection.

The studies by Streeter and Phelps along the Ohio River during 1914 and 1915 provided a better understanding of the relationship between the dissolved oxygen concentration and wastewater discharges. The reduction and/or depletion of the dissolved oxygen was found to be a result of the oxygen demand of unoxidized inorganic and organic substances. The oxygen balance method, developed by Streeter (1925), permitted a better understanding of the receiving water assimilative capacity for receiving organic wastes (Phelps, 1944). This method enabled the pollution control authorities to set effluent requirements on BOD that treated receiving water assimilative capacity as a natural resource.

Most other effluent requirements were based on dilution requirements. An example of stream standards based on dilution were those which the British Royal Commission set in 1912 for sewage and sewage effluents. By these standards, no treatment was required if the dilution ratio of the sewage flow to the receiving stream low-flow characteristics was less than 1:500 (Imhoff and Fair, 1940).

Stream standards are similar for most European countries. Their origin can be traced to the Kolkwitz–Marsson saprobien system (Chapter 4), which was then correlated to various water quality parameters. Thus, most of the stream standards in Europe recognize four water quality *classes:* Class 1 is the best water quality characterized by oligosaprobic conditions and is usually considered to be a water quality suitable for public water supply. Class 4 consists of the worst water quality with polysaprobic conditions and is unfit for most common uses. An example of French stream water quality standards is shown in Table 5.1 (Le Foll *et al.*, 1977).

TABLE 5.1 French stream water quality standards

Quality	Concentration	CLASSES				
		1A	1B	2	3	4
Temperature		≤20	20–22	22–25	25–30	>30
pH		6.5–8.5	6.5–8.5	6.5–8.5	5.5–9.5	<5.5 or >9.5
D.O.	mg/liter	>7	5–7	3–5	<3	
% Saturation		>90	70–90	50–70	<50	
BOD₅	mg/liter	≤3	3–5	5–10	10–25	>25
COD	mg/liter	≤20	20–25	25–40	40–80	>80
SO₄²⁻	mg/liter		<250		>250	
NH₄	mg/liter	≤0.1	0.1–0.5	0.5–2	2–8	>8
NO₃	mg/liter		≤44		44–100	>100
Phenols	mg/liter	≤0.001		0.001–0.05	0.05–0.5	>0.5
Orthophosphate	mg/liter	≤0.4	0.4–0.7		>0.7	
Detergent anion	mg/liter	≤0.2		0.2–0.5		>0.5
CN	mg/liter		≤0.05			>0.05
Cr	mg/liter		≤0.05			>0.05
F	mg/liter	≤0.7		0.7–1.7		>1.7
Pb	mg/liter		≤0.05			>0.05
Se	mg/liter		≤0.01			>0.01
Cu	mg/liter	≤0.05		0.05–1		>1
Zn	mg/liter				3–5	>5
As	mg/liter		≤0.05	0.05–0.1		>0.1
Fe	mg/liter	≤0.5	0.5–1	1–1.5		>1.5
Mn	mg/liter	≤0.1	0.1–0.25	0.25–0.5		>0.5
Cd	mg/liter		≤0.005			>0.005
Substances Extracted	mg/liter	≤0.2	0.2–0.5	0.5–1		>1
Escherichia	N/100 milliliter	≤2000		>2000		
Streptococcus	N/100 milliliter	≤20	20–1000	1000–10,000		>10,000
Conductance	μmhos/cm	≤400	400–750	750–1500	1500–3000	>3000
Cl⁻	mg/liter	≤100	100–200	200–400	400–1000	>1000

STREAM AND EFFLUENT WATER QUALITY STANDARDS

In order to effectively control water quality, it is necessary to describe it in precise, technical quantitative terms. Once this is accomplished, the decision for effluent discharge limitations or a beneficial use of the water can be formulated.

It is clear that statements such as "nitrates are objectionable" or "toxic metals should be negligible" are not sufficient to describe water quality requirements. For this reason, a decision must be made delineating how much is objectionable or negligible. Furthermore, any requirement for water quality must be imposed with consideration of the concomitant level of treatment requirements of wastewater effluents or water supply intakes upstream and downstream from the point of interest. Water treatment is an engineering process requiring economic resources. Therefore, if the effluent or intake water quality is to be changed, an authoritative statement must be available in order to determine whether a proposed water quality alteration is sufficient in the interests of both economy and intended use.

For the purpose of quantifying water quality, water is analyzed by various physical, chemical, and biological techniques that yield numerical values of the concentrations of various substances present in the samples. It is obvious that one sample is not sufficient to adequately quantify the status of water quality of a particular stream or wastewater effluent. Rather, the quantification should be the result of a statistical evaluation of a number of samples taken at various locations, flows, and times. Therefore, the accuracy of the water quality quantification process is related to the frequency of the data acquisition and its statistical reliability. The process of quantifying water quality then involves a comparison of the statistical water quality characteristics obtained from appropriate water quality analyses with a set of water quality criteria or standards. Therefore, it is important that the differences between "standards" and "criteria" be clearly understood. The reader is referred to Chapter 2 for a review of this topic.

The water quality standards used presently by most water pollution control engineers and pollution abatement authorities throughout the world are in the category of either stream standards or effluent standards or a combination thereof (see Chapter 2). Although the present primary emphasis in the United States is on effluent standards, stream standards are enforced where the effluent loadings exceed the waste assimilative capacity, as determined by stream standards (water quality limiting cases). In water pollution control practice, the waste assimilative capacity connotes the capability of a receiving water to assimilate a certain quantity of waste material, under the worst possible conditions, without causing deleterious effects.

In the context of this discussion, it is obvious that if the water quality goals of PL 92-500 are to be attained, adequate planning at all levels of government is mandatory. The establishment of water quality criteria, the use of these criteria in the rational enactment of water quality standards, and the use of stream and/or effluent standards must be carefully scrutinized not only in the short-term plan, but also far into the future.

Therefore, major systems for municipal, areawide, state, and regional planning have been instigated in order to clarify federal, state, and local roles in planning and management.

• Municipal planning is directly related to the building of publicly owned treatment works. Municipal facilities plans must provide ways to prevent, dispose of, and store wastes and must consider other alternatives besides conventional structural facilities to reduce municipal wastes.

• Areawide planning concentrates on comprehensive methods of controlling urban–industrial pollution. Areawide planning agencies have the responsibility for obtaining federal grants for construction of wastewater treatment plants and for managing and collecting money to maintain and upgrade existing plants.

• State planning agencies relate water quality data to permits and ensure that the compliance schedule under the permit is stringent enough to protect the quality of receiving waters.

• Regional planning—conducted by federal and state agencies—is intended to relate water pollution control and water resource management efforts (Izaak Walton League, 1973).

BENEFICIAL USES OF WATER

Inasmuch as the intended use of a water should dictate its quality requirements, delineation of water allocated for specific uses is mandatory. The traditionally accepted beneficial uses of water are (Krenkel, 1973): domestic water supply, industrial water supply, agricultural water supply, stock and wildlife watering, propagation of fish and other aquatic life, shellfish culture, swimming and bathing waters, boating and aesthetic enjoyment, water power and navigation, and transport, dispersion, and assimilation of wastes.

Although the engineering approach to water quality management is to define the intended use of the water and then ensure that the water quality satisfies that use, the philosophy of existing legislation is to require the most advanced treatment possible. However, this treatment may not cause a discernible improvement in receiving water quality. As one can see from the preceding list of beneficial water uses, a wide spectrum of water quality requirements exists. Even though the "highest" use of water is usually thought to be for drinking purposes, the most stringent quality requirements are probably for boiler-feed water. On the other hand, the "lowest" use of water is probably for waste assimilative capacity.

A discussion of the requirements for each of the beneficial uses is in order so that the water quality engineers can make decisions concerning water quality requirements concomitant with a designated use. Figures 5.1 and 5.2 show the Water Resources Council (WRC) estimates for past, existing, and future water uses for both withdrawals and consumptive use (USWRC, 1978).

Domestic Water Supply

Present and projected water withdrawals for municipal and self-supplied industrial uses represent about 20–25% of the total water use. The estimated water withdrawal for domestic and commercial use in 1975 was about 107 million m^3/day (29 billion gal/day), and for industrial use it was about 194 million m^3/day (51 billion gal/day) (USWRC, 1978). Although the National Water Commission estimated that the amount of water withdrawn for these uses should more than double by the year 2000 (Natl. Water Commis., 1973), the latest WRC report indicates a reduction in total water use by the year 2000 (however, not in domestic or commercial use).

The Federal Environmental Protection Agency issued interim primary drinking water regulations on December 10, 1975, as the first step in setting national standards for drinking water quality, under the provision of the Safe Drinking Water Act of 1974 (PL 93-523).

The standards became effective in June 1977. They apply to all public water systems, which EPA estimates to number about 240,000. The raw surface water quality regulations can be related to such surface water quality as would provide safe and potable water within the drinking water standards subsequent to treatment. The standard water treatment processes considered include coagulation, sedimentation, rapid sand filtration, and disinfection. Certain aspects of the new regulations favor groundwater sources. For example, maximum levels of turbidity are specified where surface water is the source of supply.

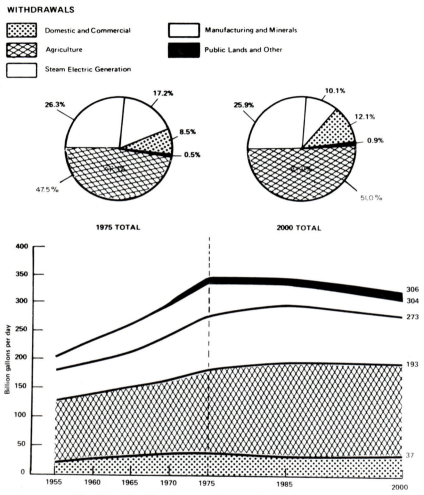

Fig. 5.1. Total freshwater withdrawals (USWRC, 1978).

The Safe Drinking Water Act defines a public water system as ''a system for the provision to the public of piped water for human consumption, if such system has at least 15 service connections or regularly serves at least 25 individuals.'' The EPA interprets service ''to the public'' to include factories and private housing developments and ''regular service'' to cover systems serving at least 25 individuals for at least 60 days out of the year. Thus, campgrounds, lodges, and other tourist accommodations

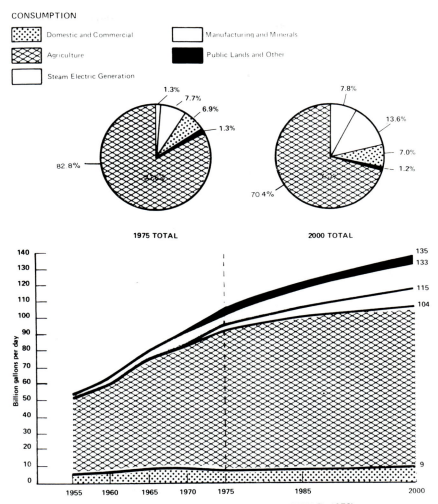

Fig. 5.2. Total freshwater consumption (USWRC, 1978).

which are open only part of the year are considered to have public water systems. These interpretations lead to classifying public water systems into two major types, those serving year-round residents and those serving transients or having intermittent use. A "community water system" is defined, then, as one which supplies at least 15 service connections used by year-round residents or serves at least 25 year-round residents. All other public water systems are classified as "noncommunity water systems."

The maximum contaminant levels for organic chemicals and for most inorganic chemicals are based on the potential health effects of long-term exposure. These levels, as pointed out by the EPA, are not necessary to protect transients or intermittent users. Therefore, the final regulations provide that maximum contaminant levels for organic chemicals and for inorganic chemicals other than nitrates are not applicable to noncommunity systems. An exception was made for nitrates because they may have adverse health effects on susceptible infants in a short period of time.

All the drinking water quality standards specified by the Act are related to water quality at a point of discharge to the distribution system. It is assumed that intake waters will be sufficiently uncontaminated so that with the application of the most effective treatment methods, a public water system would be able to protect the public health (including attainment of the recommended maximum contaminant levels).

The primary drinking water standards published by EPA in 1975 (USEPA, 1975) represent the levels of contamination which, in the judgment of EPA, "may have adverse effect on the health of persons and which specifies for each contaminant a maximum level, or for those contaminants for which it is uneconomical or technically not feasible to monitor, a treatment technique which leads to reduction in the level of the contaminant sufficient to eliminate any adverse health effects".

The secondary drinking water standards imply a regulation "which applies to public water systems and which specifies the maximum contaminant levels that are requisite to protect the public welfare, specifically contaminants that may adversely affect the odor or appearance of water and consequently may cause a substantial number of persons served by the public water system providing such water to discontinue its use."

The maximum contaminant levels of pollutants, except fluoride, are presented in Table 5.2. The bacteria levels are based on the criteria recommended by the National Technical Advisory Committee which reflect the intake water quality (FWPCA, 1968).

Industrial Water Supply

Because of the diverse nature of industrial processes, water quality requirements are industry dependent. In addition, water quality requirements may vary for various industrial processes within a single plant and for the same process in different plants (NAS/NAE, 1973). Quality considerations are similar, depending on cost of treatment, plant age, plant operating practices, and quality and quantity of the available supply (FWPCA, 1968).

TABLE 5.2 National interim primary drinking water standards[a]

Inorganic Chemicals	Concentrations, mg/liter	Organic Chemicals	Concentrations, mg/liter
Arsenic	0.1	Chlordane	0.003
Barium	1.0	Endrin	0.0002
Cadmium	0.01	Heptachlor	0.0001
Chromium	0.05	Lindane	0.004
Cyanide	0.2	Methoxychlor	0.1
Lead	0.05	Toxaphene	0.005
Mercury	0.002	2,4-D	0.1
Nitrate (as N)	10.0	2,4,5-TP Silvex	0.01
Selenium	0.01	Azodrin	0.003
Silver	0.05	Dichlorros	0.01
		Dimethonoate	0.002
		Ethion	0.02

Bacteria in Raw Water Supply[b]	Concentration	Permissible	Desirable
Coliform organisms	N/100 ml	10,000	100
Fecal coliforms	N/100 ml	2,000	20

Anticipated Secondary Contaminant Levels[c]	Concentration		
Chloride	250 mg liter	Iron	0.3 mg/liter
Color	15 units	Manganese	0.05 mg/liter
Copper	1 mg/liter	Odor	3 Thresh. Od. Numbers
Corrosivity	none	Sulfate	250 mg/liter
Foaming agents	0.5 mg/liter	Zinc	5 mg/liter
Hydrogen sulfide	0.5 mg/liter		

[a] Effective as of June 24, 1977 (USEDA, 1975).

[b] Microbiological limits are monthly averages based upon an adequate number of samples. Total coliform count may be relaxed if fecal coliform concentration does not exceed the specified limit. (NAS/NAE, 1973).

[c] From Hernandez and Barkley, 1976.

Because of the low cost of water treatment as compared to the cost of total production and marketing, industry can treat almost any water to its own specifications. However, quality characteristics that exceed those given in Table 5.3 would probably not be acceptable to industry (NAS/NAE, 1973). In addition, the Ohio River Valley Water Sanitation Commission (ORSANCO) states that all water for all uses should be (1) free from substances attributable to municipal, industrial, or other discharges, or agricultural practices that will settle to form putrescent or otherwise objectionable sludge deposits; (2) free from floating debris, oil,

TABLE 5.3 Summary of specific quality characteristics of surface waters that have been used as sources for industrial water supplies[a]

| Characteristics | Boiler Makeup Water | | Fresh | | Brackish[b] | | Textile Industry SIC-22 |
	Industrial 0 to 1500 psig	Utility 700 to 5000 psig	Once through	Makeup recycle	Once through	Makeup recycle	
Silica (SiO$_2$)	150	150	50	150	25	25	—
Aluminum (Al)	3	3	3	3	—	—	—
Iron (Fe)	80	80	14	80	1.0	1.0	0.3
Manganese (Mn)	10	10	2.5	10	0.02	0.02	1.0
Copper (Cu)	—	—	—	—	—	—	0.5
Calcium (Ca)	—	—	500	500	1200	1200	—
Magnesium (Mg)	—	—	—	—	—	—	—
Sodium + potassium (Na + K)	—	—	—	—	—	—	—
Ammonia (NH$_3$)	—	—	—	—	—	—	—
Bicarbonate (HCO$_3$)	600	600	600	600	180	180	—
Sulfate (SO$_4$)	1400	1400	680	680	2700	2700	—
Chloride (Cl)	19000	19000	600	500	22000	22000	—
Fluoride (F)	—	—	—	—	—	—	—
Nitrate (NO$_3$)	—	—	30	30	—	—	—
Phosphate (PO$_4$)	—	50	4	4	5	5	—
Dissolved solids	35000	35000	1000	1000	35000	35000	150
Suspended solids	15000	15000	5000	15000	250	250	1000
Hardness (CaCO$_3$)	5000	5000	850	850	7000	7000	120
Alkalinity (CaCO$_3$)	500	500	500	500	150	150	—
Acidity (CaCO$_3$)	1000	1000	0	200	0	0	—
pH, units	—	—	5.0–8.9	3.5–9.1	5.0–8.4	5.0–8.4	6.0–8.0
Color, units	1200	1200	—	1200	—	—	—
Organics:							
Methylene blue active substances	2[e]	10	1.3	1.3	—	1.3	
Carbon tetrachloride extract	100	100	[f]	100	[f]	100	—
Chemical oxygen demand (COD)	100	500	—	100	—	200	
Hydrogen sulfide (H$_2$S)	—	—	—	—	4	4	—
Temperature (°F)	120	120	100	120	100	120	—

[a] NAS/NAE (1973).
[b] Water containing in excess of 1000 mg/liter dissolved solids.
[c] May be ≤ 1000 for mechanical pulping operations.
[d] No particles ≥ 3 mm diameter.
[e] One mg/liters for pressures above 700 psig.
[f] No floating oil.
[g] Applies to bleached chemical pulp and paper only.
[h] 12000 mg/liter includes 6000 Fe$^+$, and 6000 Fe^{2+}.
ASTM Standards 1970[4] or Standard Methods 1971[16].

TABLE 5.3 (*Continued*)

| | | | | | Mining Industry | | Oil Recovery Injection Water | |
Lumber Industry SIC-24	Pulp and Paper Industry SIC-26	Chemical Industry SIC-28	Petroleum Industry SIC-29	Prim. Metals Industry SIC-33	Copper Sulfide Concentrator Process Water	Copper Leach Solution	Sea Water	Formation Water
—	50	—	85					
—	—	—	—	—	—	12000		
—	2.6	10	15	—	—	12000[h]	0.2	13
—	—	2						
—	—	250	220	—	1510(CaCO$_3$)	—	400	2727
—	—	100	85	—	—	12000	1272	655
—	—	—	230	—	—	—	10840	42000
—	—	—	40					
—	—	600	480	—	—	—	142	281
—	—	850	900	—	1634	64000	2560	42
—	200[c]	500	1600	500	12	—	18980	72782
—	—	—	1.2	—				
—	—	—	8					
—	1080	2500	3500	1500	2100	—	34292	118524
d	--	10000	5000	3000				
—	475	1000	900	1000	1530			
—	—	500	500	200	415			
—	—	—	—	75				
5–9	4.6–9.4	5.5–9.0	6.0–9.0	3–9	to 11.7	3–3.5	—	to 6.5
—	360	500	25					
—	—	—	—	30				
—	—	—	20					
—	95[g]	—	—	100				

scum, and other floating materials attributable to municipal, industrial, or other discharges, or agricultural practices in amounts sufficient to be unsightly or deleterious; and (3) free from materials attributable to municipal, industrial, or other discharges, or agricultural practices producing color, odor, or other conditions in such a degree as to cause a nuisance (ORSANCO, 1967).

It is significant to note that according to the Bureau of the Census (NAS/NAE, 1973), the water withdrawal of industry, including manufacturing plants and investor-owned thermal electric utilities, was some 84,000 billion gallons per year in 1968, of which 93% was used for cooling and condensing, 5% for processing, and 2% for boiler-feed water. The previously mentioned predicted reductions in water withdrawals by the year 2000, based on assumptions about water-use efficiency and recycling, amounts to 62% for the manufacturing industry. However, the more important parameter, consumptive use, is predicted to increase because of the greater evaporation resulting from the higher temperatures in the recycling process (USWRC, 1978).

For a detailed description of water quality requirements for industry, and the treatment methodology for attaining them, the reader is referred to Nordell (1961).

Agricultural Water Supply

According to the United States Water Resources Council report (USWRC, 1978), the estimated withdrawal of water for irrigation and livestock watering in 1975 was 160,655 MGD (608.1 million m³/day) with a consumptive use of 88,303 MGD (334.2 million m³/day). The same report predicts a withdrawal of 156,397 MGD (592 million m³/day) for the year 2000 with a corresponding consumptive use of 95,057 MGD (359.8 million m³/day). As noted by that report, irrigation accounted for 81% of the total water consumed in the United States, and the California region accounted for almost one-third of the total water consumed by irrigation. It may be concluded that irrigation is the largest water user in the United States and, furthermore, that the majority of the water used for irrigation is in the western United States. It is also true that irrigation is the largest water user throughout the world. Thus, consideration of water quality for agricultural purposes is quite important, not only from the quality standpoint, but from the potential adverse effects on irrigated land.

The multiple uses of water in agriculture require that streams and other irrigation supplies be of such quality that potable water can be produced economically on the farm without significant fluctuations in quality. Furthermore, the raw water supply should be satisfactory without treatment for the irrigation of vegetable and fruit crops. For the farmstead water supply, the previously mentioned drinking water standards apply.

Irrigation water quality is of particular importance in arid and semiarid regions because of potential adverse effects on the soil. Although there is concern over the pollution aspects of irrigation water, treated effluents have been successfully used for crop irrigation for many years. For example, irrigation is one of the major means of disposal of the sewage water effluent for the city of Melbourne, Australia. Since water used for irrigation is mostly lost by evapotranspiration, irrigation represents a potential means for the ultimate disposal of wastewater effluents. It is for this reason that the clean water amendments of 1977 require consideration of land treatment prior to the award of construction grant monies.

Several problems may occur in the so-called land treatment of effluents, however. A problem unique to the West, where the appropriative rights doctrine applies, is that of water rights (see Chapter 2). Contrary to the riparian rights doctrine, one cannot simply apply a wastewater to the ground without consideration of the water rights involved. In addition, the lower rainfall in arid areas increases the significance of the potential accumulation of a particular contaminant in the soil.

It is not the intent of this discussion to present a pedantic outline of land treatment procedures, but water quality requirements for agricultural water use have recently become the subject of intensive study because of the increasing application of wastewaters for irrigation. When considering the use of treated wastewaters for irrigation, Noy and Feinmesser (1977) note the following disadvantages: (a) the supply of wastewater is continuous throughout the year, while irrigation is seasonal and dependent on crop demands; (b) treated wastewater may plug nozzles in irrigation systems and clog capillary pores of heavy soils; (c) some of the soluble constituents in wastewater may be present in concentrations toxic to plants; (d) health regulations restrict the application of wastewater to edible crops; and (e) when wastewater is not properly treated, it may be a nuisance to the environment.

Regardless of the source of the water, its quality requirements for irrigation are the same. The NAE/NAS report on Water Quality Criteria (NAS/NAE, 1973) made several recommendations for irrigation water. For example, they recommended that fecal coliform concentrations should be below 1000 per 100 ml of water in order to avoid hazards from pathogenic microorganisms from use or consumption of raw crops irrigated with such waters. In addition, they recommended maximum concentrations of trace elements in irrigation waters as shown in Table 5.4. The fate of various organisms in wastewaters in groundwater is shown in Table 5.5 (USEPA, 1976a).

Water lost by evapotranspiration has no salt content; therefore, soluble salts in the water lost will be retained by the soil. In order to keep the salt levels in soils at an acceptable level without impairing plant growth, additional water must be applied. It is obvious that the salt content of irrigation water is of primary importance. Excess water application necessary to permit the use of saline irrigation water and/or for leaching the harmful

TABLE 5.4   Recommended maximum concentrations of trace elements in irrigation waters[a]

Element	For waters used continuously on all soil mg/liter	For use up to 20 years on fine textured soils of pH 6.0 to 8.5 mg/liter
Aluminum	5.0	20.0
Arsenic	0.10	2.0
Beryllium	0.10	0.50
Boron	0.75	2.0
Cadmium	0.010	0.050
Chromium	0.10	1.0
Cobalt	0.050	5.0
Copper	0.20	5.0
Fluoride	1.0	15.0
Iron	5.0	20.0
Lead	5.0	10.0
Lithium	2.5[c]	2.5[c]
Manganese	0.20	10.0
Molybdenum	0.010	0.050[d]
Nickel	0.20	2.0
Selenium	0.020	0.020
Tin[b]	—	—
Titanium[b]	—	—
Tungsten[b]	—	—
Vanadium	0.10	1.0
Zinc	2.0	10.0

[a]These levels will normally not adversely affect plants or soils.
[b]See text for a discussion of these elements.
[c]Recommended maximum concentration for irrigating citrus is 0.075 mg/liter.
[d]Only for acid fine textured soils or acid soils with relatively high iron oxide contents.

accumulation of chemicals from the soil root zone is called the "leaching requirement." This requirement will depend on several factors, including the salt tolerance of the crops, and permeability and salinity of the irrigation water. Crops should be selected on the basis of their salt tolerance and the salt content of the soil. The relative salt tolerance of some important crops is presented in Table 5.6.

The leaching requirement may be defined as

$$LR = \frac{EC_i}{EC_d}, \tag{5.1}$$

where EC_i is the conductivity of the irrigation water and EC_d is the conductivity of the drainage water past the root zone.

TABLE 5.5 Survival times of organisms

Organism	Medium	Type of Application	Survival Time
Ascaris ova	Soil	Sewage	Up to 7 years
	Vegetables	AC[a]	27–35 days
Bacillus typhosa	Soil	AC	29–70 days
	Vegetables	AC	31 days
Cholera vibrios	Spinach, lettuce	AC	22–29 days
	Nonacid vegetables	AC	2 days
Coliform	Grass	Sewage	14 days
	Tomatoes	Sewage	35 days
Entamoeba histolytica	Vegetables	AC	3 days
	Soil	AC	8 days
Hookworm larvae	Soil	Infected feces	6 weeks
Leptospira	Soil	AC	15–43 days
Polio virus	Polluted water	—	20 days
Salmonella typhi	Radishes	Infected feces	53 days
	Soil	Infected feces	74 days
Shigella	Tomatoes	AC	2–7 days
Tubercle bacilli	Soil	AC	6 months
Typhoid bacilli	Soil	AC	7–40 days

[a] AC—Artificial contamination.

The characterization of the salinity of soils is based on the conductivity of the saturation extract of the soil, the exchangeable sodium, the pH, and other chemical characteristics. Based on these characteristics, a soil can be saline, saline–alkali, or nonsaline–alkali (USDA, 1969). The basic chemical characterization of these soils is shown in Table 5.7. The conductivity expressed in μmhos/cm can be related to the concentration of dissolved salts [1000 μmhos/cm = 640 mg/liter of total dissolved solids (TDS)].

The flow of irrigation water must be sufficient to keep the concentrations of salts in the soil solution below levels harmful to the plants being grown. A salt inflow–outflow balance must be maintained, preferably with the level of salinity of the drainage water such that downstream reuse of water is possible.

The usability of water for the irrigation of crops is determined by the chemical content of the water, the sensitivity of the crops to salts and water soluble elements, and the chemical characteristics of the soil to which the water will be applied. The most important quality consideration

TABLE 5.6   Salt tolerance of crops[a]

Salt Sensitive (< 600 μmhos/cm)			
Avocado	Citrus	Strawberries	Peach
Apricot	Almond	Plum	Prune
Apple	Pear	Beans (400)	Celery
Radish	Clover		

Medium Tolerance (600–1500 μmhos/cm)			
Grape	Cantaloupe	Cucumber	Squash
Peas	Onion	Carrot	Peppers
Potato (600)	Sweet corn	Lettuce	Olive
Fig	Pomegranate	Cabbage	Broccoli
Tomato	Oats (1500)	Wheat (1600)	Rye
Alfalfa (800)	Corn (600)		

High Tolerance (> 1500 μmhos/cm)			
Asparagus	Cotton	Garden beets	Barley (1600)
	Sugar beets		

[a] USDA (1964).

in irrigation water is the total salt content expressed in mmhos/cm or in mg/liter of TDS. Based on the salt content, irrigation waters may be classified into four salinity groups as shown in Table 5.8.

The second most important parameter of irrigation water quality is the relationship of the cations of sodium to those of calcium and magnesium, called the sodium adsorption ratio (SAR), which is expressed as

$$\text{SAR} = \frac{\text{Na}^+}{[(\text{Ca}^{2+} + \text{Mg}^{2+})/2]^{1/2}}, \tag{5.2}$$

where the cations are expressed in mEq.

TABLE 5.7   Soil characteristics[a]

Parameter	Saline Soils	Saline Alkali	Nonsaline Alkali
1. Conductivity of saturation extract at 25° C	>4 mmhos/cm	>4 mmhos/cm	<4 mmhos/cm
2. Exchangeable sodium percent	< 15	> 15	> 15
3. pH	< 8.5	< 8.5	8.5–10
4. Appearance	Recognized by white crust		Deposited organics, black color

[a] From Fireman and Hayward (1955).

TABLE 5.8 Suggested guidelines for salinity in irrigation water[a]

Crop Response	TDS mg/liter	EC mmhos/cm
Water for which no detrimental effects will usually be noticed	< 500	< 0.75
Water which can have detrimental effect on sensitive crops	500–1000	0.75–1.5
Water that may have adverse effects on many crops and requiring careful management practices	1000–2000	1.5–3.0
Water that can be used for salt-tolerant plants on permeable soils with careful management practices	2000–5000	3.00–7.0

[a] FWPCA (1968).

The sodium adsorption ratio can be related to the exchangeable sodium percentage (ESP) of the soil irrigated with that water. The ESP is a fraction of the negatively charged adsorption sites in the soil occupied by Na^+ ions. If the ESP is greater than 10 to 20%, it is assumed that the soil deteriorates. Water with a high SAR causes the replacement of Ca^{2+} and Mg^{2+} ions in the soil by Na^+ ions in the water until equilibrium conditions indicated by the ESP–SAR relationship are achieved. The relationship of SAR *v.* ESP can be evaluated from this empirical equation:

$$ESP = \frac{100\,(-0.0126 + 0.01475\,SAR)}{1 + (-0.0126 + 0.01475\,SAR)}. \tag{5.3}$$

A nomogram for determining the SAR value of irrigation water and for estimating the corresponding ESP value that is at equilibrium with the water is shown in Fig. 5.3 (USDA, 1969).

Irrigation waters having SAR values of 8 or less are probably safe, those with values 12 to 15 are marginal, and continued use of water with SAR's greater than 20 could lead to serious sodium hazards. Where used on sensitive crops, SAR values above 4 may be detrimental because of sodium phytotoxicity (FWPCA, 1968).

It is significant to note that while boron is essential to plant growth, concentrations of 1 to 4 mg/liter may be toxic to plants. In addition, bicarbonates in excess of 10 to 20 mg/liter may cause iron chlorosis in plants and other problems related to the permeability and sodium-exchange capacity of soils.

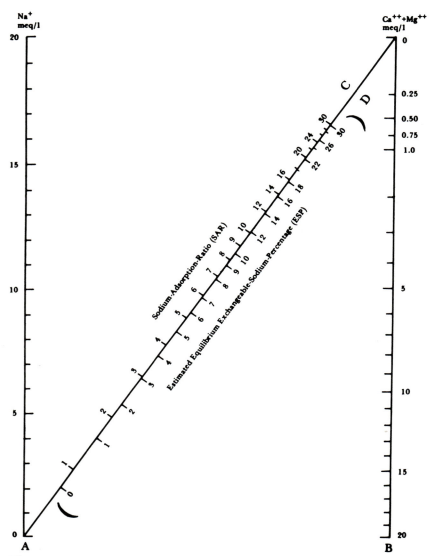

Fig. 5.3.   Nomogram for determining the SAR value of irrigation water and for estimating the corresponding ESP value of a soil that is at equilibrium with the water (USDA, 1969).

Stock and Wildlife Watering

According to the National Technical Advisory Committee (FWPCA, 1968), ''desirable quality criteria for livestock drinking water should ulti-

TABLE 5.9 NAS/NAE recommendation for livestock watering

Parameter	Limitation (mg/liter)
Arsenic	0.20
Boron	5.00
Cadmium	0.05
Chromium	1.00
Cobalt	1.00
Copper	0.50
Fluoride	2.00
Lead	0.10
Mercury	0.01
Nitrite + nitrate	100.00
Nitrite	10.00
Selenium	0.05
Vanadium	0.10
Zinc	25.00

mately be no less than for man.'' However, the problems associated with this goal are also delineated. Salinity can cause adverse physiological effects on livestock (NAS/NAE, 1973), and the NAE/NAS study recommended that waters containing 3000 mg/liter or less of soluble salts were probably satisfactory for livestock (NAS/NAE, 1973). The study also recommended limitations for individual water quality constituents, as shown in Table 5.9.

In addition, it was recommended that waters containing heavy growths of blue-green algae be avoided and that the drinking water standards for radionuclides and pesticides be applied to livestock water supplies. The possibility of livestock's contracting various diseases from contaminated water was discussed both in the NAS/NAE and the NTAC studies. However, no specific recommendations were made.

Propagation of Fish and Other Aquatic Life

The prime objective of water quality standards for fish and aquatic life is to restore and maintain environmental conditions that are essential for the survival, growth, reproduction, and general well-being of the biota.

The criteria should be based on scientific evaluation of available data by aquatic scientists, without economic considerations, and the standards should be developed with proper consideration of local conditions.

In order to establish the importance of aquatic life resources, it should be noted that the commercial freshwater and estuarine catch is in excess of $1.6 billion per year. In addition, sportfishermen spend some $2 billion per year on equipment, licenses, and trip expenses. While the recreational and aesthetic value of sportfishing cannot be estimated in monetary terms, it is undoubtedly a most significant part of aquatic resources (Bender and Jackson, 1969). Thus, the objective of Congress in establishing a goal of fishable and swimmable waters is a viable one from the economic standpoint.

In defining water quality requirements for aquatic life, it is necessary to define the extreme upper and lower limits of the various environmental factors as well as the optimum values. The extremes are governed by two important biological laws, as shown in Fig. 5.4. Liebig's law of the minimum relates the number of species to the essential material available in amounts most closely approaching the critical minimum, which then become the limiting factor. Shelford's law of tolerance states that survival of an organism can be controlled by the quantitative or qualitative deficiency or excess with respect to any one of several factors which may approach the limits of tolerance for that particular organism. According to Shelford's law, any substance or combination of substances may be toxic as soon as the concentration exceeds a certain level. The toxicity of a substance or combination of substances is usually expressed by the median tolerance limit, or TL_m, which is the concentration range at which 50% of the test organisms survive for a certain period of time. The duration of the toxicity experiments is usually 96 hours as discussed in Chapter 4.

The type and life history of a fish population is also important in determining water quality standards. Most of the freshwater fishes, such as the

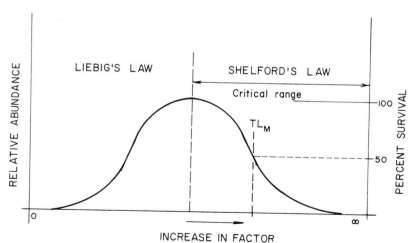

Fig. 5.4. Tolerance of organisms.

smallmouth bass, the muskie, pikes, and pickerels, spend their entire lives in fresh water. *Anadromous* fish such as the salmon, striped bass, and some of the herrings, for example, the shad, lay their eggs in fresh water and then migrate to the sea to feed and mature. These species include some of the most important food and game fishes, and the maintenance of their migration routes is a major problem. *Catadromous* fish are few in number. The eel is the best known, and lays eggs in the sea but matures in fresh water. Marine fish, as the name implies, reproduce and grow in the sea, estuaries, and brackish waters. Important species include shrimp, oysters, menhaden, and tuna.

In designating a stream for fish and wildlife protection, the following considerations should be taken into account:

1. Streams often constitute breeding migration routes. These must be reasonably clean or migration is halted and reproduction inhibited. It is a well-known axiom that a life cycle need be broken at only one point to eliminate the species.

2. Certain sites, often in streams or in shallow water in the proximity of lakes and estuaries, are favored as spawning areas. Water quality and quantity here are, and should be, particularly rigid.

3. Estuaries are some of the most biologically productive and heavily used waters in the world.

4. Wetlands are as fundamental to survival to waterfowl as water is to fish. Marshy areas are used as feeding and breeding areas in the summer and feeding and living areas in the winter.

Water quality investigations indicate that dissolved oxygen content is probably the most used and discussed water quality parameter and can be classified into three groups. The first level of oxygen concentration would just permit the fish to exist, the second level would permit the fish or aquatic organism to be active to a specified degree, and the third level would allow the organism to live, grow, and reproduce in a given area. The first level probably should not be considered. The second level may be used for specified periods of time (day or week) during which the animal should be able to feed, evade many of its enemies, and generally maintain itself in the environment. The third level would be suitable for completion of the life history of the organism.

These considerations were included and translated into the water quality criteria recommended and accepted by the Environmental Protection Agency on the recommendation of the National Technical Advisory Committee on Water Quality Criteria for Fish, Other Aquatic Life, and Wild-life (FWPCA, 1968). The criteria require that for warm-water biological systems, including game fish, the dissolved oxygen (D.O.) concentration

TABLE 5.10   Selected quality requirements for fish and other aquatic life[a]

	Fresh water		
Characteristics	Warm Water Biota	Cold Water Biota	Marine and Estuarine
pH (units)	6–9	6–9	6.7–8.5[d]
Total alkalinity (mg/liter)	> 20	> 20	
Increase in temp (°F)[b]	< 5[c]	< 5[c]	< 1.5 (June–Aug)[e] < 4 (Sept.–May)
Turbidity (Jackson Turbidity Units)	50	10	
Change in salinity			10% of natural variation
Coliform-median MPN (No./100 ml)			< 70[f]

[a] FWPCA (1968).
[b] 1° F increase = 0.56°C increase.
[c] For lakes, < 3° F.
[d] Normal pH range should not be altered by more than 0.1 pH unit.
[e] North of Long Island and in Pacific Northwest, summer limits apply July–Sept.; fall to spring limits apply October–June.
[f] For waters used for shellfish cultivation and harvesting. Maximum of 10% of samples to exceed MPN of 230/100 ml for five tube test.

should be above 5 mg/liter, assuming that the normal variations of D.O. in time are above this concentration. It is suggested, however, that under extreme conditions the dissolved oxygen concentration may range between 5 and 4 mg/liter for a short time during a 24-hour period. For cold-water biota spawning areas, the dissolved oxygen should not be below 7 mg/liter at any time. For the general well-being of trout, salmon, and associated cold-water biota, the dissolved oxygen concentration should not be below 6 mg/liter. Under extreme conditions, the concentration may range between 6 and 5 mg/liter, provided that all other water quality parameters are within the prescribed ranges. In large streams that have some stratification or that serve as migratory routes, D.O. levels may range between 4 and 5 mg/liter for periods of up to 6 hours, but they should never be below 4 mg/liter at any time or place. It should be noted, however, that considerable controversy exists over the D.O. requirements of aquatic life (Krenkel, 1979).

Recommendations regarding selected chemical, physical, and bacteriological criteria are presented in Table 5.10 (FWPCA, 1968), although

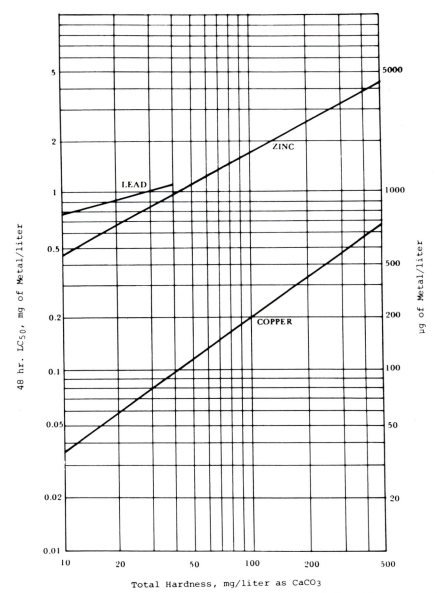

Fig. 5.5. The 48-hr lethal concentrations of three heavy metals for rainbow trout and hardness of water (NAS/NAE, 1973).

these criteria were somewhat modified by the NAS/NAE study. The reader is referred to the NAS/NAE report for recommendations on the allowable concentrations of the many contaminants that might affect aquatic life (NAS/NAE, 1973).

It is important to note that the establishment of water quality criteria for aquatic life is quite difficult because of the synergistic and antagonistic effects of various substances. For example, if D.O. levels are low or temperature is high, the metabolic activity of the organism is increased with a concomitant increase in toxicity of substances such as the heavy metals. On the other hand, if the hardness of water increases, the toxicity of the heavy metals decreases, as shown in Fig. 5.5. If more than one toxic material exists in the water, the toxicity may be synergistic, as is the case with zinc and copper. The pH may have a significant effect on the toxicity of some substances. NH_3 is an example of the effect of pH on toxicity inasmuch as free ammonia, which is highly toxic to fish, increases with increasing pH. In addition, low D.O. concentrations will also increase the toxicity of free ammonia to fish. It is also important to emphasize the synergistic effect of increasing temperature on most toxic materials.

Obviously, the factors affecting aquatic life are quite complex, and many times the agent(s) causing fish kills is difficult to identify. Thus, water quality standards for aquatic life are quite subjective, and must be, if we are to protect all aquatic life.

All effluents containing foreign material should be considered harmful and not permissible until bioassay tests have shown them otherwise. It is the obligation of the agency which produces the effluent to demonstrate that it is harmless in the concentrations to be found in the receiving waters. An appropriate application factor should then be applied in order to determine the permissible concentration of toxicants. Because of the increasing appearance of pesticides in our receiving waters, special attention has been given to their toxicity. The results of toxicity bioassay tests of various pesticides are shown in Tables 5.11 and 5.12. Finally, Table 5.13 demonstrates the variability in results and reporting of toxicity to fish (Jones, 1964).

Shellfish Culture

By definition, shellfish include mollusks, such as oysters, clams, and mussels; and crustaceans, such as lobsters, crabs, and shrimp. The importance of the shellfish industry is exemplified by the intensive study of the industry performed in Mobile Bay in 1970 (USDI, 1970). It was estimated that the economic loss in that region alone, resulting from the clo-

TABLE 5.11 Comparative toxicity of chlorinated hydrocarbon insecticides to different species of fish[a,b]

Insecticides	0.96 hour TL_m (median tolerance limit) ppm (micrograms/liter) active agent			
	Fatheads	Bluegills	Goldfish	Guppies
Aldrin	33	13	28	33
Dieldrin	16	7.9	37	22
Endrin	1.0	0.60	1.9	1.5
Chlordane	52	22	82	190
Heptachlor	94	19	230	107
Toxaphene	7.5	3.5	5.6	20
DDT	32	16	27	43
Methoxychlor	64	62	56	120
Lindane	62	77	152	138
BHC	2300	790	2300	21700

[a]Under standardized conditions—soft water as diluent. Temperature 25°C.
[b]From Henderson *et al* (1959).

sure of shellfish beds because of pollution, was between $56,700 and $227,000 annually.

Shellfish are particularly susceptible to adverse water quality because of their ingestion of water and their ability to concentrate materials in their body. For example, an active oyster passes many gallons of water

TABLE 5.12 Comparative toxicity of organic phosphorus insecticides to different species of fish[a,b]

96-hour TL_m (median tolerance limit, parts per billion (micrograms per liter) active agent.

Insecticides	Fatheads	Bluegills	Goldfish	Guppies
Chlorthion	3,000	700	2,300	1,200
Co-Ral	18,000[c]	180	> 18,000	560
Delnav	10,000	34	32,000	210
Dylox (Dipterex)	140,000	3,800	99,000	7,100
Di-Syston	3,700	63	6,500	250
EPN	250	100	450	32
Guthion	93	52	1,300	110
Malathion	23,000	90	450	840
Methylparathion	8,000	1,900	9,600	7,800
OMPA	100,000	110,000	610,000	20,000
Parathion	1,300	95	2,700	55
Systox	3,200	100	1,100	610
TEPP	1,900	1,100	21,000	1,800

[a]From Bender and Jackson (1969).
[b]Under standardized conditions—soft water as diluent—Temperature 25°C.
[c]Solubility in water less than 18 ppm.

TABLE 5.13   Lethal limits for metals as salts[a,b]

Salt	Fish Tested	Lethal Concentration (ppm)	Exposure Time (hrs)
Aluminum nitrate	stickleback	0.1 Al	144
Aluminum potassium sulfate (alum)	goldfish	100	12–96
Barium chloride	goldfish	5,000	12–17
Barium chloride	salmon	158	?
Barium nitrate	stickleback	500 Ba	180
Beryllium sulfate	fathead minnow	0.2 Be	96
Beryllium sulfate	bluegill	1.3 Be	96
Cadmium chloride	goldfish	0.017	9–18
Cadmium chloride	fathead minnow	0.9	96
Cadmium (salt?)	rainbow trout	3 Cd	168
Cadmium nitrate	stickleback	0.3 Cd	190
Calcium nitrate	goldfish	6,061	43–48
Calcium nitrate	stickleback	1,000 Ca	192
Cobalt chloride	goldfish	10	168
Cobalt (salt?)	rainbow trout	30 Co	168
Cobalt nitrate	stickleback	15 Co	160
Copper nitrate	salmon	0.18	?
Copper nitrate	stickleback	0.02 Cu	192
Copper nitrate	rainbow trout	0.08 Cu	20
Copper sulfate	stickleback	0.03 Cu	160
Copper sulfate	fathead minnow	0.05 Cu	96
Copper sulfate	bluegill	0.2 Cu	96
Copper sulfate	minnow	1.0 Cu	80
Copper sulfate	brown trout	1.0 Cu	80
Cupric chloride	goldfish	0.019	3–7
Lead chloride	fathead minnow	2.4 Pb	96
Lead nitrate	minnow	0.33 Pb	?
Lead nitrate	stickleback	0.33 Pb	?
Lead nitrate	brown trout	0.33 Pb	?
Lead nitrate	stickleback	0.1 Pb	336
Lead nitrate	goldfish	10	1–2
Lead nitrate	rainbow trout	1 Pb	100
Magnesium nitrate	stickleback	400 Mg	120
Magnesium nitrate	stickleback	50 Mn	160
Manganese (salt?)	rainbow trout	75 Mn	168
Mercuric chloride	rainbow trout	0.01 Hg	204
Mercuric chloride	rainbow trout	0.15 Hg	168
Mercuric chloride	rainbow trout	1.0 Hg	600
Nickel chloride	goldfish	10	200
Nickel chloride	fathead minnow	4 Ni	96
Nickel nitrate	stickleback	1 Ni	156
Nickel (salt?)	rainbow trout	30 Ni	168
Potassium chloride	goldfish	74.6	5–15
Potassium chloride	straw-colored minnow	373	12–29

TABLE 5.13 (*Continued*)

Salt	Fish Tested	Lethal Concentration (ppm)	Exposure Time (hrs)
Potassium nitrate	stickleback	70 K	154
Silver nitrate	stickleback	0.004 Ag	180
Sodium chloride	goldfish	10,000	240
Sodium chloride	plains killifish	16,000	96
Sodium chloride	green sunfish	10,713	96
Sodium chloride	gambusia	10,670	96
Sodium chloride	red shiner	9,513	96
Sodium chloride	fathead minnow	8,718	96
Sodium chloride	black bullhead	7,994	96
Sodium nitrate	stickleback	600 Na	180
Sodium nitrate	goldfish	1,282	14
Sodium sulfite	goldfish	100	96
Strontium chloride	goldfish	15,384	17–31
Strontium nitrate	stickleback	1,500 Sr	164
Titanium sulfate	fathead minnow	8.2 Ti	96
Uranyl sulfate	fathead minnow	2.8 U	96
Vanadyl sulfate	fathead minnow	4.8 V	96
Vanadyl sulfate	bluegill	6	96
Zinc sulfate	stickleback	0.3 Zn	204
Zinc sulfate	goldfish	100	120
Zinc sulfate	rainbow trout	0.5	64

[a] Adapted from Jones (1964).

[b] In this table the concentration values are the lowest at which defined toxic action is indicated by the data in the reference cited. It must not be assumed that lower concentrations are harmless, and for further information the works cited should be consulted as many include survival curves or tables. Most of the data are for temperatures between 15°C and 23°C. Concentrations are parts per million. Exposure times have been approximated in some cases.

through its gills daily (Maxcy, 1956). In the classic Minamata Bay incidents, shellfish were found to contain as much as 39 ppm (wet weight) of mercury (Takeuchi, 1972).

The potential for disease transmission by contaminated shellfish was confirmed by the oyster epidemics at Winchester and Southhampton, England, in 1902, where 118 of 267 guests at banquets were subjected to intestinal disorders, 21 of whom developed typhoid fever, and 5 died. (Prescott *et al.*, 1947). The last major epidemic in the United States attributed to shellfish was in the winter of 1924–1925 when some 1500 cases of typhoid fever occurred resulting in 100 deaths in different cities; all of these were traced to oysters from a single lot (Maxcy, 1956). The transmission of virus by shellfish has also been confirmed by Liu *et al.*, (1965). The problem is excacerbated by the fact that many people eat live shellfish.

The recommendations of the NAE/NAS report state that water quality for shellfish harvesting should meet the standards of the National Shellfish Sanitation Program (USDHEW, 1965).

1. Examinations are to be conducted in accordance with recommended procedures of the APHA (APHA/AWWA/WPCF, 1975).

2. No direct discharges of inadequately treated sewage are to be made.

3. Samples are to be collected under conditions of time and tide which produce maximum concentrations of bacteria.

4. The MPN must not exceed 70 per 100 ml and not more than 10% of the samples ordinarily exceed an MPN of 230 per 100 ml for a five-tube decimal dilution test in those portions of the area most probably exposed to fecal contamination during the most unfavorable hydrographic and pollution conditions.

5. The reliability of nearby wastewater treatment plants must be considered before areas for direct harvesting are approved.

TABLE 5.14   Recommended guidelines for pesticide levels in shellfish[a]

Pesticide	Concentration in shellfish (ppm-drained weight)
Aldrin[b]	0.20
BHC	0.20
Chlordane	0.03
DDT)	
DDE) Any one or all, not to exceed	1.50
DDD)	
Dieldrin[b]	0.20
Endrin[b]	0.20
Heptachlor[b]	0.20
Heptachlor Epoxide[b]	0.20
Lindane	0.20
Methoxychlor	0.20
2,4-D	0.50

[a] For USDHEW (1968).

[b] It is recommended that if the combined values obtained for Aldrin, Dieldrin, Endrin, Heptachlor, and Heptachlor Epoxide exceed 0.20 ppm, such values be considered as "alert" levels which indicate the need for increased sampling until results indicate the levels are receding. It is further recommended that when the combined values for the above five pesticides reach the 0.25 ppm level, the areas be closed until it can be demonstrated that the levels are receding.

In addition, the NAE/NAS study recommended utilization of interim guidelines for pesticides in shellfish, recommended by the Public Health Service in 1968 (USDHEW, 1968), as shown in Table 5.14.

It should also be noted that contaminated shellfish are capable of causing paralytic poisoning because of the shellfish's ability to concentrate certain dinoflagellates, genus *Gonyaulax*. No known antidote exists, and 957 cases of paralytic shellfish poisoning with at least 222 deaths were reported in the United States up to 1962 (Halstead, 1965). A chemical method for quantitative determination of the poison has been devised. However, the most commonly used method for determining the toxicity of shellfish is a bioassay using mice. The quantity of paralytic shellfish poisoning producing death is measured in "mouse units" (NAS/NAE, 1973).

Recreation

With regard to recreation and aesthetics, the National Technical Advisory Committee on Surface Waters Water Quality Criteria (FWPCA, 1968) recommended that surface waters should be free of (a) materials that will settle to form objectionable deposits; (b) floating debris, oil, scum, and other matter; (c) substances producing objectionable color, odor, taste, or turbidity; (d) materials, including radionuclides in concentrations or combinations which are toxic or produce undesirable physiological responses in human, fish, and other animal life and plants; and (e) substances and conditions or combinations thereof in concentrations which produce undesirable aquatic life.

For general recreational use, in addition to the aesthetic criteria, the average fecal coliform count should not exceed 2000 per 100 ml and the maximum should not exceed 4000 per 100 ml except in the immediate vicinity of an outfall or mixing zone. For primary contact recreation, the number of fecal coliforms should not exceed a log mean of 200 per 100 ml in a minimum of five samples for any 30-day period of the recreation season. In addition, at least 90% of the total number of samples should be less than 400 fecal coliforms/100 ml. The pH of surface waters designed for primary contact recreation should be within the range of 6.5–8.3, and the clarity of such waters should be such that the Secchi disk be visible at a minimum depth of 1.2 m (4 ft). Finally, in primary contact recreation waters, the maximum water temperature should not exceed 29.4°C (85°F), except where higher temperatures are caused by natural conditions.

The recreational use of water refers to bathing, swimming, waterskiing, and other contact sports, boating, and other aesthetic enjoyment. Fishing is a recreational use of water. However, the quality requirements are those for the protection of aquatic life. The reader is referred to Chapter 4 for a discussion of the potential of recreational use of water as a vehicle of transmission for disease.

Other Beneficial Uses of Water

Traditionally the use of water for water power, navigation and transport, and dispersion and assimilation of wastes has been considered beneficial. However, recent philosophies, as promoted by the environmentalists, have indicated that these uses of water should be considered "nonbeneficial." In the authors' opinion, nature's ability to assimilate some wastewaters without deleterious effects is an economic resource which should be considered as a beneficial use of water. It is significant to note that the National Water Commission recommended that "Federal water pollution control legislation should recognize the capacity of receiving waters to absorb heat as a valuable resource" (NWC, 1973).

With regard to quality requirements for the uses described as "other beneficial uses of water," the recommendations made by ORSANCO and listed in the industrial water supply discussion are applicable.

THE ASSIGNMENT OF STREAM USE DESIGNATIONS

An important problem which may have serious consequences for dischargers located on a stream is the intended use of the downstream water. As previously noted, waste assimilative capacity may differ significantly if the stream is assigned according to the standards for fish and wildlife propagation, recreation, drinking water, or the protection of trout or salmon. In order to avoid arbitrary decisions based on subjective and nonscientific information, a few basic rules of stream classification should be introduced. As stated by Hawkes (1975), "Recognition must be given to the existence of the different river zones and, in determining river management policies, account should be taken of their different ecological characteristics."

Huet (1949) proposed a system for the classification of streams according to key fish species. The streams are categorized in four groups and, for each group, typical fish species are correlated with the most important morphological and water quality parameters of the streams. The four groups were described as follows.

1. *Zone à Truite* (Trout Zone). Streams with steep gradients and rapid currents. Stream bed is of rock, boulders, pebbles, but sometimes of gravel and sand. Width and depth variable, often quite shallow. Water well aerated and cool, rarely exceeding 20°C.

2. *Zone à Ombre* (Grayling Zone). Larger streams with depths up to 2 m and gradient less than that in trout zone, with alternating riffles and pools; current, however, rather rapid. Stream bed finer material than in trout zone—often of washed gravel. Water, in summer, lower in oxygen than trout zone but still satisfactory. Rapids inhabited by salmonids and pools by rheophilic cyprinids.

3. *Zone à Barbeau* (Barbel Zone). Rivers of moderate gradient and current with alternating rapids and quiet water, the latter, however, being more extensive than in the grayling zone. Trout still present in rapid stretches.

4. *Zone à Breme* (Bream Zone). Includes lower stretches of rivers, canals, and ditches. Current slight, summer temperature high, oxygen depleted. Water turbid, often deep, exceeding 2 m.

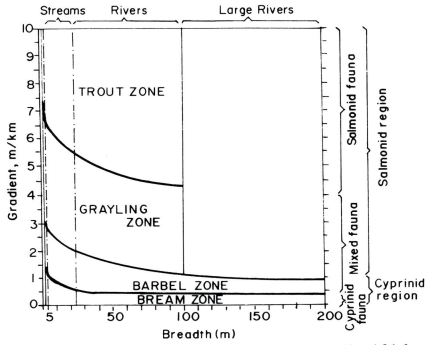

Fig. 5.6. Slope graph-showing relationship between gradient, breadth and fish faunal zone in rivers (Huet, 1949).

TABLE 5.15   Description and classification system of Maryland streams[a]

Code No.	Stream Class	Width (feet)	Depth	Temp	Water Quality	Bottom
1	Dace trickle	0 to 3	Very shallow	Cool	Usually clear, clean water except in spring	Boulders, gravel, and sand
2	Trout feeder	0 to 7	Shallow	Cold	Clear, clean water	Rubble to gravel
3	Trout stream	5 and over	Shallow with pools	Cold	Clear, clean water	Rubble to gravel
4	Sucker stream	3 to 20	Medium	Cool	Intermediate; may carry silt at times	Sand and gravel
5	Bass feeder	10 and over	Shallow with pools	Cool	Clear to slightly turbid water	Sand and gravel
6	Bass stream	20 and over	Medium with pools	Cool	Clear to slightly turbid water	Sand and gravel
7	Pickerel stream	20 and over	Moderate with pools	Cool to warm	Dark, turbid waters	Muck and sand
8	Bullhead stream	20–50	Medium	Warm to cool	Turbid water	Mud and muck
9	Catfish stream	70 and over	Moderate	Cool to warm	Turbid water	Mud, sand, and gravel
10	Carp stream	70 and over	Medium- to moderate	Warm	Muddy, sometimes highly turbid water	Mud and muck
11	Tidal stream	5 and over	Medium	Warm	Seasonally clear and muddy, More or less brackish	Sand, muck, and peat

[a]From Van Deusen (1954).

TABLE 5.15 (*Continued*)

Flow	Volume	Characteristic Forms	Shade and Cover	Miscellaneous Characters
Little	Little	Blacknose dace	Usually has some forest cover but may have none	Extreme upper reaches of most streams
Little to medium	Little to medium	Trout, muddlers, and creek chubs	Moderate shade insufficient stream cover for the large fish	Water volume not enough to support legal-sized trout population
Medium to moderate	Medium to moderate	Trout, muddlers, river, and creek chubs	Moderate amount of shade and cover for fish	With trout feeder streams flowing into main axis
Medium	Medium	Common sucker and common shiner	Lacks shade and cover	Has characteristics of a trout stream and may fall below
Medium	Medium	Smallmouth bass, crayfish, and mussels	Medium amount of cover and shade	Insufficient volume of water to support legal population of bass
Moderate	Moderate	Smallmouth bass, chubs, and crayfish	Medium amount of shade and moderate amount of cover	Sufficient volume and deep pools
Medium to slow	Medium	Chain pickerel, sunfish and crappie, and golden shiner	Moderate shade and cover	Aquatic vegetation serves as stream cover
Little to medium	Medium to little	Bullheads and variety of sunfish	Medium shade and cover	Only occasional pools
Medium	Medium and Moderate	Catfish, bullheads, and variety of panfishes	Medium shade and cover	Frequent deep pools
Medium	Little to medium	Carp, sunfishes, and catfish in marginal areas	Little shade and cover	Characterized by only a few species of fish present
Slow to medium	Medium	Variety of fresh and brackish water forms	Shade sparse and little stream cover	Shallow to deep channels with bars

Huet (1954) considered the gradient of the streams to be the primary feature characterizing the different zones. Based on his investigations of European streams, he constructed a "slope-graph" (Fig. 5.6) showing the relationship between the stream gradient, width, and fish faunal zones. For example, a river 15 m wide with a gradient of 8 m/km (ft/1000 ft) would be a trout stream, while a stream with the same width but with a slope of 2 m/km would be a barbel zone.

Van Deusen (1954) proposed a scheme for the classification of Maryland streams which was based on, and developed from, a survey conducted on Maryland's inland natural resources. The system divides a stream into eleven categories which differ in water quality, morphology, and fish population. The system of classification is presented in Table 5.15. The Maryland system appears to be adaptable to North American conditions.

It should be noted that the stream classification, as discussed previously, represents natural stream conditions without significant disruptions by man-caused pollution. The streams cannot be classified in their entirety but must be separated into stretches. Suggested criteria for dividing streams into zones include slope-width function after Huet (1954), annual or summer temperature profiles, and other ecological parameters such as current, flow, dissolved oxygen profile, dissolved nutrients and hardness, order of the stream, and shape of the valley.

Proper assignment of stream use and subsequent relation of stream use to water quality criteria represent one of the most important tasks in water quality management. Lowland sluggish streams are usually inhabited by more tolerant fish species (carp, bream, catfish). No feasible water pollution control practices will change them into a trout stream. Even the fact that the stream may have low temperatures due to the discharge of cold hypolimnetic water from an upstream reservoir will not make the stretch a trout stream. On the other hand, inhabitants of trout streams are less tolerant, and excessive levels of pollution may be damaging to fish and other biota.

Typical of water use classification is that of the state of Georgia, as shown in Table 5.16 (Georgia DNR, 1978). Under the Georgia Water Quality Control Act, the state is authorized to establish water quality standards and water use classifications. Appropriate waters may also be designated as trout streams. There are, in addition, general criteria applying to all waters of the state, including the prohibition of materials causing sludge deposits and scums, materials producing turbidity, odor, color, or other objectionable conditions, substances harmful to aquatic life, radioactive substances exceeding state or federal standards, and stream-bed alterations that may result in water quality standard violation.

TABLE 5.16 Summary of Georgia water quality standards by use classification[a]

Use Classification	Bacteria (fecal coliform) 30-day Geometric Mean (no./100 milliliter)	Bacteria (fecal coliform) Maximum (no./100 milliliter)	Dissolved Oxygen (other than trout streams) Daily Average (mg/liter)	Dissolved Oxygen (other than trout streams) Minimum (mg/liter)	Dissolved Oxygen (other than trout streams) pH	Temperature (other than estuaries or trout streams) Maximum Rise (°F)	Temperature (other than estuaries or trout streams) Maximum (°F)	Solids and Taste, Odor, and Color-Producing Substances	Remarks
Drinking water no treatment	50	—	—	—	—	—	—	None from waste discharges	b
Drinking water requiring treatment	1000	4000	5.0	4.0	6.0—8.5	5	90	—	c–e
Recreation	200	—	5.0	4.0	6.0—8.5	5	90	—	d–f
Fishing (excluding shellfishing)	1000	—	5.0	4.0	6.0—8.5	5	90	—	d–f
Agricultural	5000	—	—	3.0	6.0—8.5	5	90	—	e, g
Industrial	—	—	—	3.0	6.0—8.5	5	90	—	e, g
Navigation	5000	5000	—	3.0	6.0—8.5	5	90	—	e, g
Urban stream	2000	—	—	3.0	6.0—8.5	—	—	—	h
Wild river	No alteration of natural water quality								h
Scenic river	No alteration of natural water quality								h

[a] Georgia DNR (1978).
[b] No waste discharge.
[c] No substance in a concentration which after treatment exceeds state or federal drinking water standards.
[d] Trout streams: D.O. = 6 mg/liter daily average and greater than 5 mg/liter at all times; water temperature cannot be elevated or depressed; designated as such by the State Game and Fish Division.
[e] Estuarine waters: Maximum temperature rise limited to 1.5°F.
[f] No concentrations of toxic wastes harmful to man, fish, game, or other beneficial aquatic life.
[g] No concentrations of toxic water preventing fish survival.
[h] Designated as such by an authorized state or federal legislative branch.

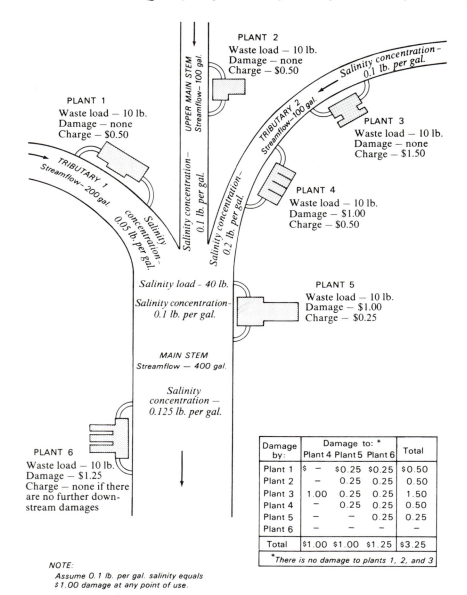

Fig. 5.7.   Schematic illustration of salinity discharges and damages on the tributaries and main stem of a river system (Kneese and Bower, 1968). (Quantities are per unit of time.)

WASTE LOAD ALLOCATION

A major question to be answered in water quality limited stream segments in the United States, or where effluent standards are not used, is how to allocate the existing waste assimilative capacity of a receiving water among several waste dischargers. The process can be quite complicated, with the objective being to minimize the total cost of attaining a particular standard at all stream locations. Hall and Dracup (1970) discuss the dynamic programming approach used by Liebmen (1965) on the Willamette River in Oregon. Another approach is that taken by Kneese and Bower (1968) in a case where upstream discharges impose a cost on downstream water users because of lower water quality. The problem posed is illustrated in Fig. 5.7 (Kneese and Bower, 1968).

While the attempts at optimizing waste assimilative capacity are quite rational, their complexity and the many assumptions inherent in their use lead most regulatory agencies to use much simpler approaches. For example, the total allowable load may be divided among the waste dischargers in some arbitrary fashion; a uniform reduction in the parameter involved based on existing discharges may be used; or each discharger may be required to attain a certain reduction based on raw wastewater characteristics.

Obviously, inequities exist in any of these approaches. For example, an industry may be penalized for voluntarily instigating wastewater treatment prior to legislation requiring it. It may cost much more for one industry to attain a given degree of treatment than another one. The fact that one industry may have been located on the river long before any of the others may not be taken into account.

Industries should provide realistic estimates of their production process outputs. At this state of negotiation between an industry and the pollution control planning agency, full collaboration must be maintained. It is not unusual that the production estimates are sometimes inflated in order to obtain a bigger piece of the allocation pie. The planning agency, on the other hand, should also be realistic in enforcing strict standards without considering natural or background water quality. In addition, a stream should not be assigned for trout fishing just because there is a small stretch of the stream with colder water caused by a reservoir discharge.

The process of waste load allocation is shown in Fig. 5.8. It usually proceeds in both directions, i.e., the planning agency (EPA or state planning agencies) must decide on the stream use in an overall planning context of the entire region. The stream use then determines the stream standards which are translated, usually using a mathematical model of the stream

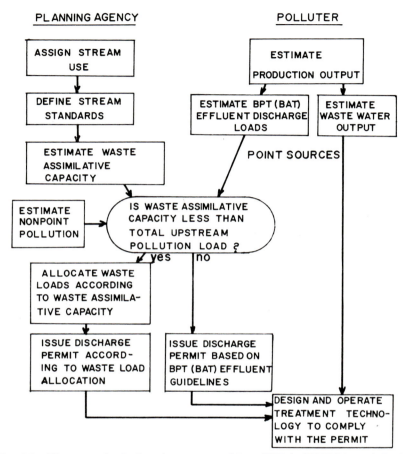

Fig. 5.8. The waste load allocation process. (Note: BPT is BPCTCA, and BAT is BATEA, both as defined in Chapter 2.)

water quality, into a waste assimilative capacity value for each important pollutant. The industries in the area of interest must meanwhile prepare realistic plans to treat their wastewaters. In many cases, these plans require treatability studies since only a few industrial wastewater effluents have a typical "textbook" quality. The effluent loadings to the receiving water body must meet the BPT or BAT limitations as described in Chapter 2. The plans are then submitted to the planning agency or to an agency responsible for water pollution control in the area.

The agency, after gathering the effluent loadings from all waste and sewage water discharges for the entire watershed, must compare the loadings to the receiving water body with its waste assimilative capacity. This must be done reach by reach since some of the pollutants can be removed or reduced during the time of residence in the receiving water body. If the loadings are below the waste assimilative capacity of the receiving water, the discharge may be allowed according to the effluent BPT or BAT guidelines.

If the waste assimilative capacity is below the total waste load input for a particular reach, the stream is considered to be water quality limited and the responsible planning agency must allocate the waste loads for each upstream discharge. As previously mentioned, the allocation may be accomplished by a uniform reduction of the pollution loads of all upstream sources for each constituent who violated the waste assimilative capacity. This process is quite sensitive since additional waste load reductions represent significant economic burdens on the dischargers. In many cases this may result in a disagreement between the planning agency and the dischargers involved. The law provides for a process of arbitration between the involved parties. In this process, it is up to the dischargers to prove that the allocation was not done properly or that the process was somewhere in violation of present scientific knowledge.

After the allocation of the waste loads is established, the dischargers must design their pollution control facilities to meet the allocated waste loads.

ANTIDEGRADATION CONCEPT

Even though current legislation does not include antidegradation requirements, the antidegradation policy set forth by former Secretary of the Interior Walter Hickel is included in state standards. Furthermore, the 1976 EPA guidelines for planning require that the policy should, as a minimum, contain the following components (USEPA 1976):

1. In all cases, existing instream beneficial water uses must be maintained and protected. Any action that would interfere with or become injurious to existing uses cannot be undertaken. Waste assimilation and transport are not recognized as beneficial uses.

2. Existing high quality waters must be maintained at their high quality unless the State decides to allow limited degradation where economically or socially justified. If limited degradation is allowed, it cannot result in violation of water quality criteria that describe the levels necessary to sustain the national water quality goals of protection and propagation of fish, shellfish, and wildlife and recreation in and on the water.

3. In all cases, high quality waters which constitute an outstanding national resource must be maintained and protected. High quality waters consist of those waters at or above the minimum levels necessary to achieve the national water quality goals.

National antidegradation rules should not be viewed as "no growth" rules. The waste loads allocated to various industrial and municipal discharges in the area of interest can be assigned in a way allowing for future growth. The planning agency and the agency responsible for issuing the discharge permits can

1. Design and allocate a part of present waste assimilative capacity for anticipated future economic growth and issue future permits only within the "surplus" of the waste assimilative capacity.

2. Issue progressively decreasing discharge permits with the initial total waste load approximately equaling the waste assimilative capacity and design waste load allocations to accommodate new sources via reduction in current source loadings.

3. Restrict any new discharges of pollutants from new and existing sources while allocating the waste load to present sources according to the total waste assimilative capacity.

4. Require land disposal of wastewaters for new projects.

REFERENCES

American Public Health Association, American Water Works Association, and Water Pollution Control Federation (1975). "Standard Methods for the Examination of Water and Wastewater", 14th ed. Washington, D.C.

Bender, M. E., and Jackson, H. W. (1969). Water quality criteria for fish and wildlife. Training Center Manual, R. A. Taft San. Eng., Cincinnati, Ohio.

Federal Water Pollution Control Administration (1968). Water quality criteria. Rep. of the Natl. Tech. Adv. Comm. on Water Qual., Washington, D.C.

Fireman, J., and Hayward, H. E. (1955). Irrigation water and saline and alkali soils. *In* "Water, The Yearbook of Agriculture." U.S. Dept. of Agric.

Georgia Department of Natural Resources (1978). Coosa River basin water quality management plan, 2nd ed. Environ. Prot. Div., Atlanta.

Hall, W. A., and Dracup, J. A. (1970). "Water Resources Systems Engineering." McGraw-Hill, New York.

Halstead, B. W. (1965). "Poisonous and Venomous Animals of the World: Vol. I, Invertebrates." U.S. Govt. Printing Office, Washington, D.C.

Hawkes, H. A. (1975). River zonation and classification. *In* "River Ecology" (B. A. Whitton, ed.). Univ. of California Press, Berkeley.

Henderson, C., Pickering, Q. H., and Tarzwell, C. M. (1959). Relative toxity of ten chlorinated hydrocarbon insecticides to four species of fish. *Trans.*, Amer. Fisheries Soc., **88**, (1), (January).

Hernandez, J. W., and Barkley, W. A. (1976). Training seminar on the National Safe Drinking Water Act. Coll. of Eng., New Mexico State Univ., Las Cruces.

Huet, M. (1949). Apercu des relations entre la pente a les populations des eaux courantes (A

153

review of relations between the gradient and populations of flowing waters). *Schweiz Z. Hydrol.* **11,** 333–351.

Huet, M. (1954). Biologie, profils en long et en travers des eaux courantes (Biology, longitudinal and cross-sectional profiles of flowing waters). *Bull. Fr. Piscic.* **175,** 41–53.

Imhoff, K., and Fair, G. M. (1940). "Sewage Treatment." Wiley, New York.

Izaak Walton League of America (1973). A citizen's guide to clean water. Arlington, Virginia.

Jones, J. R. E. (1964). "Fish and River Pollution." Butterworth, London.

Kneese, A. V., and Bower, B. T. (1968). "Managing Water Quality: Economics, Technology, Institutions." For Resources for the Future. Johns Hopkins Press, Baltimore, Maryland.

Krenkel, P. A. (1973). Some effects of wastes on natural waters. *In* "Environmental Impacts on Rivers" (H. W. Shen, ed.). Shen Publ., Fort Collins, Colorado.

Krenkel, P. A. (1979). Problems in the establishment of water quality criteria. *J. Water Pollut. Control Fed.* (September).

LeFoll, Y., Pinoit, R., and Lesouef, A. (1977). A multidimensional analysis of the results of the french 1971 surface water quality network control in the river basin Seine–Normandie. Eighth Int. Conf. on Water Pollut. Res., Sydney, Australia. *Prog. Water Technol.* **9,** 89–102.

Liebman, J. C. (1965). The optimal allocation of stream dissolved oxygen resources. Ph.D. Thesis, Cornell Univ., Ithaca, New York.

Liu, O. C., Seraichekas, H. R., and Murphy, B. L. (1965). Viral pollution and self-cleansing mechanism of hard clams. *In* "Transmission of Viruses by Water Route" (G. Berg, ed.). Wiley (Interscience), New York.

Lovett, M. H. (1958). Standards: Some pros and cons, control of river pollution. *Inst. of Sewage Purif., J. Proc.,* **4,** 409.

Maxcy, R. F. (1956). "Preventive Medicine and Public Health." Appleton, New York.

National Academy of Sciences, and National Academy of Engineering (1973). Water quality criteria—1972. EPA-R3-73-033 (March).

National Water Commission (1973). Water policies for the future. Rep. to Pres. and Congr., Washington, D.C.

Nordell, E. (1961). "Water Treatment for Industrial and Other Uses." Van Nostrand-Reinhold, Princeton, New Jersey.

Noy, J., and Feinmesser, A. (1977). The use of wastewater for agricultural irrigation. *In* "Water Renovation and Reuse" (H. I. Shuval, ed.). Academic Press, New York.

Ohio River Valley Sanitation Commission (1967). Stream quality criteria and minimum conditions. ORSANCO, Cincinnati, Ohio.

Phelps, E. (1944). The oxygen balance. *In* "Stream Sanitation." Wiley, New York.

Pickering, Q. H., Henderson, C., and Lemke, A. E. (1962). The toxicity of organic phosphorus insecticides to different species of warmwater fishes, *Trans.,* Amer. Fisheries Soc., **91,** (2), (April).

Porges, R. (1972). Wastewater quality criteria. *Proc. Int. Conf. Int. Assoc. Water Pollut. Res.,* 6th, Jerusalem, Israel.

Prescott, S. C., Winslow, C. E. A., and McCrady, M. H. (1947). "Water Bacteriology," 6th ed. J. Wiley, New York.

Streeter, H. W. (1925). A study of the pollution and natural purification of the Ohio River. *Pub. Health Bull.* No. 146, U.S. Public Health Serv., Washington, D.C.

Takeuchi, T. (1972). Biological reactions and pathological changes in human beings and animals caused by organic mercury contamination." *In* "Environmental Mercury Contamination" (R. Hartung and B. D. Ditman, eds.). Ann Arbor Science Publ., Ann Arbor, Michigan.

U.S. Department of Agriculture (1964). Irrigation water requirements. Soil Cons. Serv., Tech. Release 21, Washington, D.C.

U.S. Department of Agriculture (1969). Diagnosis and improvement of saline and alkali soils. Agric. Handbook No. 60, Washington, D.C.

U.S. Department of Health, Education and Welfare (1965). National shellfish sanitation program manual of operations. PHS Publ. No. 33. U.S. Govt. Printing Office, Washington, D.C.

U.S. Department of Health, Education and Welfare (1968). "Proceedings, Sixth National Shellfish Sanitation Workshop. (G. Morrison, ed.). U.S. Govt. Printing Office, Washington, D.C.

U.S. Department of Interior (1970). Pollution affecting shellfish harvesting in Mobile Bay, Alabama. Fed. Water Pollut. Control Admin., Southeast Water Lab., Athens, Georgia.

U.S. Environmental Protection Agency (1975). National interim primary water standards. Washington, D.C.

U.S. Environmental Protection Agency (1976a). Land treatment of municipal wastewater effluents. Design Factors-1, Washington, D.C.

U.S. Environmental Protection Agency (1976b). Guidelines for state and areawide water quality management program development. Washington, D.C.

U.S. Water Resources Council (1978). "The Nations Water Resources, 1975–2000," Vol. 1. Summary, 2nd Natl. Water Assessment. U.S. Govt. Printing Office, Washington, D.C.

Van Deusen, R. D. (1954). Maryland freshwater stream classification by watersheds. Maryland Dept. of Res. & Educ., Contribution 106, 1–30.

6 | The Dynamics of the Water Cycle

THE HYDROLOGIC CYCLE

The hydrologic cycle represents the basic source of water for all uses and is essential for maintaining life. Water is not only a building block for all living forms of life on earth, but it also supplies nutrients and accepts wastes. Undoubtedly, it is the most important component of the earth's ecological system.

Most pollutants are transported to the place of their disposal via some phase of the hydrologic cycle. The hydrologic cycle can be visualized as a permanent routing of water from its largest pool—the oceans—to dry lands and back. It starts with water evaporating from the oceans, and then cloud formation and movement carry it to the land whence precipitation occurs and causes flow on and into the dry land surfaces (Fig. 6.1). During its travel, water accepts various contaminants and salts, a part of which are ultimately deposited back into the ocean.

THE WATERSHED OR RIVER BASIN

Water quality planning, research, and analysis of many environmental problems require a working unit which can be represented in system anal-

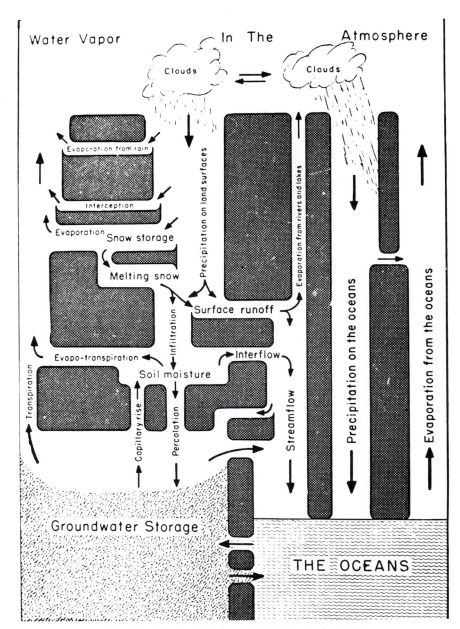

Fig. 6.1. The hydrologic cycle (Linsley *et al.*, 1949).

ysis approach terms by a "black box" concept. The "black box" represents a system for which the inputs and outputs are known or investigated
and which can be described by a set of input–output relationships. This
system should have a distinct boundary which would distinguish it from
other surrounding systems. The best representation of such a system in
hydrology and water quality management is a *watershed* or a *drainage
basin*. A watershed has distinct advantages to other hydrologic and/or
water quality system representations such as government territorial units.
These usually have hydrologic and water quality inputs which may be distributed to several output locations described by several system input–
output relationships. Even institutional arrangements and enforcement
units are usually established for a basin rather than on a territorial basis.
Examples may be seen in both Europe and the United States as shown in
Fig. 6.2.

A watershed or drainage basin is characterized by its outlet, which is
usually the lowest point in the watershed, and by its divide, which may be
represented by a connecting line of the highest points of the watershed
boundary. It is assumed that all water which falls into the area within the
divide line will either result in runoff at the outlet point, infiltrate into the
ground, or be lost by evapotranspiration.

Few simple hydrologic or geomorphic parameters can be used to characterize the watershed and its precipitation–runoff transformation (Linsley
et al., 1949; Beaumont, 1975). The complexity and development of stream
channels within a watershed is characterized by the bifurcation ratio,
which is expressed as

$$R_b = \frac{n_u}{n_{u+1}}, \tag{6.1}$$

where R_b is the bifurcation ratio, and n_u is the number of stream segments
of order u. The first-order streams are those which have no tributaries. A
second-order stream is formed by the junction of two first-order streams.
A third-order stream is formed by the confluence of two second-order
streams. Figure 6.3 shows the order distribution of streams in a hypothetical watershed.

Several important pollution transport and hydrologic watershed characteristics can be correlated to the bifurcation ratio. For example, a relationship is often found between the bifurcation ratio and the flood hydrograph (Beaumont, 1975). Basins with low bifurcation ratios produce flood
hydrographs with marked discharge peaks, while those with high ratios
give rise to low peaks over longer time periods. Also, sediment delivery is
lower for higher values of the bifurcation ratio. Normal values of the bifurcation ratio parameter range between 3 and 5 and appear to be in-

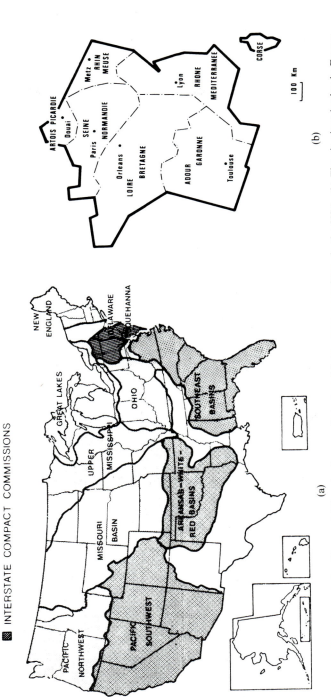

Fig. 6.2. River basin delineations: (a) River basin commissions and interagency committees in the USA; (b) The six river basins in France.

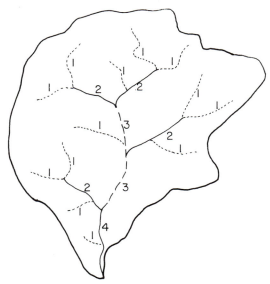

Fig. 6.3.   Order distribution of streams.

fluenced mostly by basin shapes, with elongated basins producing the highest values.

The drainage density, which is one of the most important morphometric parameters, links stream length to basin area as follows:

$$\text{Drainage density} = \frac{\text{total stream length}}{\text{total drainage area}}.$$

Other parameters which are commonly used to describe the hydrologic characteristics of a watershed include the elongation ratio and the average slope. The elongation ratio is defined as the diameter of a circle having the same area as the basin divided by the length of the longest axis of the drainage basin (Schum, 1956).

In pollution transport analysis it may be convenient to characterize the various phases of the hydrologic cycle pertaining to pollution transport. Figure 6.4 represents a schematic picture of an idealized hydrologic transport of pollutants throughout the environment. It demonstrates that the pollutant transport process has five distinctive phases.

Atmospheric transport includes air movement, wind erosion, cloud movement, dust and dirt fallout, rain and snow precipitation. The long-range transport of power plant discharges causing acid rainfall and heavy metals fallout is of major current concern.

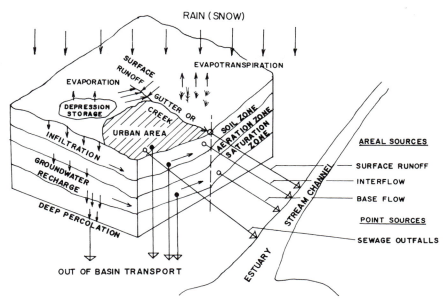

Fig. 6.4. Hydrologic transport of pollutants.

Overland flow includes rain and surface runoff, snowpack and snow-melt, erosion and sediment pickup, dilution of deposited pollutants, depression and interception, evaporation and evapotranspiration.

Groundwater flow starts with infiltration and soil water movement and includes interflow and groundwater runoff, soil absorption of pollutants and their transformation, and various phases of groundwater contamination.

Channel flow represents the primary moving mechanisms of the pollutants from their sources to the place of their final disposal, which may be a lake or the sea. Channel flow includes the following processes: convection, dispersion, sediment transport, scour, sorption, decay, and bed load transport. Some flow through reservoirs can be considered as an extension of channel flow, while other lakes with longer residence times or without significant outlets may be a final pollution disposal site.

Estuary transport is the transition between channel flow and a lake or ocean. The flow pattern is usually three-dimensional, with significant tidal backwater effects. Owing to the reduction of flow velocity, a substantial part of the suspended fraction of pollutants may be deposited in this transition zone and temporarily or permanently stored therein.

Dissolved and/or diluted pollutants move with the same velocity as the water itself (assuming that molecular diffusion is negligible when compared to convection and turbulent diffusion). Suspended particulate pollutants and sediment adsorbed fractions are subjected to sedimentation during overland and channel flow transport. Therefore, they can settle out in sections with lower velocity and be resuspended during higher flows. Obviously, the suspended and adsorbed fractions lag behind the water movement, and some of the pollutants may never reach a lake or sea but become incorporated into alluvial deposits of the stream.

Runoff Formation

The basic components of the hydrologic cycle have been shown in Figs. 6.1 and 6.4. It is obvious that the precipitation–runoff transformation is not a simple relation but, instead, is quite complex. Although runoff is related to precipitation, it is not directly proportional to it and simple statistical correlation attempts between runoff and precipitation usually fail. Runoff is a residual phenomenon which takes place only after certain demands and losses are satisfied. The most important losses include evaporation from land surfaces and interception (precipitation intercepted by vegetation and depressions and evaporated), transpiration, which refers to the water imbibed by the roots of plants from soil water and released to the atmosphere, and infiltration, which depends on antecendent rainfall. A simple mathematical formula may describe the runoff as

$$R = P - ET, \tag{6.2}$$

where R is the volume of runoff, P is precipitation, and ET is evapotranspiration which is total losses except infiltration. The total runoff can be divided into several components: *surface runoff,* which refers to the water flow on the ground surface reaching channels in the shortest time; *channel precipitation,* which is rainfall falling directly on the river surface; *interflow,* which represents the lateral movement of water in the soil zone caused by a lower permeability of subsoils; and *groundwater runoff* or *base flow,* which by definition refers to the groundwater movement recoverable by springs and wells. The movement of groundwater is very slow; sometimes the storage retention time in the groundwater zone is measured in years.

A schematic diagram of a precipitation-runoff system is shown as a block diagram in Fig. 6.5. Such flow charts serve as a basic structure of several computer models which are available for runoff estimation and

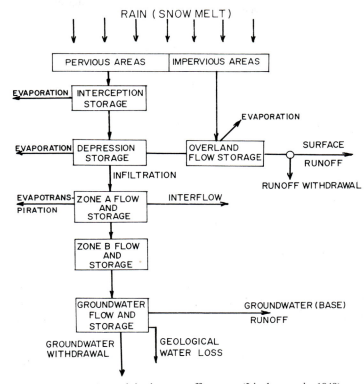

Fig. 6.5. A precipitation–runoff system (Linsley *et al.*, 1949).

modeling. The components of the runoff formation process include the following subunits.

Snowpack Formation and Snowmelt

During subfreezing temperatures, precipitation falling on the ground is in the form of snow with no direct runoff formation. The snowpack is stored on the ground surface until the heat input will result in snowmelt. The heat balance of the snowpack should include all variables such as radiation, heat convention, and wind. The simplest description of the snowmelt process is called the *temperature index* or *degree-day method,* in which the air temperature is the only variable. This approximation is possible because the air temperature is a result of the heat balance of the same inputs that act on the snowpack. The formula was presented by Gray (1973):

$$\Delta P_s = C(T_a - T_b) \quad \text{for} \quad P_s > 0.0 \quad \text{and} \quad T_a > T_b \quad (6.3)$$

and

$$P_s = P \quad \text{for} \quad T_a < T_b,$$

where P_s is the water content of the snowpack, ΔP_s is the change of the snowpack water content, T_a is the mean or maximum daily temperature, T_b is the base temperature close to 0°C, and C is a coefficient determined by trial and error, assuming lower values in early melt season and higher at a later time as shown in Fig. 6.6.

A more complex model, based on the heat budget of the snowpack, was developed by the United States Corps of Engineers in cooperation with other agencies (U.S. Army Corps of Eng., 1956).

Interception and Surface Depression Storage

On any surface there are many small depressions that must be filled before surface runoff begins. Rainfall intercepted by vegetation surfaces is removed by evaporation and that intercepted by surface depressions will

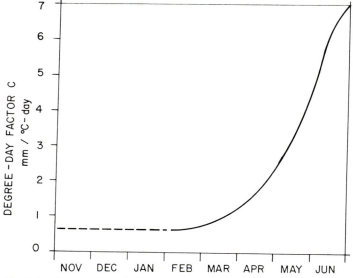

Fig. 6.6.   Snowmelt coefficient variation with time of year (Gray, 1973).

be removed by evaporation and infiltration. Based on field measurements, oak trees are able to intercept as much as 20% of rainfall (Dub, 1969) and evergreens have been reported to have interception storage up to 59% (Linsley, 1949). Many of the existing empirical equations used to estimate interception during a particular storm have the form (Gray, 1973)

$$I_{ri} = a + bP^n,\qquad(6.4)$$

where I_{ri} is the interception loss; P is the precipitation rate; and a, b, and n are constants as shown in Table 6.1.

Depression storage includes the effects of ponding and small depressions on the ground surface. Values of depression and interceptions on storage and interception estimated for the Chicago area were reported as 6.35 mm (0.25 in.) on pervious surfaces and 1.6 mm (0.0625 in.) on impervious surfaces (asphalt pavements), (Tholin and Keifer, 1960). For rural areas, Hills (1971) suggested a surface storage value of 1.25 mm (0.05 in.) for bare rural lands and woodlands. Horner and Jens (1942) defined reten-

TABLE 6.1 Evaluation of constants, a, b, and n, in an interception equation [a]

Vegetal Cover	Interception = $a + bP^n$				
	a		b		n
	1	2	1	2	
	SI[b]	US[b]	SI[b]	US[b]	
Orchards	0.10	(0.04)	0.018	(0.18)	1.00
Ash, in woods	0.05	(0.02)	0.018	(0.18)	1.00
Beech, in woods	0.10	(0.04)	0.018	(0.18)	1.00
Oak, in woods	0.13	(0.05)	0.018	(0.18)	1.00
Maple, in woods	0.10	(0.04)	0.018	(0.18)	1.00
Willow, shrubs	0.05	(0.02)	0.04	(0.40)	1.00
Hemlock and pine woods	0.13	(0.05)	0.10	(0.20)	0.50
Beans, potatoes, cabbage, and other small hilled crops	0.12h	(0.02h)	0.05	(0.15h)	1.00
Clover and meadow grass	0.03h	(0.005h)	0.03	(0.08h)	1.00
Forage, alfalfa, vetch, millet, etc.	0.06h	(0.01h)	0.032	(0.10h)	1.00
Small grains, rye, wheat, barley	0.03h	(0.005h)	0.016	(0.05h)	1.00
Corn	0.03h	(0.005h)	0.0016	(0.005h)	1.00

[a] After Horton (1938).
[b] Units

	I	P	h
1. Système International	cm	mm	m
2. United States equivalent	in.	in.	ft

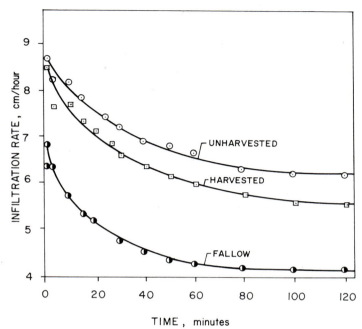

Fig. 6.7.   Variation of infiltration rate with time (Larsen *et al.*, 1972.

tion as that part of the rainfall which is permanently subtracted during the period of excess rainfall and includes interception and depression storage. For smooth cultivated lands they suggested retention values which include both depressions and interceptions of 1.25 to 2.5 mm (0.05 to 0.1 in.), and for good pasture 5 mm (0.2 in.); good forest litter was thought to be as high as 7.5 mm (0.3 in.).

Infiltration

The process of infiltration is a part of the soil water movement process and is discussed in greater detail in Chapter 16 on groundwater and pollutant movement. Infiltration is related to the gravity and capillary movement of water into the available upper soil zone storage. As can be seen in Fig. 6.7, the infiltration rate is higher at the beginning of a rain event, and then decreases to a constant infiltration rate, which approaches the saturation permeability.

Evaporation and Evapotranspiration

There are several methods of estimating evaporation including mass transfer, energy budget, water budget, empirical formulas, and evaporation measurements from pans. Most of the empirical formulas and semi-empirical formulas are based on the simple aerodynamic equation

$$E_0 = (a + bU)(e_W - e_A),\qquad(6.5)$$

in which, E_0 is the evaporation rate, U is the wind speed, e_W is the saturation vapor pressure at mean air temperature, e_A is the vapor pressure of air, and a and b are empirical coefficients.

Table 6.2 lists the most common empirical formulas for the estimation of potential evaporation.

The evapotranspiration rate is the primary factor determining water loss in the rainfall–runoff balance. Since most groundwater runoff will eventually appear on the surface as base flow or interflow, the water lost by infiltration is only temporary, but the loss by evaporation or transpiration is considered permanent in all watershed models.

Several methods have been developed for estimating evapotranspiration in cropped and forested areas (Chow, 1964; Davis and Sorensen, 1969). From the standpoint of land use–water quality modeling, two methods are recommended. The Penman method (Penman, 1949; 1956) involves the use of an equation developed in 1948 for estimating evaporation by using data for wind velocity, vapor pressures for saturated vapor in air, and estimation of crop evapotranspiration by multiplying the evaporation values by empirical constants depending on the latitude and the length of daylight. Details of this method and those following are presented in Chow (1964).

The evaporation index method, described by McDaniels (1960) is briefly summarized as

$$ET = (E_0)(KU),\qquad(6.6)$$

where ET is the crop evapotranspiration requirement on a monthly or shorter time period, E_0 is the climatic index, which is the same as the evaporation from a hypothetical shallow lake situated at the locality under consideration, and KU is the crop-use coefficient, which reflects the stage of growth of the crop (see Tables 6.3 and 6.4 for average values).

Some other methods for estimating evapotranspiration are the Blaney–Criddle method, the Modified Blaney–Criddle method, and the Lowry–Johnson method (Chow, 1964; Davis and Sorensen, 1969).

TABLE 6.2   Review of evaporation formulas of the type $E_v = (a + bU)(e_w - e_A)$

Location	Author	Original Formula for Evaporation Rate	Units
Zaikov	(from Braslavskii-Vikulina, 1963)	$\dfrac{E_v}{e_w - e_A} = 0.15[1 + 0.72\, U_A]$	E_v-mm/day U_A-m/sec e-mm Hg
Lake Hefner	(from Edinger and Geyer, 1965)	$\dfrac{E_v}{e_w - e_A} = 0.00177\ U_A^a$	E_v-in./day U_A-mi./hr e-m bars
Lake Colorado	(Edinger and Geyer, 1965)	$\dfrac{E_v}{e_w - e_A} = 0.00260\ U_A^a$	E_v-in./day U_A-mi./hr e-m bars
	Harbeck (1962) or World Metrological Organization Measurements	$\dfrac{E_v}{e_w - e_A} = 0.00338\ A^{-0.05}\,U_A$	E_v-in./day A-acre U_A-mi./hr e-m bars

[a]U measured 24 ft above the water surface.

TABLE 6.3 Crop-use coefficients for use in evapotranspiration-index method for annual crops in the Northern Hemisphere

Crop	\(KU\) Values at Listed % of Growing Season										
	0	10	20	30	40	50	60	70	80	90	100
Alfalfa	0.45	0.51	0.58	0.66	0.75	0.85	0.96	1.08	1.20	1.08	0.70
Grain sorghum	0.30	0.40	0.65	0.90	1.10	1.20	1.10	0.95	0.80	0.65	0.50
Winter wheat[b]	1.08	1.19	1.29	1.35	1.40	1.38	1.36	1.23	1.10	0.75	0.40
Cotton	0.40	0.45	0.56	0.76	1.00	1.14	1.19	1.11	0.83	0.58	0.40
Sugar beets	0.30	0.35	0.41	0.56	0.73	0.90	1.08	1.26	1.44	1.30	1.10
Cantaloupes	0.30	0.30	0.32	0.35	0.46	0.70	1.05	1.22	1.13	0.82	0.44
Potatoes (Irish)	0.30	0.40	0.62	0.87	1.06	1.24	1.40	1.50	1.50	1.40	1.26
Papago peas	0.30	0.40	0.66	0.89	1.04	1.16	1.26	1.25	0.63	0.28	0.16
Beans	0.30	0.35	0.58	1.05	1.07	0.94	0.80	0.66	0.53	0.43	0.36
Rice[c]	1.00	1.06	1.13	1.24	1.38	1.55	1.58	1.57	1.47	1.27	1.00

[a]From Davis and Sorensen (1969).
[b]Data given only for springtime season of 70 days prior to harvest (after last frost).
[c]Evapotranspiration only.

Hydrologically Active Area Concept

Some models approximate the hydrologic response of an area to a precipitation input using uniform average parameters applied throughout the entire watershed. This approximation may be acceptable if only the hydrologic response is considered. However, with different kinds of soil response to pollution loadings and different erosion rates and pollution accumulation, it is imperative to know where the pollution originates.

In urban areas, large portions of the surface are impervious, and impervious areas are almost 100% hydrologically active; i.e., almost all of the precipitation that falls on the surface will be transformed into excess rain. However, not all of the impervious areas are directly connected to a channel and not all will result in surface runoff. Roofs may be connected to underground drainage fields and part of the excess rain may overflow and percolate onto adjacent pervious areas. Therefore, to obtain surface runoff and pollution loadings from impervious areas, the excess rain must be multiplied by a factor, *DC*, which represents the fraction of impervious areas which are directly connected to a channel. This fraction will vary with the total percent of impervious areas. In high-density residential areas, almost all impervious areas are connected via either storm sewers or combined sewer overflows. In rural areas, most of the runoff from local driveways and roofs usually overflows onto adjacent pervious areas or into drainage. Figure 6.8 shows the relationship of the factor *DC* to the total percent imperviousness as used in most urban hydrologic models.

TABLE 6.4 Crop-use coefficients for use in evaporation-index method for perennial crops in the Northern Hemisphere[a]

Crop						Average KU Values by Month						
	Jan.	Feb.	Mar.	Apr.	May	June	July	Aug.	Sept.	Oct.	Nov.	Dec.
Alfalfa	0.83	0.90	0.96	1.02	1.08	1.14	1.20	1.25	1.22	1.18	1.12	0.86
Grass pasture	1.16	1.23	1.19	1.09	0.95	0.83	0.79	0.80	0.91	0.91	0.83	0.69
Grapes			0.15	0.50	0.80	0.70	0.45					
Citrus orchards	0.58	0.53	0.65	0.74	0.73	0.70	0.81	0.96	1.08	1.03	0.82	0.65
Deciduous orchards				0.60	0.80	0.90	0.90	0.80	0.50	0.20	0.20	
Sugarcane	0.65	0.50	0.80	1.17	1.21	1.22	1.23	1.24	1.26	1.27	1.28	0.80

[a]From Davis and Sorensen (1969).

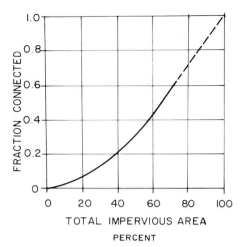

Fig. 6.8. Variation of *DC* with percent total impervious area (Hydrocomp, 1972).

Pervious surfaces on sandy or gravel soils are the least hydrologically active, inasmuch as most of the precipitation is lost by infiltration, resulting in very low-runoff potential.

The hydrologic activity of an area may have important implications as to pollution transport. It is evident that surface runoff carries most of the sediment and substances adsorbed on the sediment particles (such as phosphates, heavy metals, and some pesticides), while most of the accumulated salts in the soil and dissolved pollutants are leached downward and carried mostly by interflow and groundwater runoff.

Whether a subarea is hydrologically active or inactive depends on the magnitude of the losses which must be satisfied before surface runoff takes place. The most important losses are infiltration and interception storage. Therefore, most hydrologically active areas are impervious urban areas and/or areas which have a high groundwater table and permeability and soil storage is very low. Flat areas with poor soils adjacent to streams or swamps also may fall into this category. Sloped areas with good soil permeability are less hydrologically active. The least hydrological activity can be found in forested areas with high interception, forest litter, and soil storage. By aerial photogrammetry methods, it has been estimated that in some watersheds, almost all of the surface runoff generated during a storm originates from less than 20% of the total watershed area. The size of the hydrologically active part of a watershed depends on the intensity and volume of the rain, as shown in Fig. 6.9.

Besides the impervious areas, most hydrologically active areas are

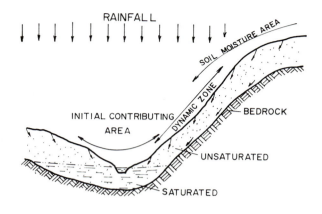

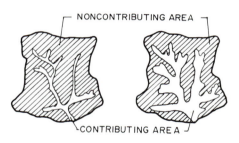

2 -YEARS RECURRENCE   10-YEARS RECURRENCE
        INTERVAL   INTERVAL

Fig. 6.9.   The dependency of hydrologically active areas on intensity and volume of rainfall (Engman, 1974).

usually located near the channel or in areas with a high groundwater table. As a storm or precipitation continues, the areas where soil storage becomes saturated increases, resulting in a reduction of the infiltration rate. Thus, the hydrologically active areas increase and locate farther from the channel.

Hydrologically active areas, especially in nonurban land use, require the most attention in their management. Erosion control practices should be applied whenever possible and the use of fertilizers should be limited. The areas with low-runoff potential do not require intensive erosion and sediment control measures as do hydrologically active areas. However, they do not possess the greater soil adsorption capability as do soils containing larger fractions of clay and organic matter. These areas may represent a danger to groundwater quality since some of the nutrients and pollutants applied or accumulated on these surfaces will penetrate into the groundwater aquifer. In spite of their good infiltration rates, use of these areas for sewage disposal should be limited if there is a danger of groundwater contamination due to an easier passage of pollutants through the soils.

Flow Estimation and Routing

After all the losses are satisfied, excess rain is routed toward the watershed outlet and becomes surface runoff. In the channel phase, the total flow is a summation of the surface runoff, interflow, and groundwater (base) flow. Figure 6.10 shows a typical runoff hydrograph and its three components.

Chow and Kulandaiswamy (1971) state that most of the hydrologic models which were developed for certain specific hydrologic problems can be considered to be a special case of a general hydrologic process if

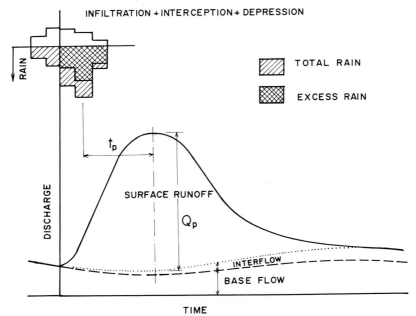

Fig. 6.10. A typical runoff hydrograph.

the hydrologic phenomenon is analyzed by a systems approach. All major hydrologic problems—and, similarly, water quality problems—are derived from the continuity equation

$$Y - X = \frac{dS}{dt}, \tag{6.7}$$

where S is the storage in the system or the storage of water in the watershed, X is the input, and Y is the output.

Transformation models of hydrologic and water quality systems can be represented mathematically by the equation

$$Y = \Phi X, \tag{6.8}$$

where Φ is a transfer function which represents the operation performed by the system on the input to transform it into output. For example, the well-known unit hydrograph is a transfer function of the watershed system. It should be noted that, in general, input and output are time functions and can be expressed by $x(t)$ and $Y(t)$, respectively, with t denoting time.

Using the system analysis approach, runoff can be considered as a response of the watershed to the precipitation input. Similarly, downstream flow is a response of the channel system to an upstream and/or lateral input. The input–output relationship for a linear system can be expressed by the convolution integral

$$Y(t) = \int_0^t h(\tau) \times (t - \tau) \, d\tau, \tag{6.9}$$

where $h(\tau)$ is the ordinate of the transform function, and τ is the lag time.

The transform function of the system is the system response to a unit pulse input which, in the case of watershed hydrology, would be a runoff response to a short-duration unit volume excess rain input. A function of this type was first proposed by Sherman (1932) and is known as the *unit hydrograph method*. The theoretical basis of the unit hydrograph was summarized as follows (Johnstone, and Cross, 1949).

1. For a given watershed, runoff producing storms of equal duration will produce a surface runoff hydrograph of approximately equivalent time bases, regardless of the intensity of the rain.

2. For a given watershed, the magnitude of the ordinates reproducing the simultaneous discharge from an area will be proportional to the volumes of surface runoff produced by storms of equal duration.

3. For a given watershed, the time distribution of runoff from a given storm period is independent of precipitation from antecendent or subsequent storm periods.

Based on the preceding assumptions, it can be deduced that the unit hydrograph resembles a transform function of a linear watershed system. As the duration of the input rainfall excess decreases to infinitesimally small values, the resulting hydrograph is called the instantaneous unit hydrograph (IUH), or the impulse response, or the kernel function of the system.

It should be also noted that the commonly used *rational formula*,

$$Q = C i A, \qquad (6.10)$$

where Q is the quantity of runoff in ft^3/sec, (liter/sec), C is the coefficient of imperviousness, i is the intensity of rainfall in inches per hour (mm/sec) during a time interval called the time of concentration, and A is the contributing area in acres (m^2) is actually a linear watershed system response to a so-called step input. The time of concentration may be defined as the time for a particle of water from the most remote part of the basin to reach the point in question.

Both the unit hydrograph system response and the rational formula are shown in Fig. 6.11.

The rational formula was developed in the late nineteenth century and is still one of the most widely used simple models for determining peak runoffs. The drawbacks of the equation and its applicability to water quality routing include uncertainities in estimating the coefficient, C, and the time of concentration, t_c. The runoff coefficient, C, includes many factors, most of which are more or less time variable and/or depend on the rainfall characteristics. These factors include infiltration, evaporation, storage, rainfall uniformity, and so on. The time of concentration, as will be seen in the subsequent portions of this chapter, depends also on the storm intensity and duration.

Tables 6.5 and 6.6 contain estimated values of the coefficient, C, as related to various land uses or surface covers. Estimation of the time of concentration, which includes the time of overland flow plus the residence time in the channel (sewer), will be discussed subsequently. (The use of the rational formula requires a storm intensity duration curve similar to those shown in Fig. 6.17.)

A linear system is defined as a system described by differential equations with all derivatives raised to the power of 1. Significant characteristics of linear systems are the principles of additivity and homogeneity. Additivity states that if Y_1 is the output from the system due to input, X_1, and Y_2 is the output from the system due to an input, X_1, and Y_2 is the output due to a second input X_2, then the input $X_1 + X_2$ will result in an output $Y = Y_1 + Y_2$. Homogeneity states that if the input is $X = A_1 X_1 +$

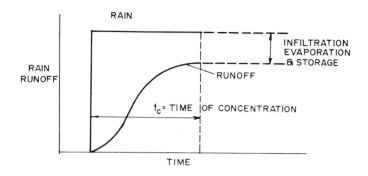

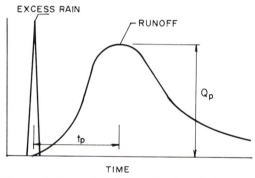

Fig. 6.11.   The unit hydrograph and the rational method response to rainfall.

$A_2 X_2$, then the output is $Y = A_1 Y_1 + A_2 Y_2$. The condition of linearity implies that the transform function, $h(\tau)$, is independent of X or Y.

The principle of linearity has long been questioned for watershed systems. Izzard (1946) showed the dependency of the hydrograph parameters upon the intensity of the effective rainfall. The deficiency of the unit hydrograph was overcome by assuming a nonlinear system (Amorocho and Brandstetter, 1971), for which the input–output relationship becomes

$$Y(t) = \int_0^t X(t - \tau)h[X(t - \tau); \tau]\, d\tau, \qquad (6.11)$$

where the variable IUH, $h[X(t - \tau); \tau]$, depends on the intensity of the effective rainfall $I(t - \tau)$ as well as the time variable of integration τ.

In spite of the limitations described, it has been emphasized that the results which have been obtained by judicious application of the unit hydrograph method have been quite satisfactory. There is no doubt that it is in-

TABLE 6.5 Runoff coefficients for urban areas (Horner and Flynt, 1936)

Description of Area	Runoff Coefficient
Flat, residential, with about 30% of area impervious	0.40
Moderately steep, residential, with about 50% of area impervious	0.65
Moderately steep, built up, with about 70% of area impervious	0.80

deed of considerable value for attempts at resolving the complex relationship between rainfall and runoff.

Since the unit hydrograph has been found to correspond to the transform function of the watershed or river channel, it should not be considered an empirical formula. However, it must be remembered that the necessary coefficients for estimation or prediction of the unit hydrograph function are of empirical nature.

Instantaneous Surface Runoff Hydrograph Equation

A watershed behaves like a reservoir system that can be represented by several basins in series. Nash (1957) proposed a model consisting of a cascade of n identical reservoirs for which

$$h(\tau) = \frac{1}{K_n} \frac{e^{-(\tau/K_n)}}{\Gamma(n)} \left(\frac{\tau}{K_n}\right)^{n-1}, \tag{6.12}$$

TABLE 6.6 Deductions from unity to obtain the runoff coefficient for agricultural areas (Bernard, 1935)

Type of Area	Values of c[a]
Topography	
Flat land, with average slopes of 1 ft to 3 ft per mi	0.30
Rolling land, with average slopes of 15 ft to 20 ft per mi	0.20
Hilly land, with average slopes of 150 ft to 250 ft per mi	0.10
Soil	
Tight impervious clay	0.10
Medium combinations of clay and loam	0.20
Open sandy loam	0.40
Cover	
Cultivated lands	0.10
Woodland	0.20

[a]The magnitude of c is obtained by adding values of c for each of the three factors: topography, soil, and cover; and then by subtracting the sum from unity.

in which K_n is the reservoir constant, and $\Gamma(n)$ is the gamma function of n.

If n approaches one, the above hydrograph function can be replaced by a single reservoir model given by

$$h(\tau) = \frac{1}{K} e^{-\tau/K}. \tag{6.13}$$

Both constants can be related to the time of travel of the water from the most remote point on the watershed to the watershed outlet. On the runoff hydrograph this time represents the time distance, t_p, between the centroid of the rain pulse and the peak of the hydrograph, as shown in Fig. 6.11. Then, according to Rao *et al.* (1972),

$$t_p = K = nK_n. \tag{6.14}$$

By numerically solving the kinematic wave equation for the overland flow portion of the rainfall–runoff transformation, Henderson and Wooding (1964) developed an equation for t_p, which, when converted to SI units, becomes

$$t_p = 6.9 \frac{L^{0.6}n_M^{0.6}}{i^{0.4}S^{0.3}}, \tag{6.15}$$

where t_p is the peak lag time in minutes, L is the length of the overland flow in meters, i is the rainfall intensity in mm/hr, S is the slope in m/m, and n_M is the Manning's roughness factor.

An almost identical formula was independently published by Morgali and Linsley (1965). Figure 6.12 presents a nomograph for the determination of t_p based on the Morgali and Linsley formula, and Table 6.7 reports the Manning roughness factor for overland flow. Rao *et al.* (1972) statistically analyzed the hydrograph curves for several urbanizing watersheds. The authors analyzed the effect of many variables on the shape of the runoff hydrograph. Only statistically significant variables were included in their final formulas. Based on the preceding work, t_p and n can be estimated as

$$t_p = 304 \frac{(AW)^{0.458}(TR)^{0.104}}{(1 + U)^{1.662}(i)^{0.269}} \tag{6.16}$$

and

$$n = 2.64 \frac{(AW)^{0.069}}{(1 + U)(i)^{0.155}}, \tag{6.17}$$

where, t_p is the lagtime in hours, AW is the watershed area in km², i is the rain intensity in mm/hour, U is a fraction of impervious areas of the total watershed, and TR is the rain duration in hours.

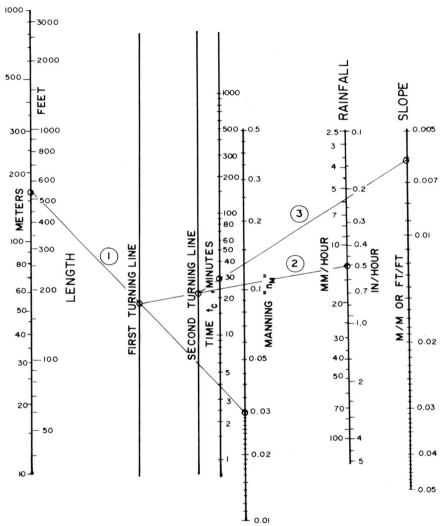

Fig. 6.12. Nomograph for determination of the time of concentration, t_c. (Ragan and Duru, 1972)

The unit hydrograph solution can be conveniently programmed for a digital computer mode. Compared to a conventional numerical finite difference solution, it offers substantial savings of computer time with adequate accuracy. There are also no problems with the convergence of the solution or its stability. A graphical representation of the algorithm is

TABLE 6.7   Typical values of the Manning's roughness factor for overland flow[a]

Ground Cover	Manning's n_m for Overland Flow
Smooth asphalt	0.012
Street pavement	0.013
Asphalt or concrete paving	0.014
Packed clay	0.03
Light turf	0.20
Dense turf	0.35
Dense shrubbery or forest litter	0.40

[a]From Crawford and Linsley, (1966).

shown in Fig. 6.13. Figures 6.14 and 6.15 show a comparison of simulated runoff by the discussed approach with actual measured data in urban and suburban sections.

Statistical Flow and Precipitation Characteristics

Precipitation and runoff are statistical quantities and therefore cannot be precisely predicted. Short-term predictions are possible; however, to predict the precipitation or flow a year or even a week ahead is still not achievable. Such phenomena that cannot be exactly predicted in the future are called stochastic or statistical as compared to deterministic events, which follow a mathematical formula or a physical law.

Both runoff and precipitation are not purely random. There are certain distinct seasonal features of the two phenomena which can be predicted with some degree of accuracy. For example, in tropical and subtropical regions, one would expect monsoon rains to appear more or less regularly

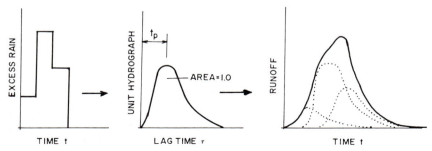

Fig. 6.13.   Unit hydrograph method.

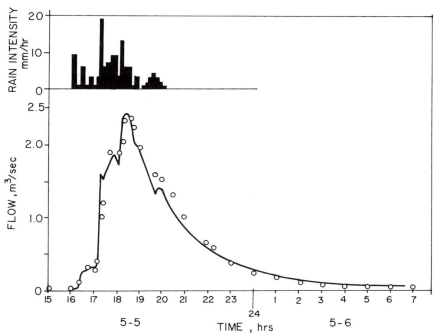

Fig. 6.14. Comparison of measured and simulated flows for urban area. (Novotny, 1977)

during a certain period of the year. Streams which are fed by mountain glaciers may be expected to have higher flows during the spring and summer thaw. These factors can be considered cyclic or harmonic; thus the precipitation or runoff pattern is a composition of cyclic and random components. Such phenomena may be characterized by statistical quantities determined from a large series of measured data.

In the United States the principal source of precipitation data is the United States Weather Bureau. Flow and most of the water quality data can be obtained from the United States Geological Survey.

Precipitation

The amount of precipitation is measured regularly at numerous United States Weather Bureau stations. Low-order stations provide precipitation data on a daily basis, while high-order stations measure precipitation with greater frequency. Some stations have continuous records. Detailed precipitation records (in hourly or even minute intervals) are usually necessary for modeling of storm water impact on water quality and for the de-

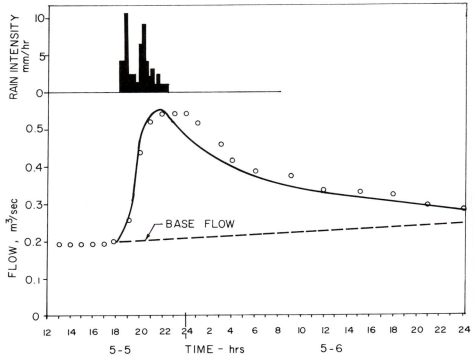

Fig. 6.15.   Comparison of measured and simulated flow for rural areas. (Novotny, 1977)

sign of storm water sewers. Precipitation is also the major input of the drainage and water quality basin systems.

The intensity of precipitation at any point depends on the amount of moisture in the clouds, the availability of nuclei, and the rate at which the vapor will condensate. Both intensity and volume are important factors in rainfall–runoff water quality systems. The total precipitation volume itself, as provided by 24-hour daily precipitation measurements, is not sufficient for estimating runoff or infiltration. Excess rain can be estimated only from the precipitation hyetograph, which is a plot of rainfall (snowfall) depth which fell during a given time interval v. time, as shown in Fig. 6.16.

The rainfall intensity varies significantly with time. High-intensity rains (storms) have a relatively short duration and are usually limited to relatively small areas of land under the storm cloud passage. Catastrophic storms are mainly caused by hurricanes.

Statistical analysis of precipitation results in the precipitation

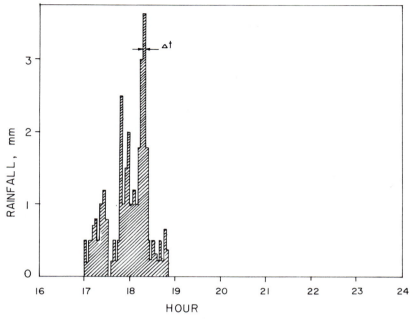

Fig. 6.16. A precipitation hyetograph.

intensity–duration curves shown in Fig. 6.17. These curves are obtained by analyzing long-term precipitation records by means of statistical frequency analysis techniques. The length of the record should be at least 50 years, and records shorter than 25 years are not acceptable. Mathematical stochastic models or data from neighboring watersheds are sometimes used to substitute for missing records. For each year of record, it is possible to determine the maximum rain intensity during a specified time interval (5, 10, 15, or 60 minutes), arrange the values in the order of magnitude and plot them on probabilistic or log-probabilistic paper, or analyze them by using, for example, a Pearson type III probabilistic distribution. (These statistical techniques will be briefly introduced in the subsequent section on runoff analysis.) From these probabilistic curves it is possible to determine rains with the specified duration which would have a recurrence interval of 1, 5, 10, 25, or more years. Cross plots are then made giving curves of intensity *v.* duration, with return periods as a parameter. Most of these curves can be approximated by a formula of the type

$$i = \frac{a}{t + b},$$

(6.18)

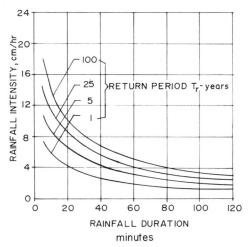

Fig. 6.17.   Precipitation intensity-duration curves

in which *i* is the rainfall intensity (in./hr or mm/hr), *t* is the duration of the rainfall (min), and *a* and *b* are empirical constants for the area under study. These curves are then used for the design of storm sewers or for storm-runoff quality impact analysis.

Mean annual precipitation, a factor which determines whether an area is arid (dry) or humid, depends on (1) latitude, (2) position and size of the continental land mass in which the area is located, (3) distance of the station from the coast, (4) temperature of the ocean and coastwise currents to adjacent land masses, (5) extent and altitude of adjacent mountain ranges, and (6) altitude of the area (U.S. Army Corps of Engineers, 1956; Dub, 1969; Davis and Sorensen, 1969).

In many areas, precipitation is the only significant source of water for various uses. Precipitation may be collected on large impervious areas and stored in underground reservoirs–cisterns, the largest of which is the system used on Gibraltar after the Spanish refused to supply water to this British land.

Runoff

Runoff is measured regularly by numerous stream gaging stations, and the results are published on a daily basis as daily average flows. These data are summarized for each water year by the United States Geological Survey on either a state or basin basis. Runoff variability is much less than that of precipitation, and, thus, daily flows may be sufficient in most water quality analysis. It is important to note that under some regulated flow conditions, average daily flow may be meaningless, however. For example, a "peaking" hydroelectric power plant may release no flow for some hours and 10,000 ft^3/sec during others.

In water quality studies, the extreme flows are of major concern. The poorest water quality is usually expected during low-flow or drought periods, while the high flood type flows carry the highest amount of pollutants to lakes and reservoirs. It is therefore necessary to characterize flows on both ends of the spectrum.

A flood is a flow which usually cannot be contained in the natural stream channels and the discharge overflows into adjacent areas called flood plains. Floods often result in extensive property damage since many developments and properties (houses, farms, and so on) are located on flood plains. Most flood discharges carry large amounts of sediment and adsorbed pollutants. Therefore, the high flow and flood pollution contribution to lakes and reservoirs is often more important than much longer periods of lower flows. A flood is expressed as either maximum daily flow or maximum stage. A flood results from a complex combination of factors affecting runoff in the watershed. Although it is the result of excess rain, other factors are also significant. The most dangerous floods occur during spring when the ground is still frozen and has very low permeability.

The flood frequency curve is also developed from long-term observations of flow data. From the records, the maximum flow observed each year is assigned a probability of being equaled or exceeded by means of the following formulation:

$$P = \frac{m}{n + 1} \times 100, \tag{6.19}$$

where P is the probability of a flow of being equaled or exceeded in percent, m is the descending order of magnitude of the flood peak, and n is the length of the record in years. The flood peaks are then plotted on probability or log-probability paper $v.$ their respective probabilities of being equaled or exceeded. The straight line or curve thus obtained is called the cummulative flood frequency curve.

An inverse function of the factor, P, i.e.,

$$T_r = \frac{P}{100}, \tag{6.20}$$

is called the recurrence interval and denotes the time in years during which one might expect a flood of a given magnitude or greater to occur. Thus a 100-year flood will have a probability of occurrence once in 100 years. However, it must be remembered that the probability of occurrence of a given flood magnitude is the same for each following year regardless of whether a similar flood recently occurred. Thus, it is possible for two 100-year floods to occur in one year.

The Pearson type III distribution curve was recommended by the Water Resources Council (1967) as a uniform technique for determining flood flow frequencies. The method requires the use of a large series of yearly floods in computations of the mean standard deviation and skew coefficients of the distribution. The Water Resources Council also recommends transforming the flow data to their logarithms and then computing the logarithmic statistical parameters. Because of this transformation, the method is called the log-Pearson type III method.

The coefficients for the log-Pearson type III distribution are computed as follows:

$$S = \left(\frac{\Sigma X^2 - (\Sigma X)^2/N}{N - 1}\right)^{1/2}, \tag{6.21}$$

coefficient of skewness

$$g = \frac{N^2\Sigma X^3 - 3N\Sigma X\Sigma X^2 + 2(\Sigma X)^3}{N(N - 1)(N - 2)S^3}, \tag{6.22}$$

where X is the logarithmic magnitude of an annual flood event, and N is the number of events on the record.

The logarithms of discharges at selected recurrence intervals or percent chance of being equaled or exceeded is computed according to the formula

$$\log Q = M + KS, \tag{6.23}$$

where M is the mean of the logarithms of the flood flows ($M = \Sigma X/N$) and K is a coefficient which can be obtained from Table 6.8 or 6.9. The flood discharge can be determined by taking the antilog of $\log Q$.

Drought

There is no adequate hydrologic definition of drought. A simple relationship of drought to rainfall shortage may not be sufficient because a significant portion of the flow may originate from groundwater sources or from melted mountain glacier ice. Therefore, in hydrology, the word drought is used to denote a period of lower than average flow.

In most water quality analysis and management problems, the drought period is usually correlated with poor quality conditions when there is insufficient flow to dilute the pollutants, or when the concentration of pollutants exceed prescribed water quality criteria. Thus, the drought period is critical for determining the waste assimilative capacity of streams for

TABLE 6.8 *K* values for negative skew coefficients

Recurrent Interval in Years (Percent Chance)

Skew Coefficient (g)	1.0101 (99)	1.0526 (95)	1.1111 (90)	1.2500 (80)	2 (50)	5 (20)	10 (10)	25 (4)	50 (2)	100 (1)	200 (0.5)
0	-2.326	-1.645	-1.282	-0.842	0	0.842	1.282	1.751	2.054	2.326	2.576
-0.1	-2.400	-1.673	-1.292	-0.836	0.017	0.846	1.270	1.716	2.000	2.252	2.482
-0.2	-2.472	-1.700	-1.301	-0.830	0.033	0.850	1.258	1.680	1.945	2.178	2.388
-0.3	-2.544	-1.726	-1.309	-0.824	0.050	0.853	1.245	1.643	1.890	2.104	2.294
-0.4	-2.615	-1.750	-1.317	-0.816	0.066	0.855	1.231	1.606	1.834	2.029	2.201
-0.5	-2.686	-1.744	-1.323	-0.808	0.083	0.856	1.216	1.567	1.777	1.955	2.108
-0.6	-2.755	-1.797	-1.328	-0.800	0.099	0.857	1.200	1.528	1.720	1.880	2.016
-0.7	-2.824	-1.819	-1.333	-0.790	0.116	0.857	1.183	1.488	1.663	1.806	1.926
-0.8	-2.891	-1.839	-1.336	-0.780	0.132	0.856	1.166	1.448	1.606	1.733	1.837
-0.9	-2.957	-1.858	-1.339	-0.769	0.148	0.854	1.147	1.407	1.549	1.660	1.749
-1.0	-3.022	-1.877	-1.340	-0.758	0.164	0.852	1.128	1.366	1.492	1.588	1.664
-1.1	-3.087	-1.894	-1.341	-0.745	0.180	0.848	1.107	1.324	1.435	1.518	1.581
-1.2	-3.149	-1.910	-1.340	-0.732	0.195	0.844	1.086	1.282	1.379	1.449	1.501
-1.3	-3.211	-1.925	-1.339	-0.719	0.210	0.838	1.064	1.240	1.324	1.383	1.424
-1.4	-3.271	-1.938	-1.337	-0.705	0.225	0.832	1.041	1.198	1.270	1.318	1.351
-1.5	-3.330	-1.951	-1.333	-0.690	0.240	0.825	1.018	1.157	1.217	1.256	1.282
-1.6	-3.388	-1.962	-1.329	-0.675	0.254	0.817	0.994	1.116	1.166	1.197	1.216
-1.7	-3.444	-1.972	-1.324	-0.660	0.268	0.808	0.970	1.075	1.116	1.140	1.155
-1.8	-3.499	-1.981	-1.318	-0.643	0.282	0.799	0.945	1.035	1.069	1.087	1.097
-1.9	-3.553	-1.989	-1.310	-0.627	0.294	0.788	0.920	0.996	1.023	1.037	1.044
-2.0	-3.605	-1.996	-1.302	-0.609	0.307	0.777	0.895	0.959	0.980	0.990	0.995
-2.1	-3.656	-2.001	-1.294	-0.592	0.319	0.765	0.869	0.923	0.939	0.946	0.949
-2.2	-3.705	-2.006	-1.284	-0.574	0.330	0.752	0.844	0.888	0.900	0.905	0.907
-2.3	-3.753	-2.009	-1.274	-0.555	0.341	0.739	0.819	0.855	0.864	0.867	0.869
-2.4	-3.800	-2.011	-1.262	-0.537	0.351	0.725	0.795	0.823	0.830	0.832	0.833
-2.5	-3.845	-2.012	-1.250	-0.518	0.360	0.711	0.771	0.793	0.798	0.799	0.800
-2.6	-3.889	-2.013	-1.238	-0.499	0.368	0.696	0.747	0.764	0.768	0.769	0.769
-2.7	-3.932	-2.012	-1.224	-0.479	0.376	0.681	0.724	0.738	0.740	0.740	0.741
-2.8	-3.973	-2.010	-1.210	-0.460	0.384	0.666	0.702	0.712	0.714	0.714	0.714
-2.9	-4.013	-2.007	-1.195	-0.440	0.390	0.651	0.681	0.683	0.689	0.690	0.690
-3.0	-4.051	-2.003	-1.180	-0.420	0.396	0.636	0.660	0.666	0.666	0.667	0.667

TABLE 6.9 K values for positive skew coefficients

Recurrence Interval in Years (Percent Chance)

Skew Coefficient (g)	1.0101 (99)	1.0526 (95)	1.1111 (90)	1.2500 (80)	2 (50)	5 (20)	10 (10)	25 (4)	50 (2)	100 (1)	200 (0.5)
3.0	-0.667	-0.665	-0.660	-0.636	-0.396	0.420	1.180	2.278	3.152	4.051	4.970
2.9	-0.690	-0.688	-0.681	-0.651	-0.390	0.440	1.195	2.277	3.134	4.013	4.909
2.8	-0.714	-0.711	-0.702	-0.666	-0.384	0.460	1.210	2.275	3.114	3.973	4.847
2.7	-0.740	-0.736	-0.724	-0.681	-0.376	0.479	1.224	2.272	3.093	3.932	4.783
2.6	-0.769	-0.762	-0.747	-0.696	-0.368	0.499	1.238	2.267	3.071	3.889	4.718
2.5	-0.799	-0.790	-0.771	-0.711	-0.360	0.518	1.250	2.262	3.048	3.845	4.652
2.4	-0.832	-0.819	-0.795	-0.725	-0.351	0.537	1.262	2.256	3.023	3.800	4.584
2.3	-0.867	-0.850	-0.819	-0.739	-0.341	0.555	1.274	2.248	2.997	3.753	4.515
2.2	-0.905	-0.882	-0.844	-0.752	-0.330	0.574	1.284	2.240	2.970	3.705	4.444
2.1	-0.946	-0.914	-0.869	-0.765	-0.319	0.592	1.294	2.230	2.942	3.656	4.372
2.0	-0.990	-0.949	-0.895	-0.777	-0.307	0.609	1.302	2.219	2.912	3.605	4.298
1.9	-1.037	-0.984	-0.920	-0.788	-0.294	0.627	1.310	2.207	2.881	3.553	4.223
1.8	-1.087	-1.020	-0.945	-0.799	-0.282	0.643	1.318	2.193	2.848	3.499	4.147
1.7	-1.140	-1.056	-0.970	-0.808	-0.268	0.660	1.324	2.179	2.815	3.444	4.069
1.6	-1.197	-1.093	-0.994	-0.817	-0.254	0.675	1.329	2.163	2.780	3.388	3.990
1.5	-1.256	-1.131	-1.018	-0.825	-0.240	0.690	1.333	2.146	2.743	3.330	3.910
1.4	-1.318	-1.168	-1.041	-0.832	-0.225	0.705	1.337	2.128	2.706	3.271	3.828
1.3	-1.383	-1.206	-1.064	-0.838	-0.210	0.719	1.339	2.108	2.666	3.211	3.745
1.2	-1.449	-1.243	-1.086	-0.844	-0.195	0.732	1.340	2.087	2.626	3.149	3.661
1.1	-1.518	-1.280	-1.107	-0.848	-0.180	0.745	1.341	2.066	2.585	3.087	3.575
1.0	-1.588	-1.317	-1.128	-0.852	-0.164	0.758	1.340	2.043	2.542	3.022	3.489
0.9	-1.660	-1.353	-1.147	-0.854	-0.148	0.769	1.339	2.018	2.498	2.957	3.401
0.8	-1.733	-1.388	-1.166	-0.856	-0.132	0.780	1.336	1.993	2.453	2.891	3.312
0.7	-1.806	-1.423	-1.183	-0.857	-0.116	0.790	1.333	1.967	2.407	2.824	3.223
0.6	-1.880	-1.458	-1.200	-0.857	-0.099	0.800	1.328	1.939	2.359	2.755	3.132
0.5	-1.955	-1.491	-1.216	-0.856	-0.083	0.808	1.323	1.910	2.311	2.686	3.041
0.4	-2.029	-1.524	-1.231	-0.855	-0.066	0.816	1.317	1.880	2.261	2.615	2.949
0.3	-2.104	-1.555	-1.245	-0.853	-0.050	0.824	1.309	1.849	2.211	2.544	2.856
0.2	-2.178	-1.586	-1.258	-0.850	-0.033	0.830	1.301	1.818	2.159	2.472	2.763
0.1	-2.252	-1.616	-1.270	-0.846	-0.017	0.836	1.292	1.785	2.107	2.400	2.670
<0>	-2.326	-1.645	-1.282	-0.842	0	0.842	1.282	1.751	2.054	2.326	2.576

various pollutants. For these reasons, an engineering definition of drought period is necessary. It would not be practical to relate waste assimilative capacity to a minimum daily discharge with a very low expectancy or a high recurrence interval. Both man and nature have the ability to tolerate small violations of water quality as long as they do not become a nuisance or a health hazard. For example, a water supply system can tolerate small increases in contaminant concentrations simply by increasing the dosage of chemicals in the treatment plant, and fish can tolerate lower oxygen levels for short periods of time without adverse effects. During recent years, it has been generally accepted in water quality management practice that a period of three to seven days of water quality violation might be tolerated, provided that the yearly expectancy of the low flow would be relatively low. Therefore, in most states, the design flow for waste assimilation capacity studies is the 7-day flow, with a recurrence interval of ten years. If the waste assimilative capacity is determined for the dissolved oxygen concentration, the low-flow characteristics should be typical for the summer period, or for the winter period in extremely cold climates.

The drought frequency analysis is performed similarly to that for floods. The previously mentioned United States Geological Survey flow data is again used; a minimum of 25 years of record is desirable. For each

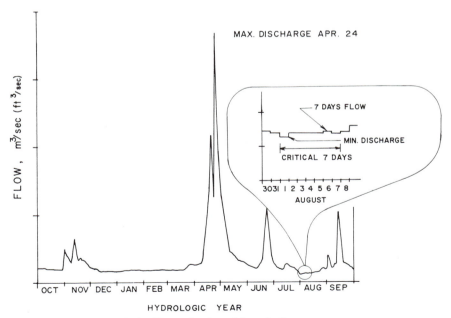

Fig. 6.18. Determination of drought-frequency curve.

year of record, the period with the lowest flow for the specified duration
(e.g., 7 days) must be determined. These low flows must be consecutive,
and the highest flow of the low-flow period then becomes the drought
characteristic of the year. The procedure is illustrated in Figs. 6.18 and
6.19. As with flood flow analysis, the drought flow characteristics are ar-
ranged according to their order of magnitude, which in this case should be
increasing. The probability of each event being equal to or less than a
given value is determined from Eq. (6.19). By plotting the low-flow characteristics with the desired duration v. their probability of being equal to
or less than a given value, the design drought flow characteristics with a
given duration and recurrence interval can be determined.

Surface runoff is the primary source of water for many uses. Because of
its inherent variability in flow and quality, it requires extensive water
management systems including reservoirs and water treatment plants.
The majority of large cities and urban areas rely on surface water sources
for both their water supply and their wastewater disposal. Surface waters
are also used for recreation to an ever-increasing extent.

Groundwater runoff is that part of the total runoff which can be recovered by springs and wells or infiltration. In hydrologic terminology, it is
referred to as base flow. The vast groundwater resources are closely related to surface water and are actually parts of one system. Very often,
groundwater flow is parallel to the flow on the surface, and a significant

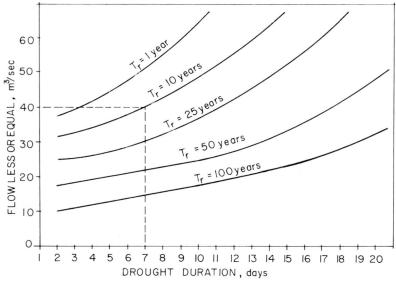

Fig. 6.19.   Drought-frequency curves.

portion of the flow of many rivers takes place in the alluvial deposits of the valleys.

The part of precipitation that infiltrates into the soil moves downward until it reaches the saturated zone (Fig. 6.20). That part of the underground geologic layers that is usually saturated by water is called an aquifer. The water table in the aquifer will fluctuate according to the water balance between the recharge from infiltration and the outflow from the aquifer. An aquifer can be confined if there is an impermeable stratum overlying it, or unconfined. The zone between the saturated zone and the surface is called the unsaturated or vadose zone.

Until the beginning of the twentieth century, groundwater aquifers were the primary source of water in the United States. Because of its long response time and usually large storage capacity, the water quality variability of a groundwater source is much less than that of surface waters. However, because of the contact of water with various geologic strata and its acid nature, the mineral content may be high.

The safe yield of an aquifer is the amount of water which can be pumped or used from the aquifer without significantly lowering its piezometric water level. The safe yield is obviously related to the infiltration input in the aquifer recharge area and to the drawdown. Obviously, seasonal fluctuations of the safe yield from shallow-well supplies are more

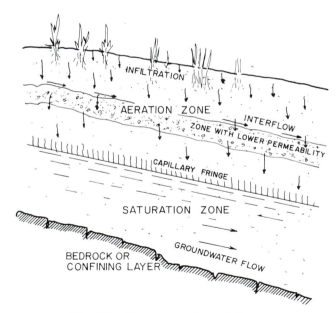

Fig. 6.20. The process of percolation.

marked than those from deep and mostly confined or artesian aquifers. A schematic diagram showing a gravity well and an artesian well is given in Fig. 6.21.

The safe yield of a well or spring also can be limited by the permeability of the geologic zones through which the water must percolate.

Springs are natural points or locations where groundwater appears on the surface. They are usually situated where the saturated water zone intercepts the ground surface. Springs formed at points where the groundwater outcrops are called gravity springs. An artesian spring is one in which the piezometric pressure of water in a confined aquifer causes water to move through cracks in the overlying confining stratum and reach the surface.

All streams originate from a spring. In fact, the water quality of the headwaters of streams during low-flow periods is almost identical to that of the underlying or adjacent aquifer. As streams accept more and more surface runoff and wastewater in their downstream sections, the water quality will change and reflect the biological, chemical, and physical processes taking place in the surface waters and the level of pollution. An overview of the role of springs in surface water contributions is shown in Fig. 6.22.

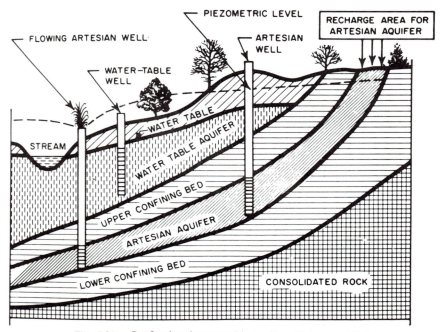

Fig. 6.21.   Confined and water table aquifers (Johnson, 1966).

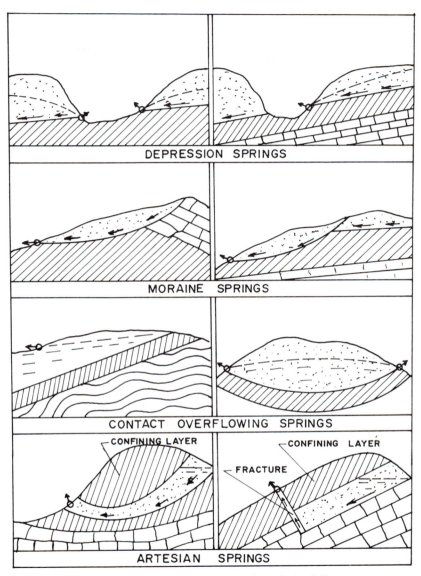

Fig. 6.22. The origination of springs (Dub, 1969).

REFERENCES

Amorocho, D., and Brandstetter, A. (1971). Determination of non-linear function—a response function in rainfall-runoff process. *Water Resour. Res.* **7,** 1087–1101.

Beaumont, P. (1975). Hydrology. *In* "River Ecology" (B. A. Whitton, ed.). Univ. of California Press, Berkeley.

Bernard, M. M. (1935). An approach to determine stream flow. *Trans.,* Amer. Soc. Civ. Eng. **100,** 347–395.

Braslavskii, A. P., and Vikulina, Z. A. (1963). "Evaporation Norms from Water Reservoirs." English translation, Israel Program for Scientific Translations, Jerusalem, Israel.

Chow, V. T. (1964). "Handbook of Applied Hydrology." McGraw-Hill, New York.

Chow, V. T., and Kulandaiswamy, V. C. (1971). General hydrologic model. *J. Hydraul. Div., Am. Soc. Civ. Eng.* **97,** 791–804.

Crawford, N. H., and Linsley, R. K. (1966). Digital simulation in hydrology, Stanford watershed model IV. Tech Rep. No. 39., Dept. Civ. Eng., Stanford Univ., Palo Alto, California.

Davis, C. V., and Sorensen, K. E. (1969). "Handbook of Applied Hydraulics." McGraw-Hill, New York.

Dub, O., and Nemev, A. (1969). Hydrologic. *Tech. Dig 34,* **LXXIV,** 378, State Publishing Company of Technical Literature, Prague, Czechoslovakia.

Edinger, J. E., and Geyer, J. C. (1965). "Heat Exchange in the Environment." Edison Electric Institute, New York.

Engman, E. T. (1974). Partial area hydrology and its application to water resources. *Water Resour. Bull.* **10,** 512–521.

Gray, D. M. (1973). "Handbook on the Principles of Hydrology." Water Info. Center, Port Washington, New York.

Harbeck, G. E. (1962). A practical field technique for measuring reservoir evaporation utilizing mass-transfer theory. *U.S. Geol. Surv. Prof. Pap.* 272-E.

Henderson, F. M., and Wooding, R. A. (1964). Overland flow and groundwater flow from a steady rainfall of finite duration. *J. Geogr. Res.* **69**, 1531–1540.

Hills, R. (1971). The influence of land management and soil characteristics of infiltration and the occurrence of overland flow. *J. Hydrol.* **13**, 163–181.

Horner, W. W., and Flynt, F. L. (1936). Relation between rainfall and runoff for small agricultural watersheds. *Trans.,* Amer. Soc. Civ. Eng. **101**, 140–206.

Horner, W. W., and Jens, S. W. (1941). Surface runoff determination from rainfall without using coefficients. *Trans. Am. Soc. Civ. Eng.,* **107**, 1039–1075.

Horton, R. E. (1938). Rainfall interception. *U.S. Monthly Weather Rev.* **47**, 603–623.

Hydrocomp International (1972). Hydrocomp simulation programming operation manual. Palo Alto, California.

Izzard, C. T. (1946). Hydraulics of runoff from developed surfaces. *Highw. Res. Board, Proc. Annu. Meet.* **26**, 129–150.

Johnson, E. E. (1966). Some basic principles of geology. *Driller's J.,* July–Aug., 11, E. E. Johnson, Inc., St. Paul, Minnesota.

Johnstone, D., and Cross, W. P. (1949). "Elements of Applied Hydrology." Roland Press, New York.

Larsen, V., Axley, J. H., and Miller, G. L. (1972). Agricultural wastewater accommodation and utilization of various forages. Water Res. Research Center-A-006-Md., 14-01-0001-790, Univ. of Maryland, College Park.

Linsley, R. K., Kohler, M. A., and Paulus, L. H. (1949). "Applied Hydrology." *McGraw-Hill,* New York.

McDaniels, L. L. (1960). Consumptive use of water by major crops in Texas. *Bulletin,* Texas Board of Water Eng., **6019**, Texas Dept. of Water Res. Austin.

Morgali, J. R. and Linsley, R. K. (1965). Computer analysis of overland flow. *J. Hydraul. Div., Am. Soc. Civ. Eng.* **91**, 81–100.

Nash, J. E. (1957). The form of the instantaneous unit hydrograph. *Bull., Int. Assoc. Sci. Hydrol.* **11**, 114–121.

Novotny, V., Tran, H., Simsiman, G., and Chesters, G. (1978). Mathematical modeling of land runoff contaminated by phosphorus. *J. Water Pollut. Control Fed.* **50**, 101–112 (January).

Penman, H. L. (1949). Natural evaporation from open water, bare soil and grass. *Proc. R. Soc. London,* Ser. A., 120–140.

Penman, H. L. (1956). Evaporation: An introductory survey. *Neth. J. Agric. Sci.* **4**, 9–29.

Ragan, R. M., and Duru, J. O. (1972). Kinematic wave nomograph for times of concentration. *J. Hydraul. Div., Am. Soc. Civ. Eng.* **98**, 1765–1772.

Rao, R. A., Delleur, J. W., and Sarma, B. S. (1972). "Conceptual hydrologic model for urbanizing basins. *J. Hydraul. Div., Am. Soc. Civ. Eng.* **98**, 1205–1220.

Schum, S. A. (1956). Evolution of drainage system and slopes in Badlands at Perth Amboy, New Jersey. *Geol. Soc. Am. Bull.* **67**, 597–646.

Sherman, L. K. (1932). Streamflow from rainfall by unit-graph method. *Eng. News-Rec.,* **108**, 501.

Tholin, A. L., and Keifer, C. J. (1960). Hydrology of urban runoff. *Trans. Am. Soc. Civ. Eng.,* Pap. No. 3061.

U.S. Army Corps of Engineers (1956). Snow hydrology. North Pacific Div., Portland, Oregon.

Water Resources Council (1967). A uniform technique for determining flood flow frequencies. Bull. No. 15, Washington, D.C.

7 | Pollution Sources, Loadings, and Wastewater Characterization

INTRODUCTION

The most important phase of water quality analysis is the identification, analysis, and characterization of wastes emanating from the cultural activities and natural sources which result in pollution inputs to receiving waters. The sources of pollution can be classified as point or diffuse (nonpoint) sources.

Point sources enter the pollution transport route at discrete and identifiable locations. They usually can be directly measured or otherwise quantified and their impact directly evaluated. Major point sources include effluents from industrial and municipal wastewater treatment plants and effluents from farm buildings or solid waste deposition sites.

Pollution from diffuse sources may originate from the weathering processes of minerals, the erosion of virgin lands and forests including residues of natural vegetation, or from artificial or semiartificial sources. The artificial categories can be related directly to human activities such as fertilizer application, the use of agricultural chemicals for controlling weeds and pests, the erosion of soil materials from agricultural farming areas and animal feedlots, and transportation and erosion in urban developments.

196

The pollution discussed in most of this chapter, both from point and nonpoint sources, is understood to be caused by man's cultural activities. This should be distinguished from water quality changes sometimes called "background pollution" caused by the contact of water with rocks, undisturbed soils and geological formations, natural erosion and elutriation of chemical and biochemical components from forest litter, migration of salt water into estuaries, or other natural causes.

In pollution transport analysis, areal sources such as dry and wet atmospheric fallout, dust and dirt accummulation in urban areas, and fertilizer and soil loss are considered to be diffuse sources. The common units for expressing the loadings from diffuse sources are in terms of mass/area/time, e.g., tons/km²/day which is identical to grams/m²/day. The English equivalent would be English tons/mi²/day. The loadings from point sources are expressed as mass/time or kg/day (lb/day).

Once the various sources and loads are identified and categorized, they must be adequately characterized. Thus, the last section of this chapter deals with the methodology used in quantitatively describing wastewater contributions and their magnitude.

POINT SOURCE POLLUTION

Municipal Effluents

A significant portion of the United States population is presently served by sewered wastewater disposal systems. These systems represent a direct pollution load to receiving waters inasmuch as they are collected at a single point. However, it must be remembered that even small household sewage disposal systems may cause severe pollution problems because of either malfunction or overloading.

As the water demand for urban areas varies with the time of day and the season, so does the sewage flow. Figure 7.1 shows a typical loading pattern for a medium size-midwestern city. The pollution load carried by the raw sewage depends on the following factors: (a) composition of the community, (b) type of sewer system, (c) living standard, (d) geography, (e) weather and season of the year, (f) presence or absence of garbage disposals.

Because of the presence of industrial and commercial establishments, it is convenient to express pollution loadings on a per capita basis. Values of per capita loadings of wastewater characteristics such as the biochemical oxygen demand (BOD), the chemical oxygen demand (COD), solids, ni-

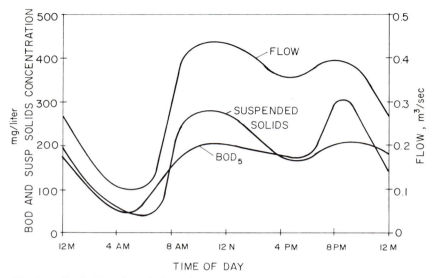

Fig. 7.1. Typical hourly variations of municipal wastewater discharges (Metcalf & Eddy, 1972).

TABLE 7.1   Per capita loadings of raw domestic sewage

Parameter	Loading	Literature Source
Flow (liter/cap/day)	220	Zanoni and Rutkowski (1972)
	290	Alexander and Stevens (1976)
	283	
BOD$_5$ (g/cap/day)	54	Fair and Geyer (1954)[a]
	46	Zanoni and Rutkowski (1972)
	47	Alexander and Stevens (1976)
Suspended solids	90	Fair and Geyer (1954)[a]
(g/cap/day)	36	Zanoni and Rutkowski (1972)
Kjeldahl nitrogen	6.5	Zanoni and Rutkowski (1972)
(as N g/cap/day)		
Ammonia nitrogen	2.6	Zanoni and Rutkowski (1972)
(as N g/cap/day)		
Total phosphate	3.7	Zanoni and Rutkowski (1972)
(as P g/cap/day)	1.8	Alexander and Stevens (1976)

[a]Reflects total municipal sewage rather than domestic.

TABLE 7.2 Per capita loadings of raw domestic sewage[a]

	Flow liter/cap/day	BOD₅ g/cap/day
Luxury homes	378	91
Better subdivisions	340	91
Average subdivisions	302	77
Low-cost housing	265	77
Apartment houses	284	77
Summer cottages	189	77

[a]After Goodman and Foster (1969).

trogen, and phosphorus have been investigated and analyzed for years. The first values of the per capita loadings appeared in literature in the first quarter of the twentieth. Tables 7.1 and 7.2 present a brief summary of per capita loadings for raw sewage. In Table 7.1, the values by Zanoni and Rutkowski (1972) reflect midwestern United States cities, while those by Alexander and Stevens (1976) are typical of European cities. If garbage disposals are installed, the values of the per capita loadings must be multiplied by the factors given in Table 7.3. Sewage volumes and BOD from various services are summarized in Table 7.4.

In determining the loadings to surface waters, it is necessary to subtract the pollution removed by treatment from the raw sewage load. A properly designed secondary sewage treatment plant should remove 85 to 90% of biodegradable pollutants expressed as BOD₅ and about the same fraction of settleable solids present in the raw sewage influent. In addition, phosphorus removal and disinfection are required in some parts of the United States. Advanced tertiary treatment should provide effluent quality which, in many aspects, should approach the quality of surface waters. Commonly accepted treatment efficiencies and effluent qualities of some secondary and tertiary treatment units are given in Table 7.5. It is useful

TABLE 7.3 Effect of home garbage disposals on per capita loadings[a]

Wastewater Characteristics	Estimated Percent Increase for 100% Disposal Use
Flow	2
BOD₅	30
Total solids	50
Suspended solids	21

[a]After Zanoni and Rutkowski (1972).

TABLE 7.4   Sewage volumes and BOD for various services[a]

Type	Volume liter/cap/day	BOD$_5$ g/cap/day
Airport		
Each employee	57	23
Each passenger	19	9
Bars		
Each employee	57	23
Each customer	7.5	4.5
Camps and resorts		
Luxury resorts	378	77
Summer camps	189	68
Factories (exclusive of industrial and cafeteria)	57	23
Hospitals		
Patients and staff av.	757	136
Hotels, motels		
Exclusive bars and restaurants	189	68
Offices	57	23
Restaurants		
Each employee	57	23
Each meal served	11.5	13.5
If garbage disposal add	4/meal	13.5/meal
Schools (including staff and students)		
Elementary	57	18
High	57	23
If cafeteria has a garbage disposal, add per person	—	4.5
Boarding schools	284	77
Swimming pools		
Employee or customer	38	13.5
Theaters (per seat or stall)	19	9
Factories (exclusive industrial waste) and cafeteria	57	23

[a]After Goodman and Foster (1969).

to know the typical concentration of different materials in domestic wastewater as shown in Table 7.6.

Industrial Effluents

In most industries, wastewater effluents result from the following water uses: Sanitary wastewater (from washing, drinking, and personal hygiene); cooling (from disposing of excess heat to the environment); process wastewater (includes water used for making goods, washing the products, waste and by-product removal and transportation); and

TABLE 7.5 Maximum effluent quality attainable from waste-treatment processes[a]

Process	BOD	COD	SS	N	P	TDS
Sedimentation, % removal	10–30	—	50–90	—	—	—
Flotation,[b] % removal	10–50	—	70–95	—	—	—
Activated sludge, mg/liter	<25	[c]	<20	[d]	[d]	—
Aerated lagoons, mg/liter	<50	—	>50	—	—	—
Anaerobic ponds, mg/liter	>100	—	<100	—	—	—
Deep-well disposal	Total disposal of waste					
Carbon adsorption, mg/liter	<2	<10	<1	—	—	—
Ammonia stripping, % removal	—	—	—	>95	—	
Denitrification and						
nitrification, mg/liter	<10	—	—	<5	—	—
Chemical precipitation, mg/liter	—	—	<10	—	<1	—
Ion exchange, mg/liter	—	—	<1	[e]	[e]	[e]

[a] USDI (1968a).

[b] Higher removals are attained when coagulating chemicals are used.

[c] $COD_{inf} - [BOD^u (removed)/0.9]$.

[d] $N_{inf} - 0.12$ (excess biological sludge), lb; $P_{inf} - 0.026$ (excess biological sludge), lb.

[e] Depends on resin used, molecular state, and efficiency desired.

cleaning (includes wastewater from cleaning and maintenance of industrial areas).

Excluding the large volumes of cooling water discharged by the electric power industry, the wastewater production from urban areas is about evenly divided between industrial and municipal sources. Therefore, the use of water by industry can significantly affect the water quality of receiving waters.

The wastewater loadings from industrial sources vary markedly with the water quality objectives enforced by the regulatory agencies. There are many possible in-plant changes, process-modifications and water-saving measures by which the industrial wastewater loads can be significantly reduced. Up to 90% wastewater reductions have been achieved recently by industries employing such methods as recirculation, operation modifications, effluent reuse, or changing to more efficient operation. As a rule, treatment of an industrial effluent is much more expensive without water-saving measures than the total cost of in-plant modifications and residual effluent treatment.

Industrial wastewater effluents are usually highly variable, with quantity and quality variations caused by bath discharges, operation start-ups and shut-downs, working hours distribution, and so on. A long-term detailed survey is usually necessary before a conclusion on the pollution impact from an industry is reached.

The population equivalent (P.E.) is defined as the number of people

TABLE 7.6 Typical composition of domestic sewage[a,b]

Constituent	Concentration		
	Strong	Medium	Weak
Solids, total	1200	700	350
Dissolved, total	850	500	250
Fixed	525	300	145
Volatile	325	200	105
Suspended, total	350	200	100
Fixed	75	50	30
Volatile	275	150	70
Settleable solids, (ml/liter)	20	10	5
Biochemical oxygen Demand, 5-day, $20°C$ ($BOD_5 - 20°$)	300	200	100
Total organic carbon (TOC)	300	200	100
Chemical oxygen demand (COD)	1000	500	250
Nitrogen, (total as N)	85	40	20
Organic	35	15	8
Free ammonia	50	25	12
Nitrites	0	0	0
Nitrates	0	0	0
Phosphorus (total as P)	20	10	6
Organic	5	3	2
Inorganic	15	7	4
Chlorides[c]	100	50	30
Alkalinity (as $CaCO_3$)[c]	200	100	50
Grease	150	100	50

[a] From Metcalf and Eddy (1972).
[b] All values except settleable solids are expressed in mg/liter.
[c] Values should be increased by amount in carriage water.

who would produce the equivalent pollution as the industrial wastewater effluent under consideration. The following per capita (cap) loadings have been commonly accepted for the population equivalent estimation:

Biochemical Oxygen Demand (BOD_5) P.E.

$$1 \text{ P.E.} = 54 \text{ g of } BOD_5/\text{day} = 0.119 \text{ lb/cap/day}$$

Suspended Solids

$$1 \text{ P.E.} = 90 \text{ g/day} = 0.198 \text{ lb/cap/day}$$

Table 7.1 can be used for P.E. estimation of nitrogen and phosphorus loadings. Population equivalents for some industrial wastes can be related to production units as shown in Table 7.7. For the reasons previously stated, these numbers represent only very rough estimates for planning purposes. For more precise numbers, the P.E. estimates should always be verified by a survey and/or by consulting the plant engineer.

Many industries discharge their wastewaters into municipal sewers, with subsequent treatment in a wastewater treatment plant, although PL-92-500 discourages this practice. Prior to discharge into a sewer, the effluent must be free of toxic substances such as cyanides and heavy metals, the wastewater pH should be close to neutral, and, generally, it should be free of substances which could impair the treatment process or cause a violation of water quality standards of the receiving waters. In many cases, however, the joint treatment of industrial and municipal wastewaters can be economically and environmentally beneficial to both the industry and the municipality.

TABLE 7.7. Flow and population equivalents for some typical industrial raw wastes[a]

Industry	Production Unit	lit, Flow one/prod. unit	P.E. BOD$_5$	P.E. SS
Meat packing				
Slaughterhouse mixed	1 hog[b]	600	18	6
Poultry	1000 kg	18,300	300	130
Milk products				
General dairy	1000 kg of raw milk	2,910	10	7
Dry milk powder	1000 kg of raw milk	2,500	6	6
Cheese factory	1000 kg of raw milk	1,660	16	6
Brewery				
Grain dewatered	bbl of beer (119.2 liters)	1,780	19	13
Beet sugar				
Flume water	ton of raw beets	9,100	22	74
Process water	ton of raw beets	2,740	40	29
Paper mill				
Unbleached	ton of paper	159,000	26	704
Bleached	ton of paper	192,000	40	282
Paper board	ton of paper	58,500	97	447
Pulp mill				
Ground wood	ton of pulp	20,400	16	—
Kraft	ton of pulp	260,000		
Sulfite	ton of pulp	244,000	1,330	
Tanning				
Vegetable	100 kg of hides	6,600	48	77
Chrome	100 kg of hides	29,000	80	520
Laundry	100 kg of clothes	12,400	24	8
Oil refining	bbl crude (159 liters)	2,910	0.6	1

[a] WPCF (1970).
[b] One hog unit = 0.4 cattle unit = 1 lamb = 1 calf.

If joint treatment is not possible, a separate treatment plant which would meet the effluent standards and guidelines must be designed and placed into operation. The treatment plant should be considered as an integral part of the production process. The treatment plant efficiency depends on the character of the wastewater, the treatment technology, variability of the influent, and other factors. A treatability study of the raw waste is usually mandatory for a workable design and/or planning purposes. Table 7.7 can serve only as a rough estimate if applied to industrial wastewater treatment units. It is more convenient to express the expected effluent water quality and quantity in terms of the best practicable technology (BPT), the best available technology (BAT), or the best conventional technology (BCT) as specified by the Environmental Protection Agency for a particular industrial wastewater effluent.

Agricultural Point Sources

Wastewaters from agriculture originate from farm buildings (dwellings and animal stables), animal feedlots, storage of manure and silage, and fertilized fields. The pollution from animal feedlots and farm fields is considered to be a nonpoint source and will be discussed subsequently. In addition to the sanitary waste from farm buildings, the wastewater contains mostly urine and silage juice leachate, both rich in organic materials. These liquid wastes also have a high content of nitrogen and phosphorus.

Farmyard manure is either in solid or liquid form. Straw or other litter materials are used to absorb urine and animal excreta. The surplus liquid from the solid manure is called dungyard water, and the liquid portion of manure is stored and used together with the solid fraction as organic fertilizer.

Silage juice is a refuse product that comes from the preservation of greencrop. It is a strong pollutant and must not be released into surface waters.

Urine, dungyard water, liquid manure, and silage juice may leak into surface waters and groundwater, and, if a concrete, impermeable dungyard and urine storage tank does not exist, the leakage may be considerable.

Table 7.8 lists the average leachate pollution from 352 farms investigated in Sweden. To obtain a comparable number, the pollution was divided by the size of the farm in hectares. The outflow quantities from animal stables and leachates were considerable and could be compared to the agricultural runoff pollution from arable lands. Locally, the leachate may be devastating to small receiving surface waters or nearby wells.

TABLE 7.8. Leachate Pollution from farms (excluding nonpoint sources)

BOD$_5$	23 kg/ha/yr
Nitrogen as N	5.3 kg/ha/yr
Phosphorus	0.2 kg/ha/yr

[a] After Brink (1975).

Combined Sewer Overflows

Most older urban areas in the United States have combined sewer systems, i.e., systems which are used for the conveyance of both sewage and storm water. When the flow becomes too large, the excess flow is diverted into the nearest water channel or receiving water body. Although only about 3 to 5% of sanitary sewage is lost from the system by overflows, untreated sewage loading can have a detrimental effect on receiving waters. In addition, the combined sewer overflows carry nonpoint surface runoff pollution, litter deposition, and catch-basin deposits.

During dry weather, when sanitary sewage alone is carried, the relatively poor flow characteristics of combined sewers encourage the settling and build-up of solids in the system until it is purged by a storm. As a result, a large sanitary pollution load may be discharged into the receiving water over a relatively short period of time, resulting in shock first-flush loadings.

Because of the variable nature of the rainfall-runoff pattern, it is impossible to list typical water quality parameters of combined sewer overflows. However, Table 7.9 illustrates some general concentration ranges. The quality may range from a very strong sanitary sewage flush during first period of the overflow to a very weak diluted sewage later in the storm. Mathematical models with acceptable degrees of reliability are available for simulation of combined sewer overflow effects (USEPA,

TABLE 7.9. Comparison of quality characteristics from first flushes and extended overflows of combined sewers, Milwaukee, Wisconsin[a]

Characteristics	First Flushes mg/liter	Extended Overflows mg/liter
COD	500–765	113–160
BOD$_5$	170–182	26–53
Susp. solids	330–848	113–174
Vol. susp. solids	221–495	58–87
Total N	17–24	3–6
Coliforms	1.5 10^5 to 310 10^5/100 ml	

[a] From Lager and Smith (1974).

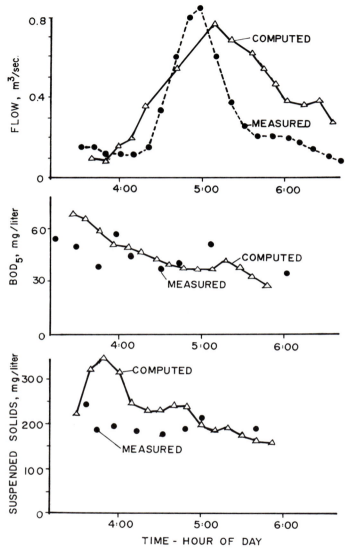

Fig. 7.2. Measured and simulated quantity and quality of a combined sewer overflow (USEPA, 1971).

1971; Hydrocomp Int., 1972). Figure 7.2 shows a comparison of measured and simulated pollutographs for a combined sewer overflow system. The differences in the observed and computed values should be noted inasmuch as they give an indication of the reliability of the model.

POLLUTION FROM DIFFUSE SOURCES

Pollution from point sources enters the transport route at a late stage of the hydrological cycle — i.e., during channel or estuary flow — or it is discharged via underwater diffusers into lakes and oceans. In contrast, nonpoint source pollution enters the hydrological cycle during its early stages, precipitation, or overland flow. The point where the pollutants enter the hydrologic transport process depends not only on the type of source and its location, but also on the form in which the pollutants occur. Gaseous, emulsified, and dispersed airborne pollutants enter the water transport route following deposition on the surface by precipitation and/or dry fallout, while soluble pollutants mix with the water directly. Relatively insoluble pollutants are either dispersed and picked up during rainfall or snowmelt events through subsequent surface runoff, or are transported by wind and subsequently redeposited. Furthermore, pollutants are adsorbed by soil particles and transported by water in the particulate phase. Figure 7.3 represents a schematic block diagram of the processes involved in the pollution transport or pickup from nonpoint sources.

With the new restrictions on point sources, such as sewage and industrial wastewaters, it has become apparent that a substantial portion of the degradation of surface waters originates from man's use of the land, i.e., from nonpoint diffuse sources.

There is a tendency to relate pollution loadings from nonpoint sources to a particular land use. In this approach, the pollution from diffuse sources is expressed as a simple value or range of unit loadings (loadings per area and time) for each land use. This approach, thought justified as a first approximation, may often lead to results which will deviate grossly from actual values. It is, therefore, necessary to delineate and analyze the basic processes and factors involved in pollution generation from diffuse sources. Loading functions of pollution from nonpoint sources typical for average conditions in the United States are summarized in a report by the Midwest Research Institute (USEPA, 1976).

It is expected that nonpoint source pollution transport processes in urban areas may be quite different from those in nonurban areas. The following factors are thought to be responsible:

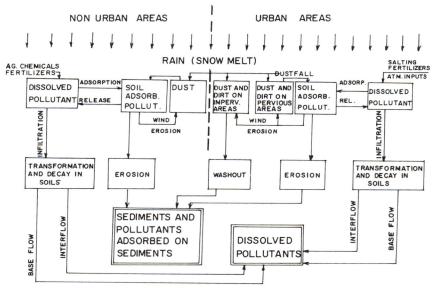

Fig. 7.3.   Transport of pollutants in urban and rural watersheds.

1. Large portions of urban areas are impervious, resulting in a much higher hydrologic activity.

2. With the exception of construction sites, most of the pervious surfaces in residential or city areas are well protected by lawns; therefore, less erosion can be expected.

3. The pollutant loadings in urban areas are affected mainly by litter cumulation, dry or wet fallout, and traffic, while in nonurban areas most of the pollution is due to erosion of soils and soil-adsorbed pollutants.

4. Over a long period of time (season) almost all of the pollutants deposited on the impervious surfaces which have not been removed by street cleaning, wind, or decay will eventually end up as surface runoff. On the other hand, in nonurban areas, soil represents an infinite pool of sediments and pollutants adsorbed by soil, and their removal rate then depends on the energy of rain or runoff which liberates the soil particles and surface protection.

Pollutant Loadings and Transport from Impervious Urban Areas

Pollutant accumulation on ground surfaces in urban areas and subsequent washout by runoff represent a major pollution contribution from nonpoint urban sources. Since impervious areas are almost 100% hydro-

logically active, most of the runoff and associated pollution in highly ur-
banized areas originates from these surfaces. The amount of deposited
pollutants depends on various factors and inputs. The major inputs are
atmospheric fallout, street litter deposition, animal and bird fecal waste,
dead vegetation, and road traffic impact. The factors which affect the
quality of the street refuse washed out to surface waters include land use,
population, traffic flow and frequency, effectiveness of street cleaning and
type of street surface, and pollution.

As previously noted, a simple unit loading value related to land use may
not provide an adequate estimation. Instead, the loading values should be
correlated to major causative factors which can be listed as follows for
various urban land uses:

1. Percent impervious area directly connected to a channel (a function
of land use or percent imperviousness).
2. Population density (a factor related to land use).
3. Dry and wet atmospheric fallout.
4. Litter accumulation (a factor related to population density and land
use).
5. Traffic density (a factor related to land use).
6. Curb height and density (factors related to land use).
7. Percent open area (a factor related to land use).
8. Average wind velocity.
9. Street cleaning practices and effectiveness.
10. Average number of dry days preceding a rain or rain intensity.
11. Depression and interception storage (a factor related to land use).

With the exception of low-density residential areas, other factors such as
slope and soil type are expected to have very little effect on pollution
loads from urban areas because most of the loading originates from impervious areas.

The magnitude of pollution generated from nonpoint urban sources is of
the same order as the raw sewage contribution. In Durham, North Carolina (Colson and Tafuri, 1975), the annual runoff pollution from nonpoint
urban sources was equal to 91% of COD load of raw sewage, the BOD
yield was 67% of that for raw sewage, and the suspended solids yield was
20 times higher than that contained in the raw municipal wastewater. In
Atlanta, Georgia, 65% of the BOD load contributed to receiving waters
can be attributed to storm water (Black, Crow, and Eidsness, 1971). This
magnitude of nonpoint pollution contribution is typical of many urban
areas.

Figure 7.4 is a schematic representation of the street surface pollutants
accumulation process. The primary sources of pollution are airborne pol-

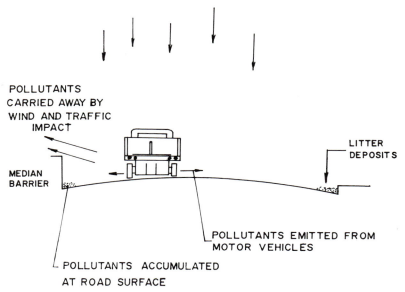

ATMOSPHERIC DUST FALLOUT

POLLUTANTS
CARRIED AWAY BY
WIND AND TRAFFIC
IMPACT

LITTER
DEPOSITS

MEDIAN
BARRIER

POLLUTANTS EMITTED FROM
MOTOR VEHICLES

POLLUTANTS ACCUMULATED
AT ROAD SURFACE

Fig. 7.4. Pollutant accumulation model schematic.

lutants deposited from the atmosphere and street litter deposited on impervious areas. These pollutants can be carried by wind and traffic and accumulate near the curb. Thus, many urban pollution studies report street pollution accumulation rates as related to the unit length of curb, as shown in Table 7.10. Reporting the street refuse loadings per unit length of curb, instead of a more meaningful areal loading, appears to be justified since it has been observed that almost 80% of the street refuse can be found within 15 cm from the curb and 97% within 1 meter from the curb

TABLE 7.10. Street refuse accumulation

Land Use Characteristic Solids Accumulation g/curb meter/day	Chicago[a] Dust and Dirt	Eight American Cities[b] Total Solids
Single family	10.4	48
Multiple family	34.2	66
Commercial	49.1	69
Industrial	68.4	127
Average of above (weighted) 22.3		

[a] American Public Works Association (1969).

[b] From Sartor and Boyd (1972). Approximate dust and dirt content of total solids = 75%.

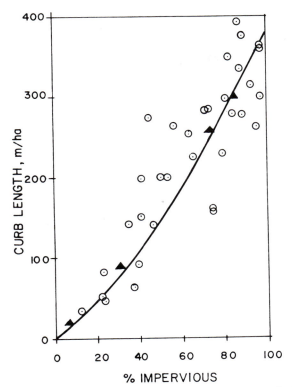

Fig. 7.5. Curb length, percent impervious relationship (Graham *et al.*, 1974; Novotny, 1978).

(Sartor *et al.*, 1974). A good correlation between curb length density and percent imperviousness of residential areas can be used for a rough approximation, as shown in Fig. 7.5.

The American Public Works Association (1969) recently developed a regression formula between curb length of urban areas v. population density based on the analysis of many American cities. The resulting regression equation was

$$CL = 311.67 - (266.07)(0.839)(2.48\ PD), \qquad (7.1)$$

where CL = curb length (m/ha) and PD = population density (persons/ha).

The street refuse washed by runoff into receiving waters contains many significant contaminants. Organic pollution, heavy metals, pesticides, and bacterial loadings are commonly associated with street refuse, mainly

TABLE 7.11.   Pollutants associated with street refuse (in μg/g of total solids)

Constitutent	Residential	Land use Industrial	Commercial	Transportation
BOD$_5$	9,160*	7,500*	8,300*	2,300**
COD	20,800*	35,700*	19,400*	54,000**
Volatile solids	71,600*	53,570*	77,000*	51,000**
Total Kjeldahl N	1,600*	1,390*	1,100*	156**
Nitrates	50*	64*	500*	80**
Phosphates	910*	1,210*	830*	610**
Lead	1,468***	1,339***	3,924***	12,000**
Total coliforms[1]	160,000*	82,000*	110,000*	NR
Fecal coliforms[1]	16,000*	4,000*	5,900*	925**

*Sarton and Boyd (1972), and Sartor *et al.* (1974)
**Shaheen (1975)
***Amy *et al.* (1974)
[1]in No/gram
NR–not reported

with its dust and dirt fraction. Tables 7.11–7.13 show typical contamination values. It must be remembered that these values, though typical, are not uniform; instead, they represent averages from a wide range of street refuse deposition and contamination for a limited number of the investigated municipalities.

Atmospheric Pollutants Deposition

Deposition of atmospheric pollutants occurs as dry or wet fallout. The deposition rates of particulate atmospheric pollutants in United States cities vary from 3.5 tons/km²/month to more than 35 tons/km²/month. Table 7.14 shows that higher deposition rates can be expected in

TABLE 7.12.   Heavy metals contamination of street refuse (μg/g of total solids)[a]

Constitutent	Residential	Land use Industrial	Commercial	Overall
Cadmium	3.45	2.83	3.92	2.82
Chromium	186	208	241	183
Copper	95	55	126	101
Nickel	22	59	59	31
Strontium	23	134	151	177
Zinc	397	283	506	338

[a]From Pitt and Amy (1973).

TABLE 7.13. Mean concentrations of chlorinated hydrocarbons and pesticides in street dust and dirt from nine United States cities[a]

	Concentration nanograms/gram of dry solids	
	Mean	Standard Deviation
Endrin	0.2	—
Dieldrin	28	28
PCB	770	770
Methoxychlor	500	1050
Lindane	2.9	7.1
Methylparathion	2	—

[a] From Amy *et al.* (1974).

congested industrial areas or business districts and lower deposition rates are common in residential and rural suburban zones.

Besides particulates, atmospheric fallout contains many other pollutants. Rainfall contributions of nitrogen are shown in Figure 7.6. The nitrogen isopleths in Fig. 7.6 indicate that the greatest nitrogen fallout from the atmosphere occurs in the midwestern areas of the United States. Atmospheric contributions of phosphorus are on the order of 10 to 100 kg/km²/yr. There is little correlation between the nitrogen and phosphorus contributions from the atmosphere and precipitation volume or intensity; therefore, the distribution of the pollutants between dry and wet fallout is more or less accidental and depends on the density and occur-

TABLE 7.14. Average annual particulate deposition rate in Milwaukee County[a,b]

Land use	1951	1957	1963	1965	1966	1967	1968	1969
Agricultural rural suburbs	4.87	5.36	7.10	8.51	9.54	10.79	8.21	6.70
Residential	7.80	6.89	7.38	8.28	8.14	7.93	7.83	6.75
Local business	12.70	9.43	10.37	8.54	9.08	10.16	10.30	8.04
Commercial	15.94	16.67	12.81	14.48	12.81	15.87	14.13	12.21
Industrial	28.57	19.59	14.37	15.80	14.51	15.00	14.76	14.20

Average monthly deposition rates in Milwaukee County in tons/km²/month, 1951–1969

Jan.	Feb.	Mar.	Apr.	May	June	July	Aug.	Sep.	Oct.	Nov.	Dec.
9.85	10.4	12.9	14.09	14.47	12.80	10.47	10.58	10.63	10.23	9.71	8.09

[a] Milwaukee County DAPC (1970).
[b] Tons/km²/month.

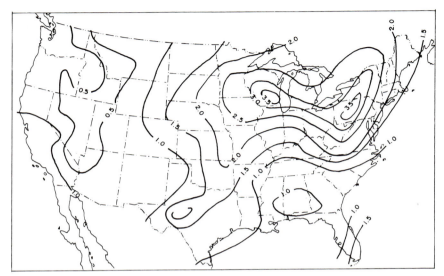

Fig. 7.6.   Nitrogen contributions.

rence of precipitation during the pollutants' fallout (Chapin and Utter-mark, 1973).

Many other pollutants are deposited from or transported by the atmosphere. Lead, which is associated with the combustion of leaded gasoline, is transported with atmospheric particulate matter. Also, cadmium, strontium, zinc, nickel, mercury, and many organic chemicals are transported with atmospheric particulates and can be associated with atmospheric fallout. The contributions of SO_2 and NO_x to the atmosphere, in particular by the Tennessee Valley Authority, have already been mentioned.

Wind Erosion

The effect of wind erosion on the surface particulate pollutant loadings seems to be significant only in certain cases. The following factors are important in assessing the effects of wind erosion: climate, soil characteristics, surface roughness, vegetation cover, and the length of the eroding surface (Beasley, 1972). The great dust storms which devastated portions of the southwestern United States during the first part of this century resulted from the worst combination of these adverse factors. Soils which are loose, dry, and finely granulated will exhibit the greatest loss by wind erosion. The velocity of the wind near the surface and, therefore, the extent of wind erosion can be significantly reduced by a roughened soil surface or a surface covered by vegetation.

TABLE 7.15. Traffic-related sources of roadway pollution[a]

Pollutant	Traffic Related Source
Asbestos	Clutch plates, brake linings
Copper	Thrust bearing, bushings, and brake linings
Chromium	Metal plating, rocker arms, crankshafts, rings, brake linings, and pavement materials
Lead	Leaded gasoline, motor oil transmission fluid, babbitt metal bearings
Nickel	Brake linings and pavement material
Phosphorus	Motor oil
Zinc	Motor oil and tires

[a]From Shaheen (1975).

In urban areas, the primary source of wind-eroded materials is open, ungrassed areas and construction sites. A variable which can be used to approximately describe the erodable area effect is the percent of unprotected open area within the vicinity of the area of interest.

Motor Vehicle Usage

Traffic can contribute significantly to pollutant deposition in urban areas. High amounts of heavy metals in storm water runoff are often attributed to motor vehicle emissions and to the breakdown of road surface materials and vehicle parts. Table 7.15 lists traffic-related sources of various pollutants. Motor vehicle usage can influence pollutant accumulation in urban areas and near high-density traffic areas by emission of pollutants, oil or gasoline spillage, mechanical impact of traffic, and tire abrasion. Therefore, in addition to traffic density, the pavement composition and conditions are significant in determining the traffic impact on pollution. Streets paved entirely with asphalt have loadings about 80% heavier than all-concrete streets (Sartor *et al.*, 1974 and Sartor and Boyd, 1972). Streets whose pavement conditions were rated "fair to poor" were found to have total solids loadings about $2\frac{1}{2}$ times more than those rated "good to excellent."

Litter Deposition

Litter deposits in urban areas include solid wastes deposited on surfaces by careless public, private, and municipal waste collection operations, animal and bird fecal droppings, fallen tree leaves, grass clippings, and other cultural deposits. The dust and dirt component of litter (materials less than 3.5 mm size) is regarded as having the greatest pollution potential of street surface refuse materials. Although most of the litter is originally greater in size than that of dust and dirt, it is possible that a significant portion of the street dust and dirt originates from the mechanical fracture of litter. The American Public Works Association study (1969) reported that residential areas had greater amounts of street surface dust and dirt, as population density increases resulted in additional pedestrian and roadway traffic. Greater street surface deposition rates by public and private waste collections would also be expected to be higher in higher density residential areas.

Effects of Vegetation

The fallout of leaves and grass clippings in urban areas contributes significantly to dust and dirt accumulation. During most of the year the dust and dirt accumulated on impervious areas orginate from the erosion of surrounding pervious areas, while atmospheric pollution and litter accumulation during the fall substantially increase the organic solids accumulated on the surface.

From the work of Carlisle *et al.* (1966), Heaney and Huber (1973) estimated an average leaf fallout of 14.5 to 26 kg/tree/year. The wooded area which was investigated was stocked with uneven-aged (about 40 to 120 years) trees with a 90 to 95% closed canopy and 155 trees/ha. The tree species were mainly oak and birch. Typical values of tree leaf fallout in Minnesota are about 380 tons/km^2/year in a forested area with about 420 trees/ha. Of these yearly values, about 65% takes place during the fall season.

The fallen leaves are 90 to 97% organic and contain about 0.04 to 0.28% phosphorus (Lutz and Chandler, 1976).

Street Dust and Dirt Deposition Rates

The deposition rate of pollutants on street and impervious area surfaces is not uniform throughout the year. Obviously, it will vary depending on many factors including source strength and emission rates, wind, traffic, and meteorology. The deposition appears to be higher in a period following a rain or street cleaning. As the pollutants accumulate on the surface, more can be removed on adjacent pervious areas by wind and traffic impact. Thus the deposition curve has a tendency to stabilize after a few days, as shown in Fig. 7.7.

The deposition rates are also affected by such factors as curb height, which acts as a natural trap for pollutants carried by wind, proximity of protected grassed areas, and exposure to major dust-carrying winds.

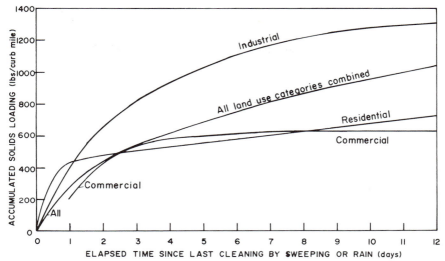

Fig. 7.7. Accumulation of pollutants by land use (Sartor *et al.*, 1974).

Pollutant Washout

Not all of the pollutants accumulated during a period preceding a rainfall will be washed off an impervious surface during the first moments of a rainfall. The rate at which rainfall washes loose particulate matter from street surfaces depends on three primary factors (Sartor and Boyd, 1972): rainfall intensity, street surface characteristics, and particle size. It may be expected that the amount of pollutants washed off will generally follow the equation

$$PL = \frac{dL}{dt} = -K_p L, \qquad (7.2)$$

where PL is the pollutant washout rate, L is amount of the pollutant present on the surface, and K_p is a coefficient depending on the rain intensity and the street surface characteristics. The coefficient, K_p, was found to be almost independent of the particle size within the range of 10 to 1000 μm. The coefficient, K_p, is approximated as $K_p = E_u R$, where E_u is the urban washout coefficient and R is the runoff rate from impervious surfaces. The values of the washout coefficient have been reported (U.S. Army Corps Eng., 1975) as $E_u = 1.81$ cm^{-1}.

Not all of the deposited litter is available for transport by surface runoff. Therefore, the sediment washout rate should be multiplied by an

TABLE 7.16. Interrelationship of sweeper efficiency and particle size[a]

Particle Size	Sweeper Efficiency
> 2000	79
840–2000	66
246–840	60
104–246	48
43–104	20
< 43	18
Overall	50

[a] From Sartor et al. (1974).

availability factor, which was proposed as (U.S. Army Corps Eng., 1975)

$$A_s = 0.57 + 0.5R^{1.1}, \qquad (7.3)$$

where R is the surface runoff rate in cm/hr. It is obvious that there must be a limit on the availability factor as the runoff rate increases. A suggested value for the maximum A_s is 0.75, which implies that about 25% of the urban litter is not available for transport.

Assuming a steady rain intensity, Eq. (7.2) can be integrated to yield the typical "decay" formula:

$$L = L_0 (1 - e^{-K_p t}), \qquad (7.4)$$

where L is the amount of litter washed out of the surface during the time period t, and L_0 is the initial amount of the litter present on the surface.

Street Sweeping Practices

Street sweeping is a common practice in American cities (as compared to some European cities which wash the streets). Most street sweeping is done mechanically, with two types of sweepers, a brush and a vacuum. Removal efficiencies with brush-type sweepers for various particle sizes

TABLE 7.17. Effectiveness of street sweeping[a]

Pollutant	Sweeping Removal Efficiency, %
Total solids	50.00
Volatile solids	42.52
COD	31.01
TKN	43.90
Phosphate	22.20

[a] From Sarton et al. (1974).

TABLE 7.18.   Percent of street pollutants in various Particle size ranges[a]

Pollutants	>2000	840–2000	Particle Size, μ 246–840	104–246	32–104	<43
Total solids	24.9	7.6	24.6	27.8	9.7	5.9
Volatile solids	11.0	17.4	12.0	16.1	17.9	25.6
COD	2.9	4.5	13.0	12.4	45.0	22.7
TKH	9.9	11.6	20.0	20.2	19.6	18.7
Phosphates	0	0.9	6.9	6.4	29.6	56.2

[a] From Sartor *et al.* (1974).

are shown in Table 7.16. Based on the data included in Table 7.17, the efficiency of sweeping to remove deposited suspended solids is about 50% per one pass of a sweeper. However, some pollutants are associated more with finer fractions (Table 7.18). The overall efficiency can be estimated by using cumulative multiplication of sweeping, efficiency for each fraction, and pollution concentrations on particles of the fraction. For example, the efficiency of sweeping for phosphate control would be only 22%, as compared to 50% for suspended sediment. The effectiveness of street sweeping to remove some common pollutants associated with street refuse is shown in Table 7.17.

Pollutant Generation from Pervious Urban and Nonurban Areas

Pollutants generated from rural areas are associated with sediment washload; transport of agricultural chemicals and nutrients; leachate of soil minerals and chemicals into the base flow and interflow; background mineral and organic natural pollution; and cultural sources such as strip-mining pollution, animal feedlots and grazing, and pollution from oil fields.

For some pollutants such as phosphorus, ammonia, toxic elements, and pesticides, soil particles are the primary carriers. These pollutants are either adsorbed readily onto clay and/or organic adsorption sites on soil particles or are precipitated in soils. Most of the traditional water quality models treat such pollutants as either conservative or nonconservative dissolved substances. Consequently, gross deviations may appear in water quality simulation runs if the particulate matter–pollutant interactions and effects are not included. The pollutants adsorbed on the soil particles may be subjected to erosion and moved with the particulate matter of the surface runoff. Soluble pollutants with a low-adsorption affinity for soil can infiltrate and reappear with interflow or can be leached into the groundwater pool and become a part of the base flow contamination.

Soil Erosion

Almost 4 billion tons of soil materials are washed into receiving waters of the United States every year (Loehr, 1972). Approximately 75% of this material originates from forested and agricultural lands. Besides the fact that the sediment particles can carry adsorbed pollutants, the sediment itself is considered to be a serious hazard to water quality.

The rates of erosion and sediment loading of surface waters have increased significantly because of man's use of lands, vegetation, and surface waters. The sediment loss from soils susceptible to erosion has been reported to be in the range of 25 to 75 tons/ha/year. Over 100 tons/ha/year can be lost from unprotected urban construction sites. Generally, cultivated soils in warmer and more humid regions are more susceptible to erosion than soils in cooler, less humid areas. Table 7.19 shows nationwide representative loadings of sediment for various land uses.

The sediment can originate and be liberated only from hydrologically active areas. Soils with high-permeability rates are much less hazardous for sediment loss than soils which have a high content of clay and fine silt fractions. The second most important factor in soil loss considerations is vegetation cover, which protects the soil surface against the impact of rain energy. Not all liberated soil particles will reach the receiving waters inasmuch as a substantial part will resettle during the overland flow phase prior to the runoff's reaching a stream. The actual sediment yield or delivery to the stream can only amount to a fraction of the total soil loss, depending on the watershed characteristics and size.

Soil erosion is a function of rainfall and runoff energy, soil characteristics, vegetation cover, length and slope of the watershed, and the extent of erosion control practices within the watershed.

TABLE 7.19.   Representative rates of erosion from various land uses[a]

Land Use	Erosion tons/ha/year
Forest	0.085
Grassland	0.085
Abandoned surface mines	8.5
Cropland	17
Harvested forest	42.5
Active surface mine	170
Construction	170

[a]USEPA (1973).

Soil Washload

Washload is the part of the total sediment load which contains most of the fine particles. Washload is usually caused by land erosion and is defined as that part of the sediment load which is composed of particles smaller than those found in appreciable quantities in the shifting portion of the stream bed (ASCE, 1976). The bed load portion is composed mostly of larger particles and sand and gravel, which originate from gulley and river bank erosion. It does not have a significant adsorptive capacity as do clay and fine soil particles and represents much less pollution hazard than the washload.

The Universal Soil Loss Equation

The universal soil loss equation is an empirical formula used for estimating soil loss from small agricultural lots. In spite of its empirical nature, it has been widely used in many water quality models, both steady state and dynamic, to estimate instantaneous sediment loadings to surface waters. Its original purpose, however, was to provide specific and reliable guides to help select adequate soil conservation practices for farm fields.

The universal soil loss equation, originally developed for areas east of the Rocky Mountains, has been applied to the rest of the United States. The equation is expressed as (Wischmeier and Smith, 1960; 1965):

$$A = (R) (K) (LS) (C) (P), \tag{7.5}$$

where A is the computed soil loss in tons per hectare during a given storm, R is the rainfall factor, K is the soil erodibility factor, LS is the slope-length gradient, C is the cropping management factor, and P is the erosion control practice factor.

Equation (7.5) does not include wind erosion but accounts only for the erosion caused by rain. In modified form, it also includes the effect of runoff energy. In order to obtain the sediment content of the runoff at an outlet point from the watershed, the soil loss estimated by Eq. (7.5) must be multiplied by a delivery ratio factor, DR, which accounts for the resettling of particles during overland flow before they reach the outlet point. Therefore,

$$AR = (DR) (A) (W), \tag{7.6}$$

where AR is the runoff sediment content at the outlet point and W is the basin area.

Rainfall factor, R The rainfall factor reflects the energy of the rain droplets falling on the surface and subsequently liberating the soil particles which become available for pickup by surface runoff. For a single storm, the rainfall factor was defined as (Wischmeier and Smith, 1960; 1965):

$$R = R_r = EI = \sum_1 [(2.29 + 1.15 \log X_i)D_i]I, \tag{7.7}$$

where R_r is the rainfall factor component due to the rain energy, E is the total kinetic energy of the rain, I is the maximum 30-minute rainfall intensity of the storm in cm/hr, X_i is the rainfall intensity in cm/hr at a time interval, i, and D_i is the rainfall in cm during the time interval, i.

Both rainfall energy and detachment of soil particles by runoff contribute to soil loss. Therefore, the rainfall factor, R, should include the effect of runoff energy as well. A modification of Eq. (7.7) has been proposed, as follows (Williams, 1972; Foster *et al.*, 1973):

$$R = aR_r + bcQq^{1/3}, \tag{7.8}$$

where a and b are weighting parameters ($a + b = 1$), c is an equality coefficient, Q is the runoff volume in cm^3, and q is the maximum runoff rate in cm/hr.

The weighting factor compares the relative amounts of erosion by rainfall and runoff under unit conditions. It was suggested (Wischmeier and Smith, 1965) that the detachment of particles by runoff and rain energy is about evenly divided ($a = b = 0.5$). The equality coefficient in SI units is about 19.3. Substituting the values of a, b, and c into the universal soil loss equation, the overall rainfall factor becomes

$$R = 0.5R_r + 7.5Qq^{1/3}. \tag{7.9}$$

Because of their empirical nature, the coefficients a, b, and c should be used with caution since they can change from storm to storm and watershed to watershed.

Soil factor, K The soil factor is a measure of the potential erodibility of a soil. The soil erodibility nomograph in Fig. 7.8 is used to find the value of the factor, K, once the soil parameters have been estimated. These parameters include percent silt plus very fine sand (particles less than 0.05 to 0.1 mm), percent sand greater than 0.1 mm, organic matter content, structure, and permeability.

Slope-length factor, LS The slope-length factor is given by the following equation (Wischmeier and Smith, 1965):

$$LS = L^{1/2} (0.0138 + 0.00974S + 0.00138S^2), \tag{7.10}$$

where L is the length in meters from the point of origin of the overland

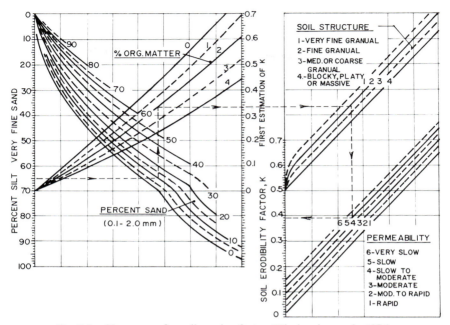

Fig. 7.8. Nomogram for soil erosion factor (Wischmeier *et al.*, 1971).

flow to the point where the slope decreases to the extent that deposition begins or to the point at which runoff enters a defined channel, and S is the average percent slope over the given overland flow length.

If the average slope is used in calculating the LS factor, the predicted erosion will be different from the actual erosion when the slope is not uniform. The equation for the LS factor shows that when the actual slope is convex, the average slope will underestimate the predicted erosion, whereas when the slope is concave, the prediction equation will overestimate the actual erosion. To minimize these errors, large eroding sites should be broken up into areas of fairly uniform slopes.

Cropping management factor, C   The cropping management factor estimates the effects of the ground cover condition of the soil and the general management practices of the area of concern. The areas with continuous fallow ground, defined as land that has been filled and kept free of vegetation and surface crusting, are assumed to have a C factor equal to 1. Table 7.20 indicates general magnitudes of the cropping management factor for various sites. These predictions indicate that erosion due to construction is on the order of 100 times greater than a completely developed urban site. The C factor for construction sites or bare vegetation free lands can

TABLE 7.20. Factor C values used in computing gross erosion[a]

Land Use	Factor C Value
Cropland	0.08
Grassland	0.01
Woodland	0.005
Construction	1.0
Urban	0.01

[a] From Brant (1972).

TABLE 7.21. C values and slope-length limits for no seeding or, if seeded, through first six weeks of growing period[a,b]

Mulch Type	T/A	Slope %	C Value	Max. Length Ft.
1. No mulch or seeding		all	1.0	—
2. Straw or hay, tied	1.0	≤5	0.20	200
down by anchoring and		6–10	0.20	100
tracking equipment used				
on slope[c]	1.5	≤5	0.12	300
		6–10	0.12	150
	2.0	≤5	0.06	400
		6–10	0.06	200
		11–15	0.07	150
		16–20	0.11	100
		21–25	0.14	75
		26–50	0.18	35
3. Crushed stone	135	≤15	0.05	200
		16–20	0.05	150
		21–33	0.05	100
		34–50	0.05	75
	240	≤20	0.02	300
		21–33	0.02	200
4. Woodchips	7	≤15	0.08	75
		16–20	0.08	50
	12	≤15	0.05	150
		16–20	0.05	100
		21–33	0.05	75
	25	≤15	0.02	200
		16–20	0.02	150
		21–33	0.02	100
5. Chemical mulches				

		C Values[d]	
Seeding	Mulch	Period 1	Period 2

C Values for Seeded Area

Seeding	Mulch	Period 1	Period 2
Temporary (grain or	None	0.70	0.10
fast-growing grass)	Straw 1T/A	0.20	0.07
	Straw 1.5T/A	0.12	0.05
	Straw 2T/A	[e]	0.05
	Stone 135T/A	0.05	0.05
	Stone 240T/A	0.02	0.02
	Woodchips 7T/A	0.08	0.05
	Woodchips 12T/A	0.05	0.02
	Woodchips 25T/A	0.02	0.02
Permanent seeding, second year		—	0.01
Sod		0.01	0.01

[a] From Ports (1975).

[b] If seeding is late in fall, these values would extend into the next spring.

[c] If the straw is not anchored to the soil, rilling may occur between the mulch. C values on moderate or steep slopes of soils having a K-value greater than 0.30 should then be taken at double the values shown in the table.

[d] The two periods are defined as follows: period 1—through first 6 weeks of growing period; period 2—after 6 weeks of growing period.

[e] Use values for no seeding for appropriate slope steepness.

be reduced if the surface is protected by seeding, hay application, or asphalt emulsion. The effects of these protection practices on the C factor are shown in Table 7.21.

Erosion control practice factor, P The P factor accounts for the erosion control effectiveness of such practices as contouring, terracing, compacting, and providing sedimentation basins. The C factors reflect protection of the soil surface against a loss of soil particles. The P factor, on the other hand, involves measures which would keep already liberated soil particles near the source and prevent further transport.

Terracing itself does not affect the P factor because the soil loss reduction from the terracing is reflected by the change in the LS factor. Values of the factor, P, for various farm erosion control practices are given in Table 7.22 and those for urban areas are presented in Table 7.23.

TABLE 7.22. Conservation practice factors P for agricultural lands[a,b]

Slope as Percentage	Up- and Down Hill	Contouring	Cross-Slope Farming without Strips	Contour Stripcropping Alternate Grain-and-Meadow Strip System	Cross-Slope Farming with Strips
1.1–2.0	1.0	0.6	—	0.3	—
2.1–7.0	1.0	0.5	0.75	0.25	0.37
7.1–12.0	1.0	0.6	0.80	0.30	0.45
12.1–18.0	1.0	0.8	0.90	0.40	0.60
18.1–24.0	1.0	0.9	0.95	0.45	0.67
>24.0		1.0	1.0	—	—

[a]From Wischmeier and Smith (1965).
[b]The conservation practice factor for terracing should equal the contour practice factor.

TABLE 7.23. Erosion control practice factor P for construction sites[a]

Surface Condition with No Cover	Factor P
1. Compact, smooth, scraped with bulldozer or scraper up and down hill	1.30
2. Same as above, except raked with bulldozer root raked up and down hill	1.20
3. Compact, smooth, scraped with bulldozer or scraper across the slope	1.20
4. Same as above, except raked with bulldozer root raked across slope	0.90
5. Loose as a disked plow layer	1.00
6. Rough irregular surface, equipment tracks in all directions	0.90
7. Loose with rough surface greater than 12 in. depth	0.80
8. Loose with smooth surface greater than 12 in. depth	0.90
Structures	
1. Small sediment basins:	
0.04 basin/acre	0.50
0.06 basin/acre	0.30
2. Downstream sediment basins:	
with chemical flocculants	0.10
without chemical flocculants	0.20
3. Erosion control structures:	
normal rate usage	0.50
high rate usage	0.40
4. Strip building	0.75

[a]From Ports (1973).

Sediment Delivery

The fraction of sediment delivered from an erosion source to any specified downstream location is affected by the size and texture of the sediment material, the size of the watershed or length of travel, the local environment, land use, and general physiographic position. In many cases, the sediment is deposited in the collecting stream and moves only during high floods.

The change (per unit area) of this downstream sediment movement from its source to any given measuring point is expressed by the delivery ratio, DR. The delivery ratio may range from a few percent for large flat watersheds to almost 100% for small steep lots.

The Sedimentation Committee recommended formula for the delivery ratio was developed by Roehl, (1962) from field investigations in the southeast Piedmont region as follows (ASCE, 1976):

$$\log DR = 4.453 - 0.23 \log W - 0.51 \log \frac{L}{R} - 2.79 \log B, \quad (7.11)$$

where DR is the sediment delivery ratio in percent, W is the drainage area of the watershed in km², L/R is the dimensionless basin length–relief ratio (watershed length measured essentially parallel to the main drainageway, divided by the elevation difference from the drainage divide to the outlet), and B is the weighted mean bifurcation ratio. (The bifurcation ratio is the ratio of the number of streams of any given order to the next higher order.)

Various other formulas have been proposed for estimation of the delivery ratio factor (USEPA, 1976; Williams and Berndt, 1972). The Midwest Research Institute study (USEPA, 1976) recommended using the delivery ratio plot based on analysis and data of the United States Soil Conservation Service as shown in Fig. 7.9. This analysis shows that for relatively homogenous basins, the delivery ratio varies inversely with the drainage basin area and drainage density, which is defined as the ratio of total channel–segment lengths (accumulated for all orders within the basin) to the basin area. The reciprocal of the drainage density factor may be thought of as an expression of the closeness of the spacing of channels, or the average distance that soil particles travel from an erosion site to the receptor water. Figure 7.9 can then be used to estimate the delivery

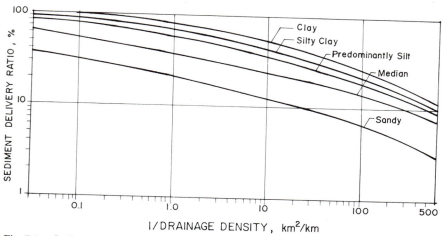

Fig. 7.9. Sediment delivery ratio for relatively homogeneous basins (McElroy *et al*, 1976).

ratios. However, as stated by the MRI study, the delivery ratio relationships need further validation.

Pollution Loading Functions

Most of the pollutants originating from pervious areas are associated with suspended soil particles. Thus, if the concentration of the pollutants adsorbed on or associated with soil sediment is known, or can be estimated, the pollution loading function then becomes

$$LP = (DR)(C_p)(A)(r_p)(W), \qquad (7.12)$$

where LP is pollutant loading to receiving stream in mass/time (tons/year), DR is the delivery ratio in percent, A is the soil loss in tons/yr/ha, C_p is the concentration of the pollutant on soil in g/g, r_p is the dimensionless enrichment ratio which accounts for the difference in the pollutant concentration in soil and runoff sediment, and W is the watershed area in km^2.

The use of Eq. (7.12) presumes knowledge of the soil pollutant concentration, C_p, and the enirchment ratio, r_p. While C_p is often measured and may be available for some pollutants such as organic matter, nutrients, and metals from agricultural sources (see Table 7.24 for metal concentrations), the exact nature of the enrichment factor, r_p, is not known and its estimation is always a rough approximation. Since most of the pollutants are associated with fine soil particles which are readily erodable, the enrichment factor may be correlated to the ratio of fine fractions in the runoff suspended sediment to that in the soil.

Equation (7.12) is static; i.e., it does not allow the prediction of future conditions and the evaluation of the effect of future management practices. However, in some cases, when a pollutant is conservative and immobile, approximate concentrations can be estimated assuming a pollutant–soil mass balance in the top soil layer. For example, if the top soil layer depth in which the pollutant remains and is uniformly distributed is 25 cm (approximately till depth) and the specific density of soil is 1.8 g/cm^3, then an application of 1 kg of the pollutant per hectare will result in a top soil concentration of

$$C_p = \frac{1(\text{kg/ha}) \times 10^9(\mu\text{g/g})}{25(\text{cm}) \times 10^8(\text{cm}^2/\text{ha}) \times 1.8(\text{g/cm}^3)} = 0.22 \frac{\mu\text{g}}{\text{g}}.$$

Agricultural Nonpoint Pollution Sources

The transport of pollutants from agricultural lands to receiving waters could take place along innumerable pathways and could involve many

TABLE 7.24.   Heavy metal concentrations in surficial materials in the United States[a,b]

	Arithmetic Analysis		Geometric Means		
Element	Average, μg/g	Range, μg/g	Conterminous United States, μg/g	West of 97th Meridian, μg/g	East of 97th Meridian, μg/g
Arsenic	—	<1,000	—	—	—
Barium	554	15–5,000	430	560	300
Cadmium	—	<20	—	—	—
Cerium	86	<150–300	75	74	78
Chromium	53	1–1,500	37	38	36
Cobalt	10	<3–70	7	8	7
Copper	25	<1–300	18	21	14
Gallium	19	<5–70	4	18	10
Germanium	—	<10	—	—	—
Gold	—	<20	—	—	—
Hafnium	—	<100	—	—	—
Indium	—	<10	—	—	—
Iron	25,000	100–100,000	18,000	20,000	15,000
Lanthanum	41	<30–200	34	35	33
Lead	20	<10–700	16	18	14
Manganese	560	<1–7,000	340	389	285
Molybdenum	3	<3–7	—	—	—
Neodymium	45	<70–300	39	36	44
Nickel	20	<5–700	14	16	13
Niobium	13	<10–100	12	11	13
Palladium	—	<1	—	—	—
Platinum	—	<30	—	—	—
Phenium	—	<30	—	—	—
Scandium	10	<5–50	8	9	7
Strontium	240	<5–3,000	120	210	51
Tantalum	—	<200	—	—	—
Tellurium	—	<2,000	—	—	—
Thallium	—	<50	—	—	—
Thorium	—	<200	—	—	—
Titanium	3,000	300–15,000	2,500	2,100	3,000
Uranium	—	<500	—	—	—
Vanadium	76	<7–500	56	66	46
Ytterbium	4	<1–50	3	3	3
Yttrium	29	<10–200	24	25	23
Zinc	54	<25–2,000	44	51	36
Zirconium	240	<10–2,000	200	170	250
Total	30,099		21,991	23,858	19,263

[a] From Shacklette *et al.* (1971).
[b] Note: "—" indicates all analyses showed element to be below detectable limits.

transport mechanisms. As discussed in the previous section, the loss of sediment from agriculture represents the primary transport vector for many pollutants. Most of the nitrate pollution reaches surface waters via groundwater. In both cases, water is the primary transporting medium, although it is recognized that nutrient loss by wind-blown particulate matter could be significant in some instances.

Agricultural lands are receiving increasing interest as a point of disposal of municipal sewage and sludges. A large pilot project is underway to dispose of all or most of the sewage sludge from Chicago on agricultural lands by means of an elaborate controlled environmental–agricultural system. The nutrient content of sewage and sludges has long been recognized as resulting in increased crop production due to the irrigation and nutrient aspects of such disposal.

A properly managed and balanced agricultural system that keeps soil loss and fertilizer excess to a minimum does not represent a great danger to water quality. In a balanced system, most of the nutrients applied to the farm fields will remain in the crops and upper soil layers and should not reach surface waters. The deficiencies are associated with the excesses from the agricultural operations. The excesses can be in the form of runoff from animal waste at animal production facilities, runoff and leachate from fertilized and manured fields, and liquid and solid wastes generated at food processing facilities connected with the agricultural operations. The excesses and losses of soil and pollutants from agricultural operations result from the increased intensity and productivity of farm production processes. The crop yields have increased markedly during the last few decades. In the early 1900s, 15 million or more hectares were required to feed a population of half as many people in the United States as existed in the latter part of this century (Loehr, 1972). Both crop production per unit area and total production are increasing even though less land is being used for farming. To maintain this production growth and intensity, more fertilizer must be used to replace the nutrients lost in the crops. Early farming used organic animal residue (manure) as a primary fertilizer. Present modern operations rely heavily on commercial chemical fertilizers, some of which are more mobile in the soil and represent a greater danger to surface waters than the older, less mobile organic fertilizers.

Large quantities of commercial fertilizers are applied annually to agricultural lands in the United States. Typical application rates range from 20 to 200 kg of N/ha and from 10 to 50 kg of P/ha. The application of fertilizers obviously increases the amount of pollutants that can be potentially lost from agricultural lands, but offsetting factors have been shown in some instances to more than compensate for the increased potential. Proper application, which includes matching the quantity and composition of fertilizers to crop needs and soil fertility, can reduce the amount of nutrient loss from croplands by increasing nutrient utilization by plants, and by increasing the crop density which reduces surface runoff and erosion. Increased root density may improve the permeability characteristics of soil, thus reducing the hydrological activity of the land. Conversely, if the addition of fertilizers creates a nutrient imbalance, or if excessive rainfall occurs shortly after application, pollution impact can be great.

It is impossible to present adequate loading figures for agricultural and forested lands. The process of pollutant–soil interaction is quite complex. The affinity of soils to retain pollutants depends on many factors, the primary ones being texture (clay content), pH, exchangeable cation content, organic matter, and the permeability and composition of the soil profile. Soils with high clay content will show a high retention capacity for many pollutants. It is known that phosphorus is mostly immobile in such soils and will remain near the layer of application. The same phenomenon applies to most of the heavy metals, pesticides, ammonia, and some other agricultural chemicals. Because of the low permeability of such soils, erosion is the primary transporting vector for these pollutants. The danger of groundwater contamination of clayey soils, with the exception of mobile components such as nitrates, is low. Under proper operating conditions, clayey soils can be used for wastewater disposal provided that the permeability is high enough to prevent excessive soil loss or sufficient erosion control practices are implemented.

Sandy soils usually have very low erosion potential; however, their pollution retention potential is lower and most of the pollutants will penetrate into the groundwater aquifer. Although the higher permeability makes them attractive for subsurface waste disposal systems, the danger of groundwater contamination should always be considered.

A summary of nutrient loading values from agricultural lands for various parts of the United States (Uttermark *et al.*, 1974) indicates an average nitrogen loading of 5.1 kg of N/ha/yr with minimum and maximum ranges of 1.2 kg/ha/yr and 13 kg/ha/yr, respectively. Average loadings for phosphorus were 0.38 kg/ha/yr with respective maximal and minimal ranges of 2.3 kg/ha/yr and 0.03 kg/ha/yr. A larger amount of the total nitrogen loading and almost all of the phosphorus lost from croplands were associated with particulate matter. Areas which are not disturbed by man have a lower pollution contribution than do areas intensively farmed or otherwise used by man.

From the discussion of the affinity of soils to adsorb pollutants, it can be deduced that most of the nutrient transport occurs only during storm events when large amounts of top soil are transported to the receiving water. The base flow and interflow pollution in rural areas can be attributed to mobile compounds such as nitrates and ions which have been leached out from soil as a result of dissolution and ion-exchange reactions. However, in some areas, the nitrate pollution of subsurface waters can be significant, and flows occurring during the dry base flow period because of a low concentration of nitrates can be deteriorating.

Pollution from Animal Feedlots

In early farming operations, small confined animal feedlots were situated in locations where natural drainage facilitated removal of the wastes. There was little concern as to the pollution potential of such operations. With advanced agricultural technology, large specialized animal feedlots are now common. In the Midwest and West, large animal feedlots produce animals for slaughter, and liquid and solid manure tends to accumulate in the feeding areas until cleanout and disposal. The feedlot pollution potential far exceeds that of other land use with the exception of urban-developing areas. Even permeable soils, due to deposited organics, become virtually impervious and greatly enhance the possibility of surface water pollution. The quantity and quality of feedlot runoff depend upon antecedent soil moisture, permeability, the number of cattle per feedlot area, the method of feedlot operation, the topography of the area, and the intensity of rainfall. The waste deposited on the feedlot surface is highly biodegradable, and its decay rate depends on temperature and moisture conditions. With increased moisture, more wastes are dissolved, thus increasing the soluble contamination of feedlot runoff. Since the biodegradation rate is proportional to temperature, less pollution is decomposed during the winter months and the pollution impact of winter storms and snow melt is higher than during the remainder of the year.

Feedlot runoff quality is not sensitive to the quantity of manure on the lot (Loehr, 1972). Once the feedlot surface is covered, the depth of manure is not an important parameter of water quality. The parameters that do affect the quality of feedlot runoff include rainfall intensity, antecedent water content of the manure pack, and type of feedlot surface (Miner *et al.*, 1966).

The concentration of pollutants in the runoff from agricultural animal feedlots is very high. The BOD_5 concentration of feedlot runoff has been reported to be in the range of 1,000 to 12,000 mg/liter.

As previously stated, runoff is a residual phenonemon, occurring after all losses (infiltration, depression storage) have been satisfied. Since animal feedlot surfaces do not have the same surface and permeability characteristics as undisturbed surfaces (the depression storage of animal feedlots being in the range of 0.55 to 0.90 cm), more than 90% of the resid-

TABLE 7.25. Pollution characteristics of manure from domestic animals

	Average Weight kg/animal	Wet Manure g/kg/day	NH_4–N g/kg/day	Total N g/kg/day	P_2O_5 g/kg/day	Total N kg/animal/yr	Phosphorus kg-P/animal/yr
Poultry	2	62	0.26	0.74	0.60	0.5	0.2
Ducks	2	—	—	8.0	0.8	5.8	0.35
Swine	125	74	0.24	0.51	0.43	23.0	8.0
Dairy cattle	450	84	—	0.23	0.34	38.0	25.0
Beef cattle	450	66	0.11	0.32	0.18	53.0	13.0
Sheep	50	72	—	0.60	0.25	11.0	2.0

[a]From Porcella *et al.* (1974).

ual net rain after the depression storage is subtracted will become runoff. The nutrient characteristics of feedlot runoff are similar to that of diluted liquid manure, as shown in Table 7.25.

Mine Drainage Water

Waters resulting from certain mining operations represent a special danger to receiving waters. Mining practices, such as those for coal, are considered quite hazardous from the standpoint of water quality impacts.

There are two different kinds of mining methods: underground or deep mining, and surface mining, including strip and open-pit mining. The principal pollution problems resulting from mining activities are erosion and acid drainage.

In strip-mining activities, bare land surfaces are exposed, and this results in high erosion yields. These can range up to 100,000 tons/km²/yr. Not only are large amounts of sediments expected, but the particles themselves may carry high concentrations of associated adsorbed metals, ore or coal residues, and chemicals associated with the mined materials (USDI, 1968b).

The acidity of coal mine drainage waters is of special concern. As water flows through a mine, it comes into contact with sulfur-bearing minerals, primary pyrites, and marcasites (chemical composition FeS_2). When pyrite is exposed to air and water, it is oxidized to ferrous sulfate ($FeSO_4$) and sulfuric acid (H_2SO_4). The reaction proceeds as:

$$2FeS_2 + 7O_2 + 2H_2O \rightleftharpoons 2FeSO_4 + 2H_2SO_4.$$

Flowing water then leaches away the ferrous sulfate and sulfuric acid, and the leachate is called acid mine water.

The ferrous sulfate can further oxidize to ferric sulfate [$Fe_2(SO_4)_3$], which hydrolyzes to form insoluble ferric hydroxide [$Fe(OH)_3$] and more sulfuric acid, as follows (Yeasted and Shane, 1976):

$$4FeSO_4 + 2H_2SO_4 + O_2 \rightleftharpoons 2Fe_2(SO_4)_3 + 2H_2O,$$
$$Fe(SO_4)_3 + 6H_2O \longrightarrow 2Fe(OH)_3 + 3H_2SO_4.$$

In addition to the acid and iron salts, various other constituents may be found in the mine drainage, such as sulfates of aluminum, calcium, magnesium, potassium, and sodium (Yeasted and Shane, 1976).

The acid mine drainage has lethal and sublethal effects on the biota of receiving streams including bacteria (Hackney and Bissonette, 1978). In addition, the streams affected by acid mine drainage exhibit a characteristic color (Cleary, 1967).

Control of pollution from mining must include both erosion control practices limiting loss of particulates from stripped lands or gob (mine refuse) piles, and control of acid mine drainage water.

Three factors are responsible for formation of acid drainage: water, air, and contact time of water with mine minerals. Thus, the control measures which have been recommended are aimed to limit the effect of these factors. They include (Yeasted and Shane, 1976; Cleary, 1967) reducing the entry of waters into mines through diversion of surface runoff and/or by sealing; minimizing the contact time of water in the mine with acid-forming minerals; equalizing the flow of water from the mines over a 24-hour period as opposed to the common practice of intermittent pumping of "slugs" of acid water; employing adequate mine-closure procedures immediately following termination of mining activities; expressing greater care in the disposal of gob and other mine refuse materials; flooding of abandoned surface mines; treating acid mine drainage by neutralizing chemicals; dilution of acid mine discharges in streams by low-flow augmentation; land reclamation and protection of strip mines and mine refuse piles.

Effects of Hydrologic Modifications

Hydrologic modifications are activities of man resulting in nonpoint source pollution that either directly or indirectly detrimentally affect the natural streamflow and associated water regime. It must be realized that almost any use or modification of land by man results in a hydrologic change of the water regime and, subsequently, results in a potential nonpoint source of pollution. Construction activities reduce permeability and surface storage; incorporation of impervious areas in a watershed increases its hydrological activity; drainage and irrigation systems in agricultural zones disrupt the groundwater regime, etc. These causes of, and effects on, nonpoint source of pollution have been previously discussed.

However, there remain other activities, mainly instream modifications, which disrupt the natural flow and result in pollution loadings similar to that from diffuse areal sources. Such activities include dredging, channelization, dam and impoundment construction and operation, and other in-water or close-to-the-stream activities.

Dredging   Dredging has a potential adverse impact on water quality because of the resuspension or redissolving of pollutants on the bottom of the waterway. This resuspension or redissolution occurs when the material is being lifted from the water or otherwise transported, and when it is being disposed of, either by dumping back into the water or onto the land.

Most current dredging activities in the United States are for the purpose of maintaining existing navigation waterways. Approximately 220 million m³ of bottom materials are dredged annually from rivers and estuaries because of navigation, and an additional 60 million m³ are dredged in the construction of navigation channels. These totals include 8 million m³ dredged annually from 115 lakes. Over 35,000 km of waterways have been modified for commercial navigation, and approximately 30,000 km of waterways and 1000 harbors are dredged each year in order to maintain waterborne commerce in the United States (Bhutani *et al.*, 1975). Figure 7.10 shows the disposition pathways of the dredged materials.

The nature and magnitude of pollution loadings introduced into surface waters by dredging depends on the character of the dredged materials. Until a few decades ago, dredging had no apparent environmental pollution problems. However, in recent years, the sediment accumulated on the bottoms of harbors and channels has become increasingly polluted as a result of point and nonpoint source pollution. As a consequence, dredged materials may range in quality from relatively clean sand and gravel to organic muck and sludge of natural or man-made origin. Examples of concentrations of pollutants in bottom samples are given in Table 7.26.

Channelization. Channel modifications are implemented primarily for flood control, erosion control, navigation, and drainage. There are seven

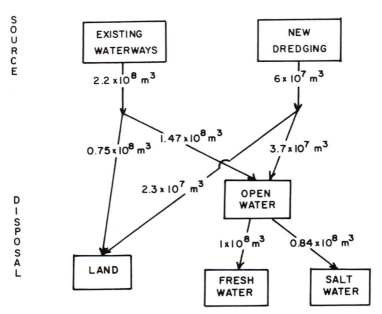

Fig. 7.10. Dredge Spoil disposal flow chart (Bhutani *et al.*, 1975; Boyd *et al.*, 1972).

TABLE 7.26. Chemical comparison of slightly and heavily polluted bottom samples[a]

Parameter	Units	Lightly Polluted		Heavily Polluted	
		Mean	Range	Mean	Range
Total volatile solids	%	2.9	0.7–5.0	19.6	10.2–49.3
Chemical oxygen demand	g/kg	21	3–48	177	39–395
Kjeldahl nitrogen	g/kg	0.55	0.01–1.31	2.64	0.58–6.80
Total phosphorus	g/kg	0.58	0.24–0.95	1.06	0.59–2.55
Grease and oil	g/kg	0.56	0.11–1.31	7.15	1.38–32.1
Initial oxygen demand	g/kg	0.50	0.08–1.24	2.07	0.28–4.65
Oxygen uptake	g/kg				
Sulfides[b]	g/kg	0.14	0.03–0.51	1.70	0.10–3.77

[a] From O'Neal and Sceva (1971).
[b] Values are conservative due to preservation method used.

different types of modifications that are potential sources of pollution (Whalen, 1977).

1. Clearing of debris and snagging of blockage operations to restore the former hydraulic capacity of a stream. These operations have the least nonpoint source pollution consequences.

2. Channel excavations which enlarge and restore an existing channel or which provide a new channel in its place. Heavy nonpoint pollution can result.

3. Channel realignment to eliminate meanders that have developed in the natural streambed. Heavy nonpoint pollution can result.

4. Construction of floodways to relieve the streambed of excessive flows of storm water. The floodways are normally dry and, if stabilized properly by vegetation, minimal pollution can result when flood flows subsequently enter the floodway.

5. Construction of flood retardation basins for the temporary storage of excess flows. These structures can be either in-stream or off-stream, with the latter ones having low pollution potential during construction.

6. Construction of debris and sediment retention basins to hold back pollution during periods of high water. The amount of nonpoint pollution is influenced by the amount of side disturbance.

7. Construction of drainage ditches or deepening of existing ditches. In-channel vegetation such as grassed waterways will limit nonpoint source pollution.

The major pollutant from channel modifications is sediment. Depending on the amount of organic matter present in the sediment (tree residuals, vegetation, etc.), decomposition materials may also be present.

In-water construction. This category includes pile driving or placement of piers into the bottom of a waterway, placement of bulkheads (vertical walls) in or adjacent to a water body, and placement of structural sections (such as pipes or tunnels) on the bottom of a water body.

The water quality effects of these activities are similar to those of dredging.

Dam construction and operation. An artificial dam or impoundment is a structural feature located on a stream which, in most cases, irreversibly changes its flow and quality regime. The most profound changes occur during the period of construction and immediately after the completion of the dam during the first flooding.

The effect is similar to dredging and other in-stream and off-stream construction activities. Heavy nonpoint source pollution can result if proper control measures are not implemented. Stream by-passes of the construction area plus control of runoff and drainage water from the construction are the most prevailing control techniques.

There are other activities that will result in nonpoint source pollution during the construction and postconstruction periods. Cutting of trees and removal of vegetation in the inundated area may somewhat improve water quality in the future lake, but it is also a significant source of pollution due to logging operations, exposing bare lands to storm action, and temporary transportation on unpaved roads.

Irreversible water quality changes caused by reservoirs include those caused by stratification, peak operation for hydropower production, accelerated eutrophication, density currents, and deposition of sediment in the reservoir. The effects of these phenomena on water quality are discussed elsewhere.

WASTEWATER CHARACTERIZATION

The previous sections of this chapter have described natural and man-caused processes resulting in potential adverse water quality changes. In order to adequately describe these changes, it is imperative that the contributing sources of contaminants be described in quantitative terms. Thus, the process of wastewater characterization is of the utmost importance.

Inasmuch as *Standard Methods* (APHA/AWWA/WPCF, 1975) describes individual analytical procedures in detail, they will not be repeated here. Instead, the methodology that should be utilized in wastewater characterization will be discussed, with emphasis on those aspects not delineated in *Standard Methods* (APHA/AWWA/WPCF, 1975).

The importance of adequate wastewater characterization becomes apparent when one attempts to model the effects of a wastewater discharge on the water quality of a river and in the design of a wastewater treatment plant. For example, one cannot predict the waste assimilation capacity of a receiving water without knowing the magnitude and characteristics of the waste load contributions. Neither can one design a wastewater treatment facility without knowledge of the materials to be treated.

The objectives of wastewater characterization have been aptly stated by Ford *et al.* (1970), as to: Provide pertinent information for the design of treatment processes; indicate waste streams with reuse or product recovery potential; serve as a basis for estimating treated effluent quality with respect to regulatory criteria; localize and identify waste streams containing refractory or potentially toxic substances.

The authors would add the objective of determining the impact of the discharges on the receiving water.

Parameters of Concern

In spite of its limitations, the biochemical oxygen demand (BOD) test is probably still the single most important parameter describing water pollution. Probably the second most important parameter is the solid content and its composition. In addition, pH, alkalinity, acidity, the nitrogen series, the phosphorus series, the heavy metals, specific organics, sodium, chloride, sulfate, oil and grease, and turbidity will probably be required for determination. An excellent format for the sequential analysis of a wastewater is shown in Fig. 7.11 (Eckenfelder and Ford, 1970). Obviously, the flow rates are quite significant and must be determined in conjunction with the parameters of concern. An excellent description of pertinent methodology for the desired flow information is contained in ORSANCO (1952).

Solids

Solids determination consists of suspended, settleable, dissolved, and total. Furthermore, it is important to know the organic and inorganic fraction of each component. For example, the biodegradable organic fraction of the solid materials discharged to a waterway will undergo oxidation and thus influence the D.O. concentration of the water. The solids found in a

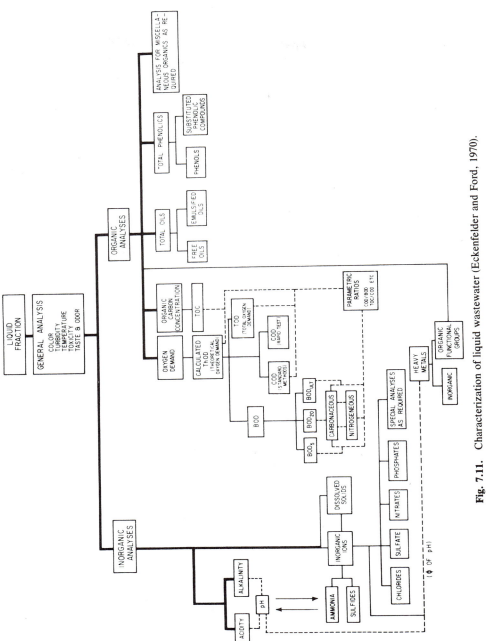

Fig. 7.11. Characterization of liquid wastewater (Eckenfelder and Ford, 1970).

typical medium strength sewage are shown in Fig. 7.12 (Metcalf and Eddy, 1972).

Nitrogen

The various forms of nitrogen existing in wastewaters each have different implications from the water quality standpoint, as elucidated in Chapter 3. Ammonia nitrogen consists of nitrogen existing as ammonia or ammonium ion and is determined by nesslerization or distillation. Organic nitrogen, which includes nitrogen associated with amino acids, amino proteins, amides, and so on, is determined by the Kjeldahl method. Total Kjeldahl nitrogen includes both organic nitrogen and ammonia nitrogen, and the Kjeldahl or organic nitrogen is determined by the difference. Nitrite nitrogen seldom exceeds 1 mg/liter and its concentration in ground and surface waters is usually much less than 0.1 mg/liter (Sawyer and McCarty, 1967). Because of its low concentration, colorimetric methods of analysis are utilized. Nitrate nitrogen analysis is most difficult, the procedures recommended are colorimetric and the "results obtained on natural samples can best be classified as semiquantitative" (Sawyer and McCarty, 1967).

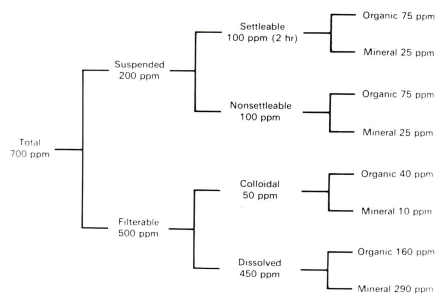

Fig. 7.12. Classification of solids Found in medium strength sewage (Metcalf and Eddy, 1972).

TABLE 7.27. Phosphorus compounds commonly encountered in sanitary engineering practice[a]

Name	Formula
Orthophosphates	
Trisodium phosphate	Na_3PO_4
Disodium phosphate	Na_2HPO_4
Monosodium phosphate	NaH_2PO_4
Diammonium phosphate	$(NH_4)_2HPO_4$
Polyphosphates	
Sodium hexametaphosphate	$Na_3(PO_3)_6$
Sodium tripolyphosphate	$NA_5P_3O_{10}$
Tetrasodium pyrophosphate	$Na_4P_2O_7$

[a]From Sawyer and McCarty (1967).

Phosphorus

As discussed in Chapter 13 on eutrophication, phosphorus is thought to be of major import in the eutrophication process. Phosphorus compounds commonly encountered in environmental engineering practice are shown in Table 7.27. It is important to note that all of the polyphosphates gradually hydrolyze in aqueous solution and revert to the ortho form (Sawyer and McCarty, 1967). The soluble orthophosphate fraction in natural waters is considered to be the biologically available form.

Oxygen Demand

Several tests are used to measure the oxygen-demanding characteristics of wastewaters, and each has advantages and disadvantages and measures different characteristics. The tests commonly used include the biochemical oxygen demand (BOD), the chemical oxygen demand (COD), total organic carbon (TOC), oxygen consumed (OC), and total oxygen demand (TOD). It is important for the water quality modeler to be familiar with each of these tests and to know their limitations and utility.

Before discussing the various measurements of oxygen demand, it is instructive to calculate the theoretical oxygen demand of an organic substance such as an amino acid. It is assumed that the oxidation is complete and occurs in two stages, carbonaceous, in which the amine is converted to ammonia, and carbonaceous plus nitrogenous, in which the amine is oxidized to nitrate. Using the method of $\frac{1}{2}$ reactions, the stoichiometry for norvalene is as follows:

1st Stage Carbonaceous oxygen demand

$$8H_2O + CH_3CH_2CH_2\overset{\overset{\displaystyle NH_2}{|}}{CH}\overset{\overset{\displaystyle O}{\|}}{-C}-OH \longrightarrow 5CO_2 + NH_3 + 24H^+ + 24e^-$$

$$6(4e^- + 4H^+ + O_2 \longrightarrow 2H_2O)$$

$$CH_3-CH_2-CH_2-\underset{\underset{\displaystyle NH_2}{|}}{CH}-\overset{\overset{\displaystyle O}{\|}}{C}-OH + 6O_2 \longrightarrow 5CO_2 + NH_3 + 4H_2O$$

Total carbonaceous plus nitrogenous oxygen demand

$$11H_2O + CH_3CH_2CH_2-\underset{\underset{\displaystyle NH_2}{|}}{CH}-\overset{\overset{\displaystyle O}{\|}}{C}-OH \longrightarrow 5CO_2 + NO_3^- + 33H^+ + 32e^-$$

$$8(4e^- + 4H^+ + O_2 + \longrightarrow 2H_2O)$$

$$CH_3CH_2CH_2\underset{\underset{\displaystyle NH_2}{|}}{CH}-\overset{\overset{\displaystyle O}{\|}}{C}-OH + 8O_2 \longrightarrow H^+ + 5H_2O + 5CO_2 + NO_3^-$$

One can easily calculate the oxygen required to oxidize a substance on a theoretical basis, and if a pure substance is completely oxidized chemically, the chemical oxygen demand (COD) will equal the theoretical oxygen demand (TOD). However, biochemical oxidation is not complete because some 10% of the original organic material usually will become a part of the nonbiodegradable residue and is thus not measured in the standard BOD test. Therefore, for these substrates Eckenfelder (1970) proposed that

$$COD = \frac{BOD_u}{0.9} = TOD,$$

where BOD_u is the ultimate biochemical oxygen demand.

BOD

The BOD test, which is still the most commonly used parameter for measuring the oxygen-depletion characteristics of wastewater, is a measure of the oxygen demand of biodegradable carbon. The methodology utilizes a laboratory technique quite similar to the natural biooxidation processes taking place in surface waters and in aerobic biological treatment units. The reaction occurs in two distinctive phases (Eckenfelder,

1970). Initially, the organic material present in the water sample is utilized by the seed microorganisms for energy and growth. This results in the utilization of oxygen and growth of new microorganisms. When the organics originally present in the sample are removed, the organisms present continue to use oxygen for autooxidation or endogenous respiration (metabolism) of their cellular mass. When the cell mass is completely oxidized, only a nonbiodegradable cellular residue remains and the reaction is complete, as shown in Fig. 7.11. The oxygen utilized is referred to as the ultimate carbonaceous BOD, BOD_u. It is assumed that nitrogenous material is not present in the hypothetical sample or in Fig. 7.13.

The removal and oxidation of the organics present in the sample is usually complete in 18 to 36 hours (phase 1), and the total oxidation of the cell mass takes at least 20 days. Obviously, the rate of reaction during the first phase is much faster than that during the second phase, as shown in Fig. 7.13.

The conventionally used bottle measurement consists of placing a known quantity of the wastewater, some microbiological "seed," and

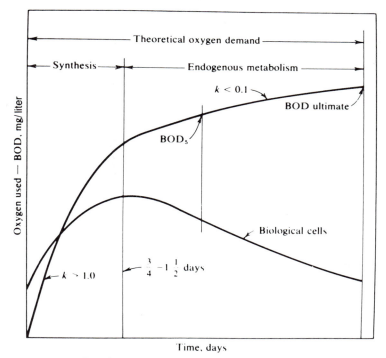

Fig. 7.13. Reactions occurring in the BOD bottle (Eckenfelder, 1970).

nutrients in well-aerated dilution water in one bottle and all but the wastewater in a second, identical bottle. The bottles, with a volume of approximately 300 ml, are placed in an incubator, with a water seal, at 20°C, for a period of five days. Both bottles are then analyzed for D.O., and the difference, accounting for dilution, comprises the standard five-day 20°C BOD. For long-term tests designed to determine the ultimate carbonaceous first-stage BOD, sufficient samples and "blanks" are prepared in order to plot the removal of oxygen each day for a designated time period.

It is assumed that when D.O. is present, the rate of oxidation is independent of the D.O. concentration. Obviously, the type and number of microorganisms (seed) present is important. The changes in D.O. can thus be related to the quantity and character of the oxidizable organic material present, and the slope of the curve is a function of the rate of oxidation. Furthermore, the oxygen demand per unit time is proportional to the amount of oxidizable material present (or remaining). The typical curve obtained is shown in Fig. 7.14.

If it is assumed that the reaction is first order, that $L = $ BOD, $L_a = $ ultimate first-stage carbonaceous BOD, $y = $ oxygen demand satisfied at time t, and $L_t = $ oxygen demand remaining at time t, then

$$\frac{dL}{dt} = -kL \quad \text{or} \quad \int_{L_a}^{L_t} \frac{dL}{L} = -\int_0^t k \, dt \tag{7.13}$$

$$\ln L \Big]_{L_a}^{L_t} = -kt \Big]_0^t ; \ln \frac{L_t}{L_a} = -kt$$

$$L_t = L_a \, e^{-kt} \tag{7.14}$$

but,
$$L_t = L_a - y$$

∴
$$L_a - y = L_a e^{-kt} \quad \text{and} \quad y = L_a(1 - e^{-kt})$$

or
$$y = L_a(1 - 10^{-k_1 t}),$$

where the k values represent the average reaction rate coefficient, k being to log base e and k_1 log base 10. It is important to note that much of the literature utilizes log base 10, while most computer models utilize log base e ($k = 2.304k_1$).

Inasmuch as the reaction-rate coefficient represents an overall reaction rate, which is an average of the first and second phases of the carbonaceous BOD reaction, the mean k may vary markedly depending on the quantity and nature of organics present in the sample. Typical values of the mean-rate coefficients are summarized in Table 7.28.

The BOD curve actually yields the oxygen consumed during the reac-

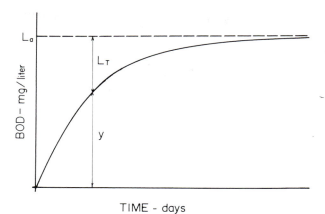

Fig. 7.14. Long-Term carbonaceous BOD results.

tion. Therefore, both the ultimate first-stage carbonaceous BOD and the reaction velocity coefficient are unknown and must be computed or determined graphically. A comprehensive description of the many methods available for determining these parameters is presented in Gaudy *et al.* (1967). Only two methods, thought by the authors to be reasonably accurate and simple, are presented here.

Log-difference Method. Differentiating the BOD reaction equation will yield

$$\frac{dy}{dt} = r = L_0 \, K \, e^{-Kt}, \tag{7.15}$$

in which *r* is the rate of oxygen utilization with time. Equation (7.15) can

TABLE 7.28. Average BOD rate coefficients, *k*, at 20 °C

Substance	$k_{\text{base } 10}$
Untreated sewage[a]	0.15–0.28
High-rate filters and anaerobic contact[a]	0.12–0.22
High-degree biological treatment effluents	0.06–0.10
Rivers with high pollution	0.12–0.3
Rivers with low pollution	0.04–0.10

[a]From Eckenfelder (1970).

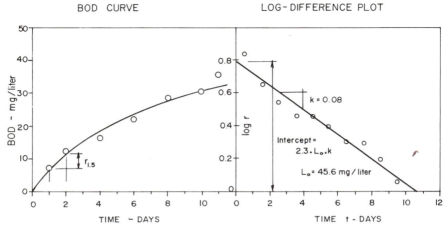

Fig. 7.15.   BOD curve and log-difference method.

be plotted on a semilogarithmic plot in this form (Fig. 7.15):

$$\log_{10} r = \log_{10}(2.3L_0k) - kt. \tag{7.16}$$

The values of the oxygen consumption must be known at daily intervals. It is convenient to read them from a smooth curve drawn through the experimental data points which will also minimize error.

Graphical Method (Thomas, 1950).   BOD determinations are made for 1, 2, 3, 4 . . . days, and $(t/y)^{1/3}$ is calculated for each day. $(t/y)^{1/3}$ is then plotted v. time and A and B are determined as shown in Fig. 7.16.

The values of L_a and k_1 are then calculated using the following equations:

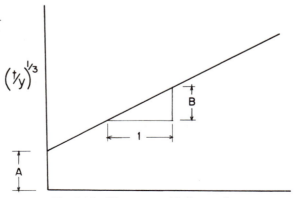

Fig. 7.16.   Thomas graphical procedure

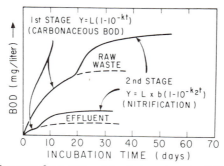

Fig. 7.17. BOD curves for a raw and treated wastewater (Malina *et al.*, 1967).

$$A = \left(\frac{1}{2.3k_1 L_a}\right)^{1/3}, \tag{7.17}$$

$$B = \frac{(2.3k_1)^{2/3}}{(6L)^{1/3}}, \tag{7.18}$$

$$k_1 = \frac{2.61B}{A}, \tag{7.19}$$

$$L_a = \frac{1}{2.3K_1 A^3}. \tag{7.20}$$

It is convenient to know the relationship between BOD_5 and BOD_u, which depends on the reaction rate and the relative stability of the substrate. For example, a raw wastewater will demonstrate a much higher rate constant than a treated wastewater, as shown in Fig. 7.17 (Malina *et al.*, 1967). The ratio of BOD_5/BOD_u and its relationship to K_{10} is shown in Fig. 7.18 (Malina *et al.*, 1967).

Nitrification. If nitrogenous material is present in the wastewater, continuation of the long-term BOD test will demonstrate another oxygen demand, the second-stage nitrogenous BOD, as shown in Fig. 7.19. The nitrifying microorganisms are autotrophic and therefore utilize the reaction products as their source of energy. Hence, their growth rate is much slower than the growth rate of the heterotrophic organisms participating in the carbonaceous BOD reaction. Therefore, with the presence of unoxidized organic carbonaceous BOD, nitrification is suppressed and will not normally begin until the carbonaceous demand is satisfied, resulting in a curve similar to that shown in Fig. 7.19.

Under some circumstances, these two oxidations can proceed simultaneously, and the resultant BOD curve will be a composite of the two reac-

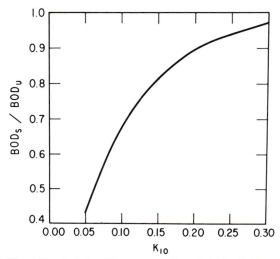

Fig. 7.18.   Relationship between K_{10} and BOD_s/BOD_u.

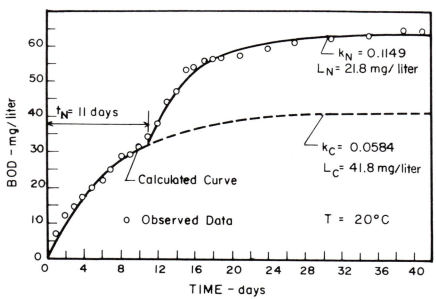

Fig. 7.19.   Observed BOD data and calculated Carbonaceous and nitrogenous deoxygenation curves (Zanoni, 1969).

tions. Mathematically the reactions can be described by this equation (Eckenfelder, 1970):

$$y = La(1 - 10^{-k_1 t}) + b(1 - 10^{-k\ t}), \tag{7.21}$$

where a = high-rate constant for carbonaceous BOD, b = low-rate constant for nitrogenous BOD, k_1 = rate constant for carbonaceous BOD, and k_n = velocity constant for nitrogenous BOD.

Nitrification occurs most often in effluents which have undergone partial oxidation of the waste components. Nitrification represents a demand on the oxygen resources of the receiving stream; therefore, it should be recognized as part of the total demand of the waste. It is not desirable in the five-day BOD test and can be eliminated by pasteurizing and reseeding the sample or by the addition of methylene blue. The rate of nitrification can be determined by a parallel set of BOD samples.

Factors Affecting the BOD. It is important to know the effect(s) of certain variables on the BOD in order that proper interpretation of the results may be made. Environmental factors may significantly affect BOD results, both in the laboratory and in the receiving water.

Temperature significantly affects the reaction rate of most biochemical reactions. It may also slightly affect the ultimate BOD value because oxidizability increases with temperature. The conventional relationship, derived from van't Hoff-Arrhenius' law, is

$$k_T = k_{20}\theta(T - 20). \tag{7.22}$$

The thermal factor, θ, is not uniform over the common temperature ranges. According to Gotaas (1948) and Zanoni (1969), it is higher at lower temperatures and decreases with higher temperatures as shown in Fig. 7.20. The commonly accepted value of $\theta = 1.047$ was proposed by Streeter and Phelps (1925). For the nitrogenous stage, Zanoni (1969) found that $\theta = 1.097$ for 10°C to 22°C, and $\theta = 0.877$ for 22°C to 30°C. It is also interesting to note that Theriault (quoted in Phelps, 1944) found the BOD_u to increase with temperature, while Gotaas (1948) demonstrated that BOD_u did not vary but the process took longer to reach the ultimate at lower temperatures.

pH The organisms which degrade the organic matter are acclimated to a narrow pH range, usually between pH 6.5–8.3. Figure 7.21 illustrates the five-day BOD values obtained from pH 4.5–9.0. Adjustment of the pH to 7.2 is necessary before reliable BOD values can be obtained on wastes which are outside the range of pH 6.5–8.3, inasmuch as the buffer solution in the standard dilution water (APHA/AWWA/WPCF, 1975) will not accommodate extreme pH values.

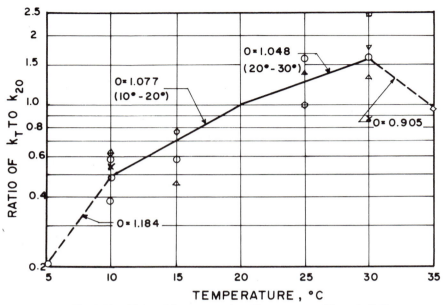

Fig. 7.20. Thermal factor at various Temperatures (Zanoni, 1969).

Essential mineral nutrients Bacteria require inorganic as well as organic nutrients for optimum metabolism. In the standard BOD test, inorganic nutrients are added to the dilution water. Nitrogen and phosphorus are two nutrients which markedly affect the results of this analysis, as shown in Fig. 7.22. Some industrial wastes lack these nutrients, and the addition of nitrogen and phosphorus significantly increases the

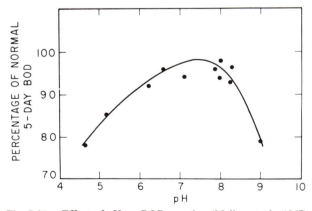

Fig. 7.21. Effect of pH on BOD reaction (Malina *et al.,* 1967).

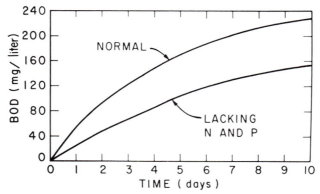

Fig. 7.22. Effects of nutrients on BOD (Malina *et al.*, 1967).

BOD. The dilution water specified for use in the standard BOD test is designed to provide the essential mineral nutrients, including nitrogen and phosphorus, at concentrations similar to those commonly found in natural streams. Since the mineral content of different streams varies markedly, a more exact estimate of the BOD of a wastewater is obtained by using water from the receiving stream as dilution water.

Microbiological population A heterogeneous and acclimatized microbial population is essential to precisely estimate the BOD of a wastewater. In most cases the streams receiving industrial wastes have adapted to the new materials through the development of acclimated populations. With certain exceptions, the organisms which may be found in the stream at some distance below a particular industrial outfall are those which are acclimated to the waste.

It is essential that the organisms present in the sample bottle during incubation be similar in type and number to those which will oxidize the waste in the receiving stream. Many industrial wastes must be "seeded" with acclimated organisms before incubation in the BOD test.

The acclimated organisms present below industrial outfalls provide an excellent source of seed for the BOD determination. With few exceptions, carefully selected seed from the receiving stream will yield the highest BOD values. When stream water is used for seed, nitrification difficulties may increase. Occasionally, it is necessary to artificially develop a microbial culture which will oxidize the industrial waste. An acclimatized heterogenous microbial culture may be developed by starting with settled domestic wastewater containing a large variety of organisms to which a small amount of the industrial effluent is added. The amount of waste added is increased until a culture develops which is adapted to the waste.

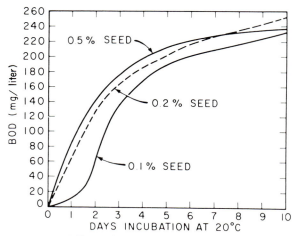

Fig. 7.23.   Effects of seed on BOD (Malina *et al.*, 1967).

The mixture of domestic wastewater and the industrial waste are aerated by bubbling air continuously through the liquid. A noticeable increase in the cloudiness or turbidity of the aerating mixture generally indicates an acclimated culture. If a D.O. probe is available, the oxygen uptake can be evaluated daily to determine when an acclimated culture has developed.

The amount of seed required to produce a normal rate of oxidation must be determined experimentally. The most frequent error is the use of insufficient seed. Figure 7.23 illustrates the effects of seed concentration on the BOD.

Large numbers of algae in stream water used for dilution water may produce significant changes in the oxygen content. When stream samples containing algae are incubated in the dark in the laboratory, the algae survive for a time. Short-term BOD determinations may show the influence of oxygen production by the algae. After prolonged lack of light, the algae die and the algal cells contribute to the total organic content of the sample and increase the BOD; therefore, samples incubated in the dark may not be representative of the deoxygenation process in the stream, since the benefits of photosynthesis are lacking. On the other hand, samples incubated in the light, under conditions of continual photosynthesis, will yield low BOD values. The influence of algae in the BOD test is difficult to evaluate, and extreme care should be taken when stream water which contains a large number of algae is used for dilution water.

Toxicity.   Various chemical elements and compounds are toxic to microorganisms. At high concentrations, some substances will kill the microbes, and at sublethal concentrations, the activity of microbes may be reduced. The effects of cyanide and some heavy metals on the BOD are illustrated in Fig. 7.24. Particular care must be used in avoiding lethal or sublethal concentration of copper in the BOD dilution water, which may originate from the water supply or the "still."

Manometric BOD determination. A technique that is quite useful in laboratory studies involves the direct measurement of oxygen uptake by use of what is called a Warburg apparatus. The process involves placing the sample, seed, and standard dilution water into a small flask which is agitated in a constant temperature bath as shown in Fig. 7.25 (Malina *et al.*, 1967). A vial in the center of the flask contains KOH which absorbs the CO_2 produced by the bacteria. The prime advantage of the test is that the manometer, which is connected directly to the sample flask, can be cali-

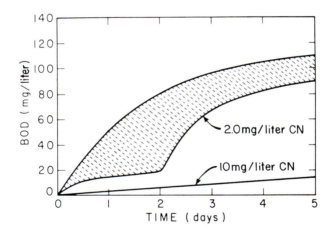

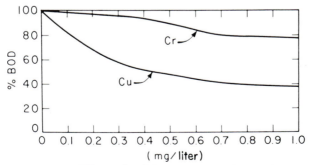

Fig. 7.24. Effects of toxicity on BOD (Malina *et al.*, 1967).

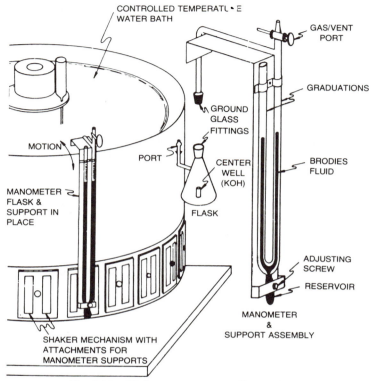

Fig. 7.25. Warburg apparatus (Malina *et al.*, 1967).

brated to read oxygen uptake. Furthermore, as many as 18 dilutions can be prepared at the same time and readings can commence within a few hours. It should be noted that the Warburg apparatus is not used for routine analysis.

The Warburg apparatus is quite expensive and requires a competent technician for its operation. However, the advantages are also contained in a relatively inexpensive manometric apparatus manufactured by Hach, as shown in Fig. 7.26. This unit, which operates on the same principle as the Warburg, can be placed directly in the incubator. The real utility of the Hach unit includes the ability to quickly determine the correct dilution, toxicity effects, and correct amount of seed. It should be noted that the manometric techniques yield a higher rate coefficient and a higher BOD value, which could be attributed to the removal of CO_2 by the KOH and the agitation, neither of which occur in the standard bottle BOD test.

Fig. 7.26. Hach BOD apparatus.

COD Test

The chemical oxygen demand is determined by a "wet" laboratory procedure employing potassium dichromate as an oxidizing agent. Some organic components are not oxidized by this method and are not included in the COD value. These organics include benzene, pyridine, toluene, and some others. On the other hand, the potassium dichromate will oxidize some nonorganics including ferrous ion, sulfides, and secondary manganese. The oxygen consumed test (OC) uses potassium permanganate as an oxidizing agent instead of potassium dichromate and is principally used in Europe.

TOD

The total oxygen demand is based on the measurement of the oxygen loss during the oxidation process which takes place at 900°C with a platinum catalyst in a laboratory analyzer. The reduction of oxygen is measured by a silver–lead fuel cell detector. In this process the total oxygen demand is measured without the shortcomings typical of the COD analysis. The results can be obtained in minutes. The TOD test also measures nitrogenous demand.

Total Organic Carbon

The TOC analysis is a measure of the amount of organic carbon in the sample and, theoretically, can be related to the oxygen demand. In the TOC analysis, the carbonaceous material is oxidized at 950°C, and the CO_2 produced is then measured. The inorganic carbon present in the sample is either removed by pretreatment or analyzed separately and then subtracted. The method is rapid and relatively simple; however, the instrument is expensive and requires an experienced technician.

For details of these and other measures of oxygen demand, the reader is referred to *Standard Methods* (APHA/AWWA/WPCF, 1975) and Eckenfelder and Ford (1970). A comparison of the techniques described is given in Table 7.29, and the relationship between oxygen and carbon parameters is shown in Fig. 7.27 (Ford *et al.,* 1970).

Relationships Among BOD, COD, and TOC

When considering routine plant control or investigational programs, the BOD is not a useful test because of the long incubation time required to obtain meaningful results. It is therefore worthwhile to attempt to develop correlations among BOD, COD, and TOC.

In attempting to correlate BOD or COD of a wastewater with TOC, one should recognize those factors which may negate the correlation. These include the following (Ford, 1968):

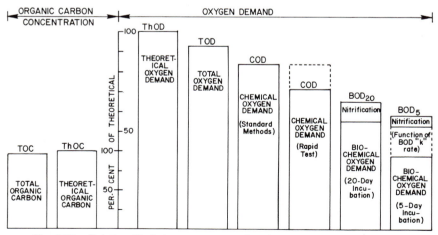

Fig. 7.27.   Relationship between oxygen and carbon parameters (Ford, 1968).

TABLE 7.29. Comparison of oxygen demand measurement techniques

	BOD Bottle	BOD Hach	BOD Warburg	TOC	COD	TOD
Equipment	Bottles, Incubator	Proprietary, Incubator	Constant Temp. Bath, Proprietary	Proprietary	Heaters, glassware	Proprietary
Time	5 days	1–5 days	1–5 days	Minutes	Hours	Minutes
Type measurement	Titration @ intervals	Continuous manometric	Continuous manometric	Infrared CO_2	Chemical oxidation	O_2 detector
Static or dynamic	Static[a]	Dynamic	Dynamic	N/A	N/A	N/A
CO_2 absorbed by KOH	No	Yes	Yes	N/A	N/A	N/A
Cost (approx.)	$150	$300	$4000	$8000	$500	$8000
Difference in results	Lower k_1, L_a	Higher k_1, L_a	Higher k_1, L_a	N/A	N/A	N/A
Biodegradable organics	directly	directly	directly	indirectly	indirectly	indirectly
Kinetics	yes[a]	yes	yes	no	no	no
Nitrogenous demand	yes[b]	yes[b]	yes[b]	no	no	yes

[a]Several bottles incubated for different periods of time are necessary for kinetics determination.
[b]Long term (over 20 to 40 days) period of analysis will reveal nitrogenous demand.

259

TABLE 7.30.   Relationship between COD and TOC for organic compounds[a]

Substance	COD/TOC (Calculated)	COD/TOC (Measured)
Acetone	3.56	2.44
Ethanol	4.00	3.35
Phenol	3.12	2.96
Benzene	3.34	0.84
Pyridine	3.33	Nil
Salicylic acid	2.86	2.83
Methanol	4.00	3.89
Benzoic acid	2.86	2.90
Sucrose	2.67	2.44

[a] After Ford (1968).

(a)   A portion of the COD of many industrial wastes is attributed to the dichromate oxidation of ferrous iron, nitrogen, sulfites, sulfides, and other oxygen-consuming inorganics.

(b)   The BOD and COD tests do not include many organic compounds that are partially or totally resistant to biochemical or dichromate oxidation. However, all the organic carbon in these compounds is recovered in the TOC analysis.

(c)   The BOD test is susceptible to variables that include seed acclimation, dilution, temperature, pH, and toxic substances.

One would expect the stoichiometric COD/TOC ratio of a wastewater to approximate the molecular ratio of oxygen to carbon ($32/12 = 2.66$). Theoretically the ratio limits would range from zero, when the organic material is resistant to dichromate oxidation, to 5.33 for methane or slightly higher when inorganic reducing agents are present (Ford, 1968).

TABLE 7.31.   Industrial waste oxygen demand and organic carbon

Type of Waste	BOD$_5$ (mg/liter)	COD (mg/liter)	TOC (mg/liter)	BOD/TOC	COD/TOC
Chemical[a]	—	4,260	640	—	6.65
Chemical	24,000	41,300	9,500	2.53	4.35
Chemical	850	1,900	580	1.47	3.28
Chemical	700	1,400	450	1.55	3.12
Olefine processing	—	321	133	—	2.40
Chemical	—	350,000	160,000	—	2.19
Synthetic rubber	—	192	110	—	1.75

[a] High concentration of sulfides and thiosulfates.

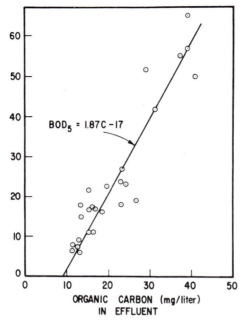

Fig. 7.28. TOC–BOD relationship

Reported BOD, COD, and TOC values for several organic compounds are given in Table 7.30, and industrial wastewaters are listed in Table 7.31. The COD/TOC ratio varies from 1.75 to 6.65 (Ford, 1968).

It has been difficult to correlate BOD with TOC for industrial wastes; however, relatively good correlation (Fig. 7.28) has been obtained for domestic wastewaters, probably because of the waste constituents. A five-day BOD–TOC ratio of 1.87 for sewage has been reported by Wurhmann (1964), and Mohlman and Edwards (1931) have reported a range of 1.35–2.62 for raw domestic waste. The calculated relationship between five-day BOD and TOC is (Ford, 1968)

$$\frac{\text{BOD}_5}{\text{TOC}} = \frac{\text{O}_2}{\text{C}} = \frac{32}{12} (0.90)(0.77) = 1.85,$$

where the ultimate BOD will exert approximately 90% of the theoretical oxygen demand, and the five-day BOD is 77% of the ultimate BOD for domestic wastes.

A decrease in the COD/TOC and BOD_5/COD ratios has been observed during the biological oxidation of both municipal and industrial wastewaters, as shown in Fig. 7.29. This can be attributed to (Ford, 1968):

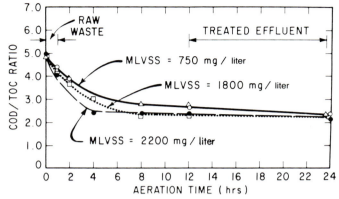

Fig. 7.29.   COD/TOC ratio at various stages of biological oxidation (Ford, 1968).

(a)   The presence of inorganic reducing substances that would be oxidized in the biological process, thereby reducing the COD/TOC ratio.

(b)   Intermediate compounds which may be formed during the biological process without significant conversion of organic matter to carbon dioxide. A reduction in COD may not be accompanied by a reduction in TOC.

(c)   The BOD reaction-rate constant k which will be greater than 0.15 in the raw waste and less than 0.1 in the treated effluent. The BOD_5/BOD_u and hence the BOD_5/COD ratio depends on this rate. This is at least in part responsible for the reduction in the BOD_5/COD or BOD_5/TOC during biological oxidation.

(d)   The concentration of nonremovable refractory materials which will account for a larger portion of the COD in the effluent than in the raw waste, thereby lowering the BOD_5/COD or the BOD_5/TOC ratio.

While the factors influencing the BOD, COD, and TOC have been observed in biological wastewater treatment plants, similar conclusions can be made in the receiving water, inasmuch as the processes causing biological degradation are similar.

REFERENCES

Alexander, G. C., and Stevens, R. J. (1976). Per capita phosphorus loading from domestic sewage. *Water Res.* **10,** 757–764.

American Public Health Association, American Waterworks Association, and Water Pollution Control Federation (1975). "Standard Methods for the Examination of Water and Wastewater," 14ed. Washington, D.C.

American Public Works Association (1969). Water pollution aspects of urban runoff. *Water Pollut. Control Res. Ser.* WP-20-15. Chicago, Illinois.

American Society of Civil Engineers (1976). Sedimentation engineering. ASCE Manual and Rep. on Eng. Practice, No. 54. New York.

Amy, G., Pitt, R., Singh, R., Bradford, W. L., and LaGraff, M. B. (1974). Water quality management planning for urban runoff. USEPA Rep. No. 44019-75-004, Washington, D.C.

Beasley, R. P. (1972). "Erosion and Sediment Pollution Control." Iowa State Univ. Press, Ames.

Bhutani, J., Holberger, R., Spewak, P., Jacobsen, W. E., and Truett, J. D. (1975). Impact of hydrologic modifications on water quality. EPA-600/2-75-007, Washington, D.C.

Black, Crow & Eidsness, Inc. (1971). Storm and combined sewer pollution sources and abatement. USEPA, Washington, D.C.

Boyd, M. B., Saucier, R. T., Kelley, J. W., Montgomery, R. L., Drown, R. D., Mathis, D. B., and Juice, C. J. (1972). Disposal of dredge spoil. Tech. Rep. H-72-8, U.S. Army Corps of Eng. Waterways Exp. Station, Vicksburg, Mississippi.

Brant, G. H. (1972). An economic analysis of erosion and sediment control methods for watersheds undergoing urbanization. Final Rep. No. 14-31-001-3392, Dow Chemical Co., Midland, Michigan.

263

Brink, N. (1975). Water pollution from agriculture. *J. Water Pollut. Control Fed.* **47,** 789–795.

Carlisle, A. A., Brown, H. F., and White, E. J. (1966). Litter fall leaf production and the effect of defoliation by Tortrix viridiana in a sissile oak Quercus petrala (querais petrala) woodland. *J. of Ecol.* **54,** 65–85.

Chapin, J. D., and Uttermark, P. D., (1973). Atmospheric contributions of nitrogen and phosphorus. Tech. Rep. WIS-WRC-73-2, Univ. of Wisconsin Water Resour. Center, Madison.

Cleary, E. J. (1967). "The ORSANCO Story—Water Quality Management in the Ohio Valley under an Interstate Compact." Johns Hopkins Press, Baltimore, Maryland.

Colson, N. V., and Tafuri, A. N., (1975). Urban land runoff considerations. *In* "Urbanization and Water Quality Control" (W. Whipple, ed.). Am. Water Resour. Assoc., Minneapolis, Minnesota.

Eckenfelder, W. W. Jr. (1970). "Water Quality Engineering for Practicing Engineers," Barnes & Noble, New York.

Eckenfelder, W. W., and Ford, D. L. (1970). "Water Pollution Control." Pemberton Press, Jenkins Publ., Austin, Texas.

Fair, G. M., and Geyer, J. C. (1954). "Water Supply and Wastewater Disposal." Wiley, New York.

Ford, D. L. (1968). Application of the total carbon analyzer for industrial wastewater evaluation. 23rd Purdue Ind. Wastes Conf., Lafayette, Indiana.

Ford, D. L., Eller, J. M., and Gloyna, E. F. (1970). Analytical parameters of petrochemical and refinery wastewaters. *Am. Chem. Soc., Div. of Petro. Chem., Prepr.,* 159th Meeting, Houston, Texas.

Foster, G. R., Mayer, L. D., and Onstad, C. A. (1973). Erosion equation derived from modeling principles. Pap. No. 73-2550, Winter Meeting Amer. Soc. Agri. Eng., Chicago, Illinois.

Gaudy, A. F., Komolrit, K., Follett, R. H., Kincannon, D. F., and Modesitt, D. E. (1967). Methods for evaluating the first order constants K_1 and L for BOD exertion. *Bioenvironmental Engr.,* Oklahoma State Univ., Stillwater (May).

Goodman, B., and Foster, J. W. (1969). Notes on activated sludge, 2nd ed. Smith & Loveless Co., Lenexa, Kansas.

Gotaas, H. B. (1948). Effect of temperature on biochemical oxidation of sewage. *Sewage Works J.* **20**(3), 441.

Graham, D. H., Costello, L. S., and Mallon, A. J. (1974). Estimation of imperviousness and specific curb lengths for forecasting stormwater quality and quantity. *J. Water Pollut. Control Fed.* **46**(4), 717–725.

Hackney, C. R., and Bissonnette, G. K. (1978). Recovery of indicator bacteria in acid mine streams. *J. Water Pollut. Control Fed.* **50**(4), 775–780.

Heaney, J. P., and Huber, W. C. (1973). Storm water management model: Refinements, testing, and decisionmaking. Dept. of Environ. Eng. Sci., Univ. of Florida, Gainesville.

Hydrocomp International (1972). Hydrocomp simulation programming operation manual. Palo Alto, California.

Lager, J. A., and Smith, W. G. (1974). Urban stormwater management and technology. USEPA Rep. No. 670/2-74-040, Washington, D.C.

Loehr, R. C. (1972). Agricultural runoff—Characteristics and Control. *J. Sanit. Eng. Div., Am. Soc. Div. Eng.* **96,** 909–924.

Lutz, H. J., and Chandler, R. I. (1976). "Forest Soils." Wiley, New York.

Malina, J. F., Ford, D. L., Eckenfelder, W. W., and Davis, E. (1967). Analytical procedures and methods, vol. I. Poland 26 Project, World Health Org., Univ. of Texas, Austin.

McElroy, A. D., Chiu, S. Y., Nebgen, S. W., Aleti, A., and Bennel, F. W. (1976). Loading functions for assessment of water pollution from nonpoint sources. EPA-600/2-76/151, USEPA, Washington, D.C.

Meinholz, T. L., Hansen, C. A., and Novotny, V. (1974). An application of the storm water management model. Proc. Natl. Symp. on Urban Runoff, Univ. of Kentucky, Lexington.

Metcalf & Eddy, Inc. (1972). "Wastewater Engineering: Collection, Treatment, Disposal." McGraw-Hill, New York.

Milwaukee County Department of Air Pollution Control (1970). Ambient air quality (particulates and sulfur oxides) in Milwaukee County. Milwaukee, Wisconsin.

Miner, J. R., Lipper, R. I., Fira, L. R., and Fink, J. W. (1966). Cattle feedlot runoff—its nature and variations. *J. Water Pollut. Control Fed.* **38**, 1582–1591.

Mohlman, F. W., and Edwards, G. P. (1931). Industrial engineering chemistry. *Anal. Ed.* **3**, 119.

Ohio River Valley Water Sanitation Commission (1952). Planning and making industrial waste surveys. Cincinnatti, Ohio.

O'Neal, G., and Sceva, J. (1971). The effects of dredging on water quality in the Northwest. USEPA, Office of Water Programs, Region X, Seattle, Washington.

Phelps, E. B. (1944). "Stream Sanitation." Wiley, New York.

Pitt, R., and Amy, G. (1973). Toxic material analysis of street surface contaminants. USEPA Rep. No. R2-73-283, Washington, D.C.

Porcella, D. B., Bishop, A. B., Anderson, J. C., Asplund, O. W., Crawford, A. B., Greeney, W. J. Jenkins, D. I. Jurinek, J. J., Lewis, W. D., Middlebrooks, E. J., and Walkinshaw, R. W. (1974). Comprehensive management of phosphorus water pollution. USEPA Rep. No. 600/5-74-010, Washington, D.C.

Ports, M. (1973). Use of the universal soil loss equation as a design standard. Am. Soc. Civ. Eng. Water Resour. Eng. Meeting, Washington, D.C.

Ports, M. (1975). Urban sediment control design criteria and procedures. Winter Meeting Amer. Soc. Agr. Eng., Chicago, Illinois.

Roehl, J. W. (1962). Sediment source areas, delivery ratios and influencing morphological factors. Publ. No. 59, Intl. Assoc. of Scientific Hydrology Commission of Land Erosion, 202–203.

Sartor, J. D., and Boyd, G. B. (1972). Water pollution aspects of street surface contamination. USEPA Rep. No. R2-72-081, Washington, D.C.

Sartor, J. D., Boyd, G. B., and Agardy, F. J. (1974). Water pollution aspects of street surface contamination. *J. Water Pollut. Control Fed.* **46**, 458–467.

Sawyer, C. N., and McCarty, P. L. (1967). "Chemistry for Sanitary Engineers," 2nd ed. McGraw-Hill, New York.

Shacklette, H. T., Hamilton, J. C., Boernagen, J. G., and Bowles, J. M. (1971). Elemental composition of surficial materials in the conterminous United States. *U.S. Geol. Surv. Prof. Pap.* 574-D, Washington, D.C.

Shaheen, D. (1975). Contributions of roadway usage to water pollution. Rep. by Biospheric, Inc., USEPA, Washington, D.C.

Streeter, H. W., and Phelps, E. B. (1925). A study of the pollution and natural purification of the Ohio River. *Public Health Bull.* No. 146, U.S. Govt. Printing Office, Washington, D.C.

Thomas, H. A. (1950). Graphical determination of BOD curve constants. *Water & Sewage Works* **97**, 123.

U.S. Army Corps of Engineers (1975). Urban storm water runoff: STORM. Hydrol. Eng. Center, Davis, California.

U.S. Dept. of Interior (1968a). ''The Cost of Clean Water,'' Vol. II, Industrial Waste Profile, No. 4, FWPCA, Washington, D.C.

U.S. Dept. of Interior (1968b). Stream pollution by coal mine drainage: Upper Ohio River basin. Ohio River Basin Proj. FWPCA, Washington, D.C. (March).

U.S. Environmental Protection Agency (1971). Storm water management model, vols. I–IV. Rep. Nos. 11024DOC07/71–11024DOC10/71, Washington, D.C.

U.S. Environmental Protection Agency (1973). Methods for identifying and evaluating the nature and extent of nonpoint sources of pollutants. Washington, D.C.

Uttermark, P. D., Chapin, J. D., and Green, K. M. (1974). Estimating nutrient loadings of lakes from nonpoint sources. USEPA Rep. 660/3-74-020, Univ. of Wisconsin Water Resour. Cent., Madison.

Water Pollution Control Federation (1970). Operation of wastewater treatment plants. Practice Manual No. 11, Washington, D.C.

Whalen, N. A. (1977). Nonpoint source control guidance–hydrologic modifications. Tech. Guid. Memo. TECG-29, USEPA, Water Planning Div., Nonpoint Sources Br., Washington, D.C.

Williams, J. R. (1972). Sediment yield computed with universal equation using runoff energy factor. Pap. Pres. at Interagency Sediment Yield Conf., USDA Sed. Lab., Oxford, Mississippi.

Williams, J. R., and Berndt, H. D. (1972). Sediment yield computer with universal equation. *J. Hydraul. Div., Am. Soc. Civ. Eng.* **98**, 2087–2098.

Wischmeier, W. H., and Smith, D. D. (1960). A universal soil-loss equation to guide conservation farm planning. *Proc. 7th Int. Congr. Soil Sci.,* Madison, Wisconsin.

Wischmeier, W. H., and Smith, D. D. (1965). Predicting rainfall-erosion losses from cropland east of the Rocky Mountains. Agric. Handbook 282, U.S. Dept. of Agric., Washington, D.C.

Wischmeier, W. H., Johnson, C. B., and Cross, B. V. (1971). A soil erodibility nomograph for farmland and construction sites. *J. Soil Water Conserv.* **26**, 189–193.

Wuhrmann, K. (1964). ''Hauptwirkungen und Wechselwirkungen einiger Betriebsparameter im Belebschlamm-system Ergebnissemehrjahriger.'' Grossversuche Verlag, Zurich, Switzerland.

Yeasted, J. C., and Shane, R. (1976). pH profiles in a river system with multiple acid loads. *J. Water Pollut. Control Fed.* **48**(1), 91–106.

Zanoni, A. (1969). Secondary effluent deoxygenation at different temperatures. *J. Water Pollut. Control Fed.* **41**(4), 640–659.

Zanoni, A. E., and Rutkowski, R. J. (1972). Per capita loadings of domestic waste water. *J. Water Pollut. Control Fed.* **44**, 1756–1762.

8 | Introduction to Water Quality Modeling

FUNDAMENTAL CONCEPTS

A system, in the broad sense, is defined as a set of objects which interact in a regular, interdependent manner. A system can be characterized by (1) its boundary, which determines if any particular object is considered to be a part of the system; (2) a statement of input–output interaction; and (3) a statement of the interrelationships between the elements of the system and the inputs and outputs, including any external interaction between the inputs and outputs (feedback).

A useful tool in representing and analyzing systems is the so-called block diagram, which is a graphical representation of the components of the system and their interactions. A component of the system can be represented by a "black box" concept indicating that the component is an isolated well-defined entity or a detailed mathematical description.

A system can be simple, consisting of one simple unit, or complex, containing several subsystems. Examples of simple systems in the water quality control field are a treatment unit, a short reach of river, and a small uniform watershed. However, most of the systems in water resources engineering and water quality control are complex, as shown in Fig. 8.1. Figure 8.1(a) shows a watershed system. The input into this

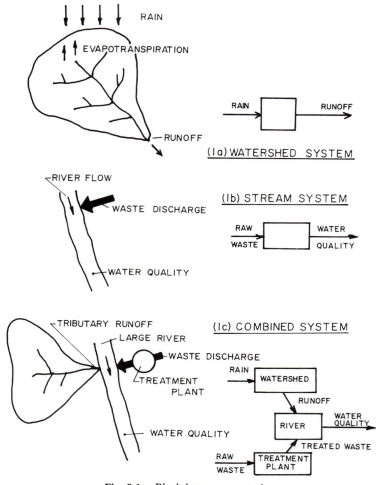

Fig. 8.1.   Black box representation.

system is precipitation (rain or snow), and the output is the quantity of flow and its quality, i.e., the amount of minerals, suspended solids, and pollution, in general, carried by the flow. The relation between the input and output is determined by the system parameters, such as surface area, roughness, slope, percent perviousness, and so on, and variables describing the state of the system, such as temperature, wind velocity, solar radiation, humidity, and crop management.

Figure 8.1(b) represents a simple river system. The input to this system is the waste load, and the output is the water quality downstream from the waste discharge. The system parameters are the cross-sectional area, the relationship between the flow velocity and magnitude, the slope of the channel, and so forth. The state variables include the temperature, the flow, and the reaction coefficients. These simple systems can be combined in a more complex unit such as shown in Fig. 8.1(c).

In system engineering terminology, the inputs are sometimes called forcing functions, and the outputs, responses. The interrelation between the inputs and outputs is named the "transform function."

Inputs to a system can be controllable, partially controllable, or uncontrollable. Those which can be completely or partially controlled are called decision variables. The outputs from each particular system are desirable, undesirable, and neutral. The relationships and possible feedback are shown in Fig. 8.2.

Each input and state variable has its constraint. When each decision variable is assigned a particular value, the resulting set of decision variables is called policy. A policy which does not violate any constraint is a feasible policy, and all possible policies form a policy space.

In order to determine if the outputs are desirable, undesirable, or neutral, it is necessary to define criteria, so the effect of any feasible policy transformed by the system into the system outputs can be evaluated. This set of criteria is called the objective.

Figure 8.3 shows a possible incorporation of the criterion concept into a simple water quality management system.

In water quality management, the objective may be stated by a regulatory agency and is usually related to downstream uses of the pollution receiving water body. The uses, such as recreation, water supply, navigation, etc., are then translated into criteria related to specific water quality requirements. For example, a warm water fishery requires that the dissolved oxygen concentration should not fall below 5 mg/liter and that the pH should be within 6 and 9, as previously discussed. The outputs from the water quality control systems are then compared to the delineated criteria.

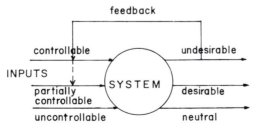

Fig. 8.2. Inputs, outputs, and feedback from a system.

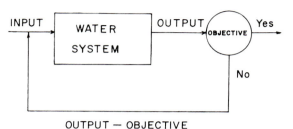

OUTPUT = OBJECTIVE

Fig. 8.3. Water quality management system.

Ecosystems are a special case of water resources and pollution control systems. They are recognizable, relatively homogenous units which include the organisms, their environment, and all of the interactions among them. Ecosystems are self-regulating; i.e., they react to stresses or to changes in inputs by reaching a new balance, which includes a modification of the system itself. Receiving water bodies can, in a sense, be considered as ecosystems.

STEPS IN SYSTEMS ANALYSIS

The basic elements of engineering systems analysis are

1. Definition of objectives.
2. Formulation of measures of effectiveness.
3. Generation of alternatives.
4. Evaluation of alternatives.
5. Selection of the best design.

In the water resources and pollution control field, the objectives are formulated by a decision team within the relevant organizational structure (national government, state government, Environmental Protection Agency, company management) with the system engineering team that will participate in the planning. The activities will also include exploratory studies, reconnaissance surveys, collection of field data, and the creation of background information such as demographic data, historical and economical development, etc. The formulation and measure of the effectiveness of the proposed action is based on the present knowledge of the technology involved, compliance with environmental laws, economic effectiveness, and other factors.

Generation and evaluation of alternatives in modern system analysis engineering require the use of models. Models can be either physically scaled or mathematical. Both types find wide application in the analysis of water resources and pollution control. In many cases, a combination of the two can be the optimal system analysis approach. Such cases include the evaluation of optimal wastewater load allocations and their effects on the receiving water. In this case, treatment unit performance is better modeled on a laboratory or pilot plant scale, while the receiving water analysis is better evaluated by means of mathematical computer models. The computer models also yield better results for modeling urban and rural runoff (quantity and quality), optimization studies, and thermal pollution. Physical models, on the other hand, are better to use for the modeling of currents, water pressures on structures, sedimentation and dispersion, and pollution technology performance.

In either case, each alternative must be developed in sufficient detail to permit its evaluation in terms of factors including system performance, quality, and ultimate cost. These evaluations are then compared for the selection of the best system, and the results of this phase are usually communicated to the decision-making body in a formal report.

After the development phase or realization of the action, the performance of the system must be monitored. The monitoring, usually continuous, should provide information on (a) compliance or noncompliance with objectives, (b) data for future corrections or additions, (c) compliance with the environmental standards dictated by the regulatory agencies, (d) environmental impacts and (e) record of performance to detect malfunctions or breakdowns.

BASIC MATHEMATICAL MODELS

The use of mathematical models of systems is the primary and most useful method of generating and evaluating various alternatives. These models represent the most economical method of system analysis since physical models are more costly and have scale and similarity problems. Mathematical models are flexible and do not have problems with time scale; i.e., alternatives which on physical models would require days and months of measurements can be generated by computers in minutes.

Mathematical models are not the only implements nor are they the only answer, in water quality management. The models must be verified by prototype measurements and surveys. A model is not valid unless it is substantiated and calibrated by field and/or laboratory measurements.

The evaluation of alternatives by mathematical models can be accomplished in either time or frequency domain. Time domain models can be either exact analytical, approximate, or stochastic (statistical).

Most of the mathematical models employed in engineering systems analysis are composited from a set of differential equations describing all units of the system and the interrelationships between them.

A general form of the equations is

$$\frac{dY_i}{dt} = f(Y_1, Y_2, \ldots, Y_K; X_1, X_2, \ldots, X_L;$$

$$P_1, P_2, \ldots, P_M; S_1, S_2, \ldots, S_N), \quad (8.1)$$

where Y is the output variable, X is the input variable, P is the system parameter, S is the state variable, and K, L, M, and N are the numbers of respective input variables, output variables, system parameters, and state variables.

For some simple systems, finding an exact analytical solution of the equations is possible, and the solved input–output relationship thus becomes the model. In most cases, an exact solution is not possible, but the use of computers enables replacement of the differential expressions by their finite difference approximations such as

$$\frac{dy}{dt} \simeq \frac{\Delta y}{\Delta t} \quad \text{or} \quad \frac{d^2y}{dx^2} \simeq \frac{\Delta y}{(\Delta x)^2}.$$

If the time increment, Δt, or the size of the element, Δx, is small enough, the solution by computers may have sufficient accuracy. However, this procedure may sometimes have severe disadvantages since the system must often be divided into very small elements. In hydrologic and hydraulic models, Δt is usually in minutes and the size of the computational element, Δx, in order to secure convergence of the solution, must be

$$\Delta x < U(\Delta t),$$

where U is the average flow velocity at time t.

Figure 8.4 shows a typical breakdown of a river reach into computational elements for a finite difference model. In such a model, the basic differential terms can be approximated by

(a) forward difference terms

$$\frac{dY}{dt} \simeq \frac{Y_i^{j+1} - Y_i^{j}}{\Delta t}, \quad (8.2)$$

where $j + 1$ denotes a future time level different from the present time level, j, by Δt, and i is the element subscript;

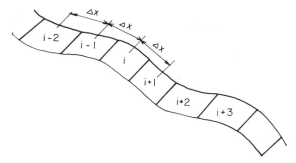

Fig. 8.4. Computational elements of a river for a finite difference model.

(b) backward difference terms

$$\frac{dY}{dx} \simeq \frac{Y_i^j - Y_{i-1}^j}{\Delta x}, \tag{8.3}$$

and (c) central difference terms

$$\frac{d^2 Y}{dx^2} \simeq \frac{Y_{i+1}^j - 2Y_i^j + Y_{i-1}^j}{(\Delta x)^2}. \tag{8.4}$$

If the differential equations describing the system are difficult to define or are enormously complex, a simpler relation can be developed from field data or measurements in the laboratory by correlating the inputs to the outputs. These models are called statistical or, on a higher level, stochastic models. Stochastic models incorporate the randomness of the input and state variables, and their results are reported and interpreted in terms of averages, variances, or standard deviations.

If the inputs and the state variables do not change with time, the system is at steady state, and simplified steady-state models can be used. In systems engineering terminology, the steady-state models are called fixed parameter models, or time invariant models. If the variables change with time, a time variable, or dynamic model must be utilized and developed.

A linear system is defined as a system described by differential equations with all derivatives raised to the power of 1. A significant characteristic of linear systems is the principle of superposition. For example, if Y_1 is the output from the system due to an input, X_1, and Y_2 is the output due to another input X_2, then if the input $X = X_1 + X_2$, it will result in the output $Y = Y_1 + Y_2$, or if

$$X = a_1 X_1 + a_2 X_2 + \ldots,$$

then the output

$$Y = a_1 Y_1 + a_2 Y_2 + \ldots.$$

The condition of linearity is that the system must be described by a set of linear differential equations. In some cases, the exact model of the system is not known but the preceding principle can be used for testing its linearity. The principle of superposition enables evaluation of responses to each input of the system separately.

The sensitivity of the system output, Y, to a variation of a system parameter, P_1, is defined by the mathematical derivative

$$s = \frac{\partial Y}{\partial P_1}. \tag{8.5}$$

The sensitivity analysis of the system is important in determining the required accuracy of the system parameters, which is, of course, related to the cost of the measurements.

Frequency domain models use the Fourier transformation to change into the frequency domain (Thomann, 1972). In a most oversimplified explanation of the Fourier transformation, the derivative symbol, d/dt, is substituted by a term $j\omega$, where $j = \sqrt{-1}$ and ω is the angular frequency. This enables one to replace integration and derivation in the time domain by simpler mathematical operations such as multiplication and division. For example, if a system in the time domain is described by the differential equation

$$a\frac{dY}{dt} + bY = X, \tag{8,6}$$

where Y = output, X = input, and a and b are coefficients (system parameters), replacing the symbol d/dt by $j\omega$, the equation can be transformed into this frequency domain form:

$$G = \frac{\hat{Y}}{\hat{X}} = \frac{1}{aj\omega + b}, \tag{8.7}$$

where $\hat{Y}$ and $\hat{X}$ are the transformed output and input, respectively, and G is the so-called frequency transform function. Note that the frequency transform function is independent of the input and output variables, which is a typical characteristic of linear systems. The frequency transform function of the linear system depends only on the system parameters and state variables. A schematic diagram of the frequency domain analysis is shown in Fig. 8.5. If the system consists of several subsystems in series, the overall transform function becomes

$$G = (G_1)(G_2)(G_3) \ \ldots \ . \tag{8.8}$$

A corresponding operation in the time domain would raise the order of the system (i.e., the order of the highest derivative in the set of the differential equations describing the system).

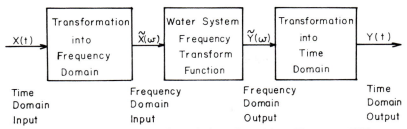

Fig. 8.5. Frequency domain analysis. (adopted from Thomann, 1972)

Since the frequency transform function is a complex quantity, it can be defined in polar form as

$$G(\omega) = \frac{Y(\omega)}{X(\omega)} = A(\omega)e^{-j\theta}, \tag{8.9}$$

where A defines the amplitude characteristics of the system, and θ is the phase shift. Application of the frequency analysis technique requires that the systems be linear and that the transformation of the inputs and outputs to and from the frequency domain, respectively, is possible.

BASIC MASS BALANCE RELATIONSHIPS— SIMPLE MODELS

A completely mixed flow unit or reactor assumes that all inputs are immediately equalized over the volume of the unit, as shown in Fig. 8.6(a). In general, two differential equations describe the process of equalization, i.e., flow equalization. Hence, flow equalization becomes

$$\frac{dV}{dt} = Q_{in} - Q, \tag{8.10}$$

and mass equalization can be described as

$$\frac{dY}{dt} V = X - QY - KVY. \tag{8.11}$$

In these equations, X is the input function which can be represented by $X = Q_{in} C_{in}$, where C_{in} = influent concentration of the substance, Q_{in} = influent flow, Q = flow from the system, Y = concentration in the system (for completely mixed systems Y also represents the effluent concentration), V = volume of the system unit, and K = decay coefficient of the substance in the system. The steady-state solution assumes that the input and the system parameters do not change with time, which means

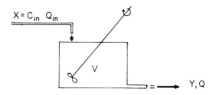

(a) COMPLETELY MIXED CONCEPT

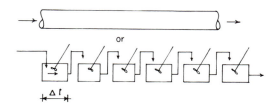

(b) PLUG FLOW CONCEPT

Fig. 8.6.   Completely (a) mixed and (b) plug flow concepts.

that $dY/dt = 0$. Then

$$\frac{Y_s}{X_s} = \frac{1}{Q + KV},$$

where X_s and Y_s are the time indifferent input and output, respectively.

A plug flow system can be represented by a series of completely mixed reactors as shown in Fig. 8.6(b). Recognizing that $Q/V = u/\Delta l$, where Δl is the length of the unit and u is the flow velocity across the reactor, and assuming that $\Delta Y = C_{in} - Y$, then if both ΔY and Δl approach very small values and the number of reactors becomes large, Eq. (8.2) can be rewritten as

$$\frac{\partial Y}{\partial t} = -u \frac{\partial Y}{\partial l} - KY. \tag{8.12}$$

In a real situation, the plug flow system formula describes the movement of a slug of the substance downstream which does not mix and keeps its original identity. Concentration changes happen only because of the decay of the substance contained in the slug if it is nonconservative.

The steady-state solution will yield

$$\frac{Y_s}{X_s} = e^{(-KL/u)}, \tag{8.13}$$

where L is the total length of the system.

In the dispersion flow system, turbulence and the velocity distribution cause particles or dissolved substances in most water systems to disperse, which, in this simplified one-dimensional case, means that the downstream moving slug is decreasing in concentration in the longitudinal sense. The diffusion or dispersion is proportional to the concentration gradient. If this concept is introduced into Eq. (8.12) one obtains

$$\frac{\partial Y}{\partial t} = D_L \frac{\partial^2 Y}{\partial l^2} - u \frac{\partial Y}{\partial l} - KY, \tag{8.14}$$

where D_L is the so-called coefficient of longitudinal dispersion, which is a hydraulic quantity depending on the turbulence intensity and velocity distribution in the system.

The steady-state solution to Eq. (8.14) will give

$$Y_s = X_s \left[\exp \left(\frac{uL}{2D_L} - L \left(\frac{u^2}{4D_L^2} + \frac{K}{D_L} \right)^{1/2} \right) \right].$$

Diffusion models are important for modeling of water bodies with a high degree of mixing, including estuaries, large water bodies, and equalization tanks. Plug flow models will usually give reasonably accurate results for shallow streams, pipe flow, or shallow overland flow, while completely mixed models are applicable for modeling flows in mixing tanks, aeration tanks, and small shallow ponds and reservoirs.

Representation of Flow Systems in the Frequency Domain

As previously stated, transformation into the frequency domain can substantially simplify the analysis, especially if the inputs are periodic and/or random. If the input is a periodic sine wave,

$$X = X_m \sin (2\pi f(t - t_0)) + \bar{X}, \tag{8.15}$$

where X_m is the amplitude of the input signal, $\bar{X}$ is the average value, f is the frequency equaling $1/T$ where T is the period, then the amplitude of the output signal will equal

$$Y_m = X_m A(f) \text{ (Note that the angular frequency } \omega = 2\pi f = 2\pi/T),$$

where $A(f)$ is the amplitude gain factor. The output signal will be shifted in the time scale by the time shift. Using the principle of superposition,

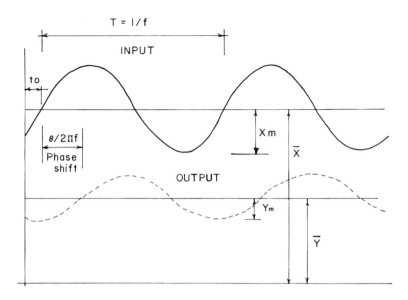

Fig. 8.7. Amplitude attenuation and phase shift.

one can also deduce that the average value of the input can be computed from the steady-state formula, as shown in Fig. 8.7.

It can be seen that both the amplitude gain factor $A(f)$ and the phase angle components of the frequency transform function, G, are significant since they relate the periodic component of the output to the input. It is therefore useful to list the formulas for their computation.

Completely Mixed Systems

Amplitude gain factor

$$A(f) = \frac{1/Q}{\left(\left(1 + K\frac{L}{u}\right)^2 + \left(2\pi f\frac{L}{u}\right)^2\right)^{1/2}} \tag{8.16}$$

Phase shift

$$\theta(f) = \tan^{-1} \frac{\frac{L}{u} 2\pi f}{1 + \frac{L}{u} K} \tag{8.17}$$

Plug Flow Systems

Amplitude gain factor

$$A(f) = e^{-KL/u} \qquad (8.18)$$

Phase shift

$$\theta(f) = 2\pi f \frac{L}{u} \qquad (8.19)$$

For plug flow systems, the amplitude gain factor is independent of frequency.

Dispersion Flow Systems

Amplitude gain factor

$$A(f) = \exp\left[\frac{Pe}{2}\left(1 - (a^2 + b^2)^{1/4}\cos\alpha\right)\right] \qquad (8.20)$$

Phase shift

$$\theta(f) = -\frac{Pe}{2}(a^2 + b^2)^{1/4}\sin\alpha, \qquad (8.21)$$

where $Pe = Lu/D_L$ is the Peclet number describing the degree of dispersion and

$a = 1 + 4D_L K/u^2,$
$b = 2\pi f D_L/u^2,$
$\alpha = \frac{1}{2}\tan^{-1}(b/a).$

It is convenient to plot the amplitude gain factor and phase shift v. frequency on a log–log scale. The result is called a Bode diagram for amplitude characteristics, as shown in Fig. 8.8. Two types of response of the

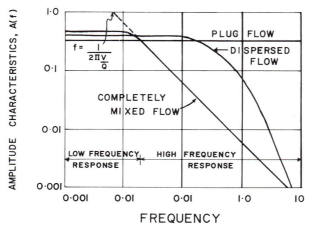

Fig. 8.8. Amplitude gain factor v. frequency.

output to a periodic harmonic input variation can be detected: a low-frequency response, in which the fluctuations are dampened only by the decay action in the system (if there is no decay $A(f) = 1.0$); and a high-frequency response, in which the dampening effect is mainly due to mixing of the input load with the volume of water present in the system. The high-frequency region is substantially more effective for dampening the fluctuations. Thus, the input variations with a high frequency will be removed at a higher rate than slow-changing fluctuations.

BASIC INPUT CHARACTERISTICS

The input characteristics are usually obtained from a survey or from surveillance programs. The data can be either discrete or continuous. Discrete data means there is a time lag between the neighboring values on the record. Depending on the method or technique used in the data acquisition, the time interval may be months, weeks, days, etc. Continuous records are produced by automatic monitoring stations in the form of charts or continuous electronic signals. In many cases, the continuous record is digitized by an analog–digital transformer, and data are transferred through further processing into digital forms, such as those on punched tape or printed tables.

The basic statistical characteristics and their evaluation have been discussed in the literature by Bendat and Piersol (1971). Data representing a physical phenomenon can be broadly characterized as being either deterministic or nondeterministic. Deterministic data are those that can be described by an explicit mathematical relationship. Because of the nature of water quality changes and the complexity of the processes affecting water or wastewater characteristics, there is no way to predict the exact value of the record at a future time. Such data are random in character and must be described in terms of probability statements and statistical averages rather than by explicit equations. However, long-term changes in the water pollution control field tend to have a functional character with random fluctuation components. Statistical evaluation techniques provide an implement to detect and quantify both the deterministic functional and the random components of a water or wastewater quality record. The deterministic component can be further categorized as being either periodic (sinusoidal) or nonperiodic (transient). In addition, the random component may be either stationary or nonstationary, as shown in Fig. 8.9.

There are an infinite number of time variable inputs and their patterns. However, since the principle of superposition can be applied to linear

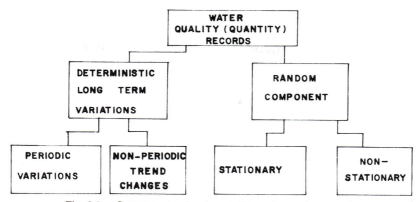

Fig. 8.9. Components of water quality and quantity records.

systems, and most of the environmental systems can be treated as linear, it is necessary to treat only the components of the inputs to which these can be decomposed. The basic components of the inputs can be categorized as:

1. Steady-state components which do not change with time.
2. Trends which are long-term changes, increasing or decreasing with a period much longer than the period of observation.
3. Time variable components—(a) pulses or step changes, periodic fluctuations, (c) random fluctuations.

In a real situation, the input changes almost always consist of a superposition of random fluctuations and a basic trend which may be either steady-state or time variable (pulses, step changes, periodic changes, etc.). Characteristic examples of basic input variations are shown in Figs. 8.10 and 8.11.

The major inputs as they may appear in environmental systems are wastewater loadings, storm water intensity, heated discharges (calorific loadings), and energy. Other variables such as temperature, meterological conditions, and flow are either system parameters or state variables.

Several statistical quantities are usually evaluated from a set of input characteristics and quality data. The following statistical values and functions are essential for their characterization.

The Mean, $\bar{X}$ is the average of all values, X_i, on the record. For a discrete series the mean is

$$\bar{X} = \frac{1}{N} \sum_{i=1}^{N} X_i, \tag{8.22}$$

where N = total number of discrete samples in the series.

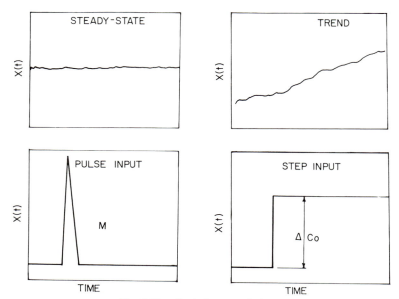

Fig. 8.10.   Basic input variations.

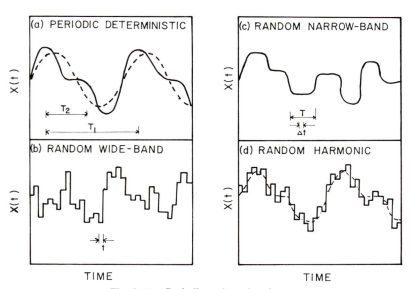

Fig. 8.11.   Periodic and random inputs.

The Variance S describes the spread of the variable, X, around the mean $\bar{X}$. For a discrete distribution the variance becomes

$$S^2 = \frac{1}{(N-1)} \sum_{i=1}^{N} (x_i - \bar{X})^2. \tag{8.23}$$

The square root of the variance is the standard deviation, $S = \sqrt{s^2}$

The Autocovariance function, R, shows the dependence between adjacent values and changes with the lag number K. Then

$$R(K) = \frac{1}{(N-K-1)} \sum_{i=1}^{N=K} (X_i - \bar{X})(X_{i+K} - \bar{X}), \tag{8.24}$$

where K = lag number. The lag time, u, becomes $u = K(\Delta t)$, where Δt is the sampling interval. From Eq. (8.24), it can be noticed that $R(0) = S^2$; therefore, it is advantageous to normalize the autocovariance function by the variance which results in the so-called autocorrelation function, which is

$$\rho(K) = \frac{R(K)}{R(0)} = \frac{R(K)}{S^2}.$$

The Spectral function, Γ, is a Fourier transformation of the autocovariance function is theoretically defined as

$$\Gamma(f) = \int_{-\infty}^{\infty} R(u)e^{-j2fu}\, du, \tag{8.25}$$

where f = frequency, and $j = \sqrt{-1}$. For discrete data, the smoothed spectral estimate $C(f)$ can be obtained by replacing the integral in Eq. (8.25) by the corresponding sum. The maximal frequency which can be detected from a discrete record is called the Nyquist frequency, f_c, given by $f_c = \frac{1}{2}(\Delta t)$. Since $\Gamma(f)$ and, therefore, $C(f)$ are even functions of frequency, it is only necessary to calculate the frequency over the range $0 \le f \le f_c$. However, to preserve the Fourier transform relationship between the series spectrum and the series autocovariance function, it is necessary to double the power associated with each frequency in the range $0 \le f \le f_c$. Thus, the formula to be used for the computation of the spectral estimate becomes

$$C(f) = 2(\Delta t) \left\{ R(0) + 2 \sum_{K=1}^{M-1} R(K)\, W(K) \cos 2\pi f K(\Delta t) \right\} \tag{8.26}$$

for $0 \le f \le f_c$, where M = maximum lag number and $W(K)$ is the lag window for smoothing the estimate such as the Tukey window, which is defined as

$$W(K) = \begin{cases} \dfrac{1}{2}\left(1 + \cos\dfrac{\pi K \Delta t}{M}\right) & \text{for} \quad K \le M \\ 0 & \text{for} \quad K > M. \end{cases}$$

The preceding approach was recommended by Jenkins and Watts (1969), although other formulas for the spectral estimate determination applicable in the water pollution control field have been published in Bendat and Piersol, 1971; Wastler, 1963; Gunnerson, 1966; and Novotny and Englande, 1974.

If the time record represents a stationary time series, the variance of the process can be decomposed into contributions for a continuous range of frequencies. Then, the following relation between variance and spectral function holds:

$$S^2 = R(0) = \int_{-\infty}^{\infty} (f)\, df = \int_{0}^{1/2(\Delta t)} D(f)\, df \qquad (8.27)$$

The plot of the autocovariation function $v.$ time of displacement is called an autocorrelogram. Figure 8.12 shows the corresponding autocorrelograms for the four basic input variations shown in Fig. 8.11. The autocorrelogram for sine-wave variations persists periodically over all time displacements with the same period as the underlying sine wave. The value of $R_x(0)$ equals $X_m^2/2$, where X_m is the amplitude of the signal. The

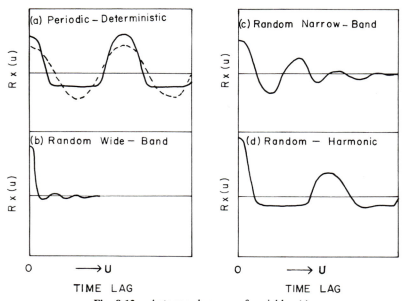

Fig. 8.12.   Autocorrelograms of variable $x(t)$.

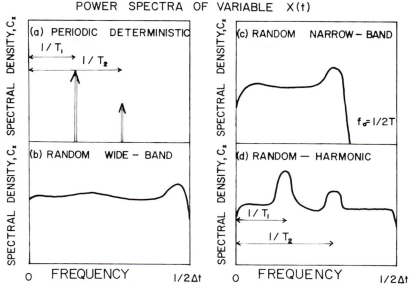

Fig. 8.13. Power spectra of variable $x(t)$.

sharply peaked autocorrelogram which diminishes rapidly to zero, as illustrated in Fig. 8.12(d), is typical of wide-band random data with a zero mean value (if the mean value were not zero, the autocorrelogram would approach a value of $\bar{X}^2$). The autocorrelogram for the sine wave plus a random component is simply the sum of their separate correlograms.

The principal application for an autocorrelation function measurement of physical data is to establish the influence of values at any time over values at a future time. It provides a tool for detecting deterministic data which might be marked in a random background.

Power spectrum estimates as shown in Eq. (8.26) describe the general frequency distribution of the data in terms of the spectral density of its variance (provided that the record is modified to a zero mean value); i.e., they describe how the variance is distributed with frequency. A plot of power spectral density function estimate v. frequency ($C_x(f)$ v. f) is called a power spectrum. Power spectra for the four typical variations of $X(t)$ as illustrated in Fig. 8.11 are shown in Fig. 8.13.

The discrete power spectrum for the deterministic harmonic signal (sine wave) is defined by a delta function at $f = 1/T$, where T is the period of the basic signal. If the signal is composed of two harmonics with frequencies $f_1 = 1/T_1$ and $f_2 = 1/T_2$, as illustrated in Fig. 8.11(a), the peaks will appear at frequencies f_1 and f_2. Although the power spectral density

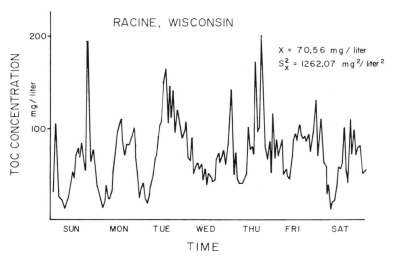

Fig. 8.14.   Time record of TOC variation, Racine, Wisconsin.

of a sine wave is infinitely large at the frequency of the sine wave and zero
at all other frequencies, the integral of the power spectrum over any frequency range has a finite value equal to the variance of the signal which,
for a simple sine wave, equals $S = X_m^2/2$.

The relatively uniform and broad spectrum, as shown in Fig. 8.13(b), is
typical for the wide-band random sample. For a theoretical random variation, the power spectrum would be uniform over all frequencies. It should
be noted that for discrete sampling, the maximum frequency which can be
resolved from the record is the Nyquist frequency, $f_c = 1/(2\Delta t)$, in which
Δt is the sampling interval.

The power spectrum for a narrow band random variation is terminated
at a frequency $f_0 = \frac{1}{2}T$, where T is the time interval at which the input
might change. Again, the area under the spectrum curve is equal to the
variance of the signal. The power spectrum of the sine wave plus a
random component is simply the sum of the power spectra for the sine
wave and random component separately, as illustrated in Fig. 8.13(d).

Figure 8.14 shows the record of the water quality variation of a midwestern wastewater discharge from a city with about 20,000 people. From the
conventional time representation, it was possible to evaluate only the
basic characteristics such as the mean and variance (standard deviation).
The time record does not allow the separation, at least not quantitatively,
of the harmonic and random components. The autocorrelogram indicates
that there is a strong 24-hour periodicity in the signal (Fig. 8.15) accounting for about 30% of the correlation in the signal. This periodicity and its

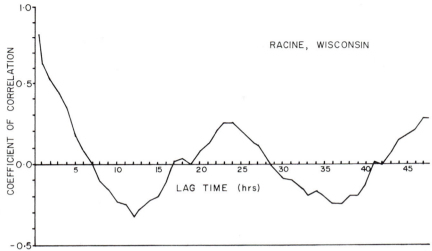

Fig. 8.15. Autocorrelogram for TOC variation, Racine, Wisconsin.

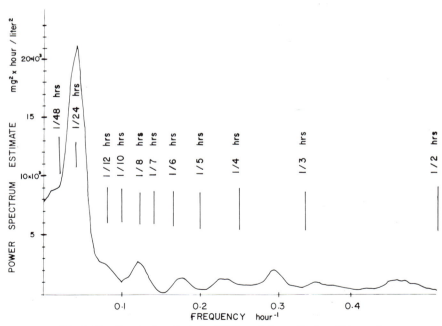

Fig. 8.16. Spectral analysis for TOC variations, Racine, Wisconsin.

influence on the variability of the signal was determined from the power spectrum plot.

The spectral analysis, as shown in Fig. 8.16, confirms the conclusion that most of the variability occurs within the frequency band $\frac{1}{48}$ hours^{-1} to $\frac{1}{16}$ hours^{-1}, with a strong peak at the frequency $\frac{1}{24}$ hours^{-1}. Almost 85% of the variance $s_x{}^2$ is caused by these low-frequency variations. There is also a less significant peak at the frequency of $\frac{1}{8}$ hours^{-1}. The remainder of the record shows the effect of random fluctuations.

INPUT–OUTPUT ANALYSIS OF LINEAR SYSTEMS

The relationship between the input and output of linear systems can be conveniently evaluated by means of the frequency transformation technique. As previously stated, the general relationship between the transformed input, $\tilde{X}$, and output, $\tilde{Y}$, is

$$\tilde{Y}(f) = \tilde{X}(f)G(f), \tag{8.28}$$

where G is the transform function of the system. For simple transient inputs, the input transformation will yield

Pulse input $X = 1,$
Step input $X = 1/(2\pi fj).$

Knowing the transform function, $G(f)$, the output can be solved and transformed back to the time domain using Fourier or Laplace transformation tables.

For periodic and random components, it is more convenient to relate the input and output spectra. The basic relationship can then be written as

$$C_y(f) = [A(f)]^2 C_x(f), \tag{8.29}$$

where $A(f)$ is the amplitude gain factor of the frequency transform function of the system, C_x is the influent spectrum estimate, and C_y is the effluent spectrum estimate.

Then the output variance, $S_y{}^2$, becomes

$$S_y = \int_0^{1/2\Delta t} [A(f)]^2 C_x(f) \, df. \tag{8.30}$$

Figure 8.17 is a graphical representation of the resulting input–output relationship. This relationship is the fundamental basis for the development of the stochastic models which can treat both steady-state average char-

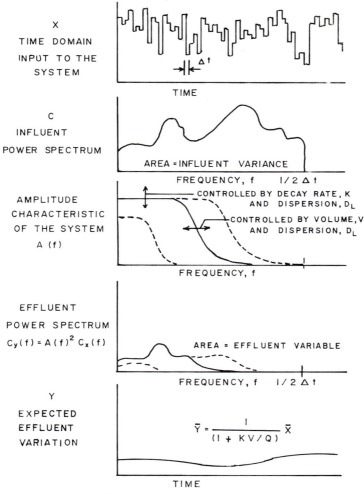

Fig. 8.17. Input-output solution.

acteristics and their residuals, i.e., in the simplest case, their variance or standard deviation.

The importance of stochastic models should not be underestimated. Long-term averages may only be of limited use to the designer of a treatment plant or to a water quality planner, since using average values would only mean designing or planning a water quality action (e.g., a treatment plant) which might be 50% overloaded. Thus, accurate knowledge of long-term averages may only be of secondary importance. Since the use of treatment units and effective control of pollution depends to a great extent on adequate knowledge of waste load, flow, or storm variability, one must be able to relate the influent and effluent variability. Some water pollution control measures such as waste assimilative capacity determination, maximum waste discharge permits, and compliance with water quality standards are based almost solely on variability and the occurrence of extremes, and are not influenced by the long-term mean.

TIME DOMAIN DETERMINISTIC MODELS FOR COMPLEX WATER RESOURCE SYSTEMS

Deterministic time domain models are probably the most widely used means of water quality systems analysis today. Their advantage lies in a simple interpretation of the model outputs which can then be directly compared to the actual real situation hydrographs and pollutographs.

A time domain model is a real representation of a system working with the same parametric variables as the prototype. However, their outputs are unique; i.e., the results are a single value or a curve which does not indicate the probabilistic range of the results or their probability of occurrence. Thus, the results may sometimes be misleading. For example, using average values of the system parameters and coefficients describing the state variables may give an excessively favorable view of the possibility of violating a water quality standard. On the other hand, worst case alternatives, when least favorable values of system parameters and state variables are used, may have an extremely low probability of occurrence. In a simple example of the dissolved oxygen concentration, which would depend only on temperature and flow when both variables are independent of each other, the combined probability of having a worst case alternative is a multiplication of probabilities for each variable separately. Thus, if both variable values have a probability of occurrence once in ten years, the combined probability would be once in every hundred years.

The problem can be partially avoided if the model is used to simulate a long time series of water quality data with statistically selected water quality parameters.

The analytical models which relied on an exact solution of simplified differential equations describing the system were the only models used in the past. With the introduction of fast computers, approximate and/or numerically more complex models are now applied almost exclusively.

These models rely primarily on a finite difference approximation of the differential equations.

A mathematical model is always an approximation of the real situation and its validity depends on the appropriate simplifications of the pertinent situation. Typical simplifications can be listed as:

(a) Steady-state assumption—assumes that all parameters and inputs to the model do not change with time and that an equilibrium state has been reached.

(b) One-dimensional assumption—implies that all parameters and variables are uniform in two directions (e.g., y and z) and changes occur only in one direction.

(c) Two-dimensional assumption—is similar. Changes occur in two dimensions. Only one direction is uniform (e.g., flow is vertically uniform).

(d) Conservative substance model—neglects the decay of the substance or its transformation to or from another form.

(e) Convective model—assumes that convection is the prevailing transport mechanism and, therefore, dispersion is neglected.

(f) Dispersion model—the opposite of the convection model; i.e., it assumes that the turbulent or tidal dispersion is the prevailing transport phenomenon.

(g) Point source models—assume that the source of pollution has no dimension.

The problem of modeling may be further complicated by the fact that some of the concentration change processes in receiving water bodies are coupled; i.e., the rate of change of one substance depends on the concentration or presence of another substance.

In many cases, the discharge of a pollutant to a water body carries significant momentum or has a different density, which means that the initial stage of mixing is influenced by the approach velocity of the waste discharge or its buoyancy.

Numerical Models

The deterministic models in water quality engineering greatly rely on numerical simultaneous solution of three basic equations: (a) equation of continuity (water conservation), (b) momentum conservation equation, and (c) mass or thermal energy conservation.

Since the differential equations are replaced by their finite difference

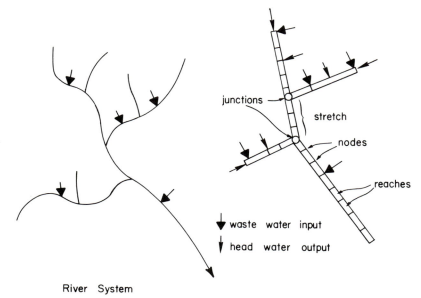

River System

Fig. 8.18.   Model representation of stream systems.

approximations, the system must be broken down into small, computational elements, as shown in Fig. 8.18. The size of the elements is dictated by the convergence requirement for the solution. The computational elements are considered uniform; i.e., no change in the cross-sectional geometry, roughness, slope, etc., can take place within the element. In stream and estuary modeling, these elements are called reaches. Several reaches in series form a stretch. Reaches are connected by nodes. Two or more stretches are connected by a junction.

In this representation, the elements are considered one-dimensional, although in their configuration they may represent a two- or three-dimensional system.

In the solution, the hydraulic mass and momentum equations are first written for each element, and the hydrodynamic behavior of the system subjected to various flow and stage inputs is then numerically solved, usually by matrix inversion. As will be seen in the following section, the set of equations describing the hydraulics of the system is nonlinear, and, therefore, their finite difference representation is more sensitive and requires finer spatial and temporal increments. Mass (water quality) conservation equations are usually linear, and the sensitivity of the solution to the magnitude of the time increment is less than that for the hydraulic portion.

In all deterministic models, the basic differential equations are represented by backward, forward, or central difference terms.

Selection of Models

The models in present use are limited in their scope and application. The early utopia of developing a "universal water quality model" has never left the stage of preliminary conception and, for all practical purposes, has been abandoned. Thus, a water quality analyst must usually limit himself to one or several water quality models describing a particular water quality problem. However, some, more complex, watershed models are available. Water quality models or segments of a watershed model can be roughly divided into five groups, as shown in Fig. 8.19.

1. *Overland flow models* simulate formation of surface runoff and may also include interflow and infiltration processes. The major inputs in such models are precipitation, atmospheric pollutant fallout, street litter accumulation, and land use characteristics. Based on the hydrologic balance

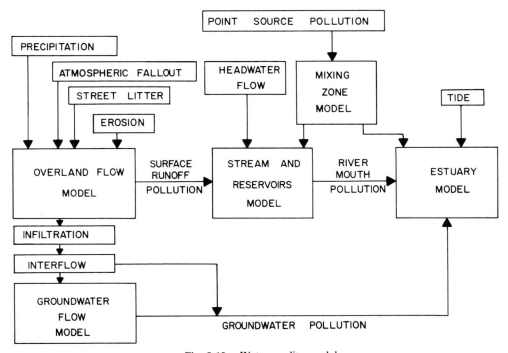

Fig. 8.19. Water quality models.

TABLE 8.1. Overview of Selected Water Quality Models

Model	Developer and/ or Source[a]	Model Category	Model Characterization	Processes Included	Parameters Modeled	Input Data and Computer Requirements
HSP-II	Hydrocomp International	overland hydrologic watershed model	dynamic	runoff, pollutants, pickup and transport	flow, sediment, most of water quality parameters	large
SWMM	Metcalf and Eddy, U. of Florida Water Res. Eng.	overland surface runoff from isolated storms	dynamic	runoff, erosion, pollutant pickup and transport	flow, sediment, most of water quality parameters	large
STORM	Wat. Res. Eng. United States Army Corps of Engineers	overland surface runoff	quasi-dynamic	surface runoff, erosion poll, pickup	flow sediment, sediment adsorbed poll.	medium
LANDRUN	Marquette U., Wisc. DNR	overland runoff	dynamic	runoff, erosion, pollut. pickup and routing	flow, sediment, sediment adsorbed pollutants	medium
DOSAG	Texas Wat. Dev. Board	stream	steady state	deoxygenation reaeration nitrification	D. Oxygen, nitrogen	small
QUAL-II	EPA	steam	semi-dynamic	stream pollutant transport	D. Oxygen, temperature, most of water quality parameters	medium
SWMM-RECEIV	Wat. Res. Eng., EPA	stream	dynamic	stream pollutant transport	D. Oxygen, nitrogen, conservative pollutants	large

294

Model	Organization	Water body	Mode	Processes	Parameters	Size
HSP-II CHANNEL QUALITY	Hydrocomp International, Palo Alto, Ca.	stream	dynamic	stream pollutant transport	D. Oxygen, nitrogen, conservative pollutant transport	large
M.I.T. Network Model	M.I.T.	stream, estuary	dynamic	pollutants transport, eutrophication, nitrification	D. Oxygen, nitrogen, conservative pollutants, temperature	large
M.I.T. Reservoir Model	M.I.T.	deep reservoir	dynamic	stratification, thermal balance, mass transfer	temperature, dissolved oxygen	large
Chen and Orlob Model	Wat. Res. Eng.	stratified estuary, lake or reservoir	dynamic	pollutants and energy balance, eutrophication	temperature, oxygen, most of wat. quality parameters	large
PLUME	Pac. Northwest, EPA	mixing zone	steady state	buoyant jet and plume mixing	conservative pollutant	medium

[a]Hydrocomp International, Palo Alto, Ca.
Metcalf and Eddy, Boston, Mass.
Water Resources Engineers, Walnut Creek, Ca.
Marquette University, Dept. of Civil Engineering, Milwaukee, Wisc.
Dept. of Natural Resources, Madison, Wisc.
University of Florida, Dept. of Environmental Eng. Sci., Gainesville, Fla.
Texas Water Development Board, Austin, TX.
Massachusetts Institute of Technology, Dept. of Civil Engineering, Cambridge, Mass.

flow chart, surface runoff hydrographs and pollutant loading time records can be generated which depend on the intensity and volume of the precipitation, quantity of pollutants accumulated on impervious surfaces, and the amount of erosion on pervious surfaces. These models are generally used for the evaluation of nonpoint source pollution loadings.

2.   *Groundwater flow models* describe the movement of water through soil and underground geological formations. Because of the very long response time of groundwater aquifers to surface inputs and also because of many unknowns, the state of the art is not as advanced as in other model categories. These models are therefore relatively coarse and are mainly steady-state modifications of the flow and mass conservation equations. The primary processes affecting groundwater quality are adsorption of pollutants on soil and chemical interaction of water with underground formations.

3.   *Mixing zone models* are models describing the spread of pollutants from the discharge point to a stream cross-section where the pollutant is laterally and vertically uniformly distributed. Because most of the stream and estuary models are one-dimensional, assuming a uniform distribution of pollutants in the vertical and lateral directions, mixing zone models are necessary for the proper design of wastewater outfalls. The mixing zone models may be two- or even three-dimensional, but one-dimensional models with the main direction along the axis of the plume or discharge are quite common. Because of the relatively short detention times of the pollutants in the mixing zone, the models are usually steady-state conservative modifications of the flow and mass conservation equations. The equations for some simplified cases can be analytically solved and/or transformed into design graphs and simple formulas. Lateral and vertical dispersion, along with the momentum and buoyancy of the discharge, are the primary mechanisms determining the shape and spread of the discharge plumes and intensity of mixing with the upstream flow.

4.   *Stream and reservoir models* are usually based on a one-dimensional approximation of the flow, momentum, and mass conservation equations. In these models, the convective transport of pollutants is more significant than dispersion, which is often neglected. These models range in complexity from simple, steady state, dissolved oxygen relationships to very complex models describing the interrelationships among pollutant removal, organic matter concentration increase, and biological life processes taking place in aquatic environments. Although the computational elements are usually one-dimensional, more complex models are capable of describing two and three-dimensional flow patterns such as stratified flow in reservoirs or lakes. The modeled parameters and processes include dissolved oxygen, most of the conservative and noncon-

servative pollutants, eutrophication, and nutrient transport under steady-state or dynamic flow conditions. A distinction should be made during the model selection process as to whether the model is to be used for quality simulation of small streams, large rivers, or lakes and reservoirs.

5. *Estuary models* do not differ significantly from stream models; in fact, both streams and estuaries can be represented by some larger water quality models. The major difference between estuarine and stream modeling is the magnitude of the tidal dispersion, which in estuaries cannot be neglected. Estuaries are also more dynamic than streams. In most cases, estuaries are modeled as uniform, nonstratified channel elements; however, two-layer stratified representation is possible.

Table 8.1 presents an overview of some models and their capabilities.

REFERENCES

Bendat, J. S., and Piersol, A. G. (1971). ''Random Data: Analysis and Measurement Procedures.'' Wiley (Interscience), New York.

Gunnerson, C. G. (1966). Optimizing sampling intervals in tidal estuaries. *J. Sanit. Eng. Div., Am. Soc. Civ. Eng.* **92,** 103–125.

Jenkins, G. M., and Watts, D. G. (1969). ''Spectral Analysis and Its Application.'' Holden-Day, San Francisco, California.

Novotny, V., and Englande, A. J., Jr. (1974). Equalization design techniques for conservative substances in wastewater treatment systems. *Water Res.* **8,** 325–332.

Thomann, R. V. (1972). ''System Analysis and Water Quality Management.'' EPA, Inc., New York.

Wastler, T. A. (1963). Application of spectral analysis to stream and estuary field surveys. Pub. No. 999-WP-7, U.S. Public Health Serv., Washington, D.C.

9 | *Transport Processes*

Water quality modeling is basically a description of the fate of pollutants in the aquatic environment. The pollutants can be either dissolved where their movement is identical with that of the water particles, or in suspended particle form, the movement of which is governed by friction, buoyancy, momentum, and gravity forces. However, the dissolved component movement may not strictly follow the hydrodynamic laws developed for pure water. The density gradients caused by dissolved pollutants, temperature differences, or saltwater intrusion may lead to stratified flow, which is one of the most complicated hydrodynamic phenomena.

The particulate matter-sediment, precipitated pollutants, organics, and pollutants adsorbed on sediments can be dispersed or be a part of moving or shifting bed load. In either case, the mathematical description and eventual solution may become quite complicated.

The pollutants themselves can be either conservative or nonconservative. The conservative pollutants do not decay or otherwise change their quantity, and their mass can be changed only by dilution, lateral inflow, and mixing. Conservative materials have no other sources or sinks than local inflows or diversions. In addition, they are not significantly affected by changes in temperature or any other biological, chemical, or physical process. Typical conservative parameters include chlorides, total dissolved solids, and sulfates.

On the other hand, nonconservative substances can undergo a transformation into another form, and their mass can be changed by chemical, biochemical, or physical processes. They are affected by temperature and chemical and/or biological phenomena, and their reaction rates are usually represented by first-order kinetics. Examples include CBOD, NBOD, and coliform bacteria.

Figure 9.1, (Texas Water Devel. Bd., 1971) shows the effects of flow on a conservative substance, and Figure 9.2 (O'Connor, 1967) demonstrates a nonconservative substance (CBOD) as it is integrated with flow.

BASIC PRINCIPLES OF WATER MOVEMENT

The description and solution of the hydrodynamic behavior of surface or groundwater systems are an essential part of every water quality model. Prime requisites of even the most simple steady-state water quality models are hydraulic quantities and relationships, such as travel time, stage-flow and stage-velocity relationships, and degree of mixing. Therefore, water quality modeling can be considered an extension of scientific hydrodynamics.

It is beyond the scope of this treatise to provide a detailed description of the hydrodynamic behavior of aquatic systems inasmuch as there are many fluid dynamic publications available to which the interested reader is referred.

The basic hydrodynamic laws which must always be included in a description of a water quality system are (1) the water conservation equation (equation of continuity) and (2) the momentum conservation equation (equation of motion).

Water conservation states that the difference (net rate) of the flow entering and leaving a control volume must equal the rate of storage in the volume. In contrast to classical fluid dynamics which mainly utilizes a fixed infinitesimally small control volume (dimensions dx, dy, and dz), the equation discussed here makes use of a control volume one unit wide in the z-direction (Fig. 9.3), with a finite height and a differential dimension in the x-direction.

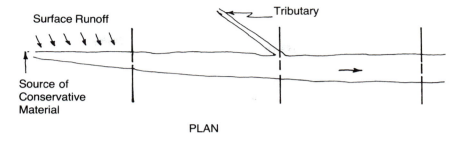

PLAN

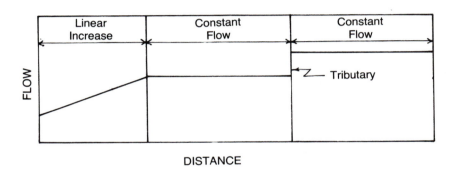

DISTANCE

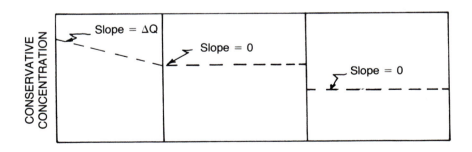

DISTANCE

Fig. 9.1. Effect of flow on conservative mineral (Texas Water Devel. Bd., 1971).

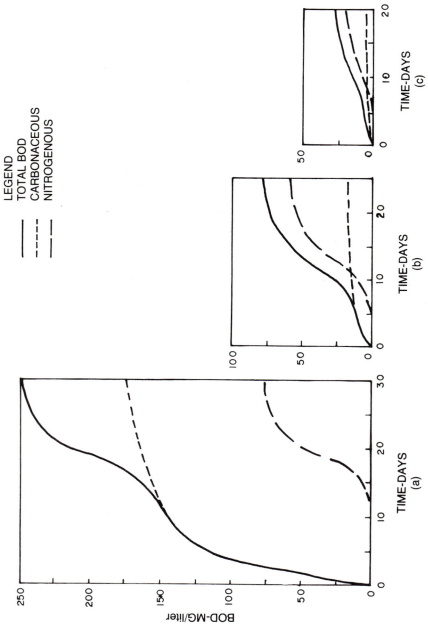

Fig. 9.2. Nonconservative substances in a river (O'Connor, 1967). A, untreated municipal wastewater; B, effluent from high rate biological treatment plant; C, river sample downstream from discharge.

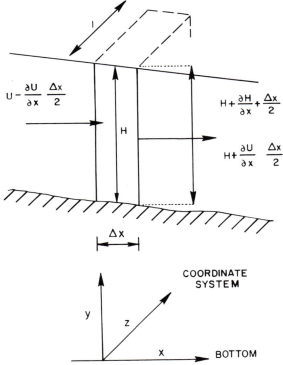

Fig. 9.3. Control volume description.

Thus, the mass balance is (Ligett,'1975)

$$\left(U - \frac{\partial U}{\partial x}\frac{\Delta x}{2} \right) \left(H - \frac{\partial H}{\partial x}\frac{\Delta x}{2} \right) - \left(U + \frac{\partial U}{\partial x}\frac{\Delta x}{2} \right)$$

inflow   outflow

$$- \left(H + \frac{\partial H}{\partial x}\frac{\Delta x}{2} \right) = \frac{\partial H}{\partial t}\Delta x, \quad (9.1)$$

change of
storage

where U is the flow velocity, Δx is the differential length of the control volume, x is the direction of flow, H is the depth, and t is time.

After multiplying and rearranging

$$\frac{\partial H}{\partial t} + \frac{\partial}{\partial x}(UH) = 0. \quad (9.2)$$

Equation 9.2 represents water conservation in a shallow uniform flow. The uniform flow approximation can be accepted for channels where $B \gg 10\,H$, in which B is the width of the channel. In a more general form, Eq. (9.2) can be replaced by

$$\frac{\partial A}{\partial t} + \frac{\partial Q}{\partial x} = q_i, \tag{9.3}$$

where A is the cross-sectional area, Q is the flow, and q_i is a lateral inflow into the segment.

Momentum conservation is based on Newton's second law of motion which states that the product of mass and acceleration equals the sum of the external forces acting on the control volume.

The original form of Newton's second law is of little use in water quality and quantity modeling. However, the momentum equation is of primary importance. The momentum equation states that the external and internal forces acting upon the control volume will result in an accumulation or increase of momentum of the body.

Based on the representation of the control volume shown in Fig. 9.3, the net increase of momentum is (Ligett, 1975)

$$\underbrace{\rho \left\{ U(UH) - \frac{\partial}{\partial x} U(UH) \frac{\Delta x}{2} \right\}}_{\text{momentum entering}} - \underbrace{\rho \left\{ U(UH) + \frac{\partial}{\partial x} U(UH) \frac{\Delta x}{2} \right\}}_{\text{momentum leaving}}$$

$$+ \underbrace{\frac{\partial}{\partial t} (\rho U H) \, \Delta x}_{\substack{\text{change of} \\ \text{momentum}}} = \underbrace{\Sigma F,}_{\substack{\text{sum of} \\ \text{the forces}}} \tag{9.4}$$

where ρ is the density of water.

Three types of forces acting upon the control volume can be considered—the forces of gravity, friction, and pressure.

The gravity force is the weight of the control volume, i.e.,

$$F_g = \rho g H \, \Delta x \sin \theta_x, \tag{9.5}$$

where θ_x is the angle of the x-axis (bottom line) with the horizontal and g is gravity acceleration. Assuming that for small angles the bottom slope, S_0, can replace $\sin \theta_x$, the gravity force component becomes

$$F_g = \rho g H S_0 \, \Delta x, \tag{9.5a}$$

The friction shear resistance force can be represented by the equation

$$F_s = \rho g H S_f \, \Delta x, \tag{9.6}$$

where S_f is the energy (friction) slope of the flow. The friction slope can be obtained from semiempirical flow formulas such as the Chezy formula

$$U = C\,(RS_f)^{1/2},$$

where C is a coefficient and R is the hydraulic radius.

The third force component is the net hydrostatic pressure of the surrounding water on the control volume, i.e.,

$$F_p = \tfrac{1}{2}\rho g \left[\left(H^2 + \frac{\partial H^2}{\partial x}\frac{\Delta x}{2} \right) - \left(H^2 - \frac{\partial H^2}{\partial x}\frac{\Delta x}{2} \right) \right]. \tag{9.7}$$

Introducing Eqs. (9.5a), (9.6), and (9.7) into Eq. (9.4) and dividing by ρ and Δx, one obtains

$$\frac{\partial}{\partial t}(UH) + \frac{\partial}{\partial x}(U^2 H) + gH\frac{\partial H}{\partial x} = gH(S_0 - S_f), \tag{9.8}$$

or in a more general form

$$\frac{\partial Q}{\partial t} + \frac{\partial}{\partial x}\left(\frac{Q^2}{A}\right) + gA\frac{\partial H}{\partial x} = gA(S_0 - S_f). \tag{9.8a}$$

The flow equations as presented here are known as the St. Venant equations after a French hydraulic scientist who in 1871 published the equations of unsteady flow in channels.

The equation of motion is often used in reduced form. Note that

$$\frac{\partial}{\partial t}(UH) + \frac{\partial}{\partial x}(U^2 H) = H\frac{\partial U}{\partial t} + U\frac{\partial}{\partial x}(UH) + UH\frac{\partial U}{\partial x}. \tag{9.9}$$

By comparison with the equation of continuity (9.2), it can be seen that the sum of the second and third terms on the right-hand side is zero and the terms therefore can be eliminated. Then, after dividing by H, the shortened form of the momentum equation becomes

$$\frac{\partial U}{\partial t} + U\frac{\partial U}{\partial x} + g\frac{\partial H}{\partial x} = g(S_0 - S_f). \tag{9.10}$$

Other Forms of the Flow Equations

The flow equations also can be written in the characteristic form

$$\frac{dQ}{dt} - \left[\frac{Q}{A} \pm (gH)^{1/2}\right]\frac{dA}{dt} = gA(S_0 - S_f) - q_i\left[\frac{Q}{A} \pm (gH)^{1/2}\right] \tag{9.11}$$

with

$$\frac{dx}{dt} = \frac{Q}{A} \pm (gH)^{1/2}.$$

Note that the characteristic form contains ordinary differential terms, not partial derivatives. It has been shown (Lighthill and Whitham, 1955; Wooding, 1965) that for Froude numbers, $Fr = U/(gH)^{1/2} < 1.0$, the preceding equations can be expressed in a kinematic form, i.e.,

$$\frac{\partial Q}{\partial x} + B \frac{\partial H}{\partial t} - q_i = 0 \tag{9.12}$$

and

$$Q = aH^m,$$

where B is the width of the channel, and a and m are coefficients which can be obtained from semiempirical flow formulas or flow rating curves. For example, the Manning flow formula would yield

$$a = \frac{1}{n_M} BS_f^{1/2} \quad \text{and} \quad m = 1.66,$$

where n_M is the Manning resistance coefficient.

The kinematic wave equation cannot be used in sections with significant backwater or tidal effects. In such cases, and for larger Froude numbers, it is necessary to utilize dynamic wave (St. Venant) flow equations. The principal advantage of the kinematic wave model is that it is easier to solve than the St. Venant or characteristic flow equations.

METHODS OF SOLUTION OF FLOW EQUATIONS

The available mathematical models and numerical techniques for flow routing utilize either a numerical solution of the St. Venant equations or one of the methods which solve an input–output transformation. Methods of flow routing similar to the unit hydrograph method for rainfall–runoff relationships have been proposed and solved by hydrologists.

Numerical Methods

The numerical solution of the flow equations can proceed in two ways. Either the original dynamic wave (St. Venant) equations are replaced by their finite difference representation and solved by using an implicit or explicit scheme, or a solution may utilize the flow equations in their characteristic form. The four most common methods used for solution of the St. Venant equations are (Price, 1974) (1) the leap-frog explicit method, (2) the two-step Lax–Wendroff explicit method, (3) the four-point Amein implicit method, and (4) the fixed-mesh characteristic method. The boundary conditions are an integral part of each method.

Usually, each method can use the input of either the cross-sectional area (or stage) or the discharge (or velocity) at the upstream boundary for the river being considered, with an initial steady-state flow throughout the reach. At the downstream boundary, two boundary conditions are required by each of the methods excluding the four-point implicit method, which requires only one boundary condition at the downstream boundary.

According to the analysis by Price (1974), the implicit method is faster with the same accuracy than the other three methods and will therefore be briefly introduced.

The four-point implicit method was developed by Amein and Fang (1970) and utilizes the Newton iteration technique. The geometric representation of the method is shown in Fig. 9.4. Let us assume that the values for all variables are known at all nodes of the network on the time line t^j, and it is desired to find the values of the variables at all nodes on the time line t^{j+1}. The flow equations are applied to a point M centered within the four-point grid formed by the intersections of the space lines x_i and x_{i+1} with the time lines t^j and t^{j+1}. At the point M, the average value and

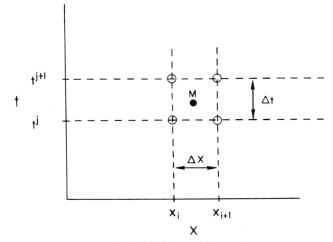

Fig. 9.4. Implicit four-point method.

the partial derivatives of a function, f, are expressed as

$$f(M) = \frac{1}{4}(f_i^j + f_{i+1}^j + f_{i+1}^{j+1} + f_i^{j+1}),$$

$$\frac{\partial f(M)}{\partial x} = \frac{1}{2\,\Delta x}\left[(f_{i+1}^j - f_i^j) + (f_{i+1}^{j+1} - f_i^{j+1})\right], \qquad (9.13)$$

$$\frac{\partial f(M)}{\partial t} = \frac{1}{2\,\Delta t}\left[(f_i^{j+1} - f_i^j) + (f_{i+1}^{j+1} - f_{i+1}^j)\right],$$

in which f represents functions such as A, Q, H, U, etc.

The finite difference form of Eqs. (9.3) and (9.8a) are

$$A_{i+1}^{j+1} + A_i^{j+1} - A_i^j - A_{i+1}^j$$
$$+ \frac{\Delta t}{\Delta x}\{[(Q_{i+1}^{j+1} - Q_i^{j+1}) + (Q_{i+1}^j - Q_i^j)]\} = 0 \quad (9.14)$$

and

$$Q_{i+1}^{j+1} + Q_i^{j+1} - Q_{i+1}^j - Q_i^j + \frac{\Delta t}{\Delta x}\left\{\left[\left(\frac{Q^2}{A}\right)_{i+1}^{j+1} - \left(\frac{Q^2}{A}\right)_i^{j+1} + \left(\frac{Q^2}{A}\right)_{i+1}^j\right.\right.$$
$$\left.\left. - \left(\frac{Q^2}{A}\right)_i^j\right]\right\} + \frac{q\,\Delta t}{4\,\Delta x}(A_{i+1}^{j+1} + A_i^{j+1} + A_{i+1}^j + A_i^j)\,[(H_{i+1}^{j+1} - H_i^{j+1})$$
$$+ (H_{i+1}^j - H_i^j)] - 2S_{0,i+1/2} + 2S_{f,i+1/2}^{j+1/2} = 0, \quad (9.15)$$

in which

$$S_{0,i+1/2} = \frac{1}{2}(S_{0,i+1} + S_{0,i}),$$

$$S_{f,i+1/2}^{j+1/2} = \frac{1}{4}(S_{f,i+1}^{j+1} + S_{f,i}^{j+1} + S_{f,i+1}^j + S_{f,i}^j),$$

and

$$S_{f,i}^j = \left(\frac{Q^2}{C^2 A^2 R}\right)_i^j,$$

where C is the Chezy coefficient and R is the hydraulic radius.

The preceding set of equations for each computational element, i, is implicit in A_i^{j+1} and Q_i^{j+1} and is nonlinear. The iteration process using Gaussian elimination procedures may be used to avoid excessive computer storage (Amein and Fang, 1970; Fread, 1971).

All of the numerical methods are quite sensitive to the magnitude of the time step, Δt, and the distance increment, Δx. As noted by Price (1974): (1) the implicit method of Amein is most accurate when $\Delta x / \Delta t$ is approximately equal to the speed of the monoclinal wave; (2) the fixed-mesh characteristic method is most accurate when $\Delta x/\Delta t$ is slightly smaller than the current speed; (3) the implicit method is computationally faster than the other methods with similar accuracy; (4) the greater the difference between the current speed and the monoclinal wave speed, the greater the advantage in using the implicit method; and (5) the time step for the implicit method must be chosen with caution if there are large differences in speeds for different parts of the flood wave, i.e., flooding over an extensive area.

Simplified Flow Routing Methods

Since the St. Venant equations are nonlinear, they represent a difficult mathematical problem, and even their finite difference representation may be unstable and time consuming.

The two partial differential equations may be simplified by (Miller and Cunge, 1975)

1. Use of the continuity equation alone;
2. Use of the momentum equation alone;
3. Simplification of the momentum equation by neglecting the resistance term, linearizing the resistance term, neglecting the local acceleration term, $\partial U/\partial t$, and neglecting the convective acceleration term, $U\partial U/\partial x$;
4. Reliance on statistical information only, i.e., without using the continuity or momentum equations directly.

In some water quality semidynamic models designed for modeling pollution during low-flow conditions, only the quality portion is dynamic, while the hydraulic portion is at steady state. The Muskingum flow routing method is a technique for approximating the modification of a flood wave in a river reach. The flow routing is performed by solving the continuity equation

$$I - Q = dS/dt, \tag{9.16}$$

after establishing a relationship between the reach outflow and storage:

$$S = f(Q),$$

where I is inflow into a reach, Q is outflow, and S is storage within the reach.

The flow storage relationship can be determined either from an empirical equation or from the flow rating curves.

Input–Output Transformation of Flow Hydrographs

It has long been recognized that a method similar to the unit hydrograph method could be utilized for flow routing in channels. Although the unit hydrograph method was proposed more than forty-five years ago, workable transform functions for channel routing appeared in the literature only very recently.

The hydrograph method is based on a numerical or graphical solution of the convolution integral

$$Y = \int_0^t h(\tau)X(t) \, d\tau, \tag{9.17}$$

where Y is the flow from the reach, X is flow into the reach, h is the transformation function, and τ is the lag time.

The transformation function can be obtained from analytical or numerical solution of the kinematic wave equations. A solution of this type presented by Harley (1967) was based upon small perturbations around an initial steady-state depth, H_0, and discharge, Q_0. The solution leads to the

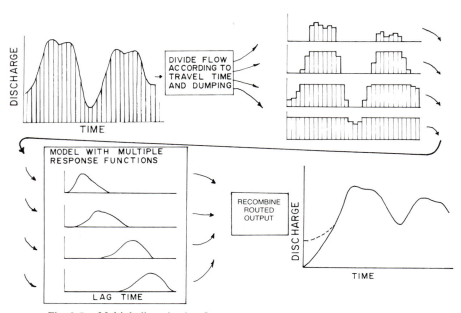

Fig. 9.5. Multiple linearization flow routing (Keefer and McQuivey, 1974).

following reach hydrograph function:

$$h_q(x, t) = \frac{1}{2(\pi D_0)^{1/2}} \times \frac{x}{t^{3/2}} \exp\left\{-\frac{(C_0 t - x)^2}{4 D_0 t}\right\}, \qquad (9.18)$$

where $D_0 = Q_0/2 S_0$, $C_0 = 1.5 Q_0/H_0$, and S_0 is the bottom slope.

If the Froude number is <0.5, the method agrees closely with the finite difference solution of the St. Venant equations.

Flow routing by this method is similar to that of the unit hydrograph for overland flow. Since the channel flow equations are highly nonlinear, it may not be possible to use only one hydrograph curve. It may be necessary to subdivide the flow range into several magnitude intervals and perform routing with appropriate hydrographs according to the flow magnitude. A multiple linearization flow routing method was developed by Keefer and McQuivey (1974) and is shown in Fig. 9.5. This method is simple to program for a digital computer or desktop calculator.

The hydrograph flow routing methods give adequate results provided that the flow Froude number is low and there are no backwater effects due to the junctions or obstacles to the river flow. They cannot be utilized for estuary flow routing where numerical methods are more appropriate.

Residence Time

Residence time or travel time of particles between points along the river flow is one of the most important factors necessary for the evaluation of pollution impact on receiving waters. The time of passage between two points on a stream system is difficult to evaluate because of the statistical nature of the phenomena involved in the process. The time of travel and mixing of particles during flow depends on the degree of turbulence, velocity distribution, configuration of the stream, energy of flow, and variability of the cross-sections of the stream.

Under steady-state conditions and in uniform channels, the average detention time between two points in a stream system is usually related to the average flow velocity as

$$\bar{\tau} = \frac{L}{\bar{U}} = \frac{L_A - L_B}{\bar{U}_{AB}}, \qquad (9.19)$$

where $\bar{\tau}$ is the detention time, L is the length of the reach (distance between points A and B), and $\bar{U}$ is the average velocity in the reach.

Even with steady-state assumptions, the real distribution of the residence time for particles of water traveling between points A and B corresponds to the velocity distribution in a typical cross-section (Fig. 9.6), which varies from zero near the bottom and banks to $u_{\max}$ near the top of

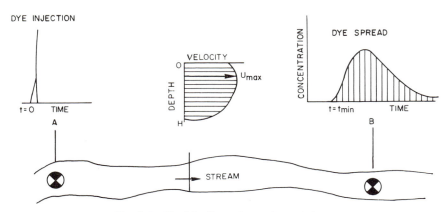

Fig. 9.6.   Particle dispersion and travel time.

the midsection. If a tracer, such as Rhodamine WT, is injected at point A and a concentration curve is obtained at point B and plotted versus time, with time zero being the time of injection at point A, the curve will approximately represent the time of travel distribution between A and B (if represented in a dimensionless form, i.e., concentration divided by the mass of the tracer). Figure 9.6 also shows a typical time of travel distribution curve. The spread of the curve depends on the amount of dispersion occurring in the reach between A and B which is a function of the level of turbulence in the reach. The flow through time may be assumed to be either the modal or peak value, or the centroid. Practically, in straight channels, the modal value will adequately represent the flow through time, while in channels with many bends, the centroid is more representative of the detention time.

The average travel velocity can be also computed using common engineering formulas such as the Chezy equation

$$\bar{U} = C(RS)^{1/2},$$

where R is the hydraulic radius, $R = A/P$, S is the slope of the energy grade line, which under the steady-state uniform flow assumption is approximately equal to the bottom slope, C is an empirical coefficient, A is the cross-sectional area, and P is the wetted perimeter.

The coefficient, C, can be evaluated from the Manning equation:

$$C = \frac{1}{n_M} R^{1/6}, \tag{9.20}$$

where n_M is a coefficient reflecting the roughness of the channel. Typical values of the coefficient n_M for overland flow were given in Chapter 6,

TABLE 9.1 Manning's roughness factors n_M for channel flow[a]

Type of Flow	n_M
Concrete lined channel	0.012
Earth canal—straight and uniform	0.017
Rock cut canals—smooth and uniform	0.025
Winding sluggish canals	0.0225
Dredged earth canals	0.025
Natural streams	
Clean, straight bank	0.025
Clean, straight bank with some weeds	0.030
Winding, some pools and shoals	0.033
Stony sections	0.045
Very weedy reaches	0.075

[a]From Brater and King (1976).

Table 6.7. These coefficients can be used for flow computations in flood plains. Values of n_M typical for channel flow are presented in Table 9.1.

Detention times can also be determined by the use of floats, using the relationship between surface and average velocity. Oranges are sometimes used for this purpose, inasmuch as they will just float on the surface of the water. In addition to cost, an added advantage is that when the tests are completed, the oranges can be eaten.

It should be emphasized that travel time determination by a dye test performed during several different flow levels is far superior to the use of any other method. The kind of curve that is most useful to stream studies is shown in Fig. 9.7.

In addition to the travel time evaluation, other hydraulic parameters are necessary for stream studies. The depth can be determined from cross-sections or computed by knowing the width and corresponding flow and velocity. Both depth and velocity (travel time) should be determined for several flows, the results of which can be conveniently plotted and correlated on log–log paper, resulting in these relationships:

$$\text{Depth} \qquad H = aQ^b$$

$$\text{Velocity} \qquad \bar{U} = \frac{L}{\tau} = cQ^d$$

$$\text{Detention time} \qquad \tau = \frac{L}{\bar{U}} = \frac{L}{c}Q^{-d}$$

$$\text{Width} \qquad W = eQ^f$$

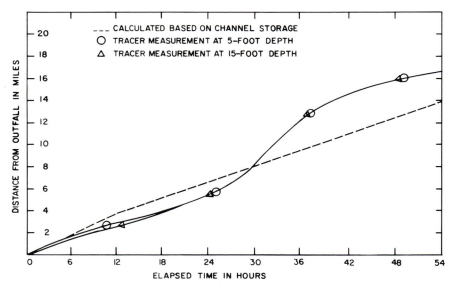

Fig. 9.7. A typical flow-through curve.

Table 9.2 shows typical values of the coefficients, and Figure 9.8 shows typical power function plots. United States Geological Survey gauging stations data can sometimes be used for determining the coefficients for reaches close to the stations.

As shown in Table 9.2, $b + d \simeq 1.0$. Furthermore, natural streams are

TABLE 9.2 Coefficients of the relation of depth and velocity (detention time) with flow[a]

| | Coefficients | | | |
| | On depth | | On velocity | |
Stream	a	b	c	d
1. Impounded waters	H (a constant)	0.0	$1/A$	1.0
2. Ideal channel flow	$\dfrac{1}{(Bn_{\mathrm{M}})^{3/5}S_f^{3/10}}$	0.6	$\dfrac{n_{\mathrm{M}}^{3/5}S_f^{3/10}}{B^{2/5}}$	0.4
3. Codorus Creek, Pa.		0.40		0.60
4. Holston River, Tenn.				
River mile 142.2–131.5		0.32		0.68
River mile 131.5–118.4		0.46		0.54
5. Scioto River, Pa.		0.30		0.70
6. Kanawha River, W. Va.				
River mile 58.6	8.541	0.03	0.0001849	0.935
River mile 46	16.468	0.028	0.0001090	0.963

[a] A = cross-sectional area, n_{M} = Manning roughness factor, B = width, and S_f = energy slope.

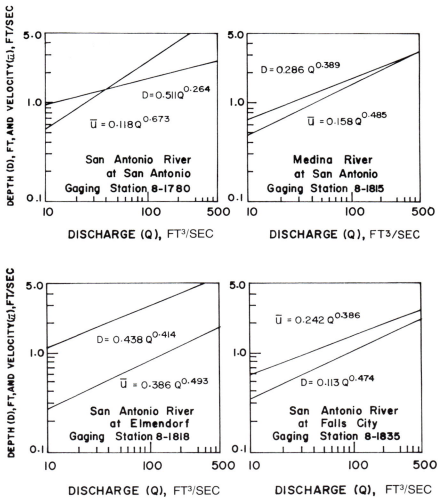

Fig. 9.8. Depth, velocity-discharge relations for selected gauging stations (Texas Water Devel. Bd., 1971).

somewhere between the two extremes of impounded waters and ideal channel flow. Therefore, it may be expected that b would be within the interval 0 to 0.6, and d within the interval 0.4 to 1.0.

Under unsteady flow conditions, the particle movement reflects the local velocity of flow at the location of the particle. The travel time determination may be especially difficult to establish for streams with rapid flow changes such as streams downstream from hydroelectric peaking power plants. Between peaks, guaranteed flow may be the only release, while during the peaks, which occur during periods of high electricity demand and last a few hours, a peak flow is released from the storage reservoir. In this case, the flow can be approximated by a square wave function of the type

$$Q = Q_{guar} + sq \text{ wave flow } (Q_{incr}, T_1, T_2), \quad (9.21)$$

where Q_{guar} is the minimum guaranteed discharge, Q_{incr} is the peak increment discharge, T_1 is the duration of the peak discharge, and T_2 is the time interval between the peaks.

The travel time of the peak flow wave is faster than the travel of the water particles, and it is therefore possible that a particle may pass through more than one consecutive peak wave during its travel downstream. Using a numerical solution of the St. Venant equations it has been found (Novotny, 1970) that the average travel velocity can be related to the duration of the peaks as

$$\frac{\bar{u} - u_{min}}{u_{max} - u_{min}} = \frac{T_1}{T_1 + T_2} \quad (9.22)$$

where $\bar{u}$ is the average travel velocity of a particle, u_{max} is a steady-state velocity corresponding to the peak flow, u_{min} is that for the minimum flow, and T_1 and T_2 are the duration of the peak wave and time between the waves, respectively.

MASS (QUALITY) TRANSPORT

Dispersion and the fate of pollutants in receiving water bodies are of major concern in environmental studies. The hydraulic behavior of the system, whether dynamic or steady, provides only a carrier mechanism. Even when the flows are at steady state, the quality routing may be dynamic.

The pollutants under consideration may be either dissolved or particulate. Dissolved pollutants are a part of the chemical composition of water and thus have the same transport pattern as the water itself. Particulate pollutants, or pollutants associated with particulate sediment movement, are carried by the turbulent drag forces and by a resultant of gravity and buoyancy forces.

In most water quality models, quality is routed and solved assuming that the hydraulic behavior of the system is at steady state for the duration of the quality time increment, Δt. Generally, the quality routing is more stable than that of quantity (hydraulics), and the time steps can be several times larger. In this case, the hydraulics of the system may be averaged over the quality time increment, Δt.

Dissolved Pollutants Transport

Water is a solvent to many components, including gases, liquids, and solids. The laws which govern the solubility and solute–solvent reactions are discussed in the chemical and physical–chemical literature. In the following analysis it will be assumed that the dissolved components are an integral part of water itself and move accordingly with the water. However, it should be remembered that the variability of the chemical composition of water can also significantly alter the hydraulics of the system. For example, waters containing varying suspended solids, dissolved salts, or temperature changes may result in stratified flow or density currents. When introduced into waters with different densities, they may behave as overflows, underflows, or interflows, and demonstrate quite complex hydrodynamic behavior.

The basic mass balance models have been discussed in Chapter 8. It was shown that under given circumstances, mass transport can be modeled by a completely mixed control volume concept, a plug flow concept resembling flow in a pipe with no longitudinal mixing, or a dispersion model which assumes both mixing and dispersion.

The completely mixed flow model is applicable to mass transfer problems in a small reservoir or impoundment and can be represented by the equation

$$dC/dt \, V = Q_{in} C_{in} - QC - KCV \pm \Sigma SV. \qquad (9.23)$$

A simple water continuity-storage equation must couple Eq. (9.23), i.e.,

$$dV/dt = Q_{in} - Q + A(P - Ev), \qquad (9.24)$$

where C is the concentration of the substance in the impoundment (mg/liter), V is the volume of the impoundment (m³), Q is the flow from the impoundment (m³/day), K is the decay coefficient (day^{-1}), ΣS denotes the sum of the sinks and sources of the nonconservative substance (mg/liter/day), Ev is the evaporation rate (m/day), P is the precipitation (m/day), A is the surface area (m²), and C_{in} and Q_{in} are influent concentration and flow, respectively.

Under steady-state conditions, dC/dt and dV/dt are zero. Assuming that ΣS also approaches zero, the relationship between the substance concentration in the impoundment (which is equal to that of the effluent) and influent mass loading becomes

$$C = \frac{Q_{in} C_{in}}{Q + KV}.$$ (9.25)

In some models, longitudinal flow of a nonconservative substance can be modeled by a series of completely mixed tanks.

A more common plug flow concept can be used for the modeling of one-dimensional transport of a nonconservative substance. Then

$$\frac{\partial C}{\partial t} = -U \frac{\partial C}{\partial x} - KC \pm \Sigma S,$$ (9.26)

where U is the flow velocity in the x-direction (main flow direction) in m/day.

Under steady-state flow conditions, the term $\partial C/\partial t$ becomes zero and the equation can be solved assuming that the boundary conditions are $C = C_0$ at $x = 0$. Then

$$C = C_0 \exp\left(-\frac{K}{U} x\right) + \frac{\Sigma S}{K}\left[1 - \exp\left(-\frac{K}{U} x\right)\right].$$ (9.27)

The plug flow model is appropriate for pollution transport in streams and channels after the waste effluent or tributary is completely mixed with the receiving water. The simplified plug flow model cannot be used for modeling mixing zones, estuaries, or where dispersion is significant.

In dynamic modeling, the plug flow mass transfer formula must be solved simultaneously with the flow equations.

Dispersion and Mixing of Pollutants in Streams and Estuaries

The simple plug flow concept of pollutant transport in streams and estuaries, which is so often used by practicing engineers in their analysis of stream pollution, can rationally be applied only to small and well-mixed streams. Frequently, such models provide adequate results only under steady-state conditions. With increased order of streams, the dispersion induced by turbulent mixing and flow distribution increases in significance. Under nonsteady flow or waste discharge conditions, dispersion cannot be neglected. For example, in case of spills or highly irregular waste discharges, dispersion and dilution may become the only effective mechanisms to reduce the peak concentration of a conservative substance.

In many estuaries, convective transport by fresh-water inflows is negligible because of a large cross-sectional area and a relatively small fresh-water flow. Therefore, tidal dispersion is the mechanism which flushes the pollutants toward the sea.

Dispersion also can be related to other hydraulic and hydrobiological transport processes such as reaeration, the uptake of contaminants by sediments and benthic biota, and sediment transport.

The term mixing zone is applied to stretches of receiving waters between the effluent outfall and a cross-section where the waste is completely mixed with the receiving water flow. The effective mechanisms which mix the waste with the ambient receiving water are the lateral and vertical dispersion which determine how long it will take before a completely mixed flow is achieved, as shown in Fig. 9.9.

The interim final *Guidelines for State and Areawide Water Quality Management Program Development,* issued by the Environmental Protection Agency in November 1976, contains the following definition of allowable mixing zones: "Water quality standards should describe the State's methodology for determining the *location, size, shape, outfall design,* and inzone quality of mixing zones. . . ."

The guidelines also include recommendations for use of various methods and techniques for defining surface area and volume of mixing zones and the recommendation that the shape of the mixing zone should be a single configuration that is easy to locate in the body of water, preferably a circle with specified radius.

Attempts to rigidly define mixing zone criteria are jeopardizing the regulatory agencies' efforts rather than assisting them. Agencies are now involved in evaluating location and design of outfalls and diffusers, model studies, and field diffusion tests—each with respect to hydrologic limitations which have no bearing on the protection of aquatic life and water uses. Mixing zones should be established on a case-by-case basis, with water quality goals and objectives as the foundation. Many states have used the term "reasonable" in statutes defining the extent of allowable mixing zones. It is a reasonable word, definable and enforceable in each unique and specific case. Specific shapes, areas, and minimization of size are not appropriate for the statutory criteria defining mixing zones. Reasonableness based on precepts such as the next ones result in a soundly based allocation for all resources in defining allowable mixing zones.

1. The allowable zone is a function of acceptable damage to the receiving stream.

2. The acceptable area of damage is related to the resource value of the affected area.

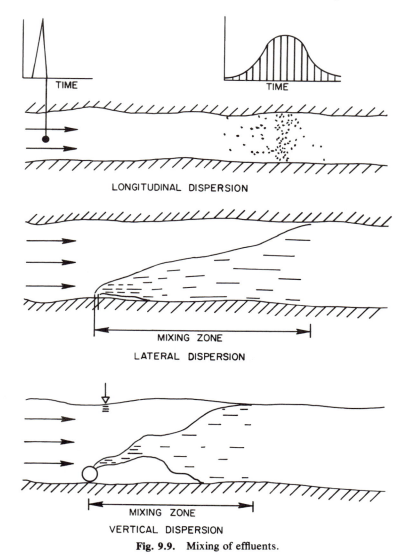

TIME

TIME

LONGITUDINAL DISPERSION

MIXING ZONE

LATERAL DISPERSION

MIXING ZONE

VERTICAL DISPERSION

Fig. 9.9.   Mixing of effluents.

3. The acceptable damage is related to the total amount of equivalent area available as a resource in the receiving water body or segment of water body.

4. Aquatic damage can be related to species generation time and/or fecundity.

5. In some cases the timing of mixing zone use may be related to seasonal utilization of impacted areas.

6. All discharges into the segment of water must be considered.

7. Zones such as spawning sites, food producing areas, and nursery areas should be avoided if possible and migratory pathways protected.

8. No shape, length, area, or cross-section should be specifically precluded from defining an allowable mixing zone.

Turbulence and Dispersion

Dispersion, as it occurs in natural water bodies, is a result of turbulent mixing by eddies and by nonuniformity of the cross-sectional velocity profiles.

Turbulence, to which dispersion is related, is an irregular motion of water particles which generally makes its appearance in fluids when they flow past solid surfaces or when neighboring streams of the same fluid flow pass over one another. The turbulent motion is superimposed on the main convective transport of fluid (Davis, 1972).

In turbulent flow, water masses move in eddies, which are statistical volumetric entities which maintain their existence for some period of time. It is assumed that the velocity vectors of all particles within the eddy volume have some degree of correlation to each other.

In stream flow, a particle moves in the general direction of the flow. As was mentioned earlier, the velocity component in the main direction of flow is U. However, because of turbulence effects, the particle will have an irregular fluctuating motion at any given instant. Thus, the velocity vector will have three components, u, v, and w. Furthermore, since the particle velocity vectors follow an irregular pattern, the Lagrangian (fixed-in-space) point velocity vectors will also fluctuate.

If we choose a fixed-in-space Lagrangian coordinate system rather than the Eulerian system which follows the particle, the velocity components will have statistical character, i.e.,

$$u = \bar{u} + u'$$
$$v = \bar{v} + v'$$
$$w = \bar{w} + w',$$

where $\bar{u}$, $\bar{v}$, and $\bar{w}$ are statistical averages of the velocity components, and u', v', and w' are deviations from the means. Since the net flow is, by definition, in the x direction, only $\bar{u}$ has a value as both $\bar{v}$ and $\bar{w}$ are zero. The fluctuating velocity components are measured in terms of root-mean-

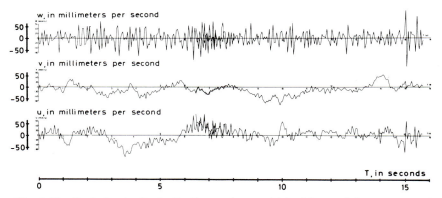

Fig. 9.10. Typical set of velocity time series as obtained by particle measurements (Hansen, 1972).

square velocities $(\bar{u}'^2)^{1/2}$, $(\bar{v}'^2)^{1/2}$, and $(\bar{w}'^2)^{1/2}$. Figure 9.10 shows a typical velocity histogram as it would be measured in an open channel flow (Hansen, 1972).

Types of Turbulence

Flow in open channels is almost always turbulent in practice, with large eddies and large fluctuations in velocity. In spite of low velocities, even lakes and reservoirs can be considered turbulent because of their large size. Obviously, geometric factors are quite important.

In homogenous turbulence, each fluctuation component is independent of its position in space; i.e., the statistical character of the velocity components at a point of flow is unaffected by the translation of the point in the space within the flow. In practice, it is difficult to find homogenous turbulence except over short distances.

In isotropic turbulence, all velocity fluctuations are equal, i.e.,

$$\bar{u}' = \bar{v}' = \bar{w}',$$

and if the fluctuations are random in space, then there is no correlation between the fluctuations in different directions. Therefore,

$$\langle u'v' \rangle_{\mathrm{av}} = 0$$

and similarly for the terms involving w'. The mathematical analysis of isotropic turbulent flow assumes that the velocity gradient is zero; however, this type of flow is not common. There is a tendency for turbulence to be isotropic near the axis of a pipe or at the edge of a boundary layer (i.e., a layer which is formed between a solid surface and a fully turbulent flow).

In a free turbulent jet, sometimes termed a submerged jet, the fluid is expelled from a nozzle into a mass of miscible fluid more or less at rest.

Flow in pipes or shallow open channels, where significant correlations exist between the velocity fluctuations along with a velocity gradient and shear stress, is called nonisotropic.

Scale of Turbulence

The average distance that a small fluid mass, or eddy, will travel before losing its identity is a measure of the average size of eddies responsible for fluid mixing. In fully developed turbulent flow, there is a wide spectrum of eddy lengths. The largest turbulence eddies are of a size comparable to the geometrical dimensions of the channel or the pipe, and the fluctuations of velocity are of the same order as the maximum velocity in the channel. These large eddies account for much of the kinetic energy of turbulent flow. The interaction of these large eddies with one another generates smaller eddies which may range down to sizes comparable to molecular motion. There is a continuous transfer of energy from larger scale eddies to small eddies, where the energy is then dissipated as heat. The rate of energy dissipation therefore may serve as a measure of the energy of turbulent flow.

The average scale of turbulent eddies can only be approximated. In the close vicinity of a solid surface (bank or bottom), Prandtl has shown that the mixing length or the average size of the eddies can be related to the distance from the solid surface:

$$\rho = \kappa y,$$

where ρ is the mixing length, y is the distance from the solid surface, and κ is a constant. Von Karman estimated the value of κ to be about 0.4. Based on Nikuradze's experimental values, Davis (1972) suggested that the value of κ is near 0.4 only in the range $0 < y \leq 0.1\ H$, where H is the depth of flow or radius of the pipe. A flow carrying suspended solids will also have reduced values of κ (Toffaleti, 1968).

Definition of Dispersion

Dispersion is analogous to but not synonymous with diffusion. Molecular diffusion is a process by which matter is transported from one part of a system to another because of the random motion of molecules. Similarly, turbulent diffusion is a result of the random motion of turbulent eddies. The equation which describes the turbulent diffusion is similar to Fick's first law of diffusion which is

$$j = -D \frac{\partial c}{\partial n}, \tag{9.28}$$

where j is the diffusive mass flow, n is a coordinate in the direction of the concentration gradient, and D is the coefficient of diffusion. The unit of D is length^2/time.

Dispersion in channel flow includes, besides molecular and turbulent diffusion, other factors such as the nonuniform velocity distribution caused by solid flow boundaries, tidal convection, wind effect, shear stress, and density differences. In addition, some minor factors also can affect the magnitude of dispersion. These factors include boat traffic, minor disturbances in the river, and biological activity. The dispersion coefficient, or the longitudinal mixing coefficient, D_L, used in water quality models includes all of these factors because of the methodology used in its measurement.

Under laminar conditions, the prevailing transport of molecules of a contaminant injected into the stream would be by molecular diffusion and would be governed by Fick's law of diffusion superimposed on the convective transport of water. In two dimensions, the governing equation becomes

$$\frac{\partial c}{\partial t} + u \frac{\partial c}{\partial x} + v \frac{\partial c}{\partial y} = D_m \frac{\partial^2 c}{\partial x^2} + D_m \frac{\partial^2 c}{\partial y^2}, \tag{9.29}$$

where D_m is the coefficient of molecular diffusion.

The same equation would be true for turbulent flow; however, in turbulent flow, the instantaneous total values of u and v, including the turbulent pulsations, must be used. If the equation could be solved, the solution would give the instantaneous, total values of c, including its fluctuations. But the instantaneous values of u and v are not known, and, furthermore, the instantaneous values of c are not required. Therefore, average values of $\bar{u}$, $\bar{v}$, and $\bar{c}$ are used. The mere introduction of $\bar{u}$, $\bar{v}$, and $\bar{c}$ would not be sufficient since this approach would ignore the transfer by u' and v'. An equation containing $\bar{u}$ and $\bar{c}$ ($\bar{v} = 0$) may be obtained by introducing $u = \bar{u} + u'$, $v = v'$, and $c = \bar{c} + c'$ into Eq. (9.29) and integrating with respect to time to obtain (Holley, 1969)

$$\frac{\partial c}{\partial t} + \bar{u} \frac{\partial c}{\partial x} = D_m \frac{\partial^2 c}{\partial x^2} + D_m \frac{\partial^2 c}{\partial y^2} + \frac{\partial \langle -u'c' \rangle_{av}}{\partial x} + \frac{\partial \langle -v'c' \rangle_{av}}{\partial y}, \tag{9.30}$$

where the angular brackets indicate the time average of the quantity within them. The last two terms represent the movement of particles by turbulent fluctuations. It has been proven experimentally that this movement has the same characteristics as the movement of molecules described by Fick's law. Therefore, by analogy to molecular diffusion

$$\langle u'c' \rangle_{av} = D_{tx} \frac{\partial c}{\partial x} \quad \text{and} \quad \langle v'c' \rangle_{av} = D_{ty} \frac{\partial c}{\partial y}.$$

Thus,

$$\frac{\partial c}{\partial t} + \bar{u} \frac{\partial c}{\partial x} = (D_m + D_{tx}) \frac{\partial^2 c}{\partial x^2} + (D_m + D_{ty}) \frac{\partial^2 c}{\partial y^2}. \tag{9.31}$$

As stated by Holley (1969), the turbulent diffusion therefore includes part of the convection because the convection written in terms of $\bar{u}$ and $\bar{v}$ does not represent the total convective transport.

The preceding development reflects the two-dimensional transport of a tracer or pollutant, assuming that the lateral velocity distribution is uniform, which would be theoretically true for an infinitesimally wide channel. For practical purposes, the average longitudinal point velocity, $\bar{u}$, must be averaged in both the vertical and lateral directions, which introduces another statistical dimension to the value of u. If the transport is averaged over the entire cross-section, a simple replacement of $\bar{u}$ and $\bar{c}$ would not be adequate since this would not account for the variability of $\bar{u}$ and $\bar{c}$ in a cross-section. Replacing $\bar{u} = U + u''$ and $\bar{c} = C + c''$, where U and C are the respective average cross-sectional velocity and concentration, will lead to a new equation for the longitudinal distribution of the average cross-sectional concentration of a tracer or pollutant as

$$\frac{\partial C}{\partial t} + U \frac{\partial C}{\partial x} = (D_m + D_{tx}) \frac{\partial^2 C}{\partial x^2} + \frac{\partial \langle \langle -u''c'' \rangle \rangle_{av}}{\partial x},$$

where the double angular brackets indicate averaging over the entire cross-section. Introducing

$$D_L \frac{\partial C}{\partial x} = \langle \langle -u''c'' \rangle \rangle_{av} + (D_m + D_{tx}) \frac{\partial C}{\partial x},$$

will lead to the most common equation describing the longitudinal transport of a pollutant in a one-dimensional flow, which is

$$\frac{\partial C}{\partial t} + U \frac{\partial C}{\partial x} = D_L \frac{\partial^2 C}{\partial x^2}, \tag{9.32}$$

where D_L is the so-called coefficient of longitudinal dispersion. Note that the coefficient includes the effect of molecular diffusion, part of turbulent convection, and lateral dispersion. As pointed out by Fisher (1968), lateral dispersion may sometimes be the most prevailing factor while molecular diffusion is almost always negligible when compared to the other two mechanisms.

When the substance is nonconservative, the decay and transformation terms must be introduced into Eq. (9.32) which then becomes

$$\frac{\partial C}{\partial t} + U \frac{\partial C}{\partial x} = D_L \frac{\partial^2 C}{\partial x^2} - KC \pm \Sigma S. \qquad (9.33)$$

For steady-state conditions, the solution of Eq. (9.33) will yield (assuming that $\Sigma S \to 0$ or that the sinks and sources term can be included in the decay term)

$$C = C_0 \exp \left[\frac{Ux}{2D_L} - x \left(\frac{U^2}{4D_L^2} + \frac{K}{D_L} \right)^{1/2} \right], \qquad (9.34)$$

(handwritten note: minus, pointing to the first term inside brackets)

where C_0 is the concentration of the substance at the beginning of the reach.

Expansion of Dispersion Formulas into Two and Three Dimensions

The dispersion equation discussed thus far can be applied to the simple one-dimensional transport of substance. However, this does not mean that the application is limited to only narrow uniform channels. In more complex systems such as estuaries, large rivers, and nonstratified lakes, the systems can be reduced to one-dimensional computation elements to which Eq. (9.32) can be applied without significant loss of accuracy. In fact, the application of one-dimensional models to the computational control volumes represents the basic structure of many water quality models. This procedure may also be applied to two-dimensional cases.

There are also some situations in which the flow and mass transport must be solved in two and three dimensions. Such cases include the modeling of mixing zones near an outfall, flow in reservoirs and lakes, the selective withdrawal of water from reservoirs, and the mixing of buoyant jets from submerged outfalls.

The expansion of Eq. (9.33) into a three-dimensional form will yield

$$\frac{\partial c}{\partial t} = \frac{\partial}{\partial x} \left(D_x \frac{\partial c}{\partial x} \right) + \frac{\partial}{\partial y} \left(D_y \frac{\partial c}{\partial y} \right) + \frac{\partial}{\partial z} \left(D_z \frac{\partial c}{\partial z} \right)$$
$$- u \left(\frac{\partial c}{\partial x} \right) - KC \pm \Sigma S. \qquad (9.35)$$

This equation assumes that the average lateral and vertical velocity equals zero and that u is not a function of x. It should be remembered that the diffusion coefficients also include the effect of the turbulent convection as previously discussed.

Numerical Solution of Dispersion

Equation (9.33) can be written for a control volume or element, V_i, as shown in Fig. 9.11 for steady-state, nonuniform hydraulics as

$$\frac{\partial C_i}{\partial t} = \frac{\left(AD_L \frac{\partial C}{\partial x}\right)_{i+1/2} - \left(AD_L \frac{\partial C}{\partial x}\right)_{i-1/2}}{V_i}$$

$$+ \frac{Q_{i-1/2}\,C_{i-1} - Q_{i+1/2}\,C_i}{V_i} - KC_i \pm \Sigma S_i, \qquad (9.36)$$

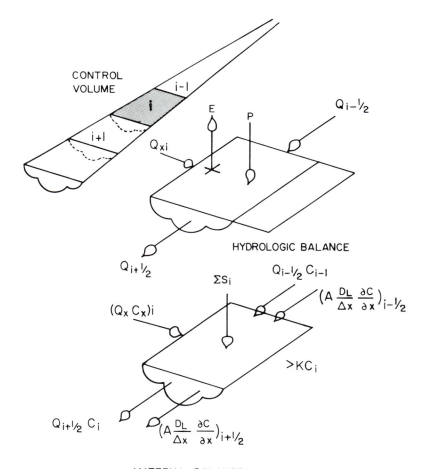

MATERIAL BALANCE

Fig. 9.11. Definition sketch for hydrologic and material balance for control volume.

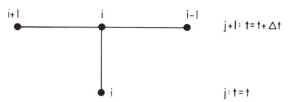

Fig. 9.12.   Computational schematics.

where $V_i = A_i \Delta x$ = volume of the element (m³ or ft³). In a dynamic situation, the hydraulic portion of the model must be solved first for each hydraulic time step. Then the water quality routing can be accomplished by writing the finite difference representation of Eq. (9.36). Since all values of the variables at a time level, t, are known, the solution for a new time level can be obtained by an implicit computational scheme. Since the quality equations are linear, the solution is more stable than that for flow, and longer time intervals can be selected.

A numerical implicit computational scheme was developed for the water quality model, QUAL II (Texas Water Devel. Bd., 1971).

The finite difference scheme was formulated by considering the constituent, C, at four points in the scheme, as shown in Fig. 9.12.

Three points are required at time $j + 1$ in order to approximate the spatial derivatives. The temporal derivative is approximated at the distance step, i, by a forward difference as

$$\frac{\partial C_i}{\partial t} = \frac{C_i^{j+1} - C_i^j}{\Delta t} \tag{9.37}$$

The terms on the right-hand side of Eq. (9.36) can be represented at a new time level, $j + 1$, by the central difference approximation of the dispersion term and by the backward difference approximation of the convection term, or

$$\frac{C_i^{j+1} - C_i^j}{\Delta t} = \frac{AD_L(C_{i+1}^{j+1} - 2C_i^{j+1} + C_{i-1}^{j+1})}{(\Delta x)^2 V_i}$$
$$+ \frac{(Q_{i-1} C_{i-1}^{j+1} - Q_i C_i^{j+1})}{V_i} \pm Q_{xi} C_{xi} \pm S_i$$
$$- K \frac{C_i^j + C_i^{j+1}}{2}. \tag{9.38}$$

The equation is solved for unknown C^{j+1} for all computational elements i. In the equation, $Q_{xi} C_{xi}$ is the local inflow or withdrawal of the pollutant and V_i is the volume of the element. Then

$$C_{i+1}^{j+1} \frac{AD_L \, \Delta t}{(\Delta x)^2 V_i} + C_i^{j+1} \left(-\frac{2AD_L \, \Delta t}{(\Delta x)^2 V_i} - \frac{Q_i \, \Delta t}{V_i} - \frac{1}{\Delta t} - \frac{K}{2} \right)$$

$$+ C_{i-1}^{j+1} \left(\frac{AD_L \, \Delta t}{(\Delta x)^2 V_i} + \frac{Q_{i-1} \, \Delta t}{V_i} \right)$$

$$= C_i^j \pm \Delta t S_i \pm \Delta t Q_{xi} C_{xi} + \Delta t \frac{K}{2} C_i^j. \quad (9.39)$$

The coefficients on the left-hand side of the equation can be given as

$$\alpha_i = \frac{AD_L \, \Delta t}{(\Delta x)^2 V_i},$$

$$\beta_i = \frac{2AD_L \, \Delta t}{(\Delta x)^2 V_i} - \frac{Q_i \, \Delta t}{V_i} - \frac{1}{\Delta t} + \frac{K}{2},$$

$$\gamma_i = \frac{AD_L \, \Delta t}{(\Delta x)^2 V_i} + \frac{Q_{i-1} \, \Delta t}{V_i},$$

$$\rho_i = -C_i^j \pm \Delta t S_i \pm Q_{xi} C_{xi} \, \Delta t + \Delta t \frac{K}{2} C_i^j.$$

The equation for all computational elements can be rearranged in a matrix:

$$
\begin{bmatrix}
\beta_1 \gamma_1 & & & & & & \\
\alpha_2 \beta_2 \gamma_2 & & & & & & \\
& \alpha_3 \beta_3 \gamma_3 & & & & & \\
& & \cdot & & & & \\
& & & \cdot & & & \\
& & & & \alpha_i \beta_i \gamma_i & & \\
& & & & & \cdot & \\
& & & & & \alpha_{n-1} \beta_{n-1} \gamma_{n-1} & \\
& & & & & & \alpha_n \beta_n
\end{bmatrix}
x
\begin{bmatrix}
C_1^{j+1} \\
C_2^{j+1} \\
C_3^{j+1} \\
\cdot \\
\cdot \\
C_i^{j+1} \\
\cdot \\
C_n^{j+1} \\
C_n^{j+1}
\end{bmatrix}
=
\begin{bmatrix}
\rho_1 \\
\rho_2 \\
\rho_3 \\
\cdot \\
\cdot \\
\rho_i \\
\cdot \\
\rho_{n-1} \\
\rho_n
\end{bmatrix}
\quad (9.40)
$$

where n is the number of computational elements.

Several solution techniques based on a matrix inversion are available (Ralson and Wilf, 1960).

The process of pollutant transformation can be coupled with some other form of the pollutant to which or from which the pollutant is transformed or consumed. Dissolved oxygen is coupled with the BOD removal, nitrogen transformation, and possibly other reactions. The movement and transformation of phosphorus is coupled with adsorption and sedimentation of sediments.

In some instances, such as water quality modeling during a low-flow period, the hydraulics of the system and most of the input can be considered to be time invariant. In this case, simplified steady-state models can be selected or developed. At steady state, the system is in equilibrium with the inputs. It should be noted that steady-state modeling may not be applicable to storm-water runoff modeling or to modeling of pollution from nonpoint sources.

Boundary Conditions   The modeling scheme outlined here is applicable both to stream and estuary systems; however, the boundary conditions may differ. The upstream boundary is usually a measured and/or inputed upstream fresh-water (headwater) hydrograph and pollutograph; hence,

$$Q_1 = \text{input upstream hydrograph,}$$

$$Q_1 C_1 = \text{input upstream pollutograph.}$$

Similarly, the downstream boundary condition for estuaries must be inputed, usually as the tide level or flow and the pollutant concentration at the seawater and of the system:

$$Q_n = f \text{ (tide level),}$$

$$C_n = \text{seawater pollutant concentration.}$$

The relationship for flow in dynamic water quality models is solved by the hydraulic portion of the model.

For stream models, the downstream boundary condition is not as important as it is for an estuary and can be simplified as

$$Q_n = Q_{n-1}$$

and

$$Q_n C_n = Q_{n-1} C_{n-1}.$$

Hydrograph Routing Models

A model schematic similar to the unit hydrograph method (Chapter 8) also can be utilized for water quality and quantity models. This method has been applied mainly to watershed hydrology models; however, it has been shown that similar methods have been applied to stream flow routing. The method is based on the equation

$$Y_{\text{outlet}} = \int_0^t h(\tau) X_{\text{inlet}}(t - \tau) \, d\tau, \tag{9.41}$$

where h is the hydrograph routing function, X is the input to the system, Y is the output, and τ is the lag time.

These models have several advantages to the numerical models. First, the solution is much faster, and, second, there is little, if any, convergence problem. Therefore, the division of reaches and the time increments can be coarser. The models, however, represent a strictly forward scheme; i.e., they cannot be used if backwater effects take place or if the system is affected by tide.

Estimating Logitudinal Dispersion

Three methods are available for estimating the longitudinal dispersion coefficient, D_L. First, there are numerous empirical and semiempirical equations primarily developed from observations in laboratory flumes. Second, the coefficient, D_L, can be estimated by injecting a tracer into the stream. A third method, which is applicable to estuaries, involves evaluation of D_L from the extent of the salinity intrusion into the estuary. Table 9.3 shows typical values of the dispersion coefficient for streams and estuaries.

Empirical Formulas for D_L Taylor (1954) investigated dispersion in pipe flow and derived a predictive equation for the longitudinal dispersion coefficient, D_L, in long straight pipes as

$$D_L = 10 \, r_0 u_* \quad (\text{m}^2/\text{sec}), \tag{9.42}$$

where r_0 is the pipe radius and u_* is the average shear stress velocity given by

$$u_* = (\tau_0/\rho)^{1/2},$$

where τ_0 is the boundary shear stress and ρ is the density of water.

A modification of Taylor's equation for streams and estuaries would yield

$$D_L = 63 \, n_{\mathrm{M}} U_t R^{5/6}, \tag{9.43}$$

TABLE 9.3 Typical values of the dispersion coefficient

System Classification	$D_L(\text{m}^2/\text{sec})$
Flumes and small rivers	3×10^{-3}
Large rivers	3×10^{-1}
	30
Estuaries	600

[a]After Gloyna (1967).

in which U_t is the stream or tidal velocity in m/sec, R is the hydraulic radius in meters, and n_M is the Manning roughness factor.

In estuaries the tidal velocity is

$$U_t = \frac{Q_f}{A} + U_r \sin \sigma t, \tag{9.44}$$

where Q_f is the fresh-water discharge at the head of the estuary, U_r is the maximum value of the tidal velocity, $\sigma = 2\pi/T$ is the frequency of the tide, and T is the tidal period.

Notable works on the theoretical determination of the dispersion coefficient are those of Elder (1959) and Fisher (1968). Elder assumed that the governing mechanism for the dispersion in streams is the vertical velocity gradient. By assuming a logarithmic velocity profile and a similarity between the momentum transfer and mass transfer, he obtained an expression analogous to Taylor's formula, which was

$$D_L = 5.93\, Hu_* . \tag{9.45}$$

Neither Elder's nor Taylor's developments account for lateral dispersion. Field measurements indicate that the two equations may give results that are low, sometimes more than an order of magnitude. Fisher (1968) concluded that this may be caused by the presence of lateral velocity gradients in addition to the vertical ones. His original equation is complicated and unusable for practical purposes. However, in a later publication, Fisher (1975) noted that his integral equation could be approximated as

$$D_L = \frac{0.07\, u'^2\, \ell^2}{\epsilon_z}, \tag{9.46}$$

in which u'^2 is the deviation of velocity from the cross-sectional mean; ℓ is the distance from the thread of the maximum velocity to the most distant bank; and ϵ_z is the transverse mixing coefficient. In his laboratory and field experiments, the ratio, u'^2/U^2 varied from 0.17 to 0.25, with a mean of 0.2. Reasonable assumptions for real streams are $u'^2 = 0.2\, U^2$ and $\ell = 0.7\, B$, in which B is the width of the stream, and $\epsilon_z = 0.6\, Hu_*$. With these substitutions, Eq. (9.46) becomes

$$D_L = \frac{0.11\, U^2 B^2}{Hu_*} . \tag{9.47}$$

Liu (1977) replaced the constant 0.11 in Eq. (9.47) by a factor

$$\beta = 0.18 \frac{u_*^{1.5}}{U} = 0.18 \left[\frac{(gRS_0)^{1/2}}{U} \right]^{1.5} . \tag{9.48}$$

Some of the predictive equations are based on the Kolmogoroff (1941) principle, which establishes a relationship between the mixing coefficient, the energy expenditure, and a measure of the scale of turbulence. This principle leads to a relationship for the dispersion of the type

$$D_L = KE^{1/3}L^{4/3}, \tag{9.49}$$

where K is a constant, E is the rate of flow energy dissipation ($E = US_0g$), and L is a measure of the scale of turbulence or, specifically, the Lagrangian eddy size. Krenkel and Orlob (1962) found excellent correlation with this equation using data obtained from a laboratory flume.

In general, the preceding equations may serve as a fair first approximation. However, it should be realized that, as noted by Fisher (1975), an order of magnitude accuracy is the best that the theoretical equations can provide. Fortunately, the concentration distribution is not excessively sensitive to the magnitude of the dispersion coefficient, and an error in estimation by a factor of 4 or slightly higher will have little effect.

Tracer Injection If a tracer or dye is injected into a stream or estuary, it will travel in a slug downstream according to the convective movement of the water. Because of turbulent dispersion and mixing, the slug's volume and extent will expand and, consequently, its concentration will decrease as shown in Fig. 9.13. If the tracer injection is instantaneous, the concentration curve at a point downstream from the tracer injection can be considered a stream system response to a pulse mass input. For a conservative tracer, Eq. (9.32) can be solved for a pulse input (Krenkel, 1960) yielding

$$C = \frac{M}{A(4\,D_L t)^{1/2}} \exp\left[-\frac{(x - Ut)^2}{4\,D_L t}\right], \tag{9.50}$$

where M is the weight of the tracer, A is the cross-sectional area of the receiving stream or estuary, and x is the distance from the point of injection.

Various analytical methods have been suggested for the determination of D_L from the tracer concentration curves using Eq. (9.50).

Levenspiel and Smith (1957) developed an equation which estimates the dispersion coefficient from the variance of the time-concentration curve, i.e.,

$$D_L = \frac{UL}{8}[(8\sigma^2 + 1)^{1/2} - 1], \tag{9.51}$$

where L is the length of flow considered and σ^2 is the variance of the time-concentration curve which can be obtained from the description of the curve. Then *should be minus*

$$\sigma_t^2 = \frac{\Sigma(t^2C)}{\Sigma C} = \left[\frac{\Sigma(tC)}{\Sigma C}\right]^2,$$

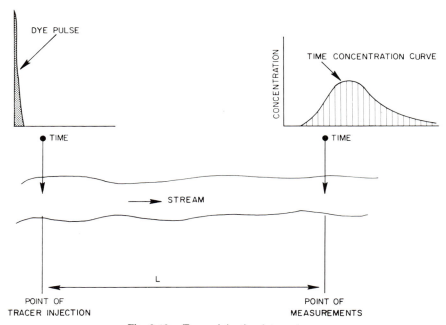

Fig. 9.13.   Tracer injection into a river.

in which the summation is taken over all uniformly spaced readings. Also,

$$\sigma^2 = \left[\frac{Q}{U}\right]^2 \sigma_t^2.$$

Levenspiel's solution requires no measure of absolute concentrations and readily lends itself to computer analysis.

Another possible method to derive the longitudinal dispersion coefficient is to use a logarithmic form of Eq. (9.50), where

$$\log Ct^{1/2} = \log \frac{M}{A(4\pi D_L)^{1/2}} - \frac{(x - Ut)^2}{4 D_L t} \log e. \qquad (9.52)$$

A plot of $\log Ct^{1/2}$ versus $(x - Ut)^2/t$ theoretically results in a straight line with a slope equaling $\log e/4D_L$ from which D_L can be determined.

The tracer used for the test should be "instantaneously" injected at a point where good mixing is ensured. The concentration curve should be measured at a point further downstream, far enough to obtain a good characteristic curve with concentrations which can be clearly measured by the techniques used for the detection of the tracer.

The time-concentration curve measured at a point downstream does

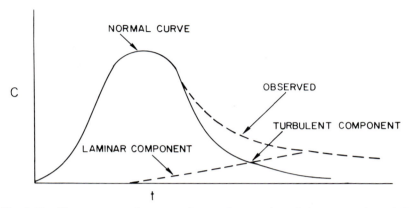

Fig. 9.14. Time-concentration curve from a dye test in turbulent open channel flow (Elder, 1959).

not usually follow a symmetrical theoretical bell-shaped curve and is skewed as shown in Fig. 9.14. The apparent deviations can be explained by the presence of laminar sublayers or "dead" zones in the turbulent flow. The release of the tracer particles from these zones is slow and primarily by molecular diffusion.

Examples of possible tracers include

1. *Rhodamine B or WT*, a fluorescent intensive dye with no known harmful or toxic effects on humans or aquatic biota. It can be detected in very low concentrations (up to μg/liter) by fluorometric techniques.

2. *Salt*, which mixes readily with the water and can also be easily measured by the conductivity of water. However, it changes the density of water and may cause density currents.

3. *Tritiated water*, which can be measured by radiometric techniques. Although it is probably the "ideal" tracer, adverse reactions to its use can be expected from public health officials and the general population.

4. *Fluorescein*, which is detected in small quantities. However, its use has been minimized in recent years because of adsorption properties and its decay when exposed to sunlight.

In estuaries the flow velocities oscillate because of relatively frequent tidal waves. It is common practice to determine the concentration at tidal slack times when the tidal component of the velocity is zero. Then, if the instantaneous injection occurs at high water slack at $t = 0$ and $x = 0$, Eq. (9.50) can be used in the following form:

$$C_{HWS} = \frac{M}{A(4\pi \, D_L NT)^{1/2}} \exp - \left[\frac{(x - U_f NT)^2}{4 \, D_L NT} \right], \tag{9.53}$$

where C_{HWS} is the concentration of the tracer at high water slack, U_f is the fresh-water velocity $(= Q_f/A)$, T is the tidal period, and $N = 1, 2, 3,$ . . . , is an integer number of slacks. The dispersion coefficient can be determined by fitting the tracer concentration curve to Eq. (9.53).

Estimation of D_L from the Salinity Distribution in Estuaries   Equation (9.53) can easily be solved for salinity intrusion into a nonstratified estuary. At the mouth of the estuary, the salinity concentration, C_0, is constant, being equal to that of the sea. Then the steady-state solution of Eq. (9.53) with boundary conditions $C = C_0$ at $x = 0$ will yield

$$C = C_0 \exp\left(Ux/D_L\right). \tag{9.54}$$

Noting that the slope of the line equals U/D_L from a semilogarithmic plot of log C versus x, D_L can be determined. Since the salinity distribution is a result of long-term averages, the average velocity, U, is approximately equal to the fresh-water velocity in the estuary. Salinity can easily be measured using conductivity, total dissolved solids, or chloride concentration.

Estimating Lateral and Vertical Dispersion

Although the majority of transport problems in water pollution control are one-dimensional, a two- or three-dimensional solution may be required for mixing zones or large stratified irregular water bodies. In these cases the intensity of the lateral and vertical diffusion must be ascertained. One would expect that there would be some correlation among the longitudinal, lateral, and vertical diffusion but the coefficients would be equal only in a rare case of isotropic turbulent flow.

From the theory of turbulence, the vertical diffusion coefficient can be related to the turbulent mixing of particles by eddies. The mixing was defined as

$$\epsilon = \ell(\overline{v'^2})^{1/2}, \tag{9.55}$$

where ϵ is the turbulent mixing coefficient and ℓ is the mean size of the eddies. According to Boussinesq the turbulent shear stress, τ, is

$$\tau = \rho\epsilon\frac{du}{dy}, \tag{9.56}$$

which can also be expressed in terms of the bottom shear stress, τ_0, as

$$\frac{\tau}{H - y} = \frac{\tau_0}{H}. \tag{9.57}$$

The velocity distribution can be approximated either by the logarithmic velocity distribution law,

$$du/dy = u_*/\kappa y \tag{9.58}$$

or by a seventh power law,

$$du/dy = u_{max}/(7\ H^{1/7}y^{6/7}). \tag{9.59}$$

The seventh power law is clearly an approximation. Since the shear stress velocity, u_*, and the bottom shear stress, τ_0, can be related as

$$u_* = \left(\frac{\tau_0}{\rho}\right)^{1/2} = (gHS_f)^{1/2},$$

then

$$\epsilon = \kappa y \left(\frac{H - y}{H}\right)(gHS_f)^{1/2}, \tag{9.60}$$

where S_f is the energy slope.

Although Eq. (9.60) does predict the value of the vertical turbulent diffusion at any point in the vertical, the values at the bottom and at the surface obtained by the equation indicate zero diffusion, which contradicts practical measurements.

An average value of the vertical mixing coefficient is

$$\bar{\epsilon} = \frac{1}{H}\int_0^H \epsilon\ dy = \frac{\kappa}{6}H(gHS_f)^{1/2}. \tag{9.61}$$

Surface (Lateral) Diffusion When a dye or pollutant is injected into a channel flow from a point or linear vertical source, the plume will spread laterally as shown in Fig. 9.15. The spread increases with time, with maximum concentrations at the centroid of the plume decreasing at a cross-section to the ambient concentration. With increasing distance from the source, the variance of concentration in a plane normal to the axis of the plume will also increase. The spread of the plume can be described by the Einstein diffusivity equation (Orlob, 1958),

$$D = \frac{U}{2}\frac{d(\sigma_z{}^2)}{dx}, \tag{9.62}$$

where D is the diffusion coefficient in any direction and σ_z is the variance of the normalized cross-sectional concentration curve.

When x is sufficiently large, the diffusion coefficient will become a constant; that is, the standard deviation will be a function of $\sqrt{x}$. From the foregoing considerations, the plume or spread of diffused matter from a point source in an infinitesimally wide two-dimensional flow system should be parabolic and symmetrical about a line through the source and

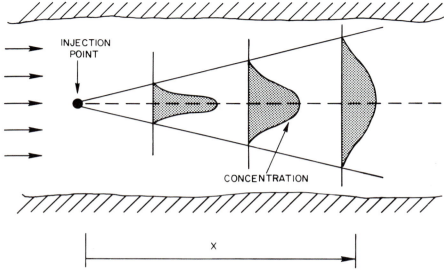

Fig. 9.15.   Lateral (surface) diffusion.

the direction of mean flow. It can readily be seen that if the standard deviations of the spread of a tracer in a two-dimensional system could be determined at any two distances, x_1 and x_2, sufficiently far downstream from the point of injection to establish a constant relation between σ_z^2 and x, the determination of the diffusion coeffient would be a simple computation. A line source in a real three-dimensional flow system would lend itself to similar analysis, assuming that no transverse velocity gradient exists.

The ratio of the longitudinal coefficient, D_L, and lateral diffusion coefficient, D_s, varied from $D_L/D_s = 30:1$ for river field data to $D_L/D_s = 80:1$ for laboratory flume data (Morris *et al.*, 1967).

Dispersion and Transport of Pollutants in Estuaries

The effects of dispersion are most important in estuaries, where dispersion is sometimes the primary moving mechanism for flushing pollutants toward the sea.

An estuary is a semienclosed coastal body of water which has a free connection with the open sea and within which seawater is mixed with fresh water from land drainage (Pritchard, 1969). A narrow embayment which extends deep inland is called a fjord. A fjord is a special case of an estuary with little fresh-water inflow.

Estuaries (including fjords) can be positive, with a net seaward flux of water; negative, with net landward flux of water; and neutral.

Large rivers form positive estuaries, and, in some cases, the fresh-water effect can be detected many miles off shore in the sea. Fjords are mainly negative estuaries.

Estuaries are hydraulically more complex than streams. Although all of the basic equations remain the same both for streams and estuaries, the boundary conditions are more complex. At one end of the estuary there is a time variable inflow of fresh water, and, at the other end, there is an oscillating, more or less regular, tidal effect bringing higher density salt water into the estuary. As a result, many estuaries are stratified. The tidal changes are much faster than most of the fresh-water inflow variations, resulting in a back and forth slug movement of water in the estuary.

Hydraulic models of estuaries usually employ a finite difference solution of the St. Venant equations along with the conservation of mass in the computational elements and nodes. In such solutions, estuaries are broken down into a network of one-dimensional channels (stretches connected by junctions) and elemental water bodies connected by nodes as shown in Fig. 9.16. The solution is performed for each element in a small time increment, Δt, subject to the boundary conditions at each inlet and the estuary outlet. Such solutions are highly dynamic if the computational time increment is less than the tidal cycle but can become steadier if Δt is increased. The one-dimensional concept assumes that the estuary is not stratified, i.e., that the density is uniform over the depth. Equations (9.35)–(9.39), along with the numerical model for unsteady flow conditions, can serve as an example of a possible computational scheme.

Estuaries are very often stratified, with the heavier, more saline, seawater near the bottom and the lighter fresh water on the top. There is usually a surface layer in which the vertical salinity gradient is small, an intermediate layer in which salinity increases very rapidly, and a lower high density layer in which the salinity gradient is again small. Although the intermediate zone is a buffer between the upper and lower zones, complete inhibition of mass transfer between the upper and lower zones is rare. In a partially mixed estuary, in addition to the tidal dispersion, there is a superimposed circulation pattern with net seaward movement in the upper layer and landward movement in the lower layer, as shown in Fig. 9.17.

In analyzing the pollutant transport of a partially mixed estuary, the estuary can be divided into longitudinal segments. Each segment is then partitioned into two vertical segments. The method of analysis is called a two-dimensional box model (Pritchard, 1969).

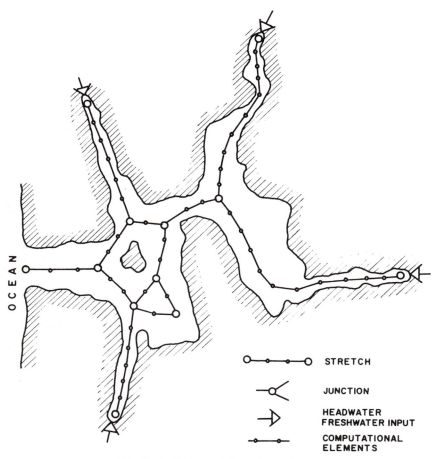

Fig. 9.16. Representation of an estuary.

A steady-state mass balance for the nth segment can be written as (Fig. 9.21)

$$(Q_u)_{n-1,n}(S_u)_{n-1,n} + (Q_v)_n(S_v)_n + E_n[(S_1)_n - (S_u)_n]$$
$$= (Q_u)_{n,n+1}(S_u)_{n,n+1}. \quad (9.63)$$

Also, from the continuity equation

$$(Q_v)_n = (Q_u)_{n,n+1} - (Q_u)_{n-1,n}, \quad (9.64)$$

it follows that

$$Q_u = Q_f \frac{S_1}{S_1 - S_u} \quad \text{and} \quad Q_1 = Q_f \frac{S_u}{S_1 - S_u}. \quad (9.65a,b)$$

In the preceding equations, Q_u and Q_1 denote the flows in the upper and lower layers, respectively, Q_f is the fresh-water inflow into the estuary, S_u and S_1 are the upper and lower layer salinity concentrations, respectively, Q_v is the volumetric rate of the net vertical flow, S_v is the salinity concentration at the boundary between the upper and lower layers, and E_n is the vertical exchange coefficient.

If the vertical salinity distribution and fresh-water inflow are known, the horizontal volume flux can be determined from Eqs. (9.65a,b). The vertical volume flux then can be computed from Eq. (9.64), and Eq. (9.63) can be solved for the only unknown, the vertical exchange coefficient, E_n, for each segment.

If a pollutant is introduced into the estuary in a segment, k, the mass balance for the segment is expanded, (Eq. 9.63), by the term q_0 which is the mass inflow of the pollutant. Then, for any segment except the one receiving waste discharge, the mass balance equation (9.63) can be written again with the concentration of the pollutant, C, replacing the salinity, S. The mass balance for the segment receiving waste becomes

$$(Q_u)_{k-1,k}(C_u)_{k-1,k} + (Q_v)_k(C_v)_k + E_k[(C_1)_k - (C_u)_k]$$
$$+ q_0 = (Q_u)_{k,k+1}(C_u)_{k,k+1}. \quad (9.66)$$

Accepting that $(S)_{n,n-1} = (S_n + S_{n-1})/2$ and similar relationships for C,

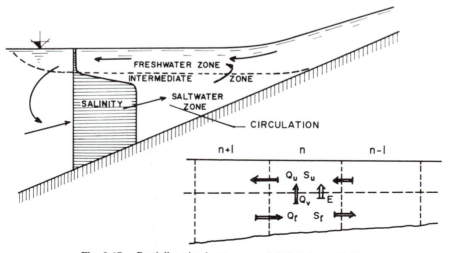

Fig. 9.17. Partially mixed estuary model (Pritchard, 1969).

the mass balance equations can be simplified to

$$(C_u)_{n-1}(Q_u)_{n-1,n} - 2(C_u)_n E_n + (C_1)_n [2E_n + (Q_v)_n] \\ - (C_u)_{n+1}(Q_u)_{n,n+1} = 0, \quad (9.67a)$$

$$(C_1)_{n+1}(Q_1)_{n,n+1} - 2(C_1)_n E_n + (C_u)_n [2E_n - (Q_v)_n] \\ - (C_1)_{n-1}(Q_1)_{n,n-1} = 0 \quad (9.67b)$$

and

$$(C_u)_{k-1}(Q_u)_{k-1,k} - 2(C_u)_k E_k + (C_1)_k [2E_k + (Q_u)_k] \\ - (C_u)_{k+1}(Q_u)_{k,k+1} + q_0 = 0. \quad (9.67c)$$

Since all variables except the concentration of pollutant are known or can be determined from the salinity distribution in the estuary, solution for the unknown C's is possible. The boundary conditions are derived from the assumption that the pollutant cannot penetrate beyond the intrusion of the ocean-derived salt wedge and that the ocean water entering the estuary has no pollutant content. Hence, $C \to 0$ at $S \to 0$ at the upstream boundary and $(C_1)_{m+1} \to 0$ at the downstream boundary. The model can be expanded to a nonconservative pollutant if a decay term is included in the mass balance equations.

Suspended (Particulate) Pollutants Transport

Sediment itself is considered a pollutant, and discharge limitations have been imposed for suspended solids. In addition, many other pollutants can be transported in the suspended solid form or adsorbed on suspended particulates. Adsorbed or suspended pollutants in channel flow include phosphates, heavy metals, ammonia, many pesticides and organics.

The suspended particles carried by streams originate from soil erosion, bank erosion, urban solids washload, and organic life processes. A substantial portion of suspended solids may be the result of industrial processes, mining, development, and agriculture.

The channel phase of the sediment transport can be divided into the suspended fraction and the fraction of sediments contained by moving streambeds. The particle size of the suspended fractions ranges from clay ($<2\mu$m), silt ($2\mu - 0.1$ mm), fine sand (>0.1 mm), and organic materials. Larger fractions are usually a part of the bed load. Primary variables known to influence sediment transport in streams are flow velocity, flow depth, slope and energy gradient, density and viscosity of the water-sediment mixture, mean fall diameter and graduation of the bed sediment, and seepage forces on the streambed. Depending on the channel hydraulic factors, the sediment can be deposited from the stream or scoured from the stream bottom.

The suspended fraction is called washload and is defined as that part of the sediment load which consists of grain sizes finer than those of the bed as determined by the available upstream supply (Shen, 1971). The part of the sediment which comprises the bed is called the bed load. Several authors (Shen, 1971; Chow, 1964) have indicated that the washload may comprise from 90 to 95% of the total sediment load. Also, almost all of the pollutants associated with suspended sediments are part of the washload, while the bed load has very little water quality relevance (with the exception of stagnant pools and reservoirs). However, much research attention has been devoted to the bed load movement of noncohesive particulates.

Hydraulics of Sediment Transport

In turbulent flow, suspended particles are supported and distributed in the flow by the mechanism of turbulent exchange, which is acting against the gravity force of the particles. The concept is similar to that of momentum exchange. Therefore, the sediment concentration equation should be analogous to that for momentum exchange.

The fall velocity of spherical particles in the turbulent flow region is given by

$$v_s = \left[\frac{4}{3}\frac{gD}{C_r}(s - 1)\right]^{1/2}, \tag{9.68}$$

where v_s is the settling or fall velocity (cm/sec), s is specific gravity of the grain (g/cm³), D is the grain diameter (cm), g is gravity acceleration (cm/sec²), and C_r is a coefficient of resistance which varies with the particle's Reynolds number.

For the laminar region,

$$v_s = \frac{gD^2}{18\nu}(s - 1), \tag{9.69}$$

where ν is the kinematic viscosity (cm²/sec).

For most suspended particles of interest to water quality, the Reynolds number of the particles, $Re = v_s D/\nu$, is in the laminar region.

In suspended sediment transport analysis, it is important to ascertain where and when a particle will settle out from the stream into the streambed or when and where the bed particles will be resuspended back into suspension. Under laminar conditions, all suspended particles with enough mass to overcome Brownian motion and buoyancy will eventually settle out. However, under turbulent conditions which prevail in nature, there will be an upflow velocity acting against the gravity of the particles. Under certain conditions there may be an equilibrium between the vertical turbulent eddy velocity component keeping the particles in suspension and the settling velocity of the particles. This results in a zero net exchange of particles between the bed load and the suspended load. In this case the suspended sediment flow is saturated. The sediment concentration profile under saturated flow conditions can be described by the equation (Vanoni, 1946)

$$C(y) = C_B \left[\frac{\dfrac{H}{y} - 1}{\dfrac{H}{\Delta} - 1} \right]^r \tag{9.70}$$

where Δ is a small distance above bottom (e.g., the roughness height), C_B is the near bottom concentration of suspended particles, $C(y)$ is the concentration of suspended sediment at a distance y from the bottom, and H is the depth.

The factor r is determined from

$$r = \frac{v_s}{\kappa (gHS_f)^{1/2}}. \tag{9.71}$$

The distribution of suspended particle concentration under saturated flow conditions is shown in Fig. 9.18.

Equation (9.70) describes the sediment concentration quite well, both in laboratory and stream conditions, and even under high sediment concentrations (above 100 g/liter) near the bed. However, problems associated with the use of this equation are due to deficiencies in the definition and changes in von Karman's coefficient, both in greater distances from the bottom, and also due to sediment content.

The distribution curve cannot be applied for the entire cross-section since it would give $C = \infty$ at $y = 0$. The magnitude of Δ is usually related to the grain size in the bed layer. The suspension of the sediment particles is only possible at a distance of approximately 2 D above the bed, where D is the mean bed particle diameter.

Since Eq. (9.70) represents the saturated sediment concentration under steady-state conditions, the total sediment flow per unit width becomes

$$g_{ss} = \int_0^H u\, C(y)\, dy. \tag{9.72}$$

The integration of the sediment flow equation requires a knowledge of the velocity distribution. As has been previously shown, two equations closely describe the vertical distribution of the velocity in the x-direction;

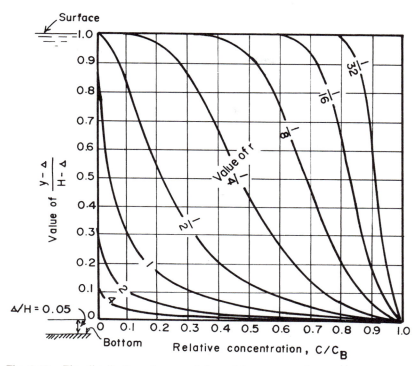

Fig. 9.18. The distribution of suspended particle concentration under saturated flow conditions (Committee on Sedimentation, 1963).

i.e., Prandtl's logarithmic velocity profile and Nikuradze's power velocity distribution.

The logarithmic profile becomes

$$\frac{u_{\max} - u}{u_*} = \frac{1}{\kappa} \ln \frac{H}{y}, \tag{9.73}$$

and, similarly, the power velocity distribution is

$$\frac{u}{u_{\max}} = \left(\frac{y}{H}\right)^{1/n}, \tag{9.74}$$

where n is an exponent with an accepted value of 7, but varying from 6 at the lowest Reynolds numbers (Re $= u_{\max} H / \nu \sim 10^3$) to 10 for the highest Reynolds numbers (Re $\sim 10^6$).

Using the logarithmic velocity profile in the suspended sediment load function would lead to a tedious integration procedure; however, Chow (1964) contains the solution and graphical nomographs. The power velocity profile provides a quicker solution; however, uncertainty as to the magnitude of n must be expected.

Toffaleti (1968) showed that the velocity distribution formula for sediment carrying streams can be written as

$$u = 1.15 \ \bar{u} \ (y/R)^{0.155}, \tag{9.75}$$

where R is the hydraulic radius. He also found that n is a function of temperature rather than the Reynolds number as suggested by Nikuradze. Toffaleti's formula is

$$1/n = 0.1213 + 0.00086 \ TDF, \tag{9.76}$$

where TDF is the stream temperature in °C.

Using the power velocity distribution law, Eq. (9.72) can be integrated to yield the saturated sediment flow formula

$$g_{ss} = \xi \ \frac{\bar{u}}{\dfrac{H}{\Delta} - 1} \ C_B H, \tag{9.77}$$

where C_B is the near bottom concentration of the suspended sediment. The magnitude of the coefficient, ξ, depends on the values of n and r as shown in Fig. 9.19 with $n = 7$.

The hydraulic roughness factor, Δ, can be approximately related to Mannings coefficient, n_M. For Δ expressed in meters, the relation has been quoted as (Mostkow, 1959):

$$n_M = 0.042 \ \Delta^{1/6}. \tag{9.78}$$

Once the saturated sediment flow is determined, estimation of the scour or deposition of the sediments can be based on the following assumptions.

Let g_s be the amount of suspended sediment with a uniform settling velocity, v_s, which is carried by a stream. If $g_s > g_{ss}$, the flow is supersaturated by the sediments and the excess $(g_s - g_{ss})$ will settle out from the stream, provided that the detention time is long enough to allow all excess particles to reach the bottom. If $g_s < g_{ss}$, the flow is unsaturated, thus no sedimentation will occur. If the bottom is covered by sediment with a similar gradation, a possible scour may take place, provided that the turbulent drag force is high enough to overcome the tolerance resistance of the deposited sediments.

If the suspended sediment load consists of different fractions with different settling velocities, the saturated sediment flow can be determined for each fraction separately. The fractions whose content is above saturation will settle out.

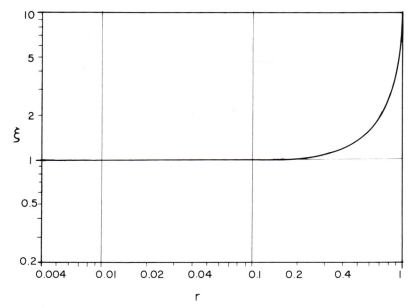

Fig. 9.19. The relationship between n and r.

Near Bottom Concentration, C_B For noncohesive sediments such as sand and gravel, the near bottom concentration, C_B, can be related to the bed load movement as has been demonstrated by Einstein (1950). However, soil particles and clay are cohesive and, therefore, the existing semi-theoretical and empirical formulas concerning erosion, deposition, and movement of cohesionless sediment cannot be used (Partheniades, 1965).

The principal difficulty lies in the fact that some of the resisting forces in cohesive soils are of an entirely different nature from those for cohesionless material. Other difficulties may be related to the flocculation effect of clay particles. Clay suspensions are not composed from isolated clay particles, but from small clusters of particles called flocs. In addition, the size of the flocs is not constant but a function of velocity gradient, turbulence intensity, soil and salt concentration, and time of agitation.

For flocculation to occur, the soil (clay) particles must be brought close enough for the surface forces to interact. The contact is caused by Brownian motion, differential settling velocity, and velocity gradients within the fluid mass.

Because of the nature of their origin and transport phenomenon, it may be anticipated that the suspended soil particles in surface waters will be mainly flocs. Thus, the turbulence intensity can be considered as the governing factor. Experiments have shown that for very long periods of agitation the maximum floc size is independent of concentration and that the average floc size increases with decreasing agitation intensity (Partheniades, 1965). A critical velocity also appears to exist above which all clay and silt particles stay in suspension and below which the clay is deposited. The critical deposition velocity was experimentally found to be 0.5 ft/sec (0.152 m/sec). The velocity was lower than the scouring velocity (approximately equal to 0.2 to 0.25 m/sec). Hence, a simultaneous interchange of suspended and bed clay may not occur (Partheniades, 1965).

Partheniades and Paaswell (1970) summarized the knowledge of scour and deposition of cohesive soils and concluded that

1. Nonsettling velocities are not necessarily scouring velocities, but there does exist a gap between lower velocities for scouring of similar sediment already deposited.

2. Under certain conditions, the gap between the two velocity limits increases with decreasing sediment size.

3. Cohesive materials with the same classification and very similar size gradation may have widely different resistances to erosion.

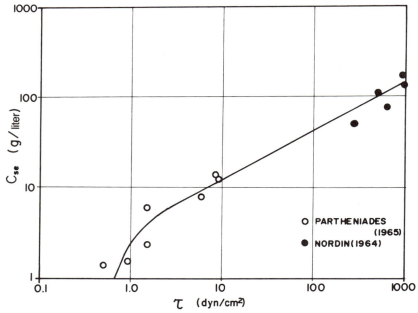

Fig. 9.20. The relationship between the equilibrium concentration and the bottom shear stress.

4. Both erosion and deposition of cohesive sediments are controlled by the bed shear stress.

Measurements by Partheniades (1965) in a laboratory flume and observations of sediment laden streams yields some preliminary information on the kinetics of scour and deposition of soil particles in turbulent water. Partheniades' experiments essentially confirmed the validity of the basic theory of sedimentation as developed for noncohesive sediments. The experiments have shown that under constant flow conditions, there is a specific sediment carrying capacity of the flow which was related by Partheniades to "the equilibrium concentration." The sediment carrying capacity of turbulent streams may be quite high. Nordin (1964) reported that recorded concentrations of cohesive sediments observed in the Rio Puerco in New Mexico were as high as 680 g/liter.

Figure 9.20 shows the relationship between the "equilibrium concentration" for the Partheniades' experiments and for the Rio Puerco and the bottom shear stress, τ_0.

A KINETIC MODEL FOR SEDIMENTATION OF COHESIVE SEDIMENTS

In this section, an attempt will be made to mathematically describe the scouring and settling processes of cohesive sediments. Very little data exist in the literature to verify the model. However, since the model is to be based on a simple classical mass balance, it is believed to be applicable.

Define M as the scour rate of sediments deposited on the bottom (in g/m^2 × hr) and K_w as a coefficient describing the settling rate of the suspended sediment. It is believed that the scour rate, M, is a function of the bed shear stress, the settling velocity of the particles, and possibly some other factors. K_w can be related to the settling velocity of the particles, v_s, and to the concentration of particles in the bottom boundary layer (C_B) and the intensity of turbulence near the bottom.

It is assumed that the concentration distribution outside the bottom boundary layer can be described by Eq. (9.70) discussed in the previous section. Both M and K_w must be determined experimentally. However, under steady-state (equilibrium) conditions, the following relation will hold:

$$M = K_w C_{Be}, \tag{9.79}$$

where C_{Be} = the equilibrium concentration at the bottom boundary layer.

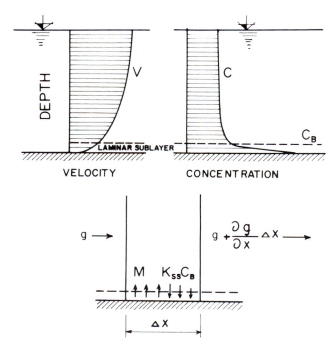

Fig. 9.21.   Mass balance of cohesive sediment analysis.

In an elemental volume with the dimensions $H \times (\Delta x) \times 1$, where H is the depth of flow (Fig. 9.21), the mass balance of the sediment carried by the flow will be

$$\frac{\partial (\int cdV)}{\partial t} = \left[-\frac{\partial g}{\partial x} + (M - K_w C_B) \right] \Delta x \times 1, \qquad (9.80)$$

where $\int cdV$ expresses the total mass of the sediment within the volume $V = \Delta x \times H \times 1$, and $g = $ the sediment discharge/unit width described by Eq. (9.81) as

$$g = \xi \bar{U} C_B \left(\frac{n}{0.042} \right)^{6r} H^{(1-r)}. \qquad (9.81)$$

Steady-State Model

Under steady-state conditions, the equation of the sediment discharge becomes

$$\frac{\partial g}{\partial x} = M - K_w C_B \qquad (9.82)$$

or

$$\frac{\partial g}{\partial x} = M - \frac{K_w}{\xi' \bar{U}} \left(\frac{0.042}{m}\right)^{6r} H^{(r-1)} g, \tag{9.83}$$

which can be integrated to yield

$$\frac{g - g_{ss}}{g_0 - g_{ss}} = e^{-ax}, \tag{9.84}$$

where g = the sediment discharge at x, g_0 = the sediment discharge at $x = 0$, and g_{ss} = the saturated sediment discharge.

The coefficient, a, becomes

$$a = \frac{K_w}{\xi' \bar{U}} \left(\frac{0.042}{n}\right)^{6r} H^{r-1}. \tag{9.85}$$

It can be seen that the saturated sediment discharges, g_{ss}, is

$$g_{ss} = \frac{M}{a} \tag{9.86}$$

and the near bottom concentration under saturated or equilibrium sediment flow conditions is

$$C_{BE} = \frac{M}{K_w}. \tag{9.87}$$

The preceding simple model can be applied only if K_w and M are independent of the concentrations. This assumption holds true only for scouring; it does not hold for the settling rate, K_w, at higher concentrations when the settling is hindered.

Effect of Hindered Settling on Sediment Carrying Capacity

At a certain concentration of suspended particles, the settling becomes hindered. The particles cease to behave as individuals and their settling is affected by the interactions and collisions among the particles. This effect will result in a reduction of the settling velocity of the particles that then becomes a function of the particle concentration.

Kynch (quoted by Tesarik and Vostrcil, 1970) introduced the assumption that at any point in the suspension, as in hindered settling, the velocity of a particle depends only on the local concentration of particles, or

$$W_s = f(c), \tag{9.88}$$

in which W_s = the velocity of settling, and c = the fraction concentration

$(0 < c \leq 1)$. The lower limit of applicability of the preceding equation is at a fraction of a packed bed, or the liquid limit of the sediment. The upper limit is the transition between discrete settling and hindered settling which occurs at a volumetric concentration of about $c = 0.005$ (Camp, 1947), which is comparable to a concentration of from 1000 to 5000 mg/liter of silt and clay particles in a turbid river water.

Tesarik and Vostrcil (1970) proposed an equation for the settling velocity of coagulated flocs as

$$W_s = \frac{\alpha}{cB},\qquad (9.89)$$

where α and B are coefficients. The values of B for chemically precipitated flocs varied between 1.7 and 2.7. For bentonite clay, the values of the coefficients were reported as $\alpha = 12.8$ and $B = 2.06$ and W_s was in mm/sec and C in g/liter.

Accepting the hypothesis that at higher concentrations the settling of particles in the near bottom layer can be hindered will have far-reaching consequences as to the sediment carrying capacity of streams. These can be summarized as:

1. At low concentrations, below the hindered settling transition (approximately $c < 0.5\%$), the sediment carrying capacity, g_{ss}, can be approximately determined from the equation

$$g_{ss} = \frac{M}{a},\qquad (9.90)$$

where

$$a = \frac{K_w}{\xi' \bar{U}} \left(\frac{0.042}{n}\right)^{6r} H^{(r-1)}$$

and the flow model is described by the equation

$$\frac{g - g_{ss}}{g_0 - g_{ss}} = e^{-ax}.$$

The coefficient K_w could be of the same order of magnitude as the settling velocity of discrete particles.

2. At higher concentrations, the reduction of the settling velocity at the near bottom layer by the hindering effect will result in a sharp increase of the sediment carrying capacity of the flow. The equilibrium concentration corresponds to the solution of the relationship

$$M = W_s C_{Be} = v_s f(C_{Be}) C_{Be}.\qquad (9.91)$$

Since W_s decreases as the concentration increases, C_{Be} increases with a higher power than M. M can be assumed to be proportional to the bottom shear stress, τ_0; thus Fig. 9.20 also represents the relationship of C_{Be} to M.

According to some models (Tesarik and Vostrcil, 1970), the hindered settling velocity may decrease with a higher power than the concentration increase. This situation may have a serious implication because it means that under certain circumstances, the sediment carrying capacity of the stream is theoretically unlimited (within the boundaries of the validity of the assumptions made in deriving the equations). In the cases in which M is greater than $K_w C_B$, any increase of C_B will result in a decrease in $(K_w C_B)$. Thus, there is no concentration, C_{Be}, to which C_B would approach. An equilibrium can be reached only if M drops to a value equalizing $K_w C_B$.

Estimation of K_w and M

The deposition of suspended soil particles and the suspension by scour of bottom deposits are processes taking place in the bottom boundary layer. The bottom boundary layer can be related to the thickness of the laminar boundary layer as shown in Fig. 9.21. The sedimentation rate of suspended particles in the boundary layer should then theoretically be equal to or proportional to the settling rate of the particles, W. However, the settling process at the bottom boundary layer does not demonstrate the characteristic of discrete settling. Because of the high concentration of the particles which may occur in the bottom layer, the settling may be hindered. For example, there will be interaction and flocculation of the particles, which may result in a reduction of the settling rate. Also, at higher flow velocities, the settling process in the boundary layer is being continuously disrupted by penetrating turbulent fluctuations, thus reducing the settling rate in the boundary layer.

The relationship of the equilibrium concentration, C_e, to bottom shear stress, τ_0, was shown in Fig. 9-20. The equilibrium concentration was obtained from the Partheniades' channel experiments and from the measurements of sediment transport of the Rio Puerco, a heavily sediment laden tributary of the Rio Grande. The Partheniades' experiments also allow the determination of the settling rate coefficient, K_w. If the concentration $(C - C_e)$ is plotted versus time on semi-log paper, the slope of the relationship will yield K_w/H. The coefficient K_w can then be related to the average concentration (Fig. 9.22). It can be seen that the coefficient K_w does decrease with the concentration. At this stage, it is not possible to evaluate whether the decrease is due to hindered settling or to the in-

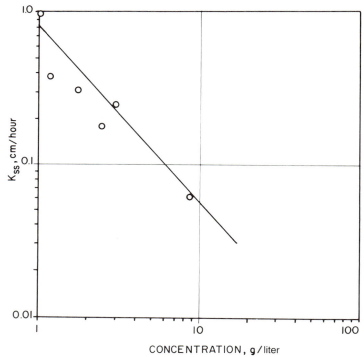

Fig. 9.22.   Relationship of K_w to average concentration of sediment.

creased turbulence in the bottom boundary layer. A thorough laboratory investigation would be necessary to resolve this question and provide more accurate data. The maximum value of K_w at the lowest velocity and concentration was in the order of 1 cm/hr, which is about the same as the settling rate for discrete clay particles. Because of the hindered settling effect and increased turbulence, the value of K_w decreases rapidly.

The sediment scouring rate, M, can be roughly estimated from $M = K_w C_e$. If M is plotted versus the bottom shear stress as in Fig. 9.23, a functional relationship is clearly indicated. Figure 9.23 also confirms that there is a certain critical value of the bottom shear stress, τ_C, below which the scour rate decreases rapidly to zero. The critical value of τ_C in the Partheniades' experiments was about $\tau_C = 1$ dyn/cm². Again, laboratory investigations, combined with field observations of sediment laden streams, may provide better and more accurate parameters of the model.

As indicated by Partheniades, the scour rate, M, may be affected by compaction of the deposited solids. Karcz and Shanmugan (1974) confirmed the hypothesis and stated that the erosion of freshly molded clay

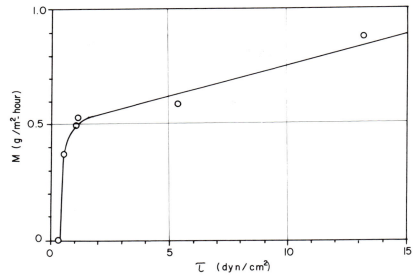

Fig. 9.23. Sediment scouring rate v. the bottom shear stress.

and mud decreases with time until a fairly steady rate of scour is attained. This rate is less dependent on the bed shear stress.

As stated before, the initiation of scour of cohesive (also noncohesive) sediments requires that the bed shear stress exceed the critical shear stress, τ_C. If $\tau_0 < C'$, no scour occurs; therefore, the equilibrium concentrations C_e and C_{Be} will approach zero. On the other hand, if $M > K_w C_B$ and C_B is greater than the transition concentration from discrete to hindered settling, there is no equilibrium concentration. The sediment concentration has then only one limit, i.e., the liquid limit of sediment at which the sediment becomes the solid mud.

REFERENCES

Amein, M., and Fang, C. S. (1970). Implicit flood routing in natural channels. *J. Hydraul. Div., Am. Soc. Civ. Eng.* **96**, 2481–2500.

Brater, E. F., and King, H. W. (1976). "Handbook of Hydraulics," 2nd ed. McGraw-Hill, New York.

Camp, T. R. (1947). Sedimentation and the design of settling tanks. *Trans. Am. Soc. Civ. Eng.* **3**, 895–958.

Chow, V. T. (ed.) (1964). "Handbook of Applied Hydrology." McGraw-Hill, New York.

Committee on Sedimentation (1963). Sediment transportation mechanics: Suspension of sediment. *J. Hydraul. Div., Am. Soc. Civ. Eng.* **89**, 45–76.

Davis, J. T. (1972). "Turbulence Phenomena." Academic Press, New York.

Einstein, H. A. (1950). The bed-load function for sediment transportation in open channel flow. Tech. Bull. No. 1026, U.S. Dept. Agric., Washington, D.C.

Elder, J. W. (1959). The dispersion of marked fluid in turbulent shear flow. *J. Fluid Mech.* **5**, 544–560.

Fisher, H. B. (1968). Dispersion prediction in natural streams. *J. Sanit. Eng. Div., Am. Soc. Civ. Eng.* **94**, 927–943.

Fisher, H. B. (1975). Discussion to simple method for predicting dispersion in streams by R. S. McQuivey and T. N. Keefer. *J. Environ. Eng. Div., Am. Soc. Civ. Eng.* **101**, 453–455.

Fread, D. F. (1971). Discussion to implicit flood routing in natural channels by M. Amein and C. S. Fang. *J. Hydraul. Div., Am. Soc. Civ. Eng.* **97**, 1156–1159.

Gloyna, E. F. (1967). Prediction of oxygen depletion and recent developments in stream model analyses. *In* Stream analysis and thermal pollution, vol. 2. Prepared for Poland Project 26, World Health Organ., Univ. of Texas, Austin.

Hansen, E. (1972). Lagrangian characteristics of surface turbulence. *J. Hydraul. Div., Am. Soc. Civ. Eng.* **98**, 1255–1273.

Harley, B. M. (1967). Linear routing in uniform open channel flow. Ph.D. Thesis, Univ. College, Cork, Ireland.

Holley, E. R. (1969). Unified view of diffusion and dispersion. *J. Hydraul. Div., Am. Soc. Civ. Eng.* **95**, 621–631.

Karcz, I., and Shanmugan, G. (1974). Decrease in scour rate of fresh deposited mud. *J. Hydraul. Div., Am. Soc. Civ. Eng.* **100**, 1735–1738.

Keefer, T. N., and McQuivey, R. S. (1974). Multiple linearization flow routing model. *J. Hydraul. Div., Am. Soc. Civ. Eng.* **100**, 1031–1046.

Kolmogoroff, A. N. (1941). Dissipation of energy in locally isotrophic turbulence. *C. R. (Dokl.) Acad. Sci. URSS* **32**, 16.

Krenkel, P. A. (1960). Turbulent diffusion and the kinetics of oxygen absorption. Ph.D. Thesis, Univ. of California, Berkeley.

Krenkel, P. A., and Orlob, G. T. (1962). Turbulent diffusion and the reaeration coefficient. *J. Sanit. Eng. Div., Am. Soc. Civ. Eng.* (March).

Levenspiel, O., and Smith, W. K. (1957). Notes on the diffusion type model for the longitudinal mixing of fluids in flow. *Chem. Eng. Sci.* **6**, 227–233.

Ligett, J. A. (1975). Basic equations of unsteady flow. *In* "Unsteady Flow in Open Channels" (K. Mahmood and V. Yevjevich, eds.). Water Resour. Publ., Fort Collins, Colorado.

Lighthill, M. J. and Whitham, G. B. (1955). On kinematic waves—I-Flood movement in long rivers. *Proc. R. Soc. London,* Ser. A: **229**, 281–316.

Liu, H. (1977). Predicting dispersion coefficient of streams. *J. Environ. Eng. Div., Am. Soc. Civ. Eng.* **103**, 59–69.

Miller, W. A., and Cunge, J. A. (1975). Simplified equations of unsteady flow. *In* "Unsteady Flow in Open Channels" (K. Mahmood and V. Yevjevich, eds.). Eater Resour. Publ., Fort Collins, Colorado.

Morris, J. S., Krenkel, P. A., and Thackston, E. L. (1967). Investigations of turbulent diffusion in inland waterways. Tech. Rep. No. 14, Dept. Environ. and Water Resour. Eng., Vanderbilt Univ., Nashville, Tennessee.

Mostkow, M. A. (1959). "Osnovy Theorii Ruslovogo Potoka" (Basic Theory of Channel Flow). Sov. Acad. Sci., Moscow.

Nordin, F. Jr. (1964). Study of channel erosion and sediment transport. *J. Hydraul. Div., Am. Soc. Civ. Eng.* **90**, 173–191.

Novotny, V. (1970). Unpublished Project Rep., Dept. Environ. and Water Res. Eng., Vanderbilt Univ., Nashville, Tennessee.

O'Connor, D. J. (1967). The temporal and spatial distribution of dissolved oxygen in streams. Dept. of Civil Eng., Manhattan College, New York.

Orlob, G. T. (1958). Eddy diffusion in open channel flow. Water Res. Center, Univ. of California, Berkeley.

Partheniades, E. (1965). Erosion and deposition of cohesive soils. *J. Hydraul. Div., Am. Soc. Civ. Eng.* **91**, 105–139.

Partheniades, E. and Paaswell, R. E. (1970). Erodibility of channels with cohesive boundary. *J. Hydraul. Div., Am. Soc. Civ. Eng.* **96**, 755–771.

Price, R. K. (1974). Comparison of four numerical methods for flood routing. *J. Hydraul. Div., Am. Soc. Civ. Eng.* **100**, 879–900.

Pritchard, D. W. (1969). Dispersion and flushing of pollutants in estuaries. *J. Hydraul. Div., Am. Soc. Civ. Eng.* **95**, 115–124.

Ralson, R., and Wilf, H. S. (1960). "Mathematical Methods for Digital Computers." Wiley, New York.

Shen, H. W. (1971). River mechanics, vol. 1, Civil Eng. Dept., Colorado State Univ., Fort Collins.

Taylor, G. I. (1954). The dispersion of matter in turbulent flow through a pipe. *Proc. R. Soc. London,* Ser. A: **223,** 446–468.

Tesarik, I., and Vostrcil, J. (1970). Hindered settling and thickening of chemically precipitated flocs. *Proc. Int. Conf. Int. Assoc. Water Pollut. Res.,* 5th, San Francisco, California.

Texas Water Development Board, (1971). Simulation of water quality in streams and canals. Rep. 128, Austin, Texas.

Toffaleti, F. B. (1968). Definitive computations of sand discharge in rivers. *J. Hydraul. Div., Am. Soc. Civ. Eng.* **95,** 225–248.

Vanoni, V. T. (1946). Transportation of suspended sediment by water. *Trans. Am. Soc. Civ. Eng.* **3,** 67–133.

Wooding, R. A. (1965). A hydraulical model for the catchment-stream problem, parts I and II. *J. Hydrol.* **3,** 21–37.

10 | The Modeling of Streams and Estuaries*

INTRODUCTION

With the passage of Public Law 92-500 (Water Pollution Control Amendments of 1972) and subsequent implementation of the planning aspects of that act, the importance of mathematical modeling has come to the forefront of water quality management. While the act purports to be concerned only with water quality, the various planning provisions clearly lead to land use planning. It is obvious that attainment of the objectives of the act are dependent on present and future control of all wastewater discharges, including those from so-called "nonpoint sources." The only practical method for delineating present and future water quality conditions is by mathematical simulation. Thus, the use of mathematical modeling is mandatory.

In addition, while the act dictates the use of effluent standards, in receiving waters where those standards will not result in the desired water quality (water quality limiting cases), allowable discharges must be determined by the use of mathematical models.

* Portions of this material were taken from "River Water Quality Model Construction," by P. A. Krenkel and V. Novotny, Chapter 17 in *Modeling of Rivers* (H. W. Shen, ed.) John Wiley & Sons, New York, 1979. Used with permission.

OVERVIEW OF MODELING

A detailed description of all of the complex water quality models would occupy a treatise. Thus, the purpose of this discussion will be to apply the approach taken in one of the most frequently used river models (DOSAG I) (Texas Water Devel. Bd., 1970) to a real situation. The emphasis will be on the parameters involved and their sensitivity and incorporation into the model. Since this is a deterministic model, the basic process consists of writing mass balance relationships for all material entering and leaving an infinitesimal volume. As will be demonstrated later, the DOSAG I model is one-dimensional and steady-state, which somewhat simplifies the model construction.

If one is dealing with the Environmental Protection Agency (EPA), the choice of model may depend on a particular region's preference. However, the model chosen for discussion herein (DOSAG I), with some modification, is probably the most frequently used for the purposes previously described. This model is primarily concerned with the oxygen balance in a river system, inasmuch as Biochemical Oxygen Demand (BOD) is still considered the single most important measure of water pollution.

Models can be classified according to whether they handle conservative or nonconservative substances, where a conservative substance is one affected only by dilution and a nonconservative substance changes its characteristics with time, i.e., BOD, coliform bacteria, temperature. Inasmuch as the conservative substances are relatively simple to model and meaningful models exist for temperature and coliform bacteria, the authors chose to use the DOSAG I model for oxygen and use other models when necessary. It is pertinent to note that while increased model accuracy usually implies increased complexity, increased complexity does not necessarily yield increased accuracy.

The range of choice for modeling is quite broad. One can choose from a number of existing models with different complexities or devise a model "from scratch." Obviously, using an existing model, where applicable, is the preferred option because the cost for development has already been paid. For an excellent discussion of existing models, their capabilities, selection, and documentation, the reader is referred to Grimsrud *et al.* (1976). The following guidelines for the selection and use of models were taken directly from Grimsrud:

1. The first step is to define the problem and to determine what information is needed and what questions need to be answered.

2. Use the simplest method that can provide the answers to your questions.

3. Use the simplest model that will yield adequate accuracy.

4. Do not try to fit the problem to a model but select a model that fits the problem.

5. Do not confuse complexity with accuracy.

6. Always question whether increased accuracy is worth the increased effort and cost.

7. Do not forget the assumptions underlying the model used, and do not read more significance into the simulation results than is actually there.

THE BASIC MODEL

The basic classical model was proposed by Streeter and Phelps (1925) and was later summarized and published by Phelps (1944). It is interesting to note that the fundamentals involved were quite sound and are still used today, with some modifications. The differential expression proposed was

$$dD/dt = K_1 L - K_2 D, \tag{10.1}$$

where $D = (C_s - C)$ = oxygen deficit; C = oxygen concentration observed and C_s = oxygen saturation concentration; L = carbonaceous biochemical oxygen demand; K_1 = deoxygenation coefficient; K_2 = reaeration coefficient; and t = time.

Equation (10.1) integrates into the familiar form of the oxygen sag equation, which is

$$D = \frac{K_1 L_0}{K_2 - K_1} (e^{-K_1 t} - e^{-K_2 t}) + D_0 e^{-K_2 t}, \tag{10.2}$$

where L_0 and D_0 are the ultimate biochemical oxygen demand and oxygen deficit, respectively.

Streeter and Phelps' approach is limited to only two phenomena, namely, deoxygenation of the water cause by bacterial decomposition of carbonaceous organic matter, and reaeration caused by the oxygen deficit and turbulence. The rate at which the BOD is exerted was assumed to be identical with that observed using the laboratory BOD test and a proportionality was assumed to exist between the reaeration rate and certain hydraulic parameters of flow.

Numerous investigators have subsequently questioned Streeter and Phelps' theory because the self-purification process in small streams does not follow the model. A classical illustration of the disagreement of field observations with the theory was published by Kittrell and Kochtitzky

(1947). These studies were made on the Holston River, Tennessee, which is used as an example in this discussion. Pertinent conclusions of that study follow.

1. The laboratory deoxygenation coefficients, K_1, varied approximately with the five-day BOD concentration of the samples collected below all sources of pollution. The range was 0.34 to 0.095 (base 10) for BOD results of 20.6 to 1.89 ppm.

2. The stream deoxygenation coefficients, K_1, were many times greater than the laboratory coefficients below the sources of pollution. Furthermore, as the distance downstream from the sources increased, the relationship was reversed. The range of the stream coefficients observed was 2.26 to 0.017 (base 10) at 20°C.

3. The rate of reduction of BOD was much more rapid in this stream (the Holston River) than in the usual deep, sluggish stream, with the reduction about seven times as great for 0.5 day and four times as great for one day.

4. The rate of the first-stage BOD reduction in the stream water followed the normal course of bacterial death rates. This indicated that the

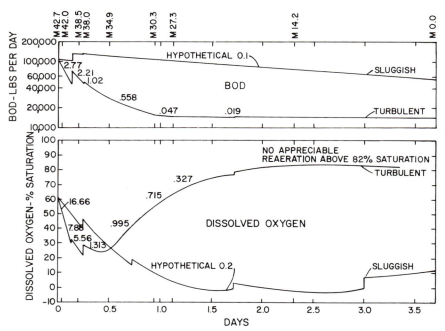

Fig. 10.1. Relative recovery capacities of turbulent and sluggish streams (Kittrell and Kochtitsky, 1947).

rate was not constant but diminished progressively with time below the entrance of pollution.

5. It was suggested that the rapid reduction in BOD may be due partially to the absorption of organic material in biological slimes on the bed of the stream.

6. The assumption that all deposited material exerted its full potential demand on the dissolved oxygen of the stream water resulted in obviously excessive reaeration coefficients, K_2, that ranged from 15.52 to 5.22 (base 10) at 20°C. It was suggested that portions of the deposited material may have decomposed anaerobically and escaped from the stream as unoxidized gases, without utilizing oxygen from the stream water.

7. A turbulent and relatively shallow river could assimilate more organic pollution without excessive dissolved oxygen depletion than could a deep sluggish stream. Figure 10.1 depicts the relative recovery capacity of the Holston River as compared to a sluggish stream, such as the Ohio River studied by Phelps ($K_1 \approx 0.25$ per day, base e).

MODIFICATIONS OF THE STREETER–PHELPS THEORY

As previously mentioned, the Streeter–Phelps theory is based on the assumption that there are only two major processes taking place in the self-purification of streams; i.e., BOD and oxygen are being removed along the stretch by the bacterial oxidation of the organic matter, and oxygen is being replaced by reaeration at the water surface. Dobbins (1964) suggested that there are several other factors which contribute to the BOD and oxygen variations in streams:

1. The removal of BOD by sedimentation or absorption.

2. The addition of BOD along the stretch by the scour of bottom deposits or by the diffusion of partly decomposed organic products from the benthal layer into the water above.

3. The addition of BOD along the stretch by local runoff.

4. The removal of oxygen from the water by diffusion into the benthal layer to satisfy the oxygen demand in the aerobic zone of this layer.

5. The removal of oxygen from the water by purging action of gases rising from the benthal layer.

6. The addition of oxygen by photosynthetic action of plankton and fixed plants.

7. The removal of oxygen by the respiration of plankton and attached plants.

8. The continuous redistribution of both BOD and oxygen by longitudinal dispersion.

Additional factors, e.g., effects of biological slimes, were also mentioned by Velz and Gannon (1962).

If unoxidized nitrogen is present (such as ammonia or unoxidized nitrogen expressed as total Kjeldahl nitrogen–TKN), it may also represent a significant sink of oxygen and must be included in the oxygen balance. Summarizing these findings, it is possible to state that the variations of water quality in rivers are caused by

1. Quantitative and qualitative variations in the volume of flowing liquid, such as deoxygenation by the microorganisms present in the water, surface reaeration, denitrification, photosynthesis and respiration of planktonic microorganisms, and, to some degree, sedimentation of suspended organics.

2. External effects, such as those of heterotrophic slimes on BOD and oxygen removal, photosynthetic action of aquatic plants and benthic algae, oxygen demand of benthal deposits, and so on.

It should be noted that the components of the oxygen balance systems are coupled; i.e., some outputs from one system represent inputs to another one. The effects of each of the inputs and the magnitudes of the inputs and parameters can best be presented by considering the generalized equation for coupled systems. Thomann (1972) considered the following equations specific to the DO problem:

BOD variation:

$$\frac{\partial L}{\partial t} = -\frac{1}{A}\frac{\partial}{\partial x}(QL) - K_r L + L_r \left[\frac{\partial Q}{\partial X}\Big/ A\right] + L_{rd}. \tag{10.3}$$

Nitrogenous oxygen demand variation:

$$\frac{\partial L^N}{\partial t} = -\frac{1}{A}\frac{\partial}{\partial x}(QL^N) - K_N L^N + L_r^N \left[\frac{\partial Q}{\partial X}\Big/ A\right] + L_{rd}^N. \tag{10.4}$$

Dissolved oxygen variation:

$$\frac{\partial C}{\partial t} = -\frac{1}{A}\frac{\partial}{\partial X}(QC) - K_d L - K_N L^N + K_a(C_s - C) - S_B(X,t)$$

$$+ P(X,t) - R(X,t) + C_r\left[\frac{\partial Q}{\partial X}\Big/ A\right]. \tag{10.5}$$

The terms are interpreted as C = dissolved oxygen concentration, mg/liter; C_r = dissolved oxygen runoff, mg/liter; C_s = dissolved oxygen saturation, mg/liter; A = cross-sectional area, m² or ft². Q = river flow, m³/day or ft³/day; L_r = BOD concentration of runoff, mg/liter; L_{rd} =

rate of addition of BOD along the stretch, mg/liter/day; L^N = nitrogenous oxygen demand, mg/liter; L_r^N = nitrogenous oxygen demand of runoff, mg/liter; L_{rd}^N = rate of addition of N–BOD along the stretch, mg/liter day; S_B = benthal oxygen demand, mg/liter day; P = DO produced by plant or phytoplankton photosynthesis, mg/liter day; R = DO removed by plant respiration, mg/liter day; K_2 = atmospheric reaeration coefficient, day^{-1}; K_r = overall BOD removal coefficient, day^{-1}; K_d = BOD deoxygenation coefficient, day^{-1}; K_N = nitrification coefficient, day^{-1}; and X = distance.

The dissolved oxygen system is thus represented by three first-order linear differential equations. It is assumed in using these equations that the effects of longitudinal dispersion usually may be neglected. However, it has been demonstrated by O'Connor (1961) that in slow-moving, highly mixed streams, such as estuaries, the longitudinal dispersion becomes important and must be included in the equations.

Assuming steady-state conditions and negligible runoff within the stretch, Eqs. (10.3)–(10.5) can be integrated into the following forms.

BOD variation:

$$L = L_0 \exp\left(-K_r \frac{x}{u}\right) + \frac{L_r d}{K_r}\left[1 - \exp\left(-K_r \frac{x}{u}\right)\right]. \tag{10.6}$$

Nitrogenous demand variation:

$$L^N = L_0^N \exp\left(-K_N \frac{x}{u}\right) + \frac{L_{rd}^N}{K_N}\left[1 - \exp\left(-K_N \frac{x}{u}\right)\right]. \tag{10.7}$$

Oxygen deficit variation: (DOSAG includes this at MU)

$$D = D_0 \exp\left[-\left(K_a \frac{x}{u}\right)\right] \qquad \text{reaeration} \qquad \text{(term a)}$$

$$+ \left[\frac{K_d}{K_a - K_r}\left\{\exp\left[-\left(K_r \frac{x}{u}\right)\right] - \exp\left[-\left(K_a \frac{x}{u}\right)\right]\right\}\right] L_0$$
$$\text{(term b)}$$

$$+ \left[\frac{K_N}{K_a - K_N}\left\{\exp\left[-\left(K_N \frac{x}{u}\right)\right] - \exp\left[-\left(K_a \frac{x}{u}\right)\right]\right\}\right] L_0^N \qquad (10.8)$$
$$\text{(term c)}$$

$$+ \frac{K_d}{K_a K_r}\left\{1 - \exp\left[-\left(K_a \frac{x}{u}\right)\right]\right\}$$

$$- \left[\frac{K_d}{(K_a - K_r)K_r}\left\{\exp\left[-\left(K_r \frac{x}{u}\right)\right] - \exp\left[-\left(K_a \frac{x}{u}\right)\right]\right\}\right] L_{rd}$$
$$\text{(term d)}$$

TABLE 10.1 Summary of differential equations to be solved by QUALII

Conservative mineral (c):
$$\frac{\partial c}{\partial t} = \frac{\partial\left(A_x D_L \dfrac{\partial c}{\partial x}\right)}{A_x \partial x} - \frac{\partial(A_x u c)}{A_x \partial x} + \frac{s_c}{A_x dx}$$

Algae (A):
$$\frac{\partial A}{\partial t} = \frac{\partial\left(A_x D_L \dfrac{\partial A}{\partial x}\right)}{A_x \partial x} - \frac{\partial(A_x u A)}{A_x \partial x} + \frac{S_A}{A_x dx} + \left(\mu - \rho - \frac{\sigma_1}{A_x}\right) A$$

Ammonia nitrogen (N_1):
$$\frac{\partial N_1}{\partial t} = \frac{\partial\left(A_x D_L \dfrac{\partial N_1}{\partial x}\right)}{A_x \partial x} - \frac{\partial(A_x u N_1)}{A_x \partial x} + \frac{S_{N_1}}{A_x dx} + \left(\alpha_1 \rho A - \beta_1 N_1 + \frac{\alpha_3}{A_x}\right)$$

Nitrite nitrogen (N_2):
$$\frac{\partial N_2}{\partial t} = \frac{\partial\left(A_x D_L \dfrac{\partial N_2}{\partial x}\right)}{A_x \partial x} - \frac{\partial(A_x u N_2)}{A_x \partial x} + \frac{S_{N_2}}{A_x dx} + (\alpha_3 N_1 - \beta_2 N_2)$$

Nitrate nitrogen (N_3):
$$\frac{\partial N_3}{\partial t} = \frac{\partial\left(A_x D_L \dfrac{\partial N_3}{\partial x}\right)}{A_x \partial x} - \frac{\partial(A_x u N_3)}{A_x \partial x} + \frac{S_{N_3}}{A_x dx} + (\beta_2 N_2 - \alpha_1 \mu A)$$

Phosphate phosphorus (P):
$$\frac{\partial P}{\partial t} = \frac{\partial\left(A_x D_L \dfrac{\partial P}{\partial x}\right)}{A_x \partial x} - \frac{\partial(A_x u P)}{A_x \partial x} + \frac{S_P}{A_x dx} + \left(\alpha_2(\rho - \mu)A - \frac{\sigma_2}{A_x}\right)$$

Biochemical oxygen demand (L):
$$\frac{\partial L}{\partial t} = \frac{\partial\left(A_x D_L \dfrac{\partial L}{\partial x}\right)}{A_x \partial x} - \frac{\partial(A_x u L)}{A_x \partial x} + \frac{S_L}{A_x dx} - (K_1 + K_3)L$$

Dissolved oxygen (ϕ):
$$\frac{\partial \phi}{\partial t} = \frac{\partial\left(A_x D_L \dfrac{\partial \phi}{\partial x}\right)}{A_x \partial x} - \frac{\partial(A_x u \phi)}{A_x \partial x} + \frac{s_\phi}{A_x dx} + \left[K_2(\phi^* - \phi) + (\alpha_3\mu - \alpha_4\rho)A - K_1 L - \frac{K_4}{A_x} - \alpha_3\beta_1 N_1 - \alpha_6\beta_2 N_2\right]$$

Coliform (F):
$$\frac{\partial F}{\partial t} = \frac{\partial\left(A_x D_L \dfrac{\partial F}{\partial x}\right)}{A_x \partial x} - \frac{\partial(A_x u F)}{A_x \partial x} - \frac{S_F}{A_x dx} - K_5 F$$

Radioactive material (R):
$$\frac{\partial R}{\partial t} = \frac{\partial\left(A_x D_L \dfrac{\partial R}{\partial x}\right)}{A_x \partial x} - \frac{\partial(A_x u R)}{A_x \partial x} - \frac{S_R}{A_x dx} - K_r R - K_a$$

$$- \left\{ 1 - \exp\left[- \left(K_a \frac{x}{u} \right) \right] \right\} \frac{P}{K_a} \qquad \text{(term e)}$$

$$+ \left\{ 1 - \exp\left[- \left(K_a \frac{x}{u} \right) \right] \right\} \frac{R}{K_a} \qquad \text{(term f)}$$

$$+ \left\{ 1 - \exp\left[- \left(K_a \frac{x}{u} \right) \right] \right\} \frac{S_B}{K_a}, \qquad \text{(term g)}$$

where subscript 0 denotes the initial conditions (initial concentrations).

The terms in the various parts of the solution are interpreted as (a) initial value of DO deficit, (b) deficit due to point source of carbonaceous BOD, (c) deficit due to point source of nitrogenous oxygen demand, (d) distributed addition of BOD along the stretch, (e) reduction of the deficit due to photosynthesis, (f) deficit due to algae and plant respiration, and (g) distributed benthal demand effect.

In order to determine the oxygen deficit for different temperatures, flow conditions, and so forth, several supporting models have been developed for evaluation of the pertinent parameters, as will be subsequently discussed. A more complicated set of differential equations has been selected in the design of the water quality model, QUAL II, as shown in Table 10.1 (Roesner *et al.*, 1973).

DETERMINATION OF PARAMETERS

Several methods exist for the determination of each parameter of the oxygen balance. In almost all cases, the values of these parameters must be checked against field data which can be obtained either from water quality monitoring stations and/or from water quality survey data. Thus, a sensitivity analysis of the model and verification with field data is a necessary step in developing an adequate water quality model. Using only theoretical equations in a complex water quality system model is always dangerous and can lead even an experienced environmental systems engineer to erroneous conclusions. In addition, the use of specialists in limnology and water chemistry is desirable.

The parameters necessary for the design of an oxygen balance model can be separated into five recognizable categories: (1) hydrologic parameters, (2) hydraulic parameters, (3) oxygen sinks, (4) oxygen sources, and (5) temperature effects.

Hydrologic Parameters

Two major hydrologic factors are required as inputs to the oxygen balance; these are flow and temperature. The design flow is usually taken as the minimum 7-day 10-year flow. For design water temperature, the average temperature of the warmest month (July or August) with a recurrence interval of once in 10 years or 20 years is probably appropriate. In many waste assimilative capacity studies, the waste load is associated with a heat load (cooling water averages). Thus, the effect of the wastewater on the thermal regime of the receiving water must be determined.

Hydraulic Parameters

Hydraulic parameters are supporting inputs related to the reaction rates, detention times, and so forth. Necessary hydraulic parameters include velocity, hydraulic radius and/or depth, slopes, and bottom roughness. Stream velocity and/or detention time is usually determined by dye tests as was the case in the Holston River studies. Fluorescent dyes such as Rhodamine WT or B are convenient since the minimum detectable concentrations are less than 1 ppb. Ideally, stretches of the stream with uniform hydraulic characteristics should be selected. In some cases, the velocity can be directly computed with a velocity equation (Mannings, Prandtl Logarithmic, or others). Depth can be determined from cross-sections or computed if one knows the width and corresponding flow and velocity. Depth and velocity should be determined for several flows, the results of which can be conveniently plotted and correlated on log–log paper, as shown in Figure 9.8. The so-called power functions are presented and discussed in Chapter 9 under Residence Time, and observed values of the coefficients are presented in Table 9.2.

As shown in Table 9.2, $b + d \approx 1.0$. Furthermore, natural streams are somewhere between the two extremes of impounded waters and ideal channel flow. Therefore, b may be expected to be within the interval 0 to 0.6 and d within the interval 0.4 to 1.0.

Oxygen Sinks

The carbonaceous deoxygenation rate is related to BOD removal and subsequent oxygen demand brought about by the BOD removal process. Figure 10.2 depicts the processes which take part in the BOD removal in small- and medium-size shallow streams. The coefficients are defined as K_r = overall BOD removal coefficient, K_d = overall BOD deoxygenation

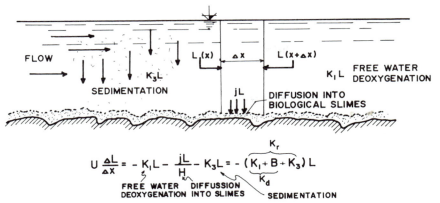

Fig. 10.2. BOD removal processes in a shallow stream (Krenkel and Novotny, 1973).

coefficient, K_1 = deoxygenation coefficient of free-flowing water ($\approx$ approx. equal to the laboratory BOD test coefficient), and K_3 = a coefficient describing sedimentation which does not exert an oxygen demand (anaerobic benthal decomposition). It follows that

$$K_r = K_d + K_3$$

and

$$K_d = B + K_1,$$

where B was defined by Velz and Gannon (1962) as a coefficient describing a boundary effect of absorption by biological slimes attached to a solid bottom (rocks, gravel, etc.). Several authors have attempted to describe the coefficient B by a formula. Since the coefficient B is thought to be related to the depth, it would express the ratio of the biological slime volume to the volume of water and possibly to the turbulence intensity, which would account for the diffusion intensity. Boško (1957, 1966) intuitively proposed the formula

$$B \approx \eta \frac{U}{H}, \qquad (10.9)$$

where the coefficient of bottom activity ranges from $\eta = 0.1$ for slow-flowing, sluggish streams to $\eta = 2.0$ (B base e) for rocky mountain streams. Novotny (1969) proposed the formula

$$B \approx \nu \frac{S_e^{1/4}}{H^{3/4}}. \qquad (10.10)$$

In Eqs. (10.9) and (10.10) the depth H is in meters and S_e is in ft/1000 ft

TABLE 10.2   Coefficients n and v for the estimation of K_d

Character of the Stream	Boško's formula n base e	Novotny's formula v base e
1. Streams with rocky bottom, higher pollution	2.0	2.0–3.0
2. Coarse gravel bottom with large rocks, higher pollution	1.5	1.1–2.0
3. Coarse gravel	1.2	0.8–1.1
4. Gravel, lower pollution	0.9	0.5–0.8
5. Sand, lower pollution	0.6	0.3–0.5
6. Slow moving rivers with muddy bottom	0.3	<0.3

(m/km), which was based on the hypothesis that the diffusion process between the free-flowing water and bottom slimes is a boundary layer type of diffusion. Thus, assuming that the diffusion rate through the boundary layer controls organic transfer from the free-flowing stream, the coefficient v is supposed to range from $v \approx 0.1$ for streams with moving bottoms (sand, mud) to $v \approx 3.0$ (B base e) for rocky streams. The possible ranges of the coefficients η and v are shown in Table 10.2.

As noted by Kittrell and Kochtitzky (1947) for the Holston River, K_1 is not constant, but decreases with decreasing BOD concentration. The variation of K_1 (base 10) in the Holston River is shown in Fig. 10.3. One might expect the same behavior to occur with the coefficient B, although to a lesser extent. Thus, the formula of Boško or Novotny yields an approximation and must be used with caution.

Both theoretical research (Tischler and Eckenfelder, 1968) and practical measurements (Monod, 1949; Wuhrman, 1955) lead to the conclusion that the coefficient of deoxygenation for the carbonaceous BOD varies (decreases) as the BOD is being removed from the solution. This phenomenon is explained by the existence of many organic compounds in common municipal and industrial wastewaters with different degrees of digestibility to the biota in streams or treatment plants. During the biodegradation process, the easily biodegradable compounds are removed linearly at a much faster rate than those with low digestibility. This can be demonstrated in a mixture of glucose (easily biodegradable), aniline, and phenol (least biodegradable), as shown by Tischler and Eckenfelder (1968) and presented in Fig. 10.4. The removal rate of glucose is highest, but linear, followed by aniline and phenol. This can be extended to a common wastewater containing a large quantity of organic compounds. As shown by Grau *et al.* (1975), with the assumption of large quantities of organic compounds being present, the character of the BOD curve and,

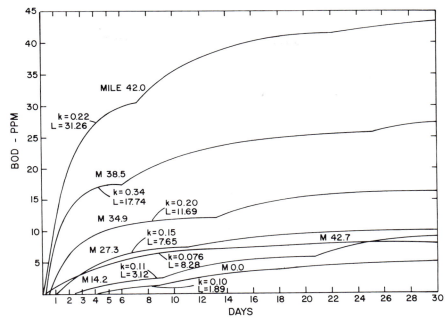

Fig. 10.3. 30-day BOD curves for the Holston River (Kittrell and Kochtitsky, 1947).

thus, the deoxygenation rate, might be considered variable with the remaining BOD resulting in the variability of the deoxygenation rate, k, as the organic waste compounds represented by the BOD parameter (or TOC, COD, etc.) are being removed from the solution.

This phenomenon can be verified by using data from the Holston River, Tennessee (Kittrell and Kochtitzky, 1947; TVA, 1969; and Krenkel and Novotny, 1973), and Mississippi River data below Minneapolis, Minnesota (Univ. of Minn., Inst. Tech., 1961). The set of data shown in Fig. 10.5 contains data from both stream and sewage treatability studies. The relationship indicates that as the waste material is being removed, either from the stream or from the wastewater by a biological process, the BOD removal coefficient decreases. In both sets of data, the K_1 value for the raw wastewater is not the highest one. This condition is probably caused by a lack of acclimated microorganisms in the sample. As the water proceeds downstream, the microorganisms are being acclimated, thus increasing the rate. However, the easily assimilated materials are oxidized first and those remaining are then removed at a reduced rate. The overall effect is a skewed bell-shaped curve as shown in Fig. 10.5.

Most of the microorganisms and fungi which take part in the self-

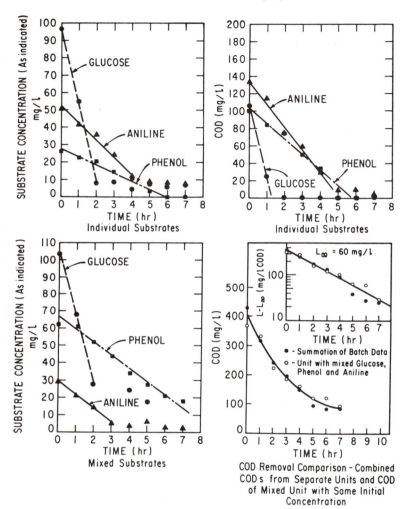

Fig. 10.4. Batch removal of simple Organic Compounds—individually and mixed, loading = 0.25 kg Chemical Oxygen Demand/Kg Mixed Liquor Volatile Suspended Solids/day (Tischler and Eckenfelder, 1968).

purification process are facultative; i.e., they metabolize the organics in either an aerobic or anaerobic environment. However, anaerobic metabolism is much slower than aerobic ($K_1 = 0.12 -$ base e). On the other hand, the most dominant species which can be found in the biological bottom slimes of polluted rivers, *Sphaerotilus natans*, is strictly aerobic. Thus, with low (near zero) oxygen concentrations, one might expect B $\rightarrow$

0. In an anaerobic environment, nitrates and possibly sulfates, become the electron acceptors in the BOD removal process, which reduces nitrates to nitrogen gas and sulphates to hydrogen sulfide.

Benthal Oxygen Demand

In polluted streams, the river bottom below a wastewater outfall may be covered by highly active biological materials, sludges, biological slimes, algae, etc. The growth and accumulation of these materials result either from deposition of suspended organics and/or from the transfer of soluble organics (BOD) from the flowing water to the biological slimes. The BOD is then metabolized to new cell growth and aerobically or anaerobically decomposed. Deposition of the suspended organics usually occurs during low flows when the velocities are low. Velz (1953) and Imhoff and Fair (1956) state that if the velocity is less than 0.6 to 1.0 ft/sec,

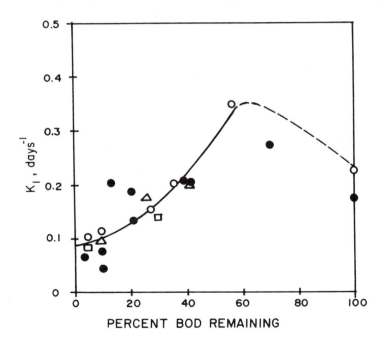

△ EPA SURVEY
□ WASTEWATER TREATABILITY DATA
○ KITTRELL-KOCHTITZKY'S RIVER DATA
● MINNEAPOLIS-ST. PAUL SEWAGE

Fig. 10.5. Percent BOD remaining *v.* K_1.

deposition takes place, and scouring occurs when the velocity increases above 1.0–1.5 ft/sec. Thus, deposited materials may be resuspended and cause a secondary increase of BOD.

Only the upper layer of deposited sludges is aerobic, with the remainder of the sludge banks anaerobic. Imhoff and Fair (1956) have reported that the thickness of the aerobic layer is approximately 5 mm. The gaseous products of the anaerobic decomposition process, CH_4 and H_2S, proceed up through the sludge layer into the overlying water. Typical values of the dissolved oxygen uptake by sludge deposits vary from 1 to 10 gm/m²/day. Thomann (1972) summarized some of the earlier works on the magnitude of the benthal demand, the results of which are presented in Table 10.3.

In order to obtain S_B as defined by Eq. (10.5), the areal benthal oxygen demand (assuming complete vertical mixing of the overlying water) must be divided by the average depth, or

$$S_B = \frac{O_2 \text{ uptake (g/m}^2\text{/day)}}{H}. \tag{10.11}$$

Soluble organic wastes support the growth of heterotrophic bottom microorganisms and fungi. As previously mentioned, filamentous bacteria, such as *Sphaerotilus natans*, are typical. Although *Sphaerotilus* may sometimes become the predominant species, many other bacteria take part in this biological "bottom" activity.

TABLE 10.3   Average values of oxygen uptake rates of river bottoms[a]

Bottom Type and Location	Uptake (gm O_2/m²/day) at 20°	
	Range	Approximate Average
Municipal sewage sludge—outfall vicinity	2–10.0	7
Municipal sewage sludge—aged downstream of outfall	1–2	1.5
Cellulose fiber sludge	4–10	7
Exclusive mud	1–2	1.5
Sandy bottom	0.2–1.0	0.5
Mineral soils	0.05–0.1	0.07
River Ivel[b]		2.15
River Hiz[b]		2.9

[a] After Thomann (1972).
[b] Anon., Water Pollution Research Laboratory, England (1967).

Milwaukee Harbor   5-6

From these considerations, it should be possible to relate the benthal oxygen demand of these aerobic bottom growths to the BOD deoxygenation rate, i.e., to the bottom deoxygenation factor, B. From the general microbiology of biological wastewater treatment processes, the following relationship has been accepted for expressing the oxygen uptake rate:

$$O_2 \text{ uptake} = a'S_r + b'X_v, \tag{10.12}$$

where S_r = BOD removed, X_v = cell concentration, and a', b' are coefficients. The ranges of a' for activated sludge and trickling filters vary within the interval 0.4 to 0.8. Applying Eq. (10.12), one can write an approximate relationship for benthal oxygen demand as

$$S_B \approx a'BL + b'X_v/H. \tag{10.13}$$

The first term on the right-hand side of Eq. (10.13) expresses the oxygen demand due to the BOD removed from the overlying water, and the second term expresses the endogenous respiration of the microbiological flora.

As a rough estimate, which also includes a safety factor, it can be assumed that

$$S_B \approx BL. \tag{10.14}$$

Thus, it may not be necessary to include a separate term for the benthic oxygen demand of the biological slimes in Eqs. (10.5) and (10.8). However, this assumption requires further investigation. For *Sphaerotilus*, values of oxygen demand up to 18 g/m²/day at 20°C have been reported (Water Pollut. Res. Lab., 1967).

Nitrification

The process of nitrification, in which organic nitrogen and/or ammonia are oxidized to a higher oxidation state, may be of significant import to the dissolved oxygen balance of streams. The oxygen requirements are over 4 mg of O_2 per mg of the total organic nitrogen, with the theoretical oxygen demand for ammonia being even higher (4.56 mg of O_2/mg of NH_4^+).

The effect of nitrification has been known to sanitary engineers for many years. Although the nitrogenous oxygen demand (NOD) of unnitrified effluents was well understood, sanitary engineers generally dismissed this potential river problem on the basis of the following factors (Wild *et al.*, 1971):

1. Nitrification is caused by specific organisms (*Nitrosomonas* and *Nitrobacter*), the population of which is minimal in surface waters.

2. The reaction constant for nitrogenous oxidation is small in relation to the constant for carbonaceous matter.

3. Oxidation of ammonia to nitrates simply converts dissolved oxygen to a form in which it is still available to prevent development of anaerobic conditions.

In addition, if nitrification was allowed to occur in the treatment plants, subsequent denitrification resulted in rising sludge and solids problems in the final clarifiers.

The philosophy that unnitrified effluents are not damaging to receiving streams has been disagreed with by biologists and conservationists, who argue that nitrates will not satisfy the oxygen requirements of fish and many other aquatic organisms requiring dissolved oxygen in higher concentrations.

A study of the Grand River by the Michigan Water Resources Commission (Mich. Water Resour. Commis., 1962) demonstrated the impact of nitrification on stream oxygen resources and clearly indicated the necessity of including organic nitrogen oxidation in oxygen balance calculations. Other examples of nitrification processes taking place in surface waters have been reported for English estuaries of the Thames and Tees rivers (Water Pollut. Res. Lab., 1971), the Delaware estuary (Tuffey *et al.,* 1974), a reach of the Truckee River below Reno, Nevada (O'Connell *et al.,* 1963), and the Passaic and Beaver Brook rivers in New Jersey (Tuffey *et al.,* 1974).

Thus, the present policy of many agencies responsible for enforcing water quality control laws is to require removal of the oxygen demanding nitrogen compounds from wastewater effluents. The nitrification process, which would follow a conventional biological treatment step removing the carbonaceous BOD, is a relatively costly process requiring long detention times in the treatment units. Furthermore, the process is quite sensitive to environmental factors and is energy intensive. Thus, including nitrification in waste assimilative capacity evaluations represents substantial impact on the cost of water quality control measures. Therefore, its inclusion must be justified by scientific findings and research. However, in a recent publication (Water Pollut. Res. Lab., 1971) from the well-known and respected British Stevenage Laboratory, investigators reported that they could find no evidence of nitrification occurring in any of Great Britain's rivers with the exception of two with very long detention times, the Mole and the Trent. They postulated that the slow growth rate of the nitrifying bacteria necessitated a long detention in the water phase. By the same reasoning, estuaries, which often have detention times on the order of months, are known to nitrify.

Nitrifying bacteria in estuaries and rivers with long detention times are probably dispersed or living in flocs. In addition to growth in the water phase, one also finds nitrifers growing in attached biological slimes (Matulewich and Finstein, 1973). The residence for these slimes is mainly a function of local hydrology, with periods of sufficient low flow duration allowing development of a nitrifying population.

An example of the nitrification effect on dissolved oxygen concentration is shown in Fig. 10.6, which is typical of a medium-size stream. A typical lag time to the beginning of nitrification is characteristic for these streams. On the other hand, in small- and medium-size streams with significant attached biological bottom growths, the nitrification lag time can be minimal or even nonexistent, as shown in Fig. 10.7.

It is generally accepted that the nitrification process is carried out in an aquatic environment by autotrophic bacteria belonging to the family Nitrobacteriaceae. These bacteria prefer solid surfaces for growth, i.e., they grow better if suspended particles or bottom slimes are present. By contrast with the heterotrophic organisms which derive their energy from the oxidation of the organic carbon components, the autotrophic nitrifying microorganisms derive their energy from the nitrification reaction itself.

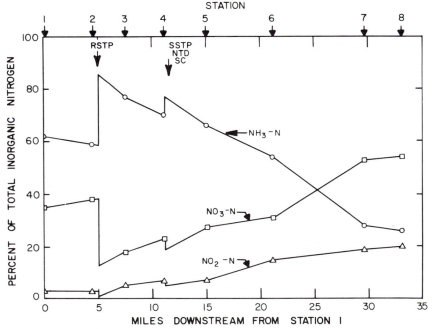

Fig. 10.6. Nitrogen conversion in the Truckee River (O'Connell *et al.*, 1963).

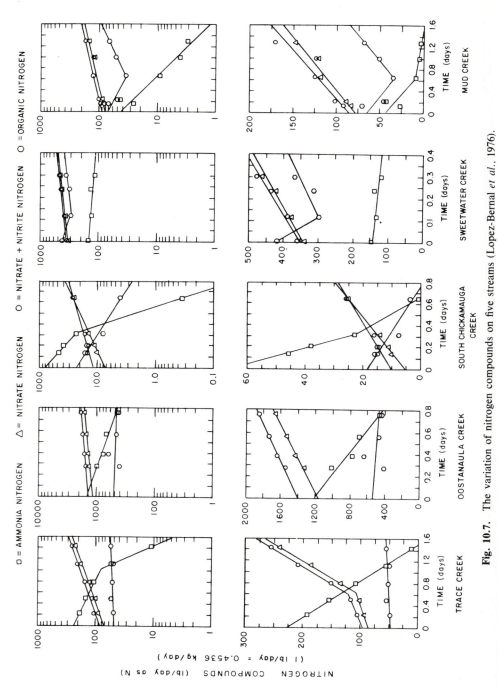

□ = AMMONIA NITROGEN   △ = NITRATE NITROGEN   ○ = NITRATE + NITRITE NITROGEN   ○ = ORGANIC NITROGEN

Fig. 10.7. The variation of nitrogen compounds on five streams (Lopez-Bernal *et al.*, 1976).

NITROGEN COMPOUNDS (lb/day as N)
(1 lb/day = 0.4536 kg/day)

By virtue of their autocatalytic nature, the growth of nitrifiers is very slow, as is the yield of cells per unit of energy source oxidized. In pure cultures, the organisms are extremely susceptible to poisons such as metal ions and other toxic materials.

Nitrification is a two-stage process involving first the autotrophic biochemical oxidation of ammonia nitrogen to nitrite by *Nitrosomonas* or similar organisms. The end product of the first reaction becomes the substrate for the second oxidative process in which a different group of bacteria, *Nitrobacter*, oxidizes nitrite to nitrate nitrogen. In simplified form, these reactions may be written as

$$NH_4^+ + \tfrac{3}{2}O_2 \xrightarrow{\textit{Nitrosomonas}} NO_2^- + 2H^+ + H_2O$$

$$NO_2^- + \tfrac{1}{2}O_2 \xrightarrow{\textit{Nitrobacter}} NO_3^-$$

The nutritional requirements of *Nitrosomonas* —NH_4^+, CO_2, Fe, Cu, etc.—are usually present in adequate supply in sewage; the energy source for *Nitrobacter*, nitrite, is usually absent but is formed by *Nitrosomonas*. However, competition with heterotrophs may cause a limitation of growth; for example, if the C:N ratio were high and the heterotrophic organisms were growing rapidly, the ammonia available for *Nitrosomonas* would tend to be severely restricted.

[handwritten margin note: If C:N ratio high, hetero-trophic organisms may use sig. quan. of NH₃]

If the nitrogen is present in organic form prior to entering the nitrification process, it must first be changed to ammonia by processes including deamination. Three ways of producing ammonia from organically bound nitrogen can be distinguished (Painter, 1970):

1. From extracellular organic nitrogen-containing compounds, chemically or biochemically,
2. From living bacterial cells during endogenous respiration, and
3. From dead and lysed cells.

The foregoing processes, i.e., heterotrophic ammonia utilization during higher C:N ratios and deamination of organically bound nitrogen, cause a time lag in the appearance of nitrification, which is usually several days, as can be seen in the typical example of the BOD reaction shown in Fig. 10.8. The time to nitrification for raw sewage has been measured, and a summary is shown in Fig. 10.8, which also contains measurements of the beginning of nitrification for secondary effluent. It can be seen that with more treatment of sewage, the time lag is reduced.

Stoichiometrically, the oxygen requirements for the overall reaction are 4.56 mg/O_2/mg of NH_4^+. However, since the reaction is autotrophic, carbon dioxide is fixed with the result that the amount of oxygen utilized in inorganic nitrogen oxidation is less than the theoretical value. As re-

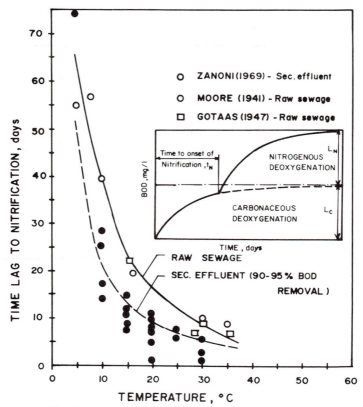

Fig. 10.8.   The effect of temperature and treatment on the onset of nitrification.

ported by <u>Wezernak and Gannon (1968), direct measurement experiments</u> <u>of oxygen utilization in nitrification indicated that the value of the oxygen</u> <u>utilization is 4.33 mg of O_2 per mg of NH_4^+.</u>

Factors Affecting Nitrification Kinetics

DeMarco *et al.* (1967) and Wild, Sawyer, and McMahon (1971) investigated the effect of environmental factors on the nitrification process. A more comprehensive literature review published by Painter (1970) summarizes the review and research conducted by the British Stevenage Water-Pollution Research Laboratory. Based on these studies it can be summarized that

1. The initial rate of ammonia oxidation is proportional to the initial concentration of nitrifying microorganisms. If nitrifying microorganisms

are not present or are present in low concentrations, a lag time of several days to the start of nitrification can occur.

2. The nitrification rate is affected by the turbulence level. Samples with lower turbulence showed lower nitrification rates.

3. The presence of organics in the samples (carbonaceous BOD) had several effects on the rate of the ammonia nitrogen oxidation:

a. If the initial ammonia nitrogen was not completely assimilated by the heterotrophic population (decomposing carbonaceous BOD) during organic oxidation, simultaneous oxidation of carbonaceous and nitrogenous matter took place without influence on the initial rate of nitrification provided that enough of the nitrifying population was present in the sample.

b. If the ammonia nitrogen was completely assimilated during organic oxidation, nitrification could not begin until some ammonia nitrogen was released back into the solution by the death and lyse of heterotrophic bacteria.

c. In either a or b, when the ammonia nitrogen was eliminated from the solution and the only source of ammonia nitrogen for nitrification was from cellular organic nitrogen, the rate of nitrification was controlled by the cell death rate, which is usually quite slow.

4. pH has a profound effect on nitrification with the optimum pH being 8.3. At pH 7.0 the rate of nitrification is reduced to 50%, and below pH 6.0, the nitrification process falls to 10% of the maximum, as shown in Fig. 10.9.

5. The nitrifying microorganisms are mainly strict aerobes; i.e., nitrification occurs only if sufficient dissolved oxygen is present. If the oxygen concentration is depleted below 2 mg/liter, the reaction rate decreases rapidly and no oxidation could be detected when the DO concentration

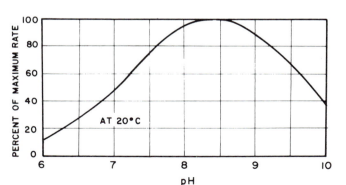

Fig. 10.9. Percent of maximum rate of nitrification at constant temperature *v*. pH (Wild *et al.*, 1971).

was less than 0.3 mg/liter (Matulewich and Finstein, 1973). Under septic or anaerobic conditions, denitrification occurs where nitrates become the source of oxygen and the oxidized nitrogen is reduced to nitrogen gas.

The original mathematical description introduced by Theriault (1927) and accepted by many authors (Thomann, 1972; Roesner *et al.*, 1973; Boško, 1966) describes nitrification similarly to the carbonaceous BOD process, i.e.,

$$\frac{\partial L^N}{\partial t} = - K_N L^N, \tag{10.15}$$

where the variables have been described by Eqs. (10.4)–(10.8). The reaction rate, K_N, has been reported to range from 0.1 to 15.8 days^{-1} (Ruane and Krenkel, 1975). While this mathematical description roughly holds for the carbonaceous BOD, its application to nitrification is suspect. It has been proposed that the nitrification of ammonia is a first-order reaction, a zero-order reaction, a second-order reaction, and an autocatalytic reaction (Wild *et al.*, 1971; Tuffey *et al.*, 1974; Wezernak and Gannon, 1968; Huang and Hopsom, 1974; Lopez–Bernal *et al.*, 1976). However, the nitrification is a function of the concentration of the microorganisms' participation in the process. Thus, the following mathematical zero-order description appears to be not only appropriate but simple:

$$\frac{\partial L^N}{\partial t} = - 4.33 K_n', \tag{10.16}$$

which, if introduced into a simplified steady-state oxygen concentration model, will yield (Lopez–Bernal, *et al.*, 1976)

$$\frac{dC}{dt} = - K_d L - 4.33 K_N' + K_2(C_s - C_t), \tag{10.16}$$

since

$$L = L_0 e^{-K_r t} = L_0 e^{-(K_d + K_3)t_d} \tag{10.18}$$

and

$$D = C_s - C_t. \tag{10.19}$$

Equations (10.18) and (10.19) can be substituted in Eq. (10.17) to form

$$\frac{dD}{dt} = K_d L_0 e^{-K_r t} + 4.33 K_N' - K_2 D. \tag{10.20}$$

Integrating Eq. (10.20) results in the following expression:

$$D = D_0 e^{-K_2 t} + \frac{K_d L_0}{K_2 - (K_d + K_3)} (e^{-K_r t} - e^{-K_2 t})$$

$$+ \frac{4.33 K'_N}{K_2} (1 - e^{-K_2 t}). \quad (10.21)$$

Equation (10.21) predicts the dissolved oxygen deficit at any point in a stream, where D = dissolved oxygen deficit at time t (mg/liter or lb/day), and D_0 = dissolved oxygen deficit at the discharge point ($t = 0$) (mg/liter or lb/day).

The conversion factor 4.33 comes from the assumption that for the oxidation of 1 mg/liter of NH_4^+—N to NO_2^-—N, 3.22 mg/liter DO are needed and for the oxidation of 1 mg/liter of NO_2^-—N to NO_3^-—N, 1.11 mg/liter DO are needed.

It should be noted that K'_N is an arithmetic coefficient, not a logarithmic coefficient.

The zero-order nitrification rate can be determined by calculating the slope of either the rate of ammonium decrease or the rate of nitrate increase, or by calculating an average of these two rates. The choice of which of these alternatives is best must be made on the basis of some knowledge of the particular stream; however, the rate of increase in nitrate usually is the more reliable rate because ammonium is affected by more factors (Ruane and Krenkel, 1975). As stated previously, the rate of increase of nitrate plus nitrite is usually about the same as the rate of increase of nitrate. This is normally true because the rate of nitrification of ammonium to nitrite nitrogen is the time limiting reaction and the rate of nitrification of nitrite to nitrate nitrogen proceeds at a relatively rapid rate.

Since no nitrate increase was seen, nitrification is assumed not occurring in Proj. III.

Since nitrification is often carried out by nitrifiers attached to the stream bed, K'_N will vary with streamflow in the same manner as benthic oxygen demand.

A zero-order oxygen demand caused by bottom organisms is constant (as long as substrate is present), requiring some mass of oxygen/unit time (i.e., g/day). To determine the effect of this demand on the overlying water, it is necessary to divide the oxygen demand by the volume of the overlying water. The volume of the overlying water is the product of the area of the streambed times the average depth. Since the area of the streambed can usually be assumed constant for relatively small changes in streamflow, the effect of change in streamflow can be accounted for by adjusting such oxygen demand rates by the ratio of depths for the respective streamflows. Thus, to adjust zero-order nitrification rates for streamflow, the following relationship can be used:

$$K'_{N@H_2} = K'_{N@H_1} \left(\frac{H_1}{H_2} \right), \quad (10.22)$$

where $K'_{N@H_2}$ and $K'_{N@H_1}$ are in terms of mg/liter/day, and H_1 and H_2 are average depths for respective streamflows. Since K'_N is derived in terms of lb/day², the following equation usually is more useful for adjusting K'_N for various streamflows:

$$K'_{N@Q_2} = K'_{N@Q_1} \left(\frac{H_1}{H_2}\right) \left(\frac{Q_2}{Q_1}\right). \tag{10.23}$$

It should be noted that a change in streamflow may also affect turbulence intensity, which would affect diffusion intensity. Boško (1957, 1966) and Novotny (1969) have proposed equations to account for the effect of turbulence on the oxidation rate of carbonaceous BOD. A similar relationship perhaps applies to the nitrification rate

$$K_N \sim K'_n \sim C\,\frac{U^m}{H^n} = C_1\,\frac{S^q}{H^p}, \tag{10.24}$$

where C and C_1 are constants for various types of streambeds and m, n, q, and p are exponents that need to be derived theoretically and verified with laboratory and field data.

Because of the complex nature of the nitrogen cycle and the undefined effects of various environmental factors on this cycle, modeling nitrification is a ''descriptive'' rather than a ''predictive'' effort at the present time. Therefore, field data need to be collected under environmental conditions similar to those for which the model will be applied.

Oxygen Sources

Atmospheric Reaeration

The process of adding oxygen from the atmosphere to water is similar to that of heat transfer; i.e., oxygen must penetrate through the air and water surface boundary layers. However, unlike heat transfer, for low-solubility gases such as nitrogen or oxygen the rate of oxygen penetration through the water surface layer is controlling. The rate of oxygen change in a water body can thus be represented by an equation which accounts only for the water surface layer process, i.e.,

$$\frac{\partial c}{\partial t}\,V \simeq A\,K_L(C_s - c) = A\,K_L\,D, \tag{10.25}$$

where c is the dissolved oxygen concentration, C_s is the oxygen saturation concentration, A is the surface area, V is the volume of the water body undergoing aeration, K_L is the surface layer oxygen transfer coefficient, and

D is the oxygen deficit ($C_s - C_t$). Note that K_L is the so-called liquid film coefficient in the chemical engineering literature.

For uniform open channel flow A/V will approach $1/H$, where H is the depth of flow and the common coefficient of reaeration K_2 is used instead of K_L. Then

$$K_2 = \frac{A}{V} K_L = \frac{K_L}{H}. \tag{10.26}$$

The coefficient, K_L, is apparently a function of the turbulence level near the water surface.

In the early work of Streeter and Phelps (1925), the reaeration coefficient, K_2, was reported in the form

$$K_2 = C \frac{U^n}{H^2}. \tag{10.27}$$

During the past fifty years, many investigators have attempted to predict the coefficient of reaeration. The resulting empirical formulas are usually based on the correlation of few hydraulic parameters with a limited amount of data. Wilson and Macleod (1974) list over twenty-five different equations describing the coefficient of reaeration. Additional equations have been listed (Bansal, 1973; Kramer, 1974; Brown, 1974; and Rathbun, 1977). The total number of predictive equations is probably over fifty.

The reaeration coefficient equations can be categorized into two groups:

1. Simple equations, which relate the reaeration coefficient to a few hydraulic parameters such as velocity, depth, roughness, or slope, are usually expressed as

$$K_2 = C \, U^m H^n S_e{}^p, \tag{10.28}$$

where C, m, n, and p are coefficients. Empirically derived coefficients for some commonly used equations are shown in Table 10.4.

2. Complex equations are based on conceptual models such as surface renewal, two-film theory, kinetic theory, energy dissipation, and turbulent diffusion. These include the equations of Dobbins (1964), Krenkel and Orlob (1962), Thackston and Krenkel (1969), Bennett and Rathbun (1972), Lau (1972), and others.

The Thackston–Krenkel (1969) formula is an example of a concept which relates the intensity of oxygen penetration through the surface to the flow energy dissipation. The formula has been reported as

$$K_2 = 0.000287(1 + F^{0.5}) \frac{u_*}{H} \quad \text{(at 20°C)}, \tag{10.29}$$

TABLE 10.4   Summary of coefficients for commonly used reaeration formulas[a] $K_2 = C\, U^m H^n S_e^p$

Author	C	m	n	p
O'Connor and Dobbins (1958)	3.73	0.5	−1.5	0.0
Krenkel and Orlob (1962)	173.6	0.408	−0.66	0.408
Owens *et al.* (1964)	5.32	0.67	−1.85	0.0
Churchill *et al.* (1962)	5.01	0.969	−1.673	0.0
Langbein and Durum (1967)	5.13	1.0	−1.33	0.0
Isaacs and Gaudy (1968)	3.87	1.0 ·	−1.5	0.0
Bennett and Rathbun (1972)	5.58	0.607	−1.689	0.0
Cadwallader and McDonnell (1969)	185.4	0.5	−1.0	0.5
Bennett and Rathbun (1972)	32.81	0.413	−1.408	0.273
Negulescu and Rojanski (1969)	10.91	0.85	−0.85	0.0

[a] K_2—reaeration coefficients (base e) in day^{-1}; U—mean stream velocity in m/sec; H—average depth in meters; and S_e—energy (bottom) slope in meters/meter.

where u_* is the friction velocity $= (gHS_e)^{1/2}$, F is the Froude number $= U/(gH)^{1/2}$, g is the gravity acceleration, and S_e is the slope of the energy gradeline. The dimension of u_* is in m/day, H is in meters, and K_2 (base e) is in days^{-1}.

Because of its wide application, the formula of O'Connor and Dobbins (1958) should be mentioned. They assumed that the reaeration coefficient was a function of oxygen molecular diffusivity, stream depth, and average velocity as

$$K_2 = \frac{(D_m \bar{U})^{1/2}}{H^{3/2}}, \tag{10.30}$$

where D_m is the molecular diffusivity of oxygen (L^2/T), H is the average stream depth, and $\bar{U}$ is the average stream velocity. The equation is based totally on theoretical considerations and has been used effectively.

With such a large number of predictive equations, it is obvious that there is a need to critically evaluate the predictive capability of at least those equations which are in common use. The need for such critique is even more apparent when it is realized that all of the predictive equations are of an empirical nature and/or the result of statistical regression analysis.

Such appraisal has been undertaken by several authors, some of whom were probably biased since they included their own formulas in the analysis and used a limited amount of data for verification. The most comprehensive analysis was done by Bennett and Rathbun (1972) and Rathbun (1977) in the United States, and Wilson and Macleod (1974) in Great Britain. The models analyzed by Bennett and Rathbun were applied to 239 data points for rivers, streams, and flumes. Wilson and Macleod compared the predictive equations with 502 data points including those used by Bennett and Rathbun.

Those who have analyzed the errors involved in determining the reaeration coefficient agree that the errors and scatter of the data are more than one would expect from experimental errors alone. The effect of wind and substances such as detergents in the stream is usually not included in the equations. Since the reaeration is controlled by the diffusion in the liquid surface film, the effect of wind is probably much less than it would be for heat transfer. However, it may become significant for quiescent water bodies such as lakes and reservoirs. It should be noted that the presence of surface active agents (Mancy *et al.*, 1960) and/or suspended sediment (Alonso *et al.*, 1975), may reduce the diffusion processes and lower reaeration may result.

The most significant of the conclusions of Bennett and Rathbun (1972) was that analysis of data from natural streams and from laboratory flumes indicated a significant difference between regression equations obtained from these different data sets. On this basis Bennett and Rathbun concluded that, since no single equation enabled accurate prediction of reaeration coefficients in both stream and laboratory flumes, existing equations were inadequate and incomplete.

Wilson and Macleod (1974) statistically evaluated 502 data points with the 16 most common predictive equations. The measured and predicted values were compared both graphically and statistically:

$$\text{Standard error} = \left(\sum_{1}^{N} (\text{Calculated value} - \text{Observed value})^2 \right)^{1/2} \bigg/ N.$$

The standard error has the same units as the parameter under investigation and may be used as a comparison of the degree of scatter for identical data and different models, but not for different data sets. Better evaluation can be performed using the normalized mean error of prediction, which is defined as

Normalized mean error
$$= \left(\sum_{1}^{N} \frac{(\text{Calculated value} - \text{Observed value})}{(\text{Observed value})} \times 100\% \right) \bigg/ N$$

Table 10.5 shows the results of the performance evaluation for the 16 most commonly used equations.

In their conclusions, Wilson and Macleod (1974) stated (among other things) that

TABLE 10.5   Performance of reaeration prediction equations[a]

Investigator	Number of Data Points	Standard Error (day^{-1})	Normalized Mean Error (%)
Simple Equations			
Dobbins (1956)	482	114.43	196.1
Churchill *et al.* (1962)	482	104.31	194.9
Churchill *et al.* (1962) (after Isaacs and Gaudy, 1968)	482	71.77	75.1
Krenkel and Orlob (1962) (after Isaacs and Gaudy, 1968)	482	81.88	14.2
Isaacs and Gaudy (1968)	482	75.91	42.9
Negulescu and Rojanski (1969)	482	99.15	54.4
Bennett and Rathbun (1972)	482	358.36	701.4
Bennett and Rathbun (1972)	482	207.31	470.4
Energy Equations			
Dobbins (1964)	382	73.43	34.5
Krenkel and Orlob (1962)	382	116.63	−34.8
Thackston and Krenkel (1969a)	382	112.36	−15.0
Thackston and Krenkel (1969a)	382	115.62	−27.6
Cadwallader and McDonnell (1969)	382	66.19	69.6
Parkhurst and Pomeroy (1972)	382	33.25	31.4
Lau (1972)	382	150.37	330.9
Bennett and Rathbun (1972)	382	157.84	298.9

[a]From Wilson and Macleod (1974).

1. Those equations for the prediction of reaeration coefficients which incorporate only the simple flow parameters of stream depth and average velocity do not permit accurate predictions over the entire range of data investigated.

2. Even the best of the available correlations gives unreliable predictions of reaeration rates. In many cases, predicted rates are many times larger or smaller than those observed.

3. Examination by a technique based on dimensional analysis strongly suggests that the scatter in the data which persists even for those equations giving the best general fit is probably mainly attributable not to experimental error but to the omission of one or more significant variables from consideration. Such missing variables can only be identified by further experimental studies.

Based on the preceding evaluations, it is evident that the use of any of the predictive equations without a good fit with the field data is suspect. Thus, several field surveys and good fit of the data to the computed model values are eminent.

Inasmuch as the method of Velz (1970) is being used in the comprehensive river water quality assessments being performed by the United States Geological Survey, such as that performed on the Willamette River, Oregon (U.S. Geol. Surv., 1975–1979), it should be discussed even though it does not fall into the mentioned categories. Velz first presented his methodology in 1939 (Velz, 1939):

> One of the first fundamental approaches to the evaluation of reoxygenation was made by Black and Phelps quoted in Phelps (1944) when they stated that reoxygenation was governed by two fundamental laws, (1) the law of solution and (2) the law of hydrodiffusion. The rate of reoxygenation is inversely proportional to the amount of dissolved oxygen present; i.e., the less oxygen present, the greater the rate of reoxygenation up to a maximum rate when no oxygen is present. Basically the relationship proposed by Phelps (1944), sometimes called Fick's law of hydrodiffusion, is for diffusion in quiescent water which must be translated into a flowing stream by means of a mix concept.

From these laws, the mathematical relationship proposed by Phelps and put into a graphical form by Velz is

$$D = 100 - \left[\left(1 - \frac{B}{100} \right) \times 81.06 \left(e^{-k} + \frac{e^{-9K}}{9} + \frac{e^{-25K}}{25} + \cdots \right) \right],$$

(10.31)

where D = final DO content (average for the depth) as percent of saturation/mix; B = initial average DO content as percent of saturation; and $K = \pi^2 \, at / 4L^2$ (this is not the same k_1 or k_r used in deoxygenation relationships); in which t = time of exposure in hours, L = depth of water in centimeters, and a = diffusion coefficient for a specific temperature (determined experimentally).

Velz solved the preceding equation assuming that the initial dissolved oxygen concentration $(B) = O$ and expressed it in graphical form as illustrated in Fig. 10.10.

Inspection indicates that it is necessary to know the mean channel depth, d (scale 2), the stream temperature, T (scale 3), and a time of mix value, t (scale 4), in order to determine the percent of oxygen saturation absorbed per mix at zero initial dissolved oxygen. For oxygen concentrations other than zero, this figure is related on a linear basis to the dissolved oxygen deficit, e.g., at 60% dissolved oxygen concentration the deficit would be 40% and, therefore, the rate of reoxygenation would be 40% of the maximum rate for 100% deficit. In addition, for a quantitative evaluation of the amount of oxygen obtained from reaeration in terms of pounds per day, it would be necessary to know the appropriate volume of river water which could be calculated from river channel information.

Of the factors entering into the reoxygenation calculations, the one

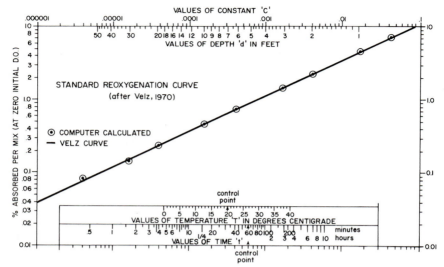

Fig. 10.10.   Standard reoxygenation curve (Velz, 1939).

most difficult to evaluate has been the time of mix. Originally it was re-lated by Velz to velocity and depth based on limited experimental channel data, but this proved inadequate for full-scale river situations. It became necessary to resort to an indirect semiempirical approach to establish a relationship between mix interval and effective depth as illustrated in Fig. 10.11, based on different flow regimes.

It is noteworthy that the United States Geological Survey's studies on the Willamette River Basin demonstrated that Velz' method yielded the most reliable results, according to their evaluations.

Finally, the technique for independent field measurement of river gas exchange capacity as proposed by Tsivoglou (1967) must be elucidated. Tsivoglou (1967) showed by laboratory experiments that the rate of ox-ygen absorption could be related to the rate of krypton gas dissolution by means of comparison of their molecular properties. The laboratory studies showed that the gas exchange coefficient of krypton-85 is 83% of that for oxygen and is fairly constant.

The field tests consist of injecting three tracers into the water simultaneously—fluorescent dye, which provides time of flow data and when to sample for other tracers; tritium, which yields information on dis-persion; and krypton-85, which undergoes the same dispersion and is also lost to the atmosphere. In any reach of the stream, the gas lost is evalu-ated on the basis of upstream and downstream concentration ratios of krypton-85 and tritium.

The assumptions made are

(a)   Tritiated water undergoes only dispersion in the stream and is not lost from the stream water in any significant amount.

(b)   Dissolved krypton-85 undergoes the same dispersion as tritiated water and is also lost to the atmosphere with no other significant losses.

(c)   The ratio of the gas transfer coefficient for krypton-85 and oxygen is 0.83 and is not significantly affected by temperature, turbulence, or the presence of pollutants.

On the basis of several field studies, Tsivoglou derived the following equation for K_2:

$$K_2 = C\frac{\Delta H}{t},\qquad(10.32)$$

where K_2 is the base e reaeration coefficient per hour, ΔH is the change in

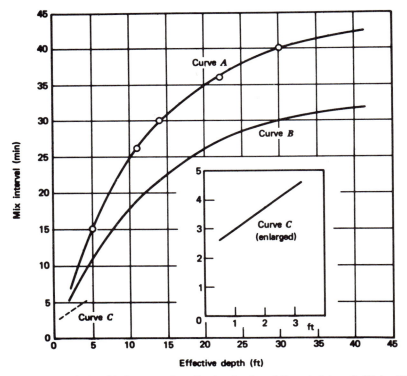

Fig. 10.11.   Relationship between the effective depth and the mix interval. (Velz, 1970). Curve A, usual freshwater streams; Curve B, tidal estuaries; Curve C, shallow, relatively high velocity streams.

water surface elevation in the stream reach in feet, t is the time of flow through the reach in hours, and C is the escape coefficient per foot. Tsivoglou and Wallace (1972) reported that $C = 0.054/\text{ft}$ may be used to estimate the reaeration coefficient for an "average" or "typical" stream that is moderately polluted ($BOD_5 = 15$ mg/liter) and reasonably well mixed at a temperature of 25°C. Tsivoglou (1974) later proposed that C varied inversely with the discharge as

$$C = C_0 - mQ, \tag{10.33}$$

which is the equation of a straight-line relationship bewtween C and Q, where m is the slope of the line and C_0 is the intercept.

The method shows promise, however. The authors attempted to use Eq. (10.32) on the data taken by Churchill *et al.* (1962) with no success. This is significant because the data taken by Churchill *et al.* (1962) are probably the only reliable field reaeration data available, and all of the parameters required by Tsivoglou are available.

Because of the environmental movement, Tsivoglou's method may be academic because introducing radioactive material into the environment is difficult if not prohibited in most states. For this reason, Rathbun *et al.* (1978) proposed the use of ethylene and propane as a tracer in lieu of krypton-85. Tsivoglou (1979), however, disputed their claims and concluded that their method was not an "acceptable alternative" to his.

Photosynthesis and Respiration of Algae and Green Plants

Organisms containing chlorophyll can assimilate atmospheric carbon dioxide and water and liberate oxygen. A simplified photosynthetic reaction can be written as

$$6CO_2 + 6H_2O \xrightarrow{\text{light}} C_6H_{12}O_6 + 6O_2.$$

Plant and algae respiration occurs continuously throughout the day, since the photosynthetic release of oxygen during the daytime during bright days is higher than the respiration, which results in a net oxygen gain. The ratio of the photosynthetic oxygen production to the respiration, P/R, has a seasonal pattern. Common ratios of P/R are equal to $1:4$. During dull or dark days, the ratio may be less than 1; i.e., respiration is greater than photosynthetic production.

The effect of aquatic plants on DO within a particular stream reach at a particular time of day is a function of the plant density, the light intensity, the water depth, the turbidity, and the DO concentration. The most important factor is apparently plant density. Oxygen production is proportional to plant density only to a certain limit; then oxygen production decreases with continuously increasing plant density and may even show a net oxygen loss. These phenomena may occur because the plants become so dense that some are shaded by other overlying plants. In such situations, even though high DO values are produced during the midday hours, extremely low DO concentration may occur at night.

Photosynthetic oxygen production is independent of the oxygen deficit; thus, it can result in oxygen supersaturation during a bright, sunny day. Photosynthetic oxygen production rates as high as 50 gm O_2/m^2/day have been reported.

The effect of plant photosynthetic activity and respiration on the oxygen balance can be determined using several methods. These include (a) the light and dark bottle techniques, (b) photosynthetic chambers, (c) evaluation of diurnal DO measurements, and (d) radiocarbon techniques. Supplementing these techniques are measurements of related variables, such as chlorophyll. The dark and light bottle technique only accounts for plankton and does not measure the influence of attached plants.

The production of oxygen and respiration rates are related to the chlorophyll content of algae or green aquatic growths; 1 mg of chlorophyll represents about 6.7 mg of O_2 production under optimal conditions.

The benefits attributed to the photosynthetic oxygen production are questionable. In deeper streams and reservoirs, only the upper 1 m depth has an oxygen production–respiration ratio P/R greater than 1 and even then it only lasts for a limited period of the year. Characteristic algal blooms with higher P/R ratios which usually occur during the spring and late summer and are more of a nuisance than a benefit to water quality. Higher photosynthetic oxygen production is also an indication of advanced eutrophication, as discussed in Chapter 13.

In deeper quiescent water bodies, photosynthetic oxygen production primarily is the result of the planktonic algae population. These algae need rather long detention times for development. Thus, reservoirs with short residence time (< 7 days) do not show appreciable oxygen gains by photosynthesis. In rivers, most of the photosynthetic oxygen is usually produced by attached algal growths and aquatic weeds which can develop in appreciable amounts only in shallow streams. Oxygen contributions or consumption by photosynthesis is probably insignificant in rivers which are deeper than 2 m.

Resulting $(P-R)$ curves can usually be represented by a sine curve function as

$$(P-R) = C_1 \sin 2\pi t + C_2. \tag{10.34}$$

Other functional representations include time series analysis and a

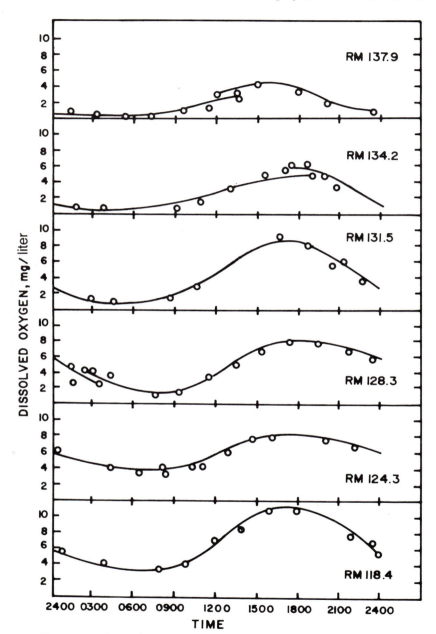

Fig. 10.12. Comparison of measured DO with TVA model (Ruane, 1972).

Fourier series function. Figure 10.12 shows observed data fitted by Eq. (10.34).

In a more complex stream oxygen model the net photosynthetic oxygen contribution is related to the concentration of algae, i.e.,

$$(P-R) = (\alpha_3\mu - \alpha_4\rho)A, \tag{10.35}$$

where A is algal biomass, μ the local specific growth rate of algae as defined in Eq. (10.36), ρ is the local respiration rate of the algae, and α_3 and α_4 are coefficients. Both μ and ρ are temperature dependent.
The local specific growth rate of algae is known to be coupled to the availability of required nutrients and light. The standard formulation for the local specific growth rate in a stream takes the form

$$\mu = \hat{\mu}\,\frac{N_3}{N_3 + K_{NS}}\,\frac{P}{P + K_{pS}}\,\frac{1}{lH}\,\ln\frac{K_{LS} + L}{K_{LS} + Le^{-lD}}, \tag{10.36}$$

where $\hat{\mu}$ is the maximum specific growth rate; N_3 is the local concentration of nitrate nitrogen; P is the local concentration of orthophosphate; L is the local intensity of light at the water surface; H is depth of flow; l is the light extinction coefficient in the river; and K_{NS}, K_{PS}, and K_{LS} are empirical half-saturation constants. Table 10.6 shows the suggested ranges of the empirical coefficients used in the photosynthetic oxygen model.

The algal growth rate to which the photosynthetic oxygen production is coupled is simulated as

$$\frac{\partial A}{\partial t} = \left(\mu - \rho - \frac{\sigma_1}{H}\right)A + \text{convection}, \tag{10.37}$$

where σ_1 is the settling rate of algae from flowing water.

It should be noted that the foregoing model relates the photosynthetic oxygen production to the algae and nutrient concentration in free-flowing water and neglects the contribution by attached aquatic weeds and algal growths. In shallow, fast moving streams, which do not usually have optimum conditions for planktonic algae development, most of the photosynthetic oxygen is produced by attached chlorophyll containing growths. In these rivers, application of the preceding model is not directly possible.

Temperature Effects

Almost all of the reaction rates in the oxygen balance are temperature dependent, with the relationships expressed as

$$K_T = K_{20}\theta^{(T-20)}, \tag{10.38}$$

TABLE 10.6 Suggested ranges of coefficients used in the photosynthetic oxygen model[a]

Coefficient	Description	Units	Range	Temperature Dependence
α_3	O_2 production per unit of algae growth	$\dfrac{\text{mg O}}{\text{mg A}}$	1.4–1.8	no
α_4	O_2 uptake per unit of algae respired	$\dfrac{\text{mg O}}{\text{mg A}}$	1.6–2.3	no
$\hat{\mu}$	Maximum specific growth rate of algae	day^{-1}	1.0–6.0	yes
ρ	Algae respiration rate	day^{-1}	0.02–0.5	yes
σ_1	Local settling rate of algae	$\dfrac{\text{m}}{\text{day}}$	0.0–0.2	no
K_{NS}	Nitrogen half saturation constant	$\dfrac{\text{mg}}{\text{liter}}$	0.1–0.4	no
K_{PS}	Phosphorus half-saturation constant	$\dfrac{\text{mg}}{\text{liter}}$	0.03–0.05	no
K_{LS}	Light half-saturation constant	$\dfrac{\text{Langleys}}{\text{min}}$	0.02–0.03	no

[a]From Roesner *et al.* (1973).

where K_T = reaction rate at temperature, T; K_{20} = reaction rate at temperature $20°C$; and θ = a thermal factor, which has the following accepted values:

Deoxygenation rates K_r, K_d, K_1	$\theta = 1.047$
Reaeration rate K_2	$\theta = 1.029$
Nitrification rate K_N	$\theta = 1.106$
Benthic oxygen demand S_B	$\theta = 1.065$
Photosynthesis–respiration	$\theta = 1.08$

Temperature has an effect on the oxygen saturation value, which can be approximated as

$$C_0 = 14.652 - 0.41022\ T + 0.007991\ T^2 - 0.000077774\ T^3. \qquad (10.39)$$

Additional Sources or Sinks of Oxygen

Oxygen can be gained by the turbulence caused by wind, which may significantly increase the water surface renewal rate; by aeration caused by weirs, sluice gate outlets, and stilling basins; and by turbine aeration. In-stream aerators have been used throughout Europe and the United States in order to introduce additional oxygen, especially during winter ice conditions, when surface aeration is blocked by ice cover. An excellent survey of supplementary aeration was prepared by King (1970).

Wind-Induced Aeration

The shear stress at the water surface induced by wind equals that of water on wind. Thus, an approximation of the surface gas transfer coefficient increase by wind action can be related to the wind velocity as

$$K_L \approx CW^\alpha, \qquad (10.40)$$

where W is the wind velocity and C and α are experimental coefficients. Based on measurements made on English rivers, the value of C may be within 0.2 to 0.35 if the wind velocity is expressed in m/sec and K_L in m/day. The value of the coefficient α for this formula is close to 1, but should theoretically approach 2. Banks and Herrera (1977) analyzed existing data on lakes and lagoons and found this correlation to exist:

$$K_w = \frac{1}{H}\,(0.384\ U^{1/2} - 0.088\ U + 0.0029\ U^2), \qquad (10.41)$$

where K_w is the reaeration coefficient due to wind action in days^{-1}, U is the wind velocity in kilometers per hour, and H is the dpeth in meters. They also concluded that oxygen added to lakes by raindrops could be significant. The reader is referred to Chapter 12 for additional discussion on wind-induced aeration.

The Effect of Hydraulic Structures on Aeration

The effect of weirs and spillways on the oxygen concentration has also been investigated. The ratio of the oxygen deficit before and after the structure can be related to the head loss. British measurements indicate that the aeration potential of weirs and spillways is primarily due to the energy of water falling to the stilling basin or tailwater and, to a much lesser extent, to the exposure of water droplets to the air. Since the energy of falling water can be related to the head difference, the following formula has been suggested (Water Pollut. Res. Lab., 1973):

$$\frac{C_s - C_u}{C_s - C_d} = 1 + 0.38abh(1 - 0.11\ h)(1 + 0.046\ T), \qquad (10.42)$$

where C_s is the oxygen saturation value, C_u is the oxygen content above the structure, C_d is the oxygen content below the structure, h is the head loss at the structure in meters, and T is the temperature in °C.

The coefficient a depends on the character of water quality and has been found to have values near 1.8 for clean water and unity for moderately polluted water. The coefficient b is a function of the weir type, and is unity for a "normal" weir (Water Pollut. Res. Lab., 1973). Some of the oxygen gained by aeration on weirs is not immediately dissolved and can be lost. The difference may be up to 1.5 mg/liter. The effect of cascades and rapids may be considered similar to weirs, but sloping passages are reported to be less efficient than free falls.

Supplementary River Aeration

Aeration using diffused air systems or surface aerators has been attempted on several rivers. However, their efficiency and the amount of oxygen they can deliver depend upon the oxygen deficit. For that reason, it is obvious that they should be located near the sag point. The reported oxygen delivery of surface aerators is between 0.5 to 1.3 kg of O_2 per kW h per mg/liter of O_2 deficit.

As discussed in Chapter 12, low oxygen concentrations may become a problem below reservoirs discharging hypolimnetic water during the summer months. Hypolimnetic water generally has a lower oxygen content, and its concentration may sometimes drop below the oxygen standard. Turbines can sometimes be used to supplement the necessary oxygen. Air is introduced through vents in the draft tubes; however, these systems result in a power loss. On the average 0.8 to 2.5 kg of O_2 can be supplied per kilowatt of power lost (Wiley and Lueck, 1960). Turbine aeration may increase the problem of cavitation; however, many dams are designed for turbine aeration. For example, about 18 dams in Wisconsin can supplement oxygen levels when the oxygen concentration drops below 3 mg/liter.

In the design of an in-stream aeration system the following procedure can be utilized (Imhoff, 1969; Maise, 1970).

1. Determine the desired oxygen supply rate for the aeration system using the formula

$$U = 3.6Q(D_u - D_d),$$ (10.43)

where U is the oxygen supply rate in kg of O_2/hr, Q is the discharge in m³/sec, D_u is the upstream (existing) oxygen deficit in mg/liter, and D_d is the desired (downstream) deficit in mg/liter.

2. Estimate the aerator efficiency, R, under standard conditions. This information is commonly supplied by the manufacturer. Standard conditions under which aerators are tested imply clean water with a temperature of 20°C.

3. Correct the aerator efficiency for field conditions.

$$R_f = R \frac{D_m}{9.02} 1.024^{T-20} \alpha,$$ (10.44)

where R_f is the aerator efficiency in kg of O_2/kW h under field conditions, R is the efficiency under standard conditions, D_m is the operating oxygen deficit in mg/liter at the aerator ($D_m \approx D_u + D_d/2$), T is the temperature of the water in °C, and α is the specific oxygen transfer rate.

The coefficient α is a quantity depending on operating conditions. Typical values for activated sludge plants range between 0.75 and 0.9 with commonly accepted values being approximately 0.85. Lower values can be expected for in-stream aeration due to lower turbulence levels. The aerator efficiency will slightly increase with an increase in stream velocity.

4. The aerator power requirement is then

$$P = \frac{U}{R_f} \quad \text{(in kW)}.$$ (10.45)

THE DOSAG I MODEL

DOSAG I (Texas Water Devel. Bd., 1970), which was developed by
the EPA and the Texas Water Development Board, is a steady-state
model which predicts dissolved oxygen concentrations in streams and
canals and results from a specified set of streamflows, wasteloads, and
temperature conditions. The model possesses other features such as
determination of the streamflow required to maintain a specific dissolved
oxygen goal, and it will search the system for available storage to achieve
that goal. The model may be used to estimate mean monthly dissolved oxygen levels over a full year. Both carbonaceous and nitrogenous oxygen
demands are included, and up to five degrees of treatment for both can be
specified. The DOSAG-I model may be adapted to any stream system,
with a major restriction that impoundments cannot be modeled.

In order to use the DOSAG I Quality Routing Model, the user must take
the stream system in question and divide it into the elements which are
used as inputs to the program. There are essentially four major elements
into which a system must be decomposed so that it can be modeled with
this program:

1. Junctions—the confluence between two streams within the river
basin being modeled.

2. Stretches—the length of a river between junctions.

3. Headwater stretches—the length of a river from its headwater to its
first junction with another stream.

4. Reaches—the subunits which comprise a stretch (headwater or
normal).

A new reach is designated at any point in the stretch where there is a significant change in the hydraulic, biologic, or physical characteristics of
the channel, including the addition of a waste load or the withdrawal of
water from the stream.

After a stream has been represented schematically, it is necessary to
specify the hydraulic, physical, and biochemical characteristics of each
reach in the stream system. The basic hydraulic and physical inputs are
flow, temperature, and coefficients, a, b, c, d, as indicated by the previously described power functions.

A Lagrangian solution technique is used to solve the dissolved oxygen
equation in the DOSAG I quality routing model. The solution used in the
model is similar to Eqs. (10.6)–(10.8) except that the DOSAG I model
does not include photosynthesis and the benthal demand.

The deoxygenation and nitrification coefficients enter the computation as direct input data. Both coefficients are expressed at a standard 20°C temperature and the model will recompute them to a desired temperature. Several options are offered by the model for computation of the reaeration coefficient, K_2. One option is to read it in for each of the reaches in the stream system. The program user also may choose to estimate K_2 values for each reach, based on the known physical and hydraulic characteristics of the stream being modeled. This option in the model applies a general formula for K_2:

$$K_2 = \frac{A_3 \, V^{B_3}}{{}_H C_3},$$ \hfill (10.46)

where A, B, and C are coefficients read as inputs.

Another technique for computing the reaeration coefficient is a direct proportionality between the reaeration coefficient and the stream discharge, or

$$K_2 = A_4 \, Q^{B_4}.$$ \hfill (10.47)

A fourth technique available for computing the reaeration coefficient for each reach is based on the investigation by Thackston and Krenkel (1969), which is identical to Eq. (10.29).

The program user may specify any of the described four methods for the prediction of the reaeration rate coefficient for a given reach.

Waste discharges are entered into the system by specifying a new reach at each location at which a discharge takes place. The model has provisions for withdrawing water at any location within the stream system. In addition, the water is withdrawn from the stream with the quality existing at the location of withdrawal as determined by the model.

ESTUARIES

The basic concepts of modeling stratified and nonstratified estuaries were discussed in Chapter 9. The reactions involved do not significantly differ from those developed for streams; however, several differences should be noted:

1. Because of the high salinity, the biological activity of heterotrophic microorganisms is reduced. Therefore, the deoxygenation coefficients found in estuaries are somewhat lower than those found in the laboratory BOD test and much lower than those in streams.

2. The effect of biological slimes on the increase of the coefficient, K_r,

is minimal, but sedimentation of organic matter may occur, and this may cause an apparent increase in the local value of K_r.

3. In most estuaries, the dispersion transport by tidal waves is more significant than the convective transport caused by fresh-water flow. Therefore, the longitudinal diffusion coefficient, D_L, must be included.

4. The oxygen saturation concentration of seawater is less than that for fresh water and must be accounted for in the determination of C_s. According to Camp (1963), the solubility of oxygen in seawater is about 82% of the solubility in fresh water.

5. Although the rate of nitrification may be retarded because of higher salinity, the detention time in most estuaries is sufficiently long to allow nitrification to occur. Therefore, the effect of nitrification cannot be neglected.

6. The reaeration coefficient should be based on the average point velocity u_t, which includes both the effect of tidal waves and the fresh-water flow, i.e.,

$$K_2 = f(u_t, H) \tag{10.48}$$

where

$$u_t = \left| \overline{\frac{Q_f}{A} + U_r \sin \sigma t} \right|. \tag{10.49}$$

U_r is the maximum value of the tidal velocity, $\sigma = 2\pi/T$ is the frequency of the tide, Q_f is the fresh-water discharge at the head of the estuary, and t is the tidal period. Note that $\left| \overline{} \right|$ denotes the mean of the absolute values.

7. In less polluted estuaries, both ammonia and nitrate nitrogen can be used by algae and may result in the undesirable proliferation of phytoplankton.

8. Because of the lower reaeration rates occurring in more polluted estuaries, anaerobic conditions may occur and nitrate can be reduced to nitrogen gas. The anaerobic conditions may further slow the deoxygenation rate.

Oxygen Balance of Estuaries

Most of the important components of the equations for the oxygen balance in estuaries remain almost the same as for streams. However, the term describing longitudinal diffusion cannot be neglected. Therefore, Eqs. (10.3)–(10.5) must be expanded to include the dispersion term. On the other hand, some factors, such as the effect of benthic slimes, benthal oxygen demand, and photosynthesis may be neglected. Equations (10.3)–(10.5) then become

BOD variation:

$$\frac{\partial L}{\partial t} = D_L \frac{\partial^2 L}{\partial x^2} - U \frac{\partial L}{\partial x} - KL. \tag{10.50}$$

Nitrogenous oxygen variation:

$$\frac{\partial L^N}{\partial t} = D_L \frac{\partial^2 L^N}{\partial x^2} - U \frac{\partial L^N}{\partial x} - K_N L^N. \tag{10.51}$$

Dissolved oxygen variation:

$$\frac{\partial C}{\partial t} = D_L \frac{\partial^2 C}{\partial x^2} - U \frac{\partial C}{\partial x} - K_r L - K_N L^N$$
$$+ P(x,t) - R(x,t) + K_2(C_s - C). \tag{10.52}$$

These equations describe the oxygen transformation occurring in a uniform, nonstratified estuary. In many cases, even irregular estuaries can be segmented into fairly uniform reaches, and a solution of Eqs. (10.50)–(10.52) can be applied.

For stratified estuaries, the vertical diffusion term, $D_y \, \partial^2/\partial_y 2$, would have to be added for all three constituents, L, L^N, and C. However, due to complex flow patterns, a solution is not feasible in most cases. Simplified methods such as Pritchard's "box" method described in Chapter 9 may give adequate results.

Over longer periods of time a steady-state distribution of pollutants can be assumed. In this case, Eqs. (10.50)–(10.52) can be integrated as in the following discussion (O'Connor 1960).

Imposing the boundary conditions

$$L = L_0; \ L^N = L_0^N; \text{ and } C = C_0 \text{ at } x = 0$$

and

$$L = 0; \ L^N = 0; \text{ and } C = C_s \text{ at } x \to \infty$$

will lead to the steady-state solutions for L, L^N, and C. Hence

$$L = L_0 e^{J_1 x} \tag{10.53}$$

$$L^N = L_0^N e^{J_3 x} \tag{10.54}$$

and

$$C = C_0 \, e^{J_2 x} + C_s \, (1 - e^{J_2 x}) - F_1 L_0 \, (e^{J_1 x} - e^{J_2 x})$$
$$- F_2 L_0^N \, (e^{J_3 x} - e^{J_2 x}) + \frac{(P - R)}{K_2} \, (1 - e^{J_2 x}), \tag{10.55}$$

where

$$J_1 = \frac{\overline{U}}{2\,D_L}\left[1 - \left(1 + \frac{4\,K_d D_L}{\overline{U}^2}\right)^{1/2}\right]$$

$$J_2 = \frac{\overline{U}}{2\,D_L}\left[1 - \left(1 + \frac{4\,K_2 D_L}{\overline{U}^2}\right)^{1/2}\right]$$

$$J_3 = \frac{\overline{U}}{2\,D_L}\left[1 - \left(1 + \frac{4\,K_N D_L}{\overline{U}^2}\right)^{1/2}\right]$$

$$F_1 = \frac{K_d}{K_2 + \overline{U}J_1 - D_L J_1{}^2} = \frac{K_d}{K_2 - \widehat{K_r}} \quad - K_d \text{ in O'Connor}$$

$$F_2 = \frac{K_N}{K_2 + \overline{U}J_3 - D_L J_3{}^2} = \frac{K_N}{K_2 - K_N}.$$

If the fresh-water runoff is very small or the cross-sectional area very large, then $\overline{U}$ becomes very small when compared to the dispersion term. Then,

$$J_1 \longrightarrow \left(\frac{K_d}{D_L}\right)^{1/2},$$

$$J_2 \longrightarrow \left(\frac{K_2}{D_L}\right)^{1/2},$$

$$J_3 \longrightarrow \left(\frac{K_N}{D_L}\right)^{1/2}.$$

It should again be noted that the long-term average velocity is $\overline{U} = Q_f/A$, while the reaeration coefficient should be based on the average absolute value of the velocity, u_t, over a complete tidal cycle.

STATISTICAL ANALYSES OF WATER QUALITY DATA

The large data banks of water quality analysis require some kind of statistical analysis for proper interpretation. For many years, attempts have been made to correlate streamflow with ionic content, ionic content with conductivity, and conductivity with streamflow. A statistical approach to the oxygen balance is presented in Chapter 11; however, inorganic and/or nonconservative substances are more amenable to statistical treatment. Models such as QUAL I and QUAL II handle conservative substances but only subject them to dilution, advection, and dispersion. Thus, factors influencing ionic concentrations in water are not accounted for.

The problem of analysis is exacerbated by the lack of quality control on existing data, the lack of purpose for collecting the data, and the lack of a uniform data storage-collection retrieval system. The discussion presented in this section is particularly useful in areas where salts are a problem.

Conductivity Relationships

Inasmuch as conductivity depends on ionic concentration, it is logical to expect a relationship to exist between conductance and, for example, TDS.

Durum (1953) found a linear relationship to exist between TDS and specific conductance (see Chapter 3), with a ratio of 0.63. He proposed this equation:

$$y = a + bx, \tag{10.56}$$

where y = concentration in mg/liter, x = specific conductance in mciromhos and $a + b$ = constants. For the particular river and conditions examined, he found the constants to be as tabulated below.

	a	b
Chloride	− 151	0.293
Sulfate	30	0.104
Calcium + magnesium	64	0.043
TDS	0	0.630

Using this relationship, he was able to approximate the relative contribution of groundwater and surface water to the TDS of the Saline River.

Ledbetter and Gloyna (1962) found that the following equation fits data taken from the Red River in Texas and the Canadian River in Oklahoma:

$$y = aK + bK^2, \tag{10.57}$$

where y = mg/liter of dissolved mineral, K = specific conductance in μmhos, and a, b = constants. Examples of their results are shown in Figs. 10.13 and 10.14. It should be noted that they could obtain no correlation between sulfate and specific conductivity.

In a comprehensive study of the relationships among flow, specific conductance, and ionic composition, Lane (1975) derived a methodology for estimating specific ion concentrations from specific conductivity measurements, assuming that the ion proportions are known or can be adequately predicted.

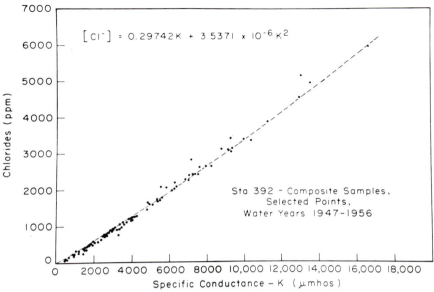

Fig. 10.13.   Chlorides *v.* specific conductance (Ledbetter and Gloyna, 1962).

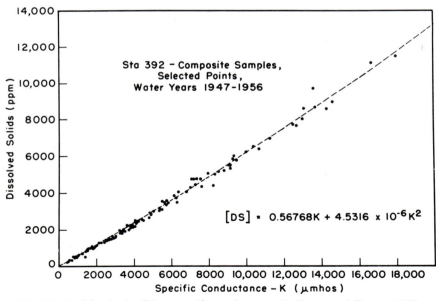

Fig. 10.14.   Dissolved solids *v.* specific conductance (Ledbetter and Gloyna, 1962).

Fair *et al.* (1968) proposed a useful relationship for estimating the TDS.

$$\text{TDS (mg/liter)} \cong (4.5 \times 10^5)(1.02)^{\text{Temp}-25}\,(\text{conductivity}), \quad (10.58)$$

where the conductivity is expressed in $\Omega^{-1}\,\text{cm}^{-1}$. They state that the equation is valid to about 20% for pH values from 5 to 9 and temperatures, T, between 10° and 40°C with TDS less than 1000 mg/liter.

Flow Relationships

Durum (1953), in the previously mentioned studies, found that the chloride concentrations in the Saline River could be represented by

$$\text{Cl}^- = \frac{34{,}500}{Q}. \quad (10.59)$$

He found variation at low and high flows; however, a correlation coefficient of 0.94 was determined.

Ward (1958), using regression analysis in studies of the Arkansas River and the Red River, derived the following equations:

$$\text{Log Cl}^-(\text{mg/liter}) = 2.20 + 0.884 \log Q - 0.188\,(\log Q)^2 \quad (10.60)$$

$$\text{Log SO}_4^{2-}(\text{mg/liter}) = 0.959 + 0.964 \log Q - 0.166\,(\log Q)^2 \quad (10.61)$$

$$\text{Log TDS (mg/liter)} = 3.272 + 0.415 \log Q - 0.112\,(\log Q)^2 \quad (10.62)$$

Ledbetter and Gloyna (1962) proposed that

$$C = KQ^b, \quad (10.63)$$

where b is given by

$$b = f + g \log Aq + hQ^n. \quad (10.64)$$

C is an inorganic pollutant concentration; Q is the flow; f, g, h, and n are regression constants, and Aq is an antecedent flow index represented by

$$Aq_k = \sum_{i=1}^{30} \frac{Qi}{i}, \quad (10.65)$$

where Q is the streamflow and i is the number of days back from the kth day.

As noted by Lane (1975) many factors contributing to the ionic concentrations in rivers must be taken into account. For example, a hysteresis effect has been noted by Hendrickson and Krieger (1964), in which concentrations apparently change according to the flow and concomitant con-

tributions from groundwater and surface water. In addition, the antecedent flow, the time of year, and the time intervals of sampling, must be accounted for.

Manczak (1971) has proposed three types of curves describing the relationship between flow and pollutant concentration, as shown in Figure
10.15. In type I, a/d corresponds to the concentration of wastes discharged from a source of pollution, with dilution as a primary factor. In
type II, characteristic of clean rivers, the curve reflects the addition of
materials by runoff, sludge resuspension, and atmospheric precipitation.
Type III represents an intermediate case in which at low flow, dilution is a
major factor, and at high flows, runoff and bottom resuspension become
important. The relationships were pedantically illustrated in rivers in Poland.

Using the proposal of Manczak (1971), Davis and Zobrist (1978) used
linear correlations, factor analysis, and regression curve analysis to estimate relative contributions of loadings from various sources. For example, their methodology would purportedly ascertain whether the major
source of phosphorus was from runoff or municipal and industrial origin.

Mineral loads were estimated by Betson and McMaster (1975) by correlating the water quality of some undisturbed streams in the Tennessee
Valley area by means of the following logarithmic functional relationship
between quality and flow:

$$C = a(Q/DA)^b \tag{10.66}$$

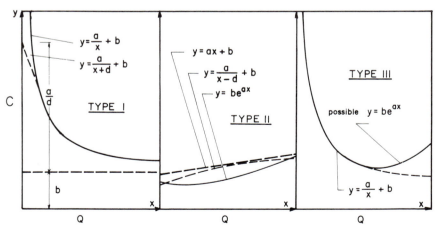

Fig. 10.15. Basic types of curves representing concentration of pollutants and rate of
flow (Manzcak, 1971). Type II, for heavily polluted rivers; type II, for clean rivers; type III,
for intermediately polluted rivers.

where C is the concentration of a mineral constituent in mg/liter, Q is the streamflow in cubic feet per second, DA is the drainage area in square miles,* and a and b are empirically determined coefficients.

The two coefficients in Eq. (10.66) were related to land use, soils, and geologic factors by the linear regression formula

$$a, b = N_1 F + N_2 C + N_3 S + N_4 I + N_5 U, \qquad (10.67)$$

in which a and b are the coefficients from Eq. (10.66), F is the fraction of the watershed area that is forested, C is the fraction of the watershed over carbonate rock, S is the drainage area fraction over shale–sandstone rock, I is the drainage area fraction over igneous rock, U is the drainage area fraction over unconsolidated rock, and $N_1 \ldots N_5$ are regression coefficients.

The four independent geological variables simply allocate the drainage area among the rock types present in the watershed, and their sum must equal 1.

Table 10.7 shows regression coefficients for fifteen water constituents obtained by analyzing 66 watersheds. As stated by the authors, the use of Eq. (10.67) requires caution since constituent rating curves (Eq. 10.66) have been found to display hysteretic effects with seasons and with rising and falling water stages. Variations among watersheds are also influenced by other factors, as previously mentioned.

TRANSPORT OF POLLUTANTS ADSORBED ON SEDIMENT

The movement of pollutants adsorbed on sediments is not well known. However, with the increasing interest in the fate of pesticides and nutrients, investigations of the process and its modeling are increasing.

The equation of continuity (water) and motion remain the same as in previous models. The mass balance equation for such pollutants as phosphorus, heavy metals, and adsorbed pesticides must be coupled with sediment transport since adsorption or release may take place between the adsorbed and dissolved pollutant phases. The adsorbed component moves with the sediment and is subject to sedimentation or scour. Another component of the system is the process of transformation of the pollutant after deposition in the benthal layers. The exchange of matter between the bottom deposits and overlying water is governed by adsorption equilibrium and limited by the diffusion velocity through the bottom boundary layer.

* If Q and DA are in m³/sec and km², respectively, multiply Q/DA by 27.83.

TABLE 10.7 Tennessee Valley Authority mineral quality model—regression with forest and geologic variables

Quality Parameter	Regression Coefficient	N^b	F^b	C^b	S^b	I^b	U^b	R	Standard Deviation	F statistic
				Regression Coefficient Value					Statistics	
SiO_2	a	64	−1.26	5.42	6.78	10.2	8.95	0.64	1.69	8.11
	b		−0.135	0.051	0.099	0	−0.927	0.40	0.114	2.26
Fe	a	29	0.035	0.020	0.009	−0.008	0.387	0.95	0.004	42.25
	b		−0.173	0.272	0.104	−0.125	0.397	0.38	0.482	0.80[a]
Ca	a	66	−8.52	53.9	13.4	8.32	8.41	0.85	9.0	31.42
	b		0.064	−0.116	−0.203	−0.229	−0.005	0.32	0.153	1.32[a]
Mg	a	66	−2.81	11.4	3.41	3.05	2.45	0.75	2.62	15.50
	b		−0.148	−0.145	−0.074	−0.104	0.513	0.67	0.197	9.99
Na	a	44	−1.79	2.23	2.84	3.00	3.74	0.74	0.50	9.45
	b		−0.318	0.079	0.122	0.110	−0.007	0.48	0.138	2.33
K	a	44	−1.08	2.51	1.94	1.58	1.80	0.72	0.47	8.12
	b		−0.152	−0.195	−0.061	0.033	−0.158	0.20	0.254	0.32[a]
HCO_3	a	66	−22.8	200	35.3	26.8	21.8	0.86	32.1	34.67
	b		0.110	−0.156	−0.294	−0.355	−0.139	0.35	0.132	1.71[a]
SO_4	a	66	−7.41	9.15	12.5	7.90	9.56	0.39	5.35	2.19
	b		−0.302	0.103	0.155	0.272	0.592	0.49	0.274	3.69

Parameter		N								
Cl	a	66	-1.86	3.21	2.95	2.58	3.81	0.60	0.93	6.68
	b		-0.171	0.010	0.088	0.099	0.067	0.24	0.14	0.72[a]
NO_3	a	63	-1.13	3.52	1.02	1.30	0.84	0.80	0.71	20.11
	b		-0.70	0.262	0.899	1.063	0.297	0.26	0.671	0.84[a]
TDS	a	66	-39.9	195.6	68.5	55.5	57.7	0.84	30.6	28.42
	b		0.016	-0.94	-0.146	-0.142	0	0.26	0.14	0.88[a]
$CaCO_3$	a	66	-33.4	182.8	48.1	34.3	31.5	0.86	29.3	33.06
	b		0.033	-0.150	-0.176	-0.222	0.131	0.49	0.151	3.78
Specific conductance	a	46	-145	357	180	142	128	0.88	54	26.74
	b		-0.015	-0.078	-0.134	-0.095	0.051	0.43	0.106	1.82[a]
pH	a	65	-0.573	8.37	7.32	7.33	6.86	0.73	0.42	13.53
	b		-0.003	-0.010	-0.003	-0.013	-0.003	0.16	0.021	0.32[a]
Color	a	63	-1.79	2.50	9.17	9.75	10.8	0.23	8.65	0.62[a]
	b		-0.376	0.211	0.339	0.204	0.448	0.32	0.364	1.26[a]

[a]Not significant at 0.9 level.
[b]N = Number of watersheds
F = fraction of watershed that is forested
C = drainage area fraction over carbonate rock
S = drainage area fraction over shale-sandstone rock
I = drainage area fraction over igneous rock
U = drainage area fraction over unconsolidated rock
R = correlation coefficient

A general mass balance equation for adsorbed pollutant movement is

Free phase:

$$\frac{\partial C}{\partial t} = - v \frac{\partial C}{\partial x} - \rho \frac{\partial S}{\partial t} \pm \Sigma N - K_d C. \tag{10.68}$$

Sorbed phase:

$$\frac{\partial S}{\partial t} \simeq K_s (S_e - S) - (K_{ss}S + M)/H. \tag{10.69}$$

C is the concentration of the dissolved pollutant (mg/liter); S is the concentration of the adsorbed pollutant (μg/g of suspended solids); S_e is the adsorption equilibrium concentration of the pollutant (μg/g of suspended solids) described by an isotherm; v is the flow velocity (m/day); ρ is the specific density of the particulate matter (g/cm³); N represents the sum of sinks and sources (g/m³/day) of the substance which includes uptake by the phytoplankton, transformation into another form, diffusion into or from benthal layers, and others; K_d is the decay coefficient describing the loss of the substance from the system (day⁻¹); K_{ss} is the settling rate of the substance (m/day); K_s is the kinetic adsorption coefficient (day⁻¹); M is the scour rate of the pollutant adsorbed on the sediment from the bottom deposits (g/m²/day); H is the depth of flow (m); x is the distance (m); and t is time (days).

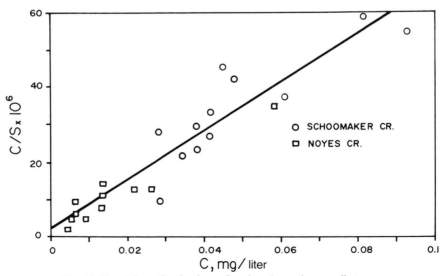

Fig. 10.16.   Adsorption isotherm for phosphorus in a small stream.

A Langmuir or Freundlich adsorption isotherm formula can be used to describe the equilibrium adsorbed concentration, S_e. The Langmuir formula is

$$S_e = \frac{bQ^\circ C}{1 + bC}.$$ (10.70)

Q° and b are the adsorption maxima and the partition coefficient, respectively, and C is the dissolved phase concentration. Figure 10.16 shows an adsorption isotherm for phosphorus in a small urban stream.

SOME MODELING CAVEATS

During the past decade, mathematical modeling has become an integral part of water resources planning and water quality management. The proper use of this technique, which is simply a method of describing a natural phenomenon by using mathematical equations, is invaluable to the planner or manager, if used properly. The alternative method would be to physically construct all possible combinations of wastewater treatment facilities and then physically measure their effect on water quality. Obviously, this would be totally impractical inasmuch as the cost would be prohibitive and the time involved would be unrealistic. The advent of high-speed computers has made the use of mathematical models even more useful because many different scenarios can be defined and their resultant effects on water quality determined in a relatively short period of time.

Obviously, using mathematical modeling does have limits because of the multitude of assumptions that must be made, the empirical nature of the coefficients involved, the uncertainty of future conditions affecting the various assumptions made, and the lack of reliable data available to verify the model. In fact, many models should only be used to provide information on the relative effects of various inputs to the model. The real crux of model output (predictions) is its interpretation by a person experienced and competent in the phenomena being modeled. Thus, the real utility of a model depends on the expertise and experience of the user and is somewhat subjective, to say the least.

The modeler should be aware of the fact that many decision makers have "not only failed to accept river-quality models as a practical tool, but often view mathematical modeling with considerable mistrust" (Hines *et al.*, 1975). It is probable that this attitude can be attributed to the failure of many models to be based on sound data or effectively applied to plan-

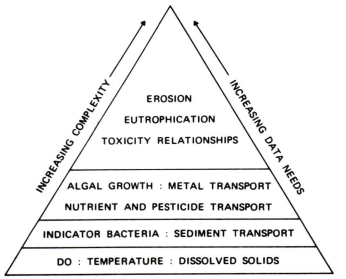

Fig. 10.17.   Complexity and data requirements of models (Hines *et al.*, 1975).

ning and management (Hines *et al.*, 1975). While the simpler models include only DO, temperature, and conservative substances, recent efforts have increased the complexity of models to include such processes as eutrophication. Obviously, increasing complexity has a concomitant increase in data requirements, as shown in Fig. 10.17.

Limitations of Models

It is imperative that the limitations of a model be known and accounted for in interpreting the results. As correctly pointed out by Hines *et al.* (1975), ''All proposed river quality models, whether conceptual or applied, should be accompanied by a statement of limitations, predictive accuracy, and suggested application.'' A similar caveat is given by the EPA in their ''208'' planning guide, in which it is stated that ''any model is, at best, only an approximation of reality, and a clear understanding of both the real situation and the approximations made by the model developer is vital to effective model use'' (USEPA, 1976).

Model Calibration and Verification

Calibration is the means by which model parameters are adjusted so that the model outputs approximate observed data or sets of data. On the other hand, verification is the process by which the true predictive capability of a model is demonstrated. The data base used for calibration of a model should not be the same as that used for model verification. Donigian and Crawford (1976) stated that "calibration should be based on several years of simulation (3 to 5 years is optimal) in order to evaluate parameters under a variety of climatic, soil moisture, and water quality conditions. . . total verification of the nonpoint source pollution model could not be performed because of insufficient water quality data . . . parameters calibrated during dry periods may not adequately represent processes occurring during wet periods . . . effects of initial conditions of soil moisture and pollutant accumulation can extend for several months resulting in biased parameter values calibrated on short simulation periods." They also noted that "The validity of the information provided by the model is a direct function of the extent of calibration and verification effort on the particular watershed. If no calibration is performed, the best that can be expected is order of magnitude estimates of annual or seasonal pollutant loadings." Even though these statements were made with respect to a dynamic simulation model, the principles are generally applicable to all models.

Model Documentation

Adequate model documentation should be an integral part of any water quality model. Documentation should convey the knowledge and experience of the model developer to the user. According to Systems Control, Inc. (1974), adequate documentation should include "a general description of the model, necessary information and direction for preparing the input deck, instruction on executing the program, and a technical discussion of the related theory on which the model is based."

Sensitivity Analysis

One of the most significant uses of models is sensitivity analysis. According to Dawdy (1969), "sensitivity analysis studies the effect on the optimal solution of changes in the input–output coefficients and in the objective function." In practical terms, sensitivity analysis will demonstrate the effect of variation of a given parameter on the output if all other factors are held constant. According to McKenzie *et al.* (1979), sensitivity analyses have at least two major uses. They identify those factors most important in model formulation and can lead to recognition of the most efficient management practices for controlling water quality problems. The results obtained from sensitivity analysis on the Willamette River Basin Study are shown in Table 10.8.

TABLE 10.8 Sensitivity analysis summary of the Willamette River study (Model similar to QUAL)

Variable Tested	Applicable River Mile Segment	Comments
Streamflow	86.5–5.0	Model is sensitive to flow, particularly at values less than 6760 ft³/sec. At 5000 ft³/sec predicted percent DO saturations are as much as 10% less than those at 6760 ft³/sec. At 3260 ft³/sec (estimated natural low flow for July 1973), predicted values are as much as 30% lower than standard conditions. At 9000 ft³/sec predicted values are higher by 6–8%.
Percent DO saturation at boundary point (River mile 86.5)	86.5–5.0	Model is sensitive to changes in initial percent of DO saturation. The major impact is near the boundary point; differences between profiles become smaller with downstream distance.
Water temperature	86.5–5.0	For the reasonably expected range of summertime water temperatures, the model is insensitive to temperature changes. Maximum predicted deviation from standard conditions are ± 3% of DO saturation.
Reaeration calculation method	86.5–26.5	Model is sensitive to the method used to calculate reaeration. Only the Velz method gave segment-by-segment reaeration inputs which resulted in good agreement of predicted and observed DO profiles.

416

Parameter	RM	Sensitivity analysis results
BOD loading	86.5–5.0	Model is relatively insensitive to the BOD_{ult} load variations. A doubling of 1974 loads (from each point source) results in deviations of 5–9% DO saturation from the standard profile. Reducing the point-source BOD load by 50% causes insignificant changes in predicted DO levels.
Rate of carbonaceous deoxygenation (k_r)	86.5–5.0	Model is relatively insensitive to changes in k_r, over a threefold range of 0.02–0.06. Predicted DO concentrations deviate no more than 6% saturation from standard profile.
Ammonia–N loading	86.5–5.0	Model is sensitive to variations in ammonia–N loading. A doubling of loads (from outfalls in the nitrifying segment RM 86.5–55) results in as much as 14% reduction in percent DO saturation values from the standard profile. Reducing the ammonia loading by 50% increases the predicted DO values by up to 8% saturation.
Rate of nitrogenous deoxygenation (k_n)	86.5–5.0	Model is insensitive to changes in k_n over a range of 0.5–0.9. Predicted DO concentrations differ from standard profile ($k_n = 0.7$) by less than 3%. Note that differences decrease with downstream distance.
Variation in water depth owing to backwater or tidal influences	24.8–5.0	Model is insensitive to expected range of changes in summertime water depth in the Tidal Reach. Predicted DO values differ from standard profile by less than 3% saturation.
Benthic demand	12.8–5.0	The model is sensitive to benthic–oxygen demand exerted between RMs 12.8–5.2. If the demand is removed, the predicted DO value at RM 5.0 is 8% higher than the standard condition.

Differences in Water Bodies

The user or designer of a water quality model should be aware of possible differences between the dissolved oxygen balance process taking place in small streams and large rivers. A model designed for large streams can give highly erroneous results if applied to a shallow stream and vice versa. It may appear obvious, but very often is neglected by modelers, that an on-site survey should precede model construction. The user of a model should obtain a "feeling" for the river, should be aware of any significant phenomenon which could affect the model, locate the stream sludge deposits or areas with attached growths, etc. This survey, which should be conducted by the modeler himself, can be done concurrently with or prior to the calibration and verification water quality surveys.

The primary factors which will affect the selection of the model include the stream depth, velocity, bottom characteristics, degree of pollution (saprobity), topographical characteristics, presence of rapids, weirs, reservoirs and dam outlets, navigation intensity, and stream currents. Some generalization and grouping of streams may be possible and can help in the selection of a model.

Large fast moving rivers with depths above 1.5 m and velocities greater than 0.5 m/sec will usually have sandy or gravel bottom with a very low potential for attached growths. Because of the higher velocity, there will be little or no sedimentation of suspended organic matter. Thus, the BOD removal coefficient may be expected to be of the same order of magnitude as the laboratory BOD rate coefficient. Since intensive developments of planktonic algae need quiescent conditions, the effect of photosynthesis may be minimal and most of the oxygen will be supplied by atmospheric aeration. The potential for nitrification is rather low, and nitrification will occur only if enough detention time is available for nitrifiers to develop. Benthal oxygen demand is minimal and often negligible. Total nutrient levels (total nitrogen and phosphorus) change only by dilution or from local nonpoint and point tributary sources. Many American and European rivers (Mississippi, Missouri, Ohio, Danube, Rhine) can be modeled using this simplified concept.

Many medium-size rivers and some large rivers in lowlands have velocities below that necessary for maintaining the suspension of organic matter. Also, impoundment of rivers for navigation and electric power production substantially slows the velocity, and deposition can occur. Typical examples of rivers in this category are the Tennessee River, the Arkansas River, the Alabama River, and some navigable tributaries of large rivers (the Kanawha River in West Virginia, the Cumberland River, and others). These rivers usually have depths above 1 to 1.5 m; however, the flow velocities are often below 0.5 m/sec, especially during low flow periods. The sedimentation of organic suspended solids may increase an apparent BOD removal coefficient, K_r, with the consequent formation of sludge bottom deposits, and benthal oxygen demand may become significant. After most of the carbonaceous BOD demand is satisfied, planktonic algae may develop in stagnant sections and photosynthetic oxygen production may be appreciable. However, the conditions for benthic slime and algae growths are not favorable. Although nitrification may occur in reaches which have sufficient detention times from a source of nitrogenous oxygen demanding material, some ammonia and phosphates can be adsorbed on suspended particles and settle out without exerting an oxygen demand. This may result in a disproportionality between the TKN removed and the nitrate formed, and the total nitrogen content will diminish. Phosphates may be reduced in the same manner.

As for the dynamics of water quality changes, shallow fast moving streams may show the greatest variability under optimal conditions. Owing to larger velocities, the bottom is mostly composed of rocks and gravel with optimal conditions for attached growths of heterotrophic slimes in polysaprobic zones or algae and aquatic weeds in mezosaprobic zones. Heterotrophic slimes contribute to significant increases of the overall BOD removal coefficient, and algae and aquatic weeds may produce large amounts of photosynthetic oxygen. Also, most of the nitrifying organisms may be located in the attached growths, and nitrification in the immediate vicinity of secondary effluent outfalls is common. Dead slime and algae can contribute large amounts of organic matter to the overall BOD mass balance and may cause problems in downstream sections, where these materials can settle out. Very little sedimentation of pollutants can be expected. Therefore, the reduction of the total nitrogen and phosphorus content is only due to their uptake by attached slimes and algae. Most of the aquatic weeds rely on bottom sediment layer nutrients, and their presence may not be reflected in a reduction of free water nutrient content.

If the velocity drops below 0.3–0.5 m/sec, the suspended organic matter may settle out and accumulate as bottom mud. Although the velocity may be low, the bottom layer is composed mainly of mud and fine sand. These fractions are constantly shifting, thus preventing any significant development of heterotrophic and autotrophic attached growths. However, planktonic algae and rooted aquatic weeds may develop in large quantities, especially in stagnant sections. The increase of the overall BOD removal coefficient is only apparent and decreases as soon as most of the suspended organics settle out. Significant amounts of nutrients and other pollutants adsorbed on suspended particles may also settle out. This reduction may be especially significant for phosphorus and ammonia.

The processes taking place in the bottom layers may significantly affect water quality. The benthic oxygen demand may be quite high during the warm summer months, but anaerobic processes may prevail in the lower layers of bottom deposits. Simultaneous nitrification and denitrification in bottom deposits have been observed, which may explain why some nitrogen is lost without corresponding oxygen demand. On the other hand, some nutrients bound into the sediments may be released back into the water.

The modeling of small shallow streams is generally more difficult than that of large rivers. Neglecting processes occurring in the bottom layers and their participation in water quality changes may lead to gross errors and misinterpretations.

REFERENCES

Alonso, C. V., McHenry, J. R., and Hong, J. C. S. (1975). The influence of suspended sediment on the reaeration of uniform streams. *Water Res.* **9,** 695–700.

Banks, R. B., and Herrera, F. F. (1977). Effect of wind and rain on surface aeration. *J. Environ. Eng. Div., Am. Soc. Civ. Eng.* **103** (EE3), 489–504.

Bansal, M. K. (1973). Atmospheric reaeration in natural streams. *Water Res.* **7**(5) 769–782.

Bennett, J. P., and Rathbun, R. E. (1972). Reaeration in open-channel flow. *U.S. Geol. Surv. Prof. Pap.* 737, U.S. Govt. Printing Office, Washington, D.C.

Betson, R. P., and McMaster, W. M. (1975). Nonpoint source mineral water quality model. *J. Water Pollut. Control Fed.,* **47**(10), 2461–2473.

Boško, K. (1957). Hydraulicke' parametre ako akasivatelia samočisticej schopnosti recipientov. *Vodohospod. Čas. Slovak Acad. of Science* **5**(1), 57–70.

Boško, K. "An explanation of the difference between the rate of BOD progression under laboratory and stream conditions" by A. Nejedly *In* "Advances in Water Pollution Research," Vol. **I,** Water Poll. Contr. Fed., Washington, D.C.

Brown, L. C. (1974). Statistical evaluation of reaeration prediction equations. *J. Environ. Eng. Div., Am. Soc. Civ. Eng.* **100**(EE5), 1051–1068.

Cadwallader, T. E., and McDonnell, A. J. (1969). A multivariate analysis of reaeration data. *Water Res.* **3,** 731–742.

Camp, T. R. (1963). "Water and Its Impurities." Van Nostrand-Reinhold, Princeton, New Jersey.

Churchill, M. A., Elmore, H. L., and Buckingham, R. A. (1962). The prediction of stream reaeration rates. "Advances in Water Pollution Research," Vol. 1. Proc. of the Int. Conf., London. Pergamon, Oxford.

Davis, J. S., and Zobrist, J. (1978). The interrelationships among chemical parameters in rivers—analyzing the effect of natural and anthropogenic sources. *Prog. Water Technol.* **10**(5/6), 65–78.

Dawdy, D. R. (1969). Considerations involved in evaluating mathematical modeling of urban hydrologic systems. *U.S. Geol. Surv. Water Supply Pap.* 1591-D, Washington, D.C.

DeMarco, J., Kurbiel, J., and Symons, J. M. (1967). Influence of environmental factors on the nitrogen cycle in water. *J. Am. Water Works Assoc.* **59**(5), 580–592.

Dobbins, W. E. (1956). The nature of the oxygen transfer coefficient in aeration systems. *In* "Biological Treatment of Sewage and Industrial Wastes," pp. 141–253. VanNostrand-Reinhold, Princeton, New Jersey.

Dobbins, W. E. (1964). BOD and oxygen relationships in streams. *J. Sanit. Eng. Div., Am. Soc. Civ. Eng.* **90**, 53–78.

Donigian, A. S., and Crawford, N. H. (1976). Modelling nonpoint pollution from the land surface. Hydrocomp, Inc., USEPA 600/3-76-083, Athens, Georgia.

Durum, W. H. (1953). Relationship of the mineral constituents in solution to stream flow: Saline River near Russell, Kansas. *Trans. Am. Geophys. Union* **34**(3), 435–442.

Fair, G. M., Geyer, J. C., and Okun, D. A. (1968). "Water and Wastewater Engineering," Vol. 2. Wiley, New York.

Gotaas, H. B. (1947). Effect of temperature on biochemical oxidation of sewage. Sewage Works J. **20**(3), 441.

Grau, P., Dohanyos, M., and Chudoba, J. (1975). Kinetics of multicomponent substrate removal by activated sludge. *Water Res.* **9**(7), 637–642.

Grimsrud, G. P., Finnemore, E. J., and Owen, E. J. (1976). Evaluation of water quality models: A management guide for planners. Office of Res. & Develop. USEPA-600/5-76-004, Washington, D.C.

Hendrickson, G. E., and Krieger, R. A. (1964). Geochemistry of natural waters of the blue grass region, Kentucky. *U.S. Geol. Surv. Pap.* 1700., Washington, D.C.

Hines, W. G., Rickert, D. A., McKenzie, S. W., and Bennett, J. P. (1975). Formulation and use of practical models for river-quality assessment. *U.S. Geol. Surv. Circ.* 715-B, Reston, Virginia.

Huang, C. S., and Hopson, N. E. (1974). Nitrification rate in biological processes. *J. Environ. Eng. Div., Am. Soc. Civ. Eng.* **100**(EE2) 409–422.

Imhoff, K., and Fair, G. M. (1956). "Sewage Treatment," 2nd ed. Wiley, New York.

Imhoff, K. R. (1969). Wie berechnet man eine Künstliche Gewässerbelüftung? (How to Compute Artificial Aeration of surface waters?). *Gas Wasserfach* **110**(20), 543–545.

Isaacs, W. P., and Gaudy, A. F. (1968). Atmospheric oxygenation in a simulated stream. *J. Sanit. Eng. Div. Am. Soc. Civ. Eng.* **94**, (SA2) 319–344.

King, D. L. (1970). Reaeration of streams and reservoirs, analysis and bibliography. Eng. and Res. Cent., U.S. Bur. of Reclamation, Rep. No. REC-OCE-70-75, Denver, Colorado.

Kittrell, F. W., and Kochtitzky, O. W. Jr. (1947). Natural purification characteristics of a shallow turbulent stream. *Sewage Works J.* **19**(6), 1032–1049.

Kramer, G. R. (1974). Predicting reaeration coefficients for polluted estuaries. *J. Enviriron. Eng. Div., Am. Soc. Civ. Eng.* **100**(EE1), 77–92.

Krenkel, P. A., and Novotny, V. (1973). The assimilative capacity of the South Fork Holston River and Holston River below Kingsport, Tennessee. Rep. to Tennessee Eastman Co., Kingsport.

Krenkel, P. A., and Novotny, V. (1979). *In* "Modeling of Rivers" (H. W. Shen, ed.), 18-1-18.40. Wiley, New York.

Krenkel, P. A., and Orlob, G. T. (1962). Turbulent diffusion and the reaeration coefficient. *J. Sanit. Eng. Div., Am. Soc. Civ. Eng.* **88**(3), 53–83.

Lane, W. L. (1975). Extraction of information on inorganic water quality. Hydrol. Pap. No. 73, Colorado State Univ., Fort Collins.

Langbein, W. B., and Durum, W. H. (1967). The aeration capacity of streams. *U.S. Geol. Surv. Circ.* No. 542, U.S. Dept. of the Interior, Washington, D.C.

Lau, L. Y., (1972). Prediction equation for reaeration in open-channel flow. *J. Sanit. Eng. Div., Am. Soc. Civ. Eng.* **98**(SA6), 1063–1068.

Ledbetter, J. O., and Gloyna, E. F. (1962). Predictive techniques for water quality — inorganics. Civ. Eng. Dept., Univ. of Texas, Austin.

Lopez-Bernal, F., Krenkel, P. A., and Ruane, R. J. (1976). Nitrification in free-flowing streams, 8th IAWPR Int. Conf., Sydney, Australia, *Progr. in Water Technol.* **9**(4), 821, 832.

Maise, G. (1970). Scaling methods for surface aerators. *J. Sanit. Eng. Div., Am. Soc. Civ. Eng.* **96**(SA5), 1099–1114.

Mancy, K. H., McKeown, J. J., and Okun, D. A. (1960). Effects of surface active agents on bubble aeration. Conf. on Biol. Waste Treatment, Manhattan College, New York.

Manczak, H. (1971). A statistical model of the interdependence of river flow rate and pollution concentrations. *In* Proc. of the Specialty Conference on Automatic Water Quality Monitoring in Europe (P. A. Krenkel, ed.). Tech. Rep. No. 28, Dept. of Environ. and Water Res. Eng., Vanderbilt Univ., Nashville, Tennessee.

Matulewich, V. A., and Finstein, M. S. (1973). Distribution of nitrifying bacteria in a polluted river. Pres. at Am. Soc. for Microbiology, Miami Beach, Florida.

McKenzie, S. W., Hines, W. G., Rickert, D. A., and Rinella, F. A. (1979). Steady-state dissolved oxygen model of the Willamette River, Oregon. *U.S. Geol. Surv. Circ.* 715-J, Arlington, Virginia.

Michigan Water Resources Commission (1962). Oxygen relationships of Grand River Lansing to Grand Ledge — 1960 survey. Lansing, Michigan.

Monod, J. (1949). The growth of bacterial cultures. *Annu. Rev. Microbiol.* **3**, 371–94.

Moore, E. W. (1941). Long-time biochemical oxygen demands at low temperatures. *Sewage Works J.* **13**(5), 561.

Negulescu, M., and Rojanski, V. (1969). Recent research to determine reaeration in natural streams. *Water Res.* **3**(3), 189–202.

Novotny, V. (1969). Boundary layer effect on the course of the self-purification of small streams. *In* "Advances in Water Pollution Research." Pergamon, Oxford.

O'Connell, R. L., Thomas, N. E., Godsil, F. J., and Hirth, C. R. (1963). Report of the survey of the Truckee River. Div. Water Supply and Pollut. Control, U.S. Dept. Health, Educ. and Welf., Public Health Serv.

O'Connor, D. J., and Dobbins, W. E. (1958). Mechanism of reaeration in natural streams. *Trans. Am. Soc. Civ. Eng.* **123**, 641–666.

O'Connor, D. J. (1961). Oxygen balance of an estuary. *Trans. Am. Soc. of Civ. Eng.* **126**, 641–684.

Owens, M., Edwards, R. W., and Gibbs, J. W. (1964). Some reaeration studies in streams. *Int. J. Air Water Pollut.* **8**, 469.

Painter, H. A. (1970). A review of literature on inorganic nitrogen metabolism in microorganisms. *Water Res.* **4**, 393–450.

Parkhurst, J. D. and Pomeroy, R. D. (1972). Oxygen absorption in streams. J. Sanit. Eng. Div., Am. Soc. Civ. Eng. **98**(SA1), 101–124.

Phelps, E. B. (1944). The oxygen balance. *In* "Stream Sanitation." Wiley, New York.

Rathbun, R. E. (1977). Reaeration coefficients of streams — state of the art. *J. Hydraul. Div., Am. Soc. Civ. Eng.* **103**(HY4), 409–424.

Rathbun, R. E., Stephens, D. W., Shultz, D. J., and Tai, D. Y. (1978). Laboratory studies of gas tracers for reaeration. *J. Environ. Eng. Div., Am. Soc. Civ. Eng.* **104,** 215–229.

Roesner, L. A., Mouser, J. R., and Evenson, D. E. (1973). Computer programs documentation for the stream quality model QUAL-II. EPA Cont. No. 68-01-0739, Water Res. Eng., Walnut Creek, California.

Ruane, R. J. (1972). Unpublished TVA reports, Chattanooga, Tennessee.

Ruane, R. J., and Krenkel, P. A. (1975). Nitrification and other factors affecting nitrogen in the Holston River. Proc. of IAWPR Conf. on Nitrogen as a Water Pollutant, Copenhagen, Denmark.

Streeter, H. W., and Phelps, E. B. (1925). A study of the pollution and natural purification of the Ohio River. *Public Health Bull.* 146, U.S. Public Health Serv., Washington, D.C.

Systems Control, Inc. (1974). Use of mathematical models for water quality planning. Water Resour. Info. System. Dept. of Ecology, State of Washington.

Tennessee Valley Authority (1969). Comprehensive plan for water quality management in the Tennessee Valley, vol. II. Chattanooga, Tennessee.

Texas Water Development Board (1970). DOSAG I, simulation of water quality in streams and canals. Report No. PB-202-974. Austin, Texas.

Thackston, E. L., and Krenkel, P. A. (1969). Reaeration prediction in natural streams. *J. Sanit. Eng. Div., Am. Soc. Civ. Eng.* **95**(SA1), 65–94.

Theriault, E. J. (1927). The oxygen demand of polluted waters. *Public Health Bull.* 173, U.S. Govt. Printing Office, Washington, D.C.

Thomann, R. V. (1972). System analysis and water quality management. Environ. Res. and Appl., Inc., New York.

Tischler, L. F., and Eckenfelder, W. W. (1968). Linear substrate removal in the activated sludge process. Pap. Pres. 4th Int. Conf. on Water Pollut. Res., Prague, Czechslovakia.

Tsivoglou, E. C. (1974). The reaeration capacity of Canandaigua Outlet, Canandaigua to Clifton Springs, New York. State Dept. of Environ. Cons., Project No. C-5402, Atlanta, Georgia.

Tsivoglou, E. C. (1979). Discussion of Laboratory studies of gas tracers for reaeration. *J. Environ. Eng. Div., Am. Soc. Civ. Eng.* **105,** 426–428.

Tsivoglou, E. C. (1967). Measurement. of stream reaeration. Fed. Water Pollut. Control Adm., U.S. Dept. of the Interior, Washington, D.C.

Tsivoglou, E. C., and Wallace, J. R. (1972). Characterization of stream reaeration capacity. USEPA-R3-72-012, Washington, D.C.

Tuffey, T. J., Hunter, J. V., and Matulewich, V. A. (1974). Zones of nitrification. *Water Resour. Bull.* **10**(3), 555–565.

U.S. Environmental Protection Agency (1976). Areawide assessment procedures manual. Cincinnati, Ohio.

University of Minnesota, Institute of Technology (1961). Pollution and recovery characteristics of the Mississippi River. Dept. of Civ. Eng., Sanit. Eng. Div.

U.S. Geological Survey (1979). River-quality assessment of the Willamette River Basin, Oregon. *U.S. Geol. Surv. Circ.* Ser. 715, 1975–1979.

Velz, C. J. (1939). Deoxygenation and reoxygenation. *Trans. Am. Soc. Civ. Eng.* **104,** 560–577.

Velz, C. J. (1953). Recovery of polluted streams. *Sewage Works J.* **100**(12), 495.

Velz, C. J. (1970). ''Applied Stream Sanitation.'' Wiley (Interscience), New York.

Velz, C. J., and Gannon, J. J. (1962). Biological extraction and accumulation in stream self-purification. *In* ''Advances in Water Pollution Research.'' Pergamon, Oxford.

Ward, J. C. (1958). Correlation of stream flow quantity with quality. M.E. Thesis, Univ. of Oklahoma, Norman.

Water Pollution Research Laboratory (1967). Water Pollution Research, 1967. Stevenage, Herts, England.

Water Pollution Research Laboratory (1971). Nitrification in the BOD test. *In* Notes on water pollution, No. 52. Stevenage, Herts, England.

Water Pollution Research Laboratory (1973). Aeration at weirs. *In* Notes on water pollution. Dept. of the Environ., Stevenage, Herts, England.

Wezernak, C. T., and Gannon, J. J. (1968). Evaluation of nitrification in streams. *J. Sanit. Eng. Div., Am. Soc. Civ. Eng.* **94**(SA5), 883–895.

Wild, H. E., Sawyer, C. N., and McMahon, T. C. (1971). Factors affecting nitrification kinetics. *J. Water Pollut. Control Fed.* **43**(9), 1845–1854.

Wiley, A. J., and Lueck, B. F. (1960). Turbine aeration and other methods of reaerating streams. *Tech. Assoc., Pulp and Paper Inst.* **43**, 241.

Wilson, G. T., and Macleod, N. (1974). A critical appraisal of empirical equations and models for the prediction of reaeration of deoxygenated water. *Water Res.* **8**(6), 341–366.

Wuhrman, K. (1955). Factors affecting efficiency and solids production in the activated sludge process. *In* ''Biological Treatment of Sewage and Industrial Wastes,'' p. 49. Van Nostrand-Reinhold, Princeton, New Jersey.

Zanoni, A. E. (1969). Secondary effluent deoxygenation at different temperatures. *J. Water Pollut. Control Fed.* **41**(4) 640–659.

11 Applications of Water Quality Models*

INTRODUCTION

As previously stated, computer models play an important role in water quality management decisions. These models are primarily used by two groups: (1) regulatory agencies in determining the assimilative capacity of receiving streams in order to set effluent limits on industrial, municipal, and wastewater discharges, and (2) industrial wastewater dischargers and water quality management planners in exploring various alternatives to meet water quality objectives. Alternatives usually considered by the latter group include various levels of wastewater treatment and flow augmentation. Other alternatives sometimes considered include moving the point of discharge to a larger stream, artificial in-stream aeration, and appealing to regulatory authorities for lower water quality objectives because of unreasonable economic impact. One of the more important uses of models by both groups includes the determination of the sensitivity of model outputs to estimated reaction coefficients (i.e., K_d, K_2, K_n) and

* This chapter was adapted from "Basic Approach to Water Quality Modeling," by Peter A. Krenkel and R. J. Ruane, Chapter 18 *in Modeling of Rivers* (H. W. Shen, ed.) John Wiley & Sons, New York, 1979. Used with permission.

other variables that possibly may have been estimated (i.e., S_B, H, U, H, P, R).

This chapter presents the general approach to determining the assimilative capacity of a stream and gives several case histories. In the first case, DOSAG I will be used on a free-flowing river in which both carbonaceous and nitrogenous BOD is oxidized. This will illustrate the sensitivity analyses for K_d, K_2, T, H, and the level of wastewater treatment. In the second case, a modified version of the Streeter–Phelps model is applied to a river in which photosynthesis and respiration take place and alternative control measures such as level of wastewater treatment, flow augmentation, artificial in-stream aeration, temperature, and DO in the stream above the waste discharge are considered.

GENERAL APPROACH TO WATER QUALITY MODELING

The determination of receiving wastewater assimilative capacity consists of a series of steps.

Step 1. The first step involves defining the objectives of the study. This includes a determination of the standards for water quality and applicable critical minimal streamflow, the planning period for 10-, 20-, or 30- year waste load projections and consideration of the allocation of assimilative capacity for other wastewater discharges.

Step 2. The next step is to contact those who may have existing information in order to minimize effort and save costs. In many cases, the United States Geological Survey will have streamflow information, critical minimum flows, stream geometry, and contour maps. In addition, water quality and time of travel data, and even existing computer models may be obtained from the local EPA region, the state regulatory agency, and regional river control agencies, such as the Tennessee Valley Authority, Corps of Engineers, United States Geological Survey, local industries, areawide planning agencies, and universities.

Step 3. If information is not available to construct a model, estimating procedures may be used to determine the sensitivity of water quality management decisions to the assimilative capacity of the stream. If assimilative capacity is found to be unimportant, decisions can be made without conducting a detailed stream study.

Three estimating procedures are considered here: (1) Fair *et al.* (1968); (2) an approach using two statistically derived equations (Ruane, 1971); and (3) an approach using estimated values of U, K_2, K_n, and K_d in the Streeter–Phelps equation.

The Fair, Geyer, and Okun procedure has been used for many years by practicing engineers. The reader is referred to their reference books.

The statistical equations of procedure 2 were derived by multiple-regression analysis of 20 assimilative capacity studies of streams in the Tennessee Valley. The following equation was developed on the basis of all 20 streams analyzed:

$$Y = 398{,}700 \frac{DO_{mix}^{0.951} Q^{1.026} S^{0.580}}{T^{1.474} (DO_{sag})^{1.434}}, \tag{11.1}$$

where Y = assimilative capacity of the stream in pounds per day of 5-day BOD at 20°C (includes upstream BOD); Q = streamflow, in cubic feet per second, immediately downstream from points of waste discharges; T = stream temperature, in °C; S = streambed slope, in feet per foot (determined from quadrangle maps with a scale of 1 : 24,000); DO_{mix} = DO of the stream in milligrams per liter, after waste mixes with stream; and DO_{sag} = minimum allowable DO of the stream in milligrams per liter. A correlation coefficient of 92% was obtained with this relationship.

By eliminating four streams with excessive atmospheric aeration and one stream with an excessively high oxidation rate from the original 20 streams, this equation was derived:

$$Y = 10{,}138 \frac{DO_{mix}^{1.094} Q^{0.864} S^{0.06}}{T^{1.423} (DO_{sag})^{1.474}}, \tag{11.2}$$

where the variables are the same as those defined in Eq. (11.1). The correlation coefficient obtained with this relationship was 95%. For the latter case, all the computed values were within 50% of the observed values, as shown in Fig. 11.1.

These equations should be used only if the independent variables fall within the range of the original data. For both equations, the ranges for temperature, DO_{mix}, and DO_{sag} are 20° to 30°C, 1 to 5 mg/liter and 3 to 5 mg/liter, respectively. For Eq. (11.1), the range of slope was from 0.00037 to 0.00675, and the range of streamflows was from 0.8 to 148 ft³/sec (1.4 to 252 m³/min). For Eq. (11.2), the range of slopes was from 0.00037 to 0.0062, and the range of streamflows was from 0.8 to 83 ft³/sec (1.4 to 141 m³/min). It should be noted that the range of slopes for Eq. (11.2) included only one value significantly greater than 0.003, while the four data points that were eliminated from Eq. (11.2) and that represent relatively shallow, rapid streams were for streams having slopes equal to or greater than 0.003.

Elevation above mean sea level was not included in either equation because of its high correlation with slope. However, the effect of elevation on DO saturation remains a part of the estimated values for assimilative

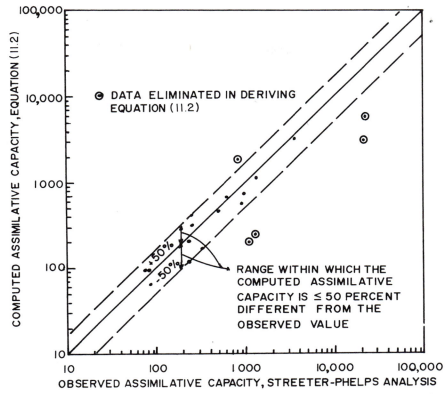

Fig. 11.1. Computed v. observed assimilative capacities for the streams used to derive Eq. (11.2) (TVA, 1973). Conditions: $DO_{sag} = 4$ mg/liter; $DO_{mix} = 7$ mg/liter; Temperature = 25°C.

capacity because it influenced the DO saturation values used in deriving the equations. Elevations for the 20 streams investigated ranged from 300 to 2300 ft (90 to 700 m) for Eq. (11.1) and from 300 to 2000 ft (90 to 619 m) for Eq. (11.2), and the higher values of slope were associated with higher elevations.

Equation (11.2) is recommended for computing the assimilative capacity of streams that have a slope less than 0.003. All but one data point for Eq. (11.2) are for streamflows less than 25 ft³/sec (42 m³/min); however, a preliminary comparison of the prediction equation with the results from two larger rivers [155 ft³/sec (264 m³/min), and 800 ft³/sec (1360 m³/min)], suggests that Eq. (11.2) may also be representative of larger streams.

For rapid, shallow streams, use of Eq. (11.1) should probably be considered. Although it yields results considerably higher than those from Eq. (11.2), these results are lower than the observed data for rapid streams and should therefore be conservative.

The 5-day BOD values resulting from the preceding equations usually are converted to ultimate carbonaceous BOD using a deoxygenation coefficient of 0.11 days^{-1} at 20°C (base 10). This ultimate carbonaceous BOD is the allowable stream waste load, and it is assumed that the UOD (ultimate oxygen demand including the nitrogenous demand) of the waste load cannot exceed this value.

If water quality data are not available for a stream, the following assumptions usually can be made to yield conservatively low results: (1) stream temperature is equal to the maximum allowed by water quality standards, (2) DO in the stream above the outfall is 85% of saturation concentration, (3) natural 5-day BOD in the stream above the outfall is 1.5 mg/liter, and (4) when the wastewater discharge exceeds the minimum critical flow in the receiving stream, the DO in the stream immediately downstream from the waste discharge is equal to the minimum concentration allowed in the stream by water quality standards. Assumptions 2 and 3 are based on the premise that water quality above the outfall is not affected by upstream pollution.

To account for uncertainties in estimating the assimilative capacity of streams, a range of "probable" error is used to test sensitivity to the water quality management decisions. From the results of the statistical analysis, this probable range of error was estimated to be 150%. It should be noted that the actual assimilative capacity could be more than 50% greater than the calculated value, but it is less likely that it should be less than 50% of the calculated value.

The third method considered here involves the use of the Streeter–Phelps equation which has been modified to account for nitrification and coefficients for this equation as estimated on the basis of easily measured hydraulic parameters such as streamflow, slope of the streambed, and approximate width of the stream.

As stated earlier, streamflow information is available from the United States Geological Survey or watershed authorities. For the purposes of estimating stream water quality coefficients, streambed slopes can be determined from quadrangle maps on which elevation contours are included. The approximate width of a stream also can be estimated from quadrangle maps if the width is on the order of 200 feet or more. For smaller streams, a site visit is necessary; however, less than one man-day is required to obtain sufficient information.

The estimation of K_d using Bosko's and Novotny's equations were discussed earlier. Bosko's formula requires a determination of the velocity and depth, and Novotny's equation requires slope and depth. Foree (1976) developed empirical relationships for estimating velocity using information on drainage area, the specific discharge, and slope of the streambed. For $0.30 \leq q < 0.90$ and $5 < A < 250$,

$$U = 14 + 5.6(q - 0.08)S_e^{0.5}, \tag{11.3}$$

where q = the specific discharge in the stream in cubic feet per second per square mile, A = drainage basin area in square miles, and S_e = slope of streambed in feet per mile. For $q < 0.30$ and $5 < A < 250$,

$$U = 0.40 + 4.1qS_e^{0.5}. \tag{11.4}$$

The data used to develop these equations were for streams having $0.45 < Q < 4.0$ and ≈ 0 (pool) $< S_e < 41.8$. Foree stated that caution should be used in applying these equations outside the range for which they were developed. Stall and Yang (1970) have also presented relationships between velocity and hydraulic variables.

Using the continuity equation and assuming an approximately rectangular channel section for the stream, depth can be estimated from the equation

$$H = Q/UW,$$

where W = width of the stream.

Foree (1976) also developed a relationship between K_2 and S_e:

$$K_2 = 0.30 + 0.19S_e^{1.2}, \tag{11.5}$$

where K_2 is to the base e at 25°C and S_e is in feet per mile. However, the other expressions for K_2 given in Chapter 10 can also be used if velocity and depth can be estimated as shown in Eq. (11.5). It should be noted that sensitivity analysis on the estimated velocity is especially critical since an error in velocity determination compounds the error in estimating the depth, and K_2 is especially sensitive to these errors since for most of the formulas, velocity is in the numerator and depth is in the denominator.

In using the foregoing approaches to model water quality, extreme caution should be exercised. The user should refer to the referenced literature to determine the applicability of the equations to specific cases, and should perform sensitivity analyses on the estimated variables. A considerable amount of research is needed to improve on the preceding relationships for U, K_2, and K_n.

Step 4. If the approximate analysis in step 3 indicates that a valid model is needed for the receiving stream, select a model as suggested by Grimsrud *et al.* (1976) and recommended in our Chapter 10.

Step 5. If the information obtained in step 2 is not sufficient for developing the water quality model, a stream survey for water quality will be needed. It is preferred that survey data be taken during a summer low-flow, steady state. A time-of-travel study during this period using tracer dyes is also highly recommended. If water quality data do exist on the stream in question, factors that may influence the coefficients should be considered. For example, if the wastewater discharge characteristics have improved since the date of the last survey, it is conceivable that the nitrification rate may have increased, and the carbonaceous BOD oxygenation rate may have decreased. Also, if the stream was heavily burdened with macrophytes or attached algae growths, these biological growths may have decreased significantly and affected the velocity in the stream as well as photosynthesis and respiration.

Step 6. The need for sensitivity analyses on critical variables that may affect management decisions is again needed. An example of the use of sensitivity analyses will be discussed in the case histories that will be described later.

Step 7. Following the completion of the sensitivity analyses, the investigator may want to consider the need for additional data or studies to resolve important issues that may be economically justified on a basis of high-cost considerations for wastewater discharges.

Step 8. The final step in this process is to make the water quality management decision.

APPLICATION OF DOSAG I MODEL

A schematic picture of the stream network is shown in Fig. 11.2. The stream system was divided into three stretches:

1. Main River River mile 142.2–147.8
2. Tributory River River mile 0.0–1.0
3. Main River River mile 128.9–142.2

The system has 10 stream reaches plus 5 wastewater discharges for a total of 15 reaches.

The hydraulic characteristics were determined from survey data, which indicated that the value of the exponent on the velocity equation is 0.674 for river miles above RM 131.5, and 0.539 for river miles below RM 131.5. The multipliers were determined from known velocities and discharges and the exponent as described in Chapter 10. The coefficients for depth determination were unknown. Therefore, the exponent on the depth equation was determined by means of the known velocity exponent from

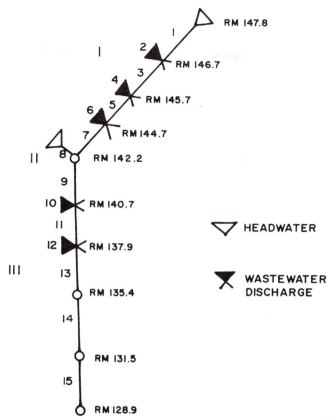

Fig. 11.2. Schematic diagram of DOSAG (Krenkel and Novotny, 1973).

the relationship

$$b + d = 1,$$

and the final values of the multipliers were obtained from the model delineation and verification.

Model Verification

For the DOSAG I model verification, two sets of survey data were available, 1946 and 1969.

Delineation and Verification of the DOSAG Model Using the 1946 Data

The 1946 data offer an excellent opportunity for the development of a valid model for these reasons:

(a)   The reported data are representative because of the care with which they were taken.

(b)   The effect of weeds on the oxygen balance was not reported and therefore assumed to be nonexistent.

(c)   The character of the wastewater loads contributed to the river has not significantly differed to date from the 1946 conditions.

(d)   The flow conditions were very close to the critical low flow.

The BOD$_5$ profile of the river is shown in Fig. 11.3. During the delineation procedure, the deoxygenation coefficients were varied within acceptable ranges until a good fit was obtained. A similar procedure was followed with the dissolved oxygen data. The only parameter by which the reaeration rate can be adjusted is depth. Therefore, since the first attempt, using EPA and TVA depths, produced results that were not in accord with the observed data, the depths in each reach were varied until a reasonable fit was obtained. The plot finally derived is shown in Fig. 11.4. The depths used in Fig. 11.4 were close to those reported during a 1969 survey. In addition, a survey in 1973 reported depths in RM 142.2 to 131.5

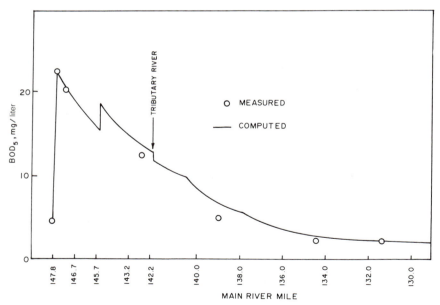

Fig. 11.3.   Verification of DOSAG model BOD$_5$, 1946 survey, temperature 23°C (Krenkel and Novotny, 1973).

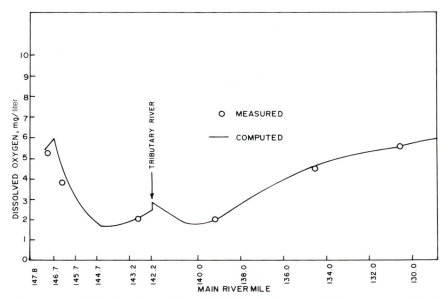

Fig. 11.4. Verification of DOSAG model for dissolved oxygen, 1946 survey (Krenkel and Novotny, 1973).

to be in the order of 1 to 2.5 feet, with some locations showing depths below 1 foot. It is interesting to note that the depths used herein were verified in a subsequent study by the EPA (USEPA, 1978).

Verification of the DOSAG Model Using 1969 Survey Data

The model, as delineated and verified using 1946 survey data, was then compared with the 1969 survey data. The basic coefficients were not changed. The discharges, temperatures, and wastewater loads were adjusted to those reported for the 1969 survey. The computed BODs, DO values, and nitrogenous demand were then plotted and compared with the observed values. The BOD plot (Fig. 11.5) shows excellent agreement, as does the plot of the nitrogenous demand (Fig. 11.6).

As shown in Fig. 11.7, the computed DO profile is somewhat lower than the observed values. This deviation may be explained by noting that the 1969 survey was conducted during bright summer days when the aquatic weeds contributed substantially to the DO variations.

Since the DOSAG I model does not take photosynthesis and respiration into account, computed DO represents conditions without the aquatic weeds. With this in mind, the computed oxygen profile is thought to be satisfactory.

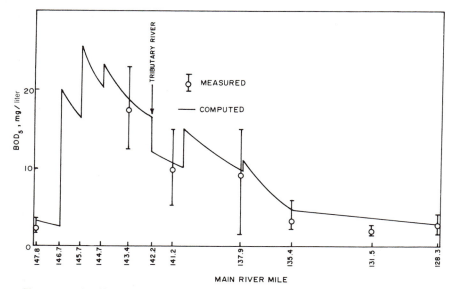

Fig. 11.5. Verification of DOSAG model BOD$_5$, 1969 Survey, Temperature 26°C (Krenkel and Novotny, 1973).

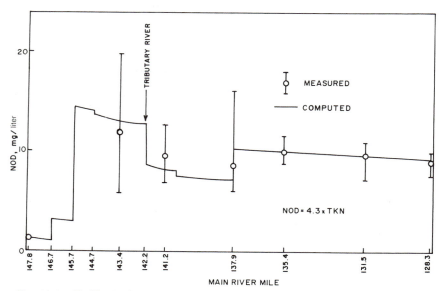

Fig. 11.6. Verification of DOSAG model, nitrogenous demand, 1969 survey, temperature 26°C (Krenkel and Novotny, 1973).

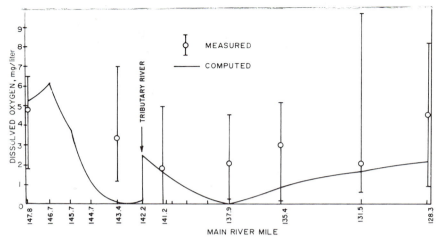

Fig. 11.7. Verification of DOSAG model, dissolved oxygen, 1969 survey, temperature 26°C (Krenkel and Novotny, 1973).

The verification of the DOSAG I model on both surveys, i.e., the 1946 survey and 1969 survey, demonstrated that the DOSAG I model can be successfully applied (excluding the effects of photosynthesis). It is obvious that proper care and effort must be devoted to the delineation of the model, selection of adequate parameters, etc., however. Only then can the model be used for the determination of waste assimilative capacity, as is the case with the model derived herein.

Verification of the Estimated Deoxygenation Coefficients

The results obtained by the model enable a comparison of the theoretically determined parameters with those applied by the model and verified by the two surveys. The results of the comparison of the theoretical deoxygenation, B, with the actual value of $(K_d - K_1)$ are shown in Table 11.1.

Determination of the Waste Assimilative Capacity

The model was constructed on the 1946 water quality investigation, which was deemed to be the best approximation of water quality conditions which would occur during a low-flow period without aquatic weeds. The solution matrix was designed with the following vectors:

(1) Wastewater load in the Main River
   50,000, 25,000, and 14,000 lbs of BOD/day
(2) Water temperature
   26°C, 30°C, and 34°C

TABLE 11.1 Comparison of estimated and model deoxygenation factors B

Reach RM	Character of Bottom	Velocity (ft/sec)	Depth (ft)	Slope per Miles	Coefficient		Computed B		Measured $K_d - K_1$ (B)
							Bosko Ev 10.9	Novotny Ev 10.10	
145.7–144.7	gravel	0.63	2.5	0.62	1.2	1.0	0.30	1.08	1.0
144.7–142.2	sand	0.42	2.5	0.62	0.6	0.4	0.10	0.43	0.25
142.2–140.7	gravel rock	1.70	1.9	0.5	1.5	1.5	1.34	1.90	2.12
140.7–137.9	gravel rock	1.04	2.4	0.8	1.5	1.5	0.65	1.79	1.60
137.9–135.4	gravel rock	0.94	2.4	0.234	1.5	1.5	0.59	1.34	1.14
135.4–131.5	gravel lower pollution	1.36	2.3	0.234	0.9	0.6	0.41	0.55	0.55
131.5–128.9	gravel lower pollution	0.89	6.1	0.234	0.8	0.5	0.12	0.22	0.15

438

TABLE 11.2 Wastewater discharges corresponding to 50,000 lbs BOD₅/day

| | Sewage and Industrial Flows | | |
Number of Reach	Flow Rate (ft³/sec)	Dissolved Oxygen (mg/liter)	Carbon BOD (mg/liter)
1	0.0	0.0	0.0
2	573.0	6.3	28.6
3	0.0	0.0	0.0
4	16.1	5.0	390.0
5	−27.1	0.0	0.0
6	26.3	5.0	56.0
7	0.0	0.0	0.0
8	0.0	0.0	0.0
9	0.0	0.0	0.0
10	0.0	0.0	20.0
11	−159.3	0.0	0.0
12	159.3	5.9	22.7
13	0.0	0.0	0.0
14	0.0	0.0	0.0
15	0.0	0.0	0.0

(3) Deoxygenation coefficient, K_r, in the river
     100% of the present, 75% of the present, 40% of the present, and
     $K_r = 0.23$ day^{-1}
(4) Average depth in the Main River between RM 142.2 and 128.9
     1.8 ft, 2.55 ft, and 4.2 ft.

The upstream dissolved oxygen concentration at RM 147.8 was selected to be 5.0 mg/liter, which was similar to the reported average concentration of oxygen in the discharge from an upstream dam. It was observed that because of high reaeration occurring between RM 147.8 and 146.7 (the first waste discharge), the dissolved oxygen concentration increased about 1 mg/liter.

It was also assumed that the DO concentration of all wastewaters discharged would not fall below 5.0 mg/liter as stipulated by the local regulatory authority. The wastewater flows and withdrawals corresponding to 50,000 lbs BOD/day are presented in Table 11.2.

The headwater water quality characteristics were as follows:

No. of headwater	River	Percent DO Saturation	Carbon BOD mg/liter	Nitrogen BOD mg/liter
1	Main RM 5.6	65.7	2.3	1.1
2	Trib. RM 1.0	85.0	2.5	0.9

Other BOD loads were computed by the model assuming 50% and 72% waste reduction. Dissolved oxygen profiles were computed and plotted for various BOD loads in the river for temperatures of 26°C and 30°C, and for different reductions of the deoxygenation coefficient, K_r. Two typical plots are shown in Figs. 11.8 and 11.9. The Thackston and Krenkel formula for K_2 was used.

A similar analysis was performed for nitrogenous BOD but is not presented here. The information obtained in the various plots similar to Figs. 11.8 and 11.9 was extracted and plotted as shown in Figs. 11.10 and 11.11. Similar plots were also developed for 25 and 75% reduction in K_r. These plots were then used to develop Fig. 11.12, which shows the effect of K_r reduction on the assimilative capacity of the river. It should be noted that for the hypothetical case where $K_r = 0$, the waste assimilative capacity converges to infinity.

Based on long-term BOD curves of the wastewater effluents with existing treatment and the BOD curves of a pilot plant activated sludge effluent, it was concluded that the activated sludge treatment would reduce the deoxygenation coefficient by 50%. This same reduction (above 50%) can be expected for the river deoxygenation rate.

The effect of depth on the waste assimilative capacity (DO standard =

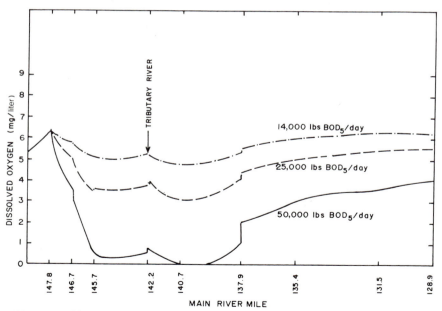

Fig. 11.8.   Dissolved oxygen profiles for various BOD_5 loads in the Main river, temperature 26°C, existing K_r (Krenkel and Novotny, 1973).

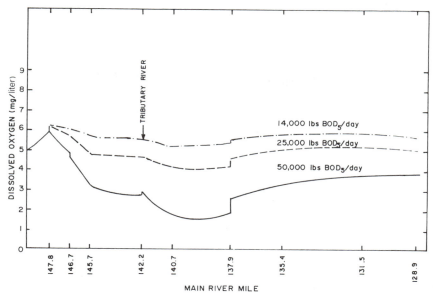

Fig. 11.9. Dissolved oxygen profiles for various BOD₅ loads in the Main river, temperature 30°C, K_r 60% reduced (Krenkel and Novotny, 1973).

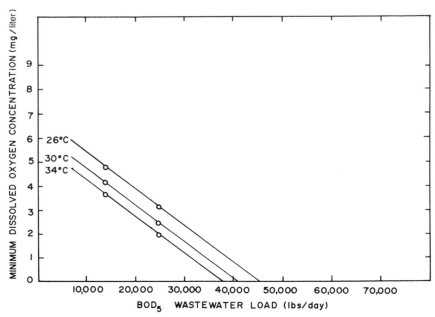

Fig. 11.10. Relationship of minimum dissolved oxygen concentration to BOD₅ load in the Main river, existing K_r (Krenkel and Novotny, 1973).

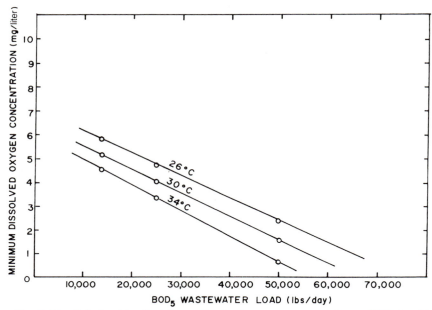

Fig. 11.11. Relationship of minimum dissolved oxygen concentration to BOD_5 load in the Main River, K_r 60% reduced (Krenkel and Novotny, 1973).

3.0 mg/liter) is shown in Fig. 11.13. It is significant to note that a deviation in the depth estimation of 1 foot represents a change in the waste assimilative capacity of 5000 pounds of BOD per day.

The effect of using different formulas for K_2 also was evaluated, and the resulting assimilative capacity was as follows (assuming K_r is reduced 50%):

K_2 Formula	K_2 Day^{-1} Base e	Assimilative Capacity (lbs/day)
Churchill *et al.* (1962)	0.95	2,300
O'Connor and Dobbins (1958)	1.7	7,300
Thackston and Krenkel (1969)	2.4	13,500

Best Estimate of the Waste Assimilative Capacity

The estimate of the waste assimilative capacity is based on the following assumptions:

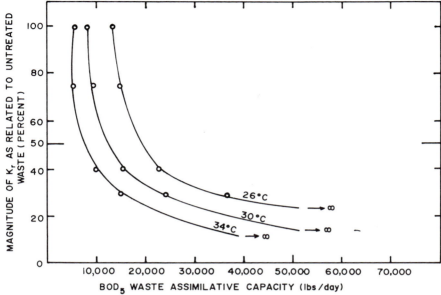

Fig. 11.12. Waste assimilative capacity related to magnitude of K_r for a minimum dissolved oxygen of 5.0 mg/liter (Krenkel and Novotny, 1973).

Flow conditions:	750 ft³/sec in Main River above tributary
	800 ft³/sec in Main River after the confluence
Temperature:	30°C in all reaches
Depth:	2.52 ft
K_r reduction:	50%
K_2:	Thackston and Krenkel formula

From Figure 11.12, the corresponding waste assimilative capacity which would satisfy the minimum DO requirement of 5.0 mg/liter of DO in all reaches, the waste assimilative capacity would be 13,500 lbs BOD/day.

It may be argued that, although depth and K_2 significantly affected the assimilative capacity of the river, the range of assimilative capacity values did not in turn significantly affect the actual DO in the river. This was deduced from plots similar to those in Fig. 11.9. If this line of argument was unacceptable to regulatory authorities, additional field surveys would have been justified to obtain a more accurate measure of depth and additional information for calibration and verification of the model.

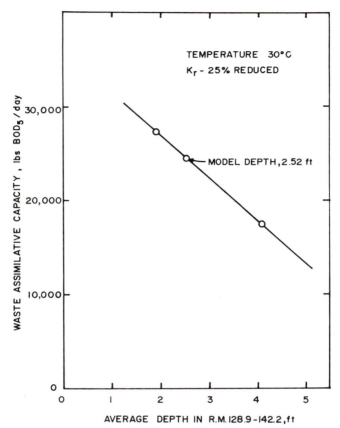

Fig. 11.13. Effect of depth variation on waste assimilative capacity of Main River (Krenkel and Novotny, 1973).

MODELING EFFECTS OF AQUATIC WEEDS

This case considers a river in which aquatic weeds significantly affect DO. The river reach is downstream from a large industrial complex that has a total discharge of 100,000 lbs/day of BOD's and 50,000 lbs/day of nitrogenous BOD. Factors that have been analyzed and modeled mathematically in the 40-mile river reach include carbonaceous BOD, nitrogenous BOD, benthic oxygen demand, aquatic plant photosynthesis and respiration, and reaeration. Results showed that carbonaceous BOD was important only for the first 10 miles, nitrogenous BOD was appreciable in the lower 31 miles, benthic oxygen demand was least important, and aquatic weeds exerted the greatest influence on DO. The mathematical model used was the same as that given in Eq. (10.7) in Chapter 10. The model indicated that aquatic weeds exerted such a tremendous demand that some solution other than secondary waste treatment and low-flow augmentation from upstream reservoirs was needed to alleviate the DO problem in the 40-mile reach.

For the purpose of this discussion, emphasis will be placed on P and R in the terms (10.8e) and 10.8f) presented in Chapter 10.

Photosynthetic oxygen production and respiration are discussed together because only their net effect is evaluated in this analysis. For all practical purposes, photosynthetic activity occurs only during daylight hours, as expressed by

$$CO_2 + H_2O + \text{minor nutrients} \xrightarrow{\text{light}} \text{plant cells} + O_2. \qquad (11.6)$$

Plant respiration occurs continuously throughout the day; however, it may vary in magnitude in proportion to the DO concentration. These two factors are combined into one factor called $(P-R)$ which is actually the net amount of oxygen produced and consumed by the plants.

The effect of aquatic plants on DO within a particular stream reach at a particular time of day is a function of the plant density and distribution, light intensity, water depth, turbidity, temperature, and DO. The most important factor is plant density and distribution. Oxygen production is proportional to plant density only to a certain limit. Then, with increasing density in the plants, oxygen production decreases and may even become a net oxygen consumer if the density becomes too great. This occurs because the plants become so dense that some of them are shaded by other overlying plants. Light intensity also is an important factor affecting photosynthetic oxygen production, and, as expected, oxygen production is proportional to light intensity. Figure 11.14 illustrates how these two factors affect DO in a river. A high density plant growth will cause low DO concentrations, even on bright days. Even though high DO values are produced during the afternoon, extremely low DO concentrations result at night. Lower density plant growths result in net addition of DO to the stream only on bright days. On dull days, even low density plant growths consume more DO than they produce. Based on these considerations, it is obvious that aquatic plants must be regarded as a disadvantage in any stream. Westlake (1966) reported that net DO production is not only affected by the average plant density, which is the density normally reported for a stream reach, but also by the distribution of the plants within

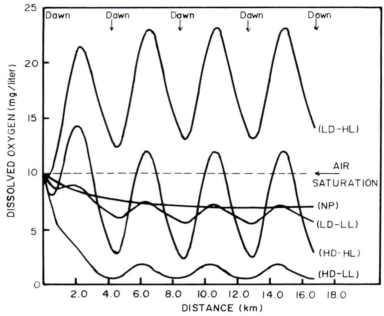

Fig. 11.14.   Calculated effect of differences in plant density and daily sunlight distributions on the distribution of dissolved oxygen on a river 0.66 m deep with a velocity of 5 cm/sec and a temperature of 15°C (Owens *et al.*, 1969). The oxygen distribution with no plants is also shown. HD, high plant density, 500 gm dry wt/m²; LD, low plant density, 100 gm dry wt/m²; HL, bright day, 650 cal/cm²-day; LL, dull day, 57 cal/cm²-day; NP, no plants.

the reach. In other words, even though the average density may be relatively low, if the weeds are concentrated within a small area, the net effect of the weeds may be to consume more DO than that produced.

The significance of water depth, turbidity, and temperature effects on photosynthetic oxygen production have not been quantitatively evaluated. Water depth and turbidity have the effect of influencing light intensity. The weed growths in this river were apparently very much influenced by water depth. As shown in Fig. 11.15, plant density is inversely proportional to water depth, as would be expected. The effect of temperature on plant respiration has been reported by Bedick (1966). For this analysis it was assumed that temperature affected photosynthetic oxygen production to the same degree. Bedick found that respiration increased by a factor of 2.16 for a change in temperature from 10° to 20°C. Using this result and assuming that plant respiration and photosynthetic

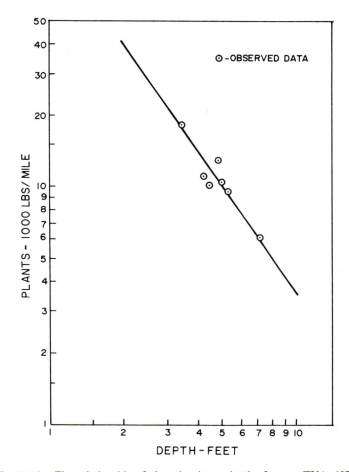

Fig. 11.15. The relationship of plant density to depth of water (TVA, 1973).

oxygen production follow the van't Hoff–Arrhenius relationship, Eq. (11.7) describes the effects of temperature on $(P-R)$:

$$(P-R)_T = (P - R)_{T_1}[1.080^{(T-T_1)}], \qquad (11.7)$$

where T is the temperature at which $(P-R)$ is wanted and T_1 is the temperature at which $(P-R)$ is known. The factor 2.16 is in close agreement with a factor of 2.3 reported by Westlake (1966) for both respiration and photosynthesis and by Rabinowitch (1956) and Talling (1957) for phytoplankton.

Respiration by potamogeton has been reported by Bedick (1966) to vary with DO concentration according to the relationship

$$R_2 = R_1 \left[\frac{0.658 + 0.0693(DO_2)}{0.658 + 0.0693(DO_1)} \right]. \tag{11.8}$$

Without incurring significant error, this relationship may be simplified to

$$R_2 = R_1 \left[\frac{10 + DO_2}{10 + DO_1} \right]. \tag{11.9}$$

Similar relationships for other species of aquatic weeds have been developed by Bedick (1966) and Owens and Maris (1964).

The effect of plant photosynthetic activity and respiration on the DO balance may be determined in one of two ways. One method is the light and dark bottle technique. This method could not be used on this river because it only accounts for plankton and does not measure the influence of attached plants. Other disadvantages of this method have been reported by O'Connor and DiToro (1968), Symons *et al.* (1967), and Pratt and Berkson (1959).

The second method involves the analysis of the diurnal DO variation in the stream. This method was first proposed by Odum (1956). In flowing streams where the diurnal DO variation differs from station to station, Odum's upstream–downstream method should be used. Basically, this method is a solution of the Streeter–Phelps equation with various terms added to account for [in addition to $(P-R)$] benthic oxygen demand and nitrogenous BOD; hence $(P-R)$ is determined by trial and error. In this analysis, all other variables were either calculated or estimated and $(P-R)$ was calculated to make the mathematical model fit the observed DO conditions of the survey. The resulting $(P-R)$ curves for each stream reach analyzed are presented in Fig. 11.16. To simplify and reduce calculations, the following sine curve was used to represent the curved portions of $(P-R)$:

$$(P-R) = C_1 \sin 2\pi t + C_2, \tag{11.10}$$

where C_1 is the amplitude of the sine curve; C_2 is the offset between the axis where $(P-R)$ equals 0 and the sine curve axis where $\sin 2\pi t$ equals 0; and t is the difference in time between the time of day where $\sin 2\pi t$ equals 0 at dawn and the time of day in question. This relationship is illustrated in Fig. 11.17, and the values of C_1 and C_2 for each respective reach are shown in Fig. 11.16. In most cases, the sine function was used to represent $(P-R)$ from a period approximately one hour before to two hours after the curve intersected with its axis. Normally, two hours after the curve intersected with the axis, the maximum respiration rate was

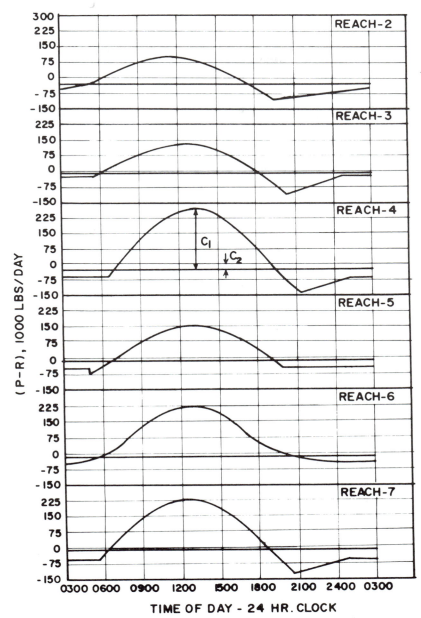

Fig. 11.16. Computed $(P-R)$ curves to represent river reaches 2–7 (TVA, 1973).

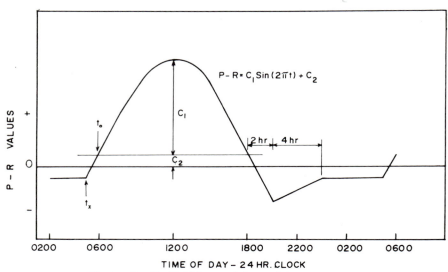

Fig. 11.17.   (*P*–*R*) Values and times of day (TVA, 1973).

reached. Following this point, the respiration gradually decreased for a period of approximately four hours and then remained constant for the remainder of the night. Minor changes from this normal procedure had to be made for reaches 2, 5, and 6 to obtain better solutions for predicting the DO concentration. The reasons for these necessary changes are not well understood. The maximum respiration rate was measured at dusk for most of the reaches. It probably occurs at this time because dissolved oxygen concentration is higher during this time than it is during the remainder of the night.

The relative magnitudes of the positive and negative values shown in Figure 11.16 are mainly a function of plant density; however, as discussed previously, the magnitudes of (*P*–*R*) are not directly proportional to plant density but may be inversely proportional, depending upon the distribution of the plants within the reach, i.e., whether the plants are concentrated in one area or spread evenly throughout the reach.

Another method for modeling photosynthetic oxygen production and respiration has been proposed by O'Connor and DiToro (1968). They suggest using a Fourier series for representing photosynthetic oxygen production and assume respiration to be constant. The Fourier series, while adequate in most cases, is not very flexible and, therefore, is hard to adjust for cases in which it will not fit the observed diurnal DO curves. The major disadvantage of the model used for this river is that plant respiration cannot be treated separately; i.e., the effects of DO concentration on respiration cannot be incorporated very readily. Kartchner *et al.* (1969) proposed a Fourier series and a time series index approach for representing the dissolved oxygen variation in a stream. These models could be used to represent photosynthetic oxygen production. The time series index approach has the advantage of adequately representing irregular diurnal DO curves. Kartchner *et al.* (1969) used these approaches in analyzing data accumulated by water quality monitors.

O'Connor and DiToro (1968) used an unsteady-state term for representing photosynthetic oxygen production; i.e., they evaluated not only the diurnal variation but also the variation of photosynthetic oxygen production within the reach as a function of time. In this analysis it was assumed that $(P-R)$ was constant within the respective reaches and, therefore, did not vary with time of travel within the reach. Hence, in the integration of the differential equation for DO in a stream reach, $(P-R)$ was treated as a constant. Although $(P-R)$ varies with time and position within each reach, it may be considered as a constant for relatively short intervals of flow time within the diurnal variation without causing significant error.

The peak respiration which occurs around dusk for most of the $(P-R)$ curves determined for the reaches investigated is similar to results obtained by other investigators (Armstrong *et al.*, 1968; O'Connell and Thomas, 1965), while they are not similar to the results reported by others (Odum, 1956; Gunnerson and Bailey, 1963; Odum, 1957). Possibly, the reason the latter investigators did not find increased respiration rates near dusk is that Odum assumed respiration did not vary throughout the day, and Gunnerson and Bailey analyzed a diurnal DO curve which varied from only 10 mg/liter to 11 mg/liter, yielding a DO difference of only 1 mg/liter throughout the day. Copeland (1965) presented evidence that the community metabolism increases as labile organic material is produced during photosynthesis, which results in increased respiration during the photosynthetic period. However, studies by the TVA (1973) show that plant respiration does not vary with light intensity. Several investigators (Gessner and Pannier, 1958; Owens and Maris, 1964; Bedick, 1966) have found plant respiration, or possibly the total biota respiration, to be a function of DO concentration. O'Connell and Thomas (1965) determined $(P-R)$ in the Truckee River by the upstream–downstream technique and a field laboratory analysis using an "algal chamber." The results for $(P-R)$ were very similar except that peak respiration resulted from the upstream–downstream analysis and not for the "algal chamber" analysis. The authors did not give the DO concentration for the "algal chamber" and did not state whether this DO concentration varied as it did in the river. Hence, it may be possible that the peak respiration did not occur because the DO was not varied.

The net result of photosynthetic oxygen production and respiration is dependent on streamflow:

$$(P-R)_2 = (P-R)_1 \left(\frac{Q_2}{Q_1}\right)^d, \qquad (11.11)$$

where $(P-R)$ is expressed in pounds per day[2], Q is the streamflow, and d is the inverse of the slope of the line resulting from a plot of log Q vs. log(Detention time)$^{-1}$.

Using the mathematical model for DO that was developed for this river, it was determined that the aquatic weeds exerted such a large respirational demand on the DO that even if no organic or nitrogenous wastes were discharged to the river, the DO would still be below the state standards. Similar problems have been reported by other investigators (McDonnell and Kountz, 1966, and Fisher, 1971). This problem exists even on "bright" days but is especially critical on "dull" days, as shown by DO data collected during another survey. Using the DO simulation model, it was estimated that the minimum DO would be less than 1 mg/liter for a minimum flow of 760 ft^3/sec, 27°C, and 7 mg/liter of DO in the river above the waste discharges, even if no wastes were discharged to the river.

Thus, it was necessary to explore solutions other than waste treatment for increasing DO in the river. Using the model, the following factors and combinations thereof were evaluated for their effect on the minimum DO concentration: advanced or tertiary waste treatment, increased DO above the waste discharges, streamflow augmentation, decreased temperature, artificial in-stream reaeration, and control of the aquatic weed growths.

The first combination of solutions considered was percent waste removal, streamflow augmentation, upstream DO, and decreased temperature. A range of possible values was selected for each alternative, and three values were chosen for each alternative within the respective range. The values considered for each alternative were as follows: percent waste removal—85, 90, and 95; streamflow—760, 1150, and 1500; upstream DO—5, 6, and 7; and stream temperature—17°, 21°, and 25°C. The consideration of three values for each alternative required computations for 81 separate DO profiles similar to the one shown in Fig. 11.18. Since the minimum DO is the critical condition for which the alternatives must be designed, this DO concentration was summarized for each of the 81 profiles. To best summarize the results of the foregoing calculations, the following linear multiple regression equation was developed between minimum DO and the controllable variables:

Minimum DO = -6.49 + 0.104 (percent removal of wastes)
+ 0.00241 (streamflow) $-$ 0.253 (stream temperature)
+ 0.379 (initial DO concentration).   (11.12)

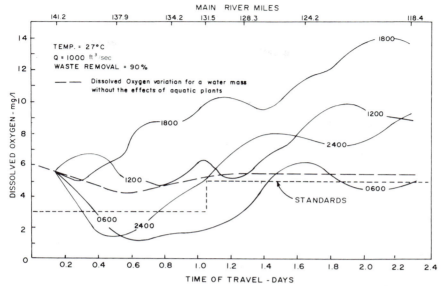

Fig. 11.18. Computed dissolved oxygen concentrations for various times of the day (TVA, 1973).

This relationship has a standard error of estimate equal to 0.2 mg/liter and a multiple correlation coefficient equal to 0.983 indicating the relationship is statistically significant at the 1% level. The independent variables were shown to be truly independent by using an analysis of moment matrix, covariance matrix, and correlation matrix.

Using the preceding model for minimum DO as a function of the four independent variables, it may be shown that on "bright" days, the state stream standards may be maintained by varying only these variables. This is important because these "bright" day conditions are expected to be prevalent, and, therefore, an adequate DO level can be obtained without removing the aquatic weeds or using artificial in-stream reaeration. One combination of values that may be used to attain 3 mg/liter DO is 1500 ft³/sec, 92% removal of wastes, 25°C, and 7 mg/liter of DO at RM 1.2. Of course, other combinations of these variables may be used to attain 3 mg/liter on "bright" days; but it is necessary to use optimization techniques to determine the best combination.

The only alternative solution that can be used to maintain state stream standards without increasing the other variables beyond the minimum requirements, i.e., 85% removal of wastes, 760 ft³/sec, and a temperature increase equal to the maximum allowable, is artificial in-stream reaeration. In addition, on "dull" days, either artificial in-stream reaeration or essentially complete removal of aquatic weeds is necessary to meet state stream standards. To determine the cost of in-stream reaeration, it is necessary to determine (1) how much oxygen is needed during critical conditions, and (2) how long the aerators would be required to operate. The latter requirement cannot presently be calculated and can only be evaluated by determining how DO in the river is affected by solar radiation, turbidity, and temperature.

To estimate the maximum amount of artificial in-stream reaeration required, the DO model was changed to measure the effects of "dull" days and determine how much artificial in-stream reaeration is required to maintain certain DO concentrations. To simulate DO conditions during critical conditions, the model was altered to fit the minimum DO concentrations observed during "dull" day conditions. With the assumption that $(P-R)$ was a constant value equal to some fraction of the respiration rate, it was determined that 0.65 times the observed respiration rates for the remaining downstream reaches yielded a good fit of the minimum DO observed under "dull" day conditions.

Artificial in-stream aeration was included in the model by automatically increasing the DO in the increments of 0.5 mg/liter as required to maintain the desired DO level.

Based on this modified DO model, the required pounds of oxygen added by artificial in-stream reaeration and the location of the necessary aerators were determined for maintaining various DO levels under various stream conditions. The results of a typical analysis are plotted in Fig. 11.19. These results are based on a streamflow equal to 760 ft³/sec, DO at RM 1.2 equal to 7 mg/liter, percent waste removal equal to 95, and a temperature of 27°C. To determine how each of these variables affected the total pounds of artificial aeration needed and how the variables affected location of aeration, each variable was varied within its respective limits.

For 85% waste reduction, 760 ft³/sec, 27°C, and 5 mg/liter DO, the total amount of required artificial in-stream aeration was 146,000 pounds of oxygen per day. If the stream temperature was lowered or raised 2°C, the amount of oxygen required was decreased by 28,000 or increased by 34,000 pounds per day, respectively. The required in-stream reaeration decreased by 6000 pounds per day if the upstream DO was increased from 5 to 7 mg/liter and decreased by 11,000 pounds per day if the streamflow was increased from 760 to 1500 ft³/sec. If percent waste removal was increased to 90%, the required in-stream aeration was 15,000 pounds per day. Obviously, many combinations of these variables may be considered to reduce the amount of in-stream aeration required. For this reason, the following equation was developed to determine the necessary amount of in-stream aeration to maintain the present state stream standards:

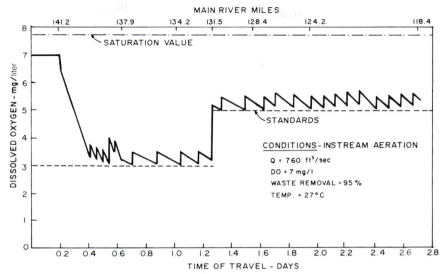

Fig. 11.19.   The use of artificial stream reaeration (TVA, 1973).

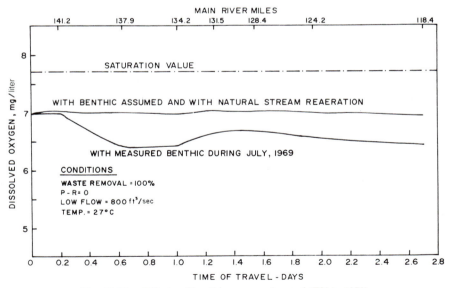

Fig. 11.20.   Effects of benthic oxygen demand (TVA, 1973).

$$I = 73,580 + 15,600T - 3588r - 19.18Q - 5296 \, \text{DO}_{1.2.} \quad (11.13)$$

where I is the total amount of supplemental aeration required from RM 1.2 to RM 106.3. This relationship has a standard error of estimate of 1500 lbs/day and a multiple correlation coefficient equal to 0.997, indicating statistical significance at greater than the 1% level. The independent variables were shown to be truly independent by using an analysis of moment matrix, covariance matrix, and correlation matrix.

The DO model was checked to see if it would simulate expected "natural" DO concentrations if the effects of wastes and $(P-R)$ were eliminated from the model. Using the rather high benthic oxygen demand measured in the river, the model predicted the lower curve shown in Fig. 11.20. This DO profile is somewhat lower than what would be expected "naturally." Most unpolluted streams have a benthic oxygen demand of about 1 g O_2/m^2/day. When this value was used, the upper DO profile resulted, yielding DO concentrations approximately 90% of the saturation concentration, which is normal for large streams.

REFERENCES

Armstrong, N. E., and Gloyna, E. F. (1968): Ecological aspects of stream pollution. *In* Advances in water quality improvement (Gloyna and Eckenfelder, eds.). Water Resour. Symp. No. 1, Austin, Texas.

Bedick, T. M. (1966). Effect of oxygen concentration on the respiration of aquatic macrophytes. M.S. Thesis, Pennsylvania State Univ. (December).

Churchill, M. A., Elmore, H. L., and Buckingham, R. A. (1962). The prediction of stream reaeration rates. *In* "Advances in Water Pollution Research," Vol. 1. Pergamon, Oxford.

Copeland, B. J. (1965). Evidence for regulation of community metabolism in a marine ecosystem. *Ecology* **46**(4), 563–564.

Fair, G. M., Geyer, J. C., and Okun, D. A. (1968). Water purification and wastewater treatment and disposal. *In* "Water and Wastewater Engineering," Vol 2. Wiley, New York.

Fisher, S. (1971). The Cross-Florida Barge Canal: A lesson in ecology. *Civ. Eng.* (April). 44–48.

Foree, E. G. (1976). Reaeration and velocity prediction for small streams. *J. Environ. Eng. Div., Am. Soc. Civ. Eng.* **102**(EE5). 937–952.

Gessner, F., and Pannier, F. (1958). Influence of oxygen tension on respiration of phytoplankton. *Limnol. Oceanogr.* **3**, 478–480.

Grimsrud, G. P., Finnemore, E. J., and Owen, E. J. (1976). Evaluation of water quality models: A management guide for planners. Office of Research & Development, USEPA-600/5-76-004, Washington, D.C. (July).

Gunnerson, C. G., and Bailey, T. E. (1963). Oxygen relationships in the Sacramento River. *J. Sanit. Eng. Div. Am. Soc. Civ. Eng.* **89**(SA4). 95–124.

Kartchner, A. D., Dixon, N., and Hendricks, D. W. (1969). Modeling diurnal fluctuations in stream temperature and dissolved oxygen. 24th Annu. Purdue Ind. Waste Conf. (May). Lafayette, Indiana.

Krenkel, P. A., and Novotny, V. (1973). The assimilative capacity of the South Fork Holston River and Holston River below Kingsport, Tennessee. Rep. to Tennessee Eastman Co., Kingsport.

Krenkel, P. A., and Ruane, R. J. (1979). *In* "Modeling of Rivers" (H. W. Shen, ed.), 17-1–17-22. Wiley, New York.

McDonnell, A. J., and Kountz, R. R. (1966). Algal respiration in a eutrophic environment. *J. Water Pollut. Control Fed.* (May). 841–847.

O'Connell, R. L., and Thomas, N. A. (1965). "Effect of benthic algae on stream dissolved oxygen. *J. Sanit. Eng. Div. Am. Soc. Civ. Eng.*, Pap. **91**(4345), 1–16.

O'Connor, D. J., and DiToro, D. M. (1968). The distribution of dissolved oxygen in a stream with time varying velocity. *Water Resour. Res.* **4**(3), 639–646.

O'Connor, D. J., and Dobbins, W. E. (1958). Mechanism of reaeration in natural streams. *Trans. Am. Soc. Civ. Eng.* **123**, 641–666.

Odum, H. T. (1956). Primary production in flowing waters. *Limnol. Oceanogr.* **1**(2), 102–117.

Odum, H. T. (1957). Trophic structure and productivity of Silver Springs, Florida. *Ecol. Monogr.* **27**, 55–112.

Owens, M., and Maris, P. J. (1964). Some factors affecting the respiration of some aquatic plants. *Hydrobiologia* **23**, 533–543.

Owens, M., Knowles, G., and Clark, A. (1969). The prediction of the distribution of dissolved oxygen in rivers. *In* "Advances in water pollution research." Proc., 4th Int. Conf. on Water Pollut. Res., Prague, Czechoslovakia.

Pratt, D. M., and Berkson, H. (1959). Two sources of error in the oxygen light and dark bottle method. *Limnol. Oceanogr.* **4**(3), 328–334.

Rabinowitch, E. I. (1956). Kinetics of photosynthesis. *In* "Photosynthesis and related processes, II, Part 2." pp. 1209–2088. Wiley (Interscience), New York.

Ruane, R. J. (1971). Statistical equation for estimating the assimilative capacity of a stream for BOD. Unpublished Rep. Tennessee Valley Authority, Water Quality Branch.

Stall, J. B., and Yang, C. T. (1970). Hydraulic geometry of 12 selected stream systems of the United States. Res. Rep. No. 32, Univ. of Illinois, Water Resour. Center, Urbana (July).

Symons, J. M., Irwin, W. H., Clark, R. M., and Robeck, G. G. (1967). Management and measurement of DO in impoundments. *J. Sanit. Eng. Div. Am. Soc. Civ. Eng.* **93**(SA6), 181–209.

Talling, J. F. (1957). Photosynthetic characteristics of some freshwater plankton diatoms in relation to underwater radiation. *New Phytol.* **56**, 29–50.

Tennessee Valley Authority, Water Quality Branch (1973). Unpublished Reports. Chattanooga, Tennessee.

Thackston, E. L., and Krenkel, P. A. (1969). Reaeration prediction in natural streams. *J. Sanit. Eng. Div., Am. Soc. Civ. Eng.* **95**(SA1), 65–94.

U.S. Environmental Protection Agency (1978). Holston River study. Surveillance and Analysis Div., Athens, Georgia (October).

Westlake, D. F. (1966). A model for quantitative studies of photosynthesis by higher plants in streams. *Int. J. Air Water Pollut.* **10**, 883–896.

12 | *Lakes and Reservoirs*

CHARACTERIZATION OF RESERVOIRS AND LAKES

Impoundments including both natural lakes and man-made reservoirs can be divided broadly into three categories:

1. A deep reservoir which is characterized by horizontal isotherms. Seasonal thermal or density stratification is typical for these water bodies;

2. A weakly stratified reservoir characterized by isotherms which are tilted along the horizontal axis of the reservoir; and

3. A vertically mixed reservoir whose temperature and density distribution are roughly uniform with depth during both summer and winter periods.

Both deep and vertically mixed reservoirs can be represented by a one-dimensional flow concept; however, the major temperature and concentration gradients in deep reservoirs take place along the vertical axis, while those in vertically mixed reservoirs take place mainly along the longitudinal axis.

Vertically mixed reservoirs are those that do not exhibit appreciable thermal or water quality variation with depth. Hydraulically and qualitatively, these reservoirs resemble large rivers and nonstratified estuaries. Therefore, the same representation can be used for both. On the other hand, weakly stratified or deep reservoirs are stratified during summer and sometimes during winter periods. The primary parameter utilized to categorize reservoirs is the densimetric Froude number, N_F, which may be written as

$$N_F = \frac{LQ}{HV}\left(\frac{\rho_0}{g\beta}\right)^{\frac{1}{2}},\tag{12.1}$$

where L is the reservoir length, Q is the volumetric discharge through the reservoir, H is the depth, V is the volume, ρ_0 is the reference density, β is the average density gradient in the reservoir, and g is gravity acceleration.

For the purpose of classifying reservoirs, β and ρ_0 can be approximated as having respective values of 10^{-3} kg/m^{-4} and 1000 kg/m^3. Substituting these values into Eq. (12.1) leads to the following expression for N_F:

$$N_F = 320\,\frac{LQ}{HV},\tag{12.2}$$

where L and H are in meters, V is in m^3, and Q is in m^3/sec.

Theoretical and experimental investigations of stratified flow indicate that flow separation occurs when N_F is less than $1/2\pi$ (Long, 1962). It then can be deduced that stratified deep reservoirs would have an $N_F \ll 1/2\pi$; weakly stratified reservoirs would have an $N_F \sim 1/2\pi$, and completely mixed reservoirs, $N_F \gg 1/2\pi$.

Inasmuch as most reservoirs have multiple uses and optimization is often necessary to avoid water shortages and deteriorated water quality, it is important that water quality predictive techniques be available.

RESERVOIR DYNAMICS

Stratification and the Annual Thermal Cycle

Lake and reservoir stratification introduces a new dimension into the process of water quality analysis. Stratification is basically a layering of fluid masses caused by density differences which, in turn, may be caused by temperature, TDS, or suspended solids differences. In deep reservoirs, temperature differences between the top and bottom layers may reach more than 15°C during a warm summer period.

In most reservoirs and lakes, stratification is a result of the heat balance between the water impounded in the reservoir and external inputs. Heat inputs include solar and atmospheric radiation, conductive heat exchange between the atmosphere and water, and heat from tributaries.

Water can emit heat back to the atmosphere through surface radiation.

In addition, heat is also lost by evaporation, conduction, and through reservoir (lake) discharges. Most of the warming and cooling processes take place in a relatively thin surface layer. Therefore, if vertical mixing is not sufficient to equalize the surface heat gradient, stratification may occur.

The process of stratification of deep reservoirs and lakes and its annual cycle is shown in Fig. 12.1. If the beginning of the annual stratification cycle is considered to be in midwinter, the vertical temperature distribution in northern lakes would follow a basic physical density distribution with heavier 4°C cold water near the bottom and lighter, near freezing water (0°C) at the top. This phase of the annual stratification cycle is called winter or negative stratification and is typical for northern and/or colder climatic conditions.

As spring approaches, the upper layers receive more heat than can be returned, resulting in an increase of the surface temperature. With the surface temperature near 4°C, the thermal and density distribution becomes more uniform. At this stage the resistance to vertical mixing is at its lowest level, and even small turbulent perturbations can achieve partial or complete vertical mixing and circulation of water in the reservoir.

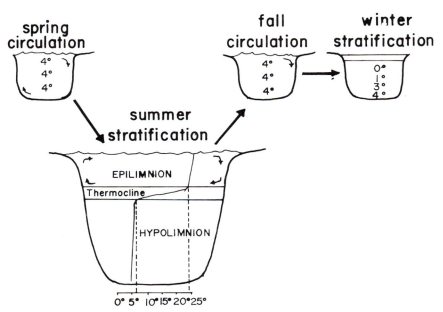

Fig. 12.1. Seasonal temperature variations in a temperate lake and the generalized thermal stratification during the summer season (Greeson, 1969).

During the late spring and early summer, the thermal input increases the surface temperature while colder, heavier water remains near the bottom. Because of a larger angle of incidence, solar radiation can penetrate into deeper layers below the surface. If the mixing is not sufficient to equalize the vertical distribution of temperature and/or density, a summer or positive stratification will develop. The thermal stratification of a reservoir that occurs during the summer period is classified in three distinct zones. The upper, warmer and less dense zone is called the epilimnion; the lower, colder and denser zone is called the hypolimnion. The thermocline is the transition zone between the two. The temperature and density gradients are lower in both the epilimnion and the hypolimnion and relatively high in the thermocline. The thermocline zone is sometimes called the mesolimnion.

Although a particular course of stratification depends on climatic conditions, mixing levels, and geographical and geometrical factors, a definite thermocline should develop by the end of spring. In late spring and early summer, the heat input reaches its highest point. At that time, strong thermal and density stratification exists, with the thermocline limiting the downward flux of energy and mass from the epilimnion to the hypolimnion (Water Res. Eng., 1966).

In the late summer, surface temperatures begin to decrease and the denser surface water causes mixing of a portion of the epilimnion, thus reducing the thermal gradient in this zone to nearly zero. With the coming of fall, colder temperatures, and decreasing radiation input, the surface water temperature continues to decline. As the thermocline becomes less definitive, its resistance to vertical mixing diminishes until a stage of relatively uniform vertical temperature and density distribution is reached.

Finally, with the continued heat loss occurring during late fall and early winter, the surface temperature may again drop below 4°C, resulting in the initially described stratification, and the annual cycle is closed. The complete annual cycle may have two stratification periods — winter and summer — and two periods in which the temperature and quality distribution are nearly uniform — the spring and fall circulations or turnovers. The complete annual cycle, as described, with two circulation periods may take place in colder climatic conditions where water temperatures drop below 4°C. In more temperate zones, where the temperature does not fall to 4°C, the period of stratification occurs from spring to fall, and only one turnover occurs.

From the water quality standpoint the summer stratification period is the most critical time. In the stagnant, warm epilimnion layer, usually corresponding to the euphotic zone, there is a danger of the proliferation of algae, which contribute to acceleration of the eutrophication process and general water quality deterioration. Dead organic matter and residues from the abundant epilimnetic growths will settle into the hypolimnion, exerting significant demand on the oxygen resources. Since oxygen resupply through the thermocline is limited because of mass transfer inhibition, the dissolved oxygen concentrations in the hypolimnion may be significantly reduced, sometimes leading to anaerobic conditions near the bottom. In the absence of oxygen, the conditions are favorable for reducing chemical reactions such as the conversion of sulfates to sulfides or the reduction of iron and manganese to more soluble forms.

Stratification also may have an effect on the hydraulic behavior of reservoirs. The vertically mixed reservoir can be described using a one-dimensional plug flow concept similar to streams. In stratified reservoirs, however, the velocity distribution depends on the stratification. The incoming water has a tendency to move into a reservoir layer which has a density similar to its own. Such flows are called density currents.

If water is withdrawn from a stratified reservoir, the flow layer is centered around the elevation of the outlet and its thickness is inversely proportional to the density gradient. This hydraulic phenomenon happens because of the stabilizing effect of the stratification since vertical motion is inhibited by buoyant forces caused by the density differences in adjacent parcels of water (Huber *et al.*, 1972).

Stratified reservoirs are not the only instances in which density currents can occur. As previously mentioned, estuaries may become stratified because of saltwater intrusion. Similar cases have been reported in some urban reservoirs receiving runoff from street salting (Cherkauer, 1977). Another cause of stratified density flows is cooling water discharges from cooling operations. The stratified density flow is usually quite stable and can persist in rivers and estuaries for very long distances. For a comprehensive treatise on the thermal properties of lakes, the reader is referred to Hutchinson (1957).

Transport and Mixing in Reservoirs

The transport and mixing processes occurring in deep reservoirs and lakes are closely associated with thermal stratification. Under nonstratified conditions, which usually occur during spring and fall, the flow regime in the reservoirs is similar to a ''plug'' channel flow or a completely mixed basin. This will depend on whether the reservoir is a narrow valley impoundment or a large surface basin. However, under summer conditions, when the flow turbulence is not capable of equalizing the heat inputs throughout the reservoir volume, the stratified flow can be separated into two different flow regimes, as shown in Fig. 12.2 (Elder, 1964). The phenomenon depicted is called an underflow.

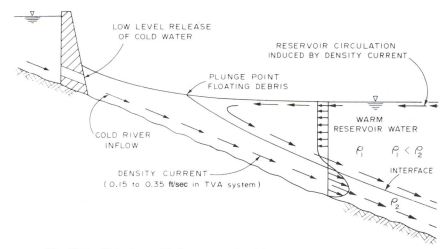

Fig. 12.2.   Velocity profile in a reservoir with a density current (Elder, 1964).

Underflows can be caused by the discharge of colder, denser water from an upstream stratified reservoir or by water containing excessive suspended or dissolved solids. Underflows caused by highly turbid waters were first noted on Lake Mead, while the underflow caused by the upstream release of hypolimnetic waters frequently occurs in TVA reservoirs. Fish kills purportedly caused by the low dissolved oxygen content resulting from the release of cold hypolimnion water have been reported downstream from Fort Loudon Dam (Jones, 1964). The same phenomenon has been observed in the Moldava River reservoir cascade in Europe by Buliček (1960–1965). Two different flow regimes are noted; the lower strata, which is bounded by the channel bottom and the interface with the velocity profile approximating that of distorted pipe flow, and the upper layer, which is bounded by the interface and the atmosphere and behaves like free surface flow.

One immediate problem that may occur is caused by the possibility of a flow reversal in the upper layers of the reservoir. Under these conditions, should the water pollution control plant discharge be located downstream or upstream from the water treatment plant intake, or what should be done to prevent pollution of upstream facilities? It should also be noted that if organic settleable suspended solids are present in the underflow, one would expect to find a benthal oxygen demand near the plunge point.

A schematic representation of a reservoir system and the major flow components occurring during summer stratification is shown in Fig. 12.3. The figure also shows the major inputs affecting the thermal balance of the reservoir. Both heat and mass throughout the reservoir volume are

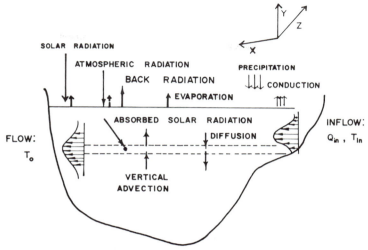

Fig. 12.3. Schematic diagram of reservoir system.

responses to external inputs subject to system parameters which include the level of mixing and diffusion, among others. On the other hand, the level of mixing and vertical diffusion are strongly affected by the temperature and density distribution.

Dispersion and Mixing

Vertical dispersion in reservoirs, or the lack thereof, is the primary factor causing stratification. As demonstrated in Chapter 9, turbulent mixing is analogous to molecular diffusion and can be described by Fick's law. For a substance, S, the mass transfer by diffusion is [see also Eq. (9.28)]

$$G = -D_y \frac{\partial S}{\partial y}, \tag{12.3}$$

where G is the mass transfer of the substance, S is the substance concentration, and D_y is a coefficient of diffusion.

In turbulent systems, which also include reservoirs, D_y is a function of the turbulence level of the system. For reservoirs, D_y is a temporal and spatial variable, and, thus, it will be referred to as $D(y,t)$. For heat transfer along the reservoir vertical axis, Eq. (12.3) can be rewritten as

$$H = -\rho\, c_p\, D(y,t) \frac{\partial T}{\partial y}, \tag{12.4}$$

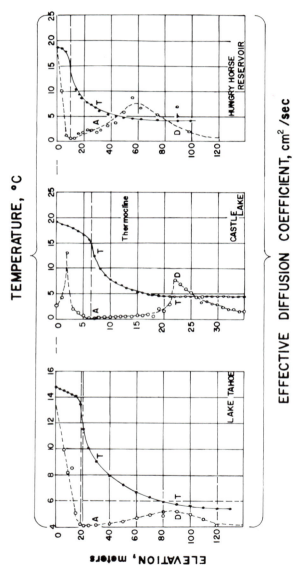

EFFECTIVE DIFFUSION COEFFICIENT, cm²/sec

Fig. 12.4.   Effective diffusion and temperature profiles for selected impoundments (Orlob and Selna, 1968).

where H is the heat flux, ρ is the water density, c_p is the heat capacity of water, T is the temperature, y is the axis of the vertical temperature gradient, and t is time.

Unlike uniform channel flow, in which the theoretical development of the magnitude of the vertical diffusion coefficient was possible using basic turbulent flow theory, an analytical or theoretical estimation of $D(y,t)$ for reservoirs is not presently possible.

During stratified flow conditions, three separate flow regimes exist. Therefore, one would expect that the level of mixing and diffusion in the three zones would also be different. Figure 12.4 shows the temperature and diffusion coefficient distribution for three reservoirs. Each of the examples presented corresponds to a summer period when the respective water bodies were well stratified. Although each particular distribution reflects the climatic and hydrologic characteristics of the reservoir, some common features can be detected. The following list contains the general properties of the vertical diffusion coefficient distribution (Orlob and Selna, 1968; Water Res. Eng., 1966).

1. In the epilimnion, the diffusion coefficient and effective mixing are greater near the surface but decrease rapidly with depth as the thermocline is approached.

2. The diffusion coefficient has its minimum value at or near the thermocline. Furthermore, the thermocline diffusion coefficient is at least an order of magnitude less than that at the surface.

3. The distribution of the diffusion coefficient in the hypolimnion is similar to that for pipe or open channel flow with minimum values near or at the bottom and near the thermocline, with the maximum value occurring near the midpoint of the hypolimnion.

It may be difficult to explain precisely why the diffusion coefficient behaves as shown. As illustrated in Fig. 12.5, the diffusion coefficient is a product of several energy inputs. In open channel flow, the vertical momentum diffusivity equation would be

$$\Sigma_m = \frac{\tau}{\rho}\frac{1}{\partial u/\partial y}, \tag{12.5}$$

where τ is the shear stress of the fluid, Σ_m is the turbulent momentum diffusivity, and $\partial u/\partial y$ is the vertical flow velocity gradient.

If Eq. (12.5) is solved as in Chapter 9, Eq. (9.60), the vertical diffusivity distribution is a curve with a maximum at the midpoint of the reservoir and zero values at the upper and lower boundaries of flow.

Under stratified flow conditions, a force acts on the momentum that is caused by turbulent mixing. That is, if two particles of different densities

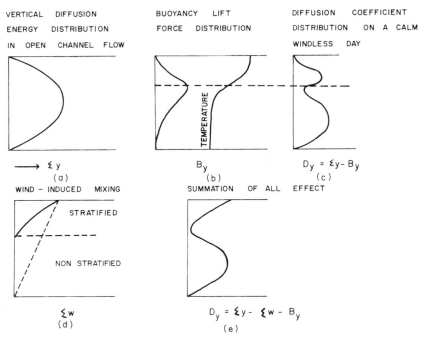

Fig. 12.5.   Theoretical distribution of diffusivities in a reservoir.

are displaced by turbulent pulsations, the buoyancy forces resulting from the different densities of the particles will have a tendency to return the particle with lower density above the particle with higher density. The lift force acting against the displacement of the two particles is proportional to the temperature gradient, as shown in Fig. 12.5. Finally, there is a water surface stress, τ_s, induced by the wind which may be described as

$$\tau_s \sim \rho_A \, W^3, \tag{12.6}$$

where ρ_A is the air density and W is the wind speed which causes an increase in surface diffusivity.

By summing up all these effects, the shape of the vertical diffusion distribution can be explained. However, actual distributions and values of the diffusion coefficient will depend on local conditions and factors characterizing the reservoir. Therefore, one would expect that the diffusion coefficient should be evaluated from field measurements of stratified reservoirs. However, as pointed out by Wunderlich (1968), it is very difficult to compute diffusivities from field data. Harleman and Huber (1968) found that neglecting the turbulent portion of the vertical diffusion coefficient and retaining only the molecular heat diffusivity ($D_t \approx 5.15$ cm²/hr at 20°C) would not lead to significant error since for all depths, excepting those near the surface, the convective transport prevails over diffusion. The values of the thermal diffusion coefficient at and below the thermocline suggested by Orlob and Selna (1968) were in the order of 10 to 100 cm²/hr, i.e., approximately ten times higher than that for molecular diffusion. As noted in Water Res. Eng. (1966), most of the numerical solutions employed for modeling reservoirs introduce errors which behave in the same fashion as an additional diffusive term. The magnitude of this pseudodiffusion was found to be in the order of ten times the molecular diffusion. Since the governing equations are not greatly sensitive to the diffusion term, an order of magnitude estimation is quite sufficient.

CONCEPTUAL REPRESENTATION OF A RESERVOIR

A deep reservoir or lake is characterized by the very slow, convective movement of water toward the reservoir outlet and great vertical water quality variations during the summer stratification period. In general, a three-dimensional representation might be considered necessary if a truly accurate theoretical model is required.

However, enough is known about the hydraulic and dynamic behavior of deep reservoirs to permit reasonable simplifications. As in the case of streams and estuaries, where one-dimensional representation with major variations along the main flow direction is adequate for most situations, a one-dimensional concept is also possible for reservoirs. In this case, the major variations in temperature and quality and their strongest gradients take place in the vertical direction.

As streams and estuaries were segmented along the x-axis, reservoirs will be segmented along the y-axis. The segmentation involves a separation of the reservoir volume into a set of horizontal control volumes or "slices" having the depth, Δy, and horizontal boundary areas, $A(y)$ and $A(y + \Delta y)$, as shown in Fig. 12.6. Each control volume accepts a portion of the tributary flow, q_{in}, and part of the downstream reservoir discharge, q_{out}, is withdrawn from the control volume at the downstream end. In addition, an average vertical flow, Q_v, passes through the slice.

The water conservation equation for each control volume then becomes

$$(q_{in} + Q_v)_i = \left(q_{out} + \left(Q_v + \frac{\partial Q_v}{\partial y}\right)(\Delta y)\right)_i. \qquad (12.7)$$

This equation reduces to

$$\left(\frac{\partial Q_v}{\partial y}\right)_i = \left(\frac{q_{in} - q_{out}}{\Delta y}\right)_i. \qquad (12.8)$$

Each slice is
characterized by
density and
temperature

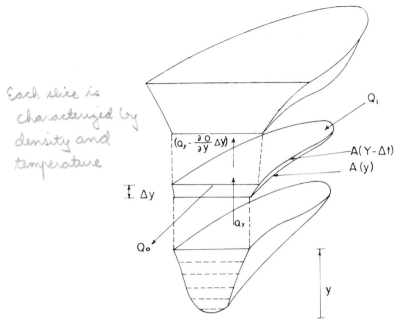

Fig. 12.6.   Conceptual representation of a reservoir.

In addition, the continuity equation for the entire reservoir must be satisfied as

$$\frac{\partial V}{\partial t} = \sum_{0}^{N} (q_{\text{in}} - q_{\text{out}})_i + P - E_v = Q_{\text{in}} - Q_{\text{out}} + P - E_v, \quad (12.9)$$

where V is the reservoir volume, P is the precipitation, E_v is evaporation, and N is the number of control volume slices.

It can be seen that at any section, i, the value of the vertical flow rate $(Q_v)_i$ can be essentially found by summing the inflows and outflows in all control volumes below the desired elevation. When the reservoir volume is determined, the water surface elevation can be estimated from a known volume–depth relationship for the reservoir. Because of very low flow velocities in the reservoir, the momentum equation can be neglected.

ENERGY BUDGET OF A RESERVOIR

A deep reservoir has only a limited capability to equalize its heat inputs. As the heat inputs increase, the surface and near surface layers become warmer and a nonuniform temperature profile develops as previously described. The thermocline can then be located in the zone where

$$\partial^2 T / \partial y^2 = 0, \tag{12.10}$$

which means that the thermocline is located in the zone containing the inflection point of the temperature profile.

The heat balance of a reservoir includes the following components, as shown in Fig. 12.3: solar (short-wave) radiation, reflected solar radiation, atmospheric (long-wave) radiation, reflected atmospheric radiation, back radiation of the water surface, conduction, evaporation, and precipitation. In addition to the surface heat (energy) inputs and outputs, the reservoir heat balance includes the influent heat content and the effluent heat loss.

All surface inputs with the exception of solar radiation are significant only on a relatively thin surface layer. Solar radiation can penetrate into deeper layers. A general heat balance equation for an ith slice below the surface can be written as

$$\frac{\partial T}{\partial t} V_i + \frac{\partial}{\partial y} (Q_v T) \, \Delta y = \frac{\partial}{\partial y} \left[A \, D_y \frac{T}{y} \right] \Delta y + q_{in} T_{in} - q_{out} T$$

$$- \frac{1}{\rho c_p} \frac{\partial (Q_s(y) A)}{\partial y} \Delta y, \tag{12.11}$$

where T is the average temperature in the slice, T_{in} is the influent temperature, $Q_s(y)$ is the solar radiation at elevation y, $V_i = A(\Delta y)$ is the volume of the slice, i, and A is the average mid-section horizontal cross-sectional area of the slice. It should be noted that Eq. (12.11) implies complete horizontal mixing of the water in the slice.

Figure 12.7 shows the ratio of the solar energy at depth $y' = y_s - y$ to the solar radiation at the surface, Q_{so}. The solar radiation can be separated into βQ_{so} absorbed at the surface and $Q_s(y) = (1 - \beta) Q_{so} e^{-\nu y'}$ penetrating into the depth y'. ν is the extinction coefficient for solar radiation which is a function of turbidity. From Fig. 12.7, the extinction coefficient for pure seawater would be about $\nu = 0.000245$ cm^{-1}, while that for the more turbid Fontana reservoir water was 0.0074 cm^{-1}.

Then for all layers except the surface

$$\frac{\partial Q_s(y)}{\partial y} = - \frac{(1 - \beta)}{\nu} Q_{so} e^{-\nu(y_s - y)}, \tag{12.12}$$

where y_s is the water surface elevation. Note that y' is measured from the surface down.

At the water surface layer, the heat diffused away from the layer should equal the absorbed solar radiation in the layer, Q_{so}, plus the heat inputs and outputs from the surface which include atmospheric radiation, Q_a,

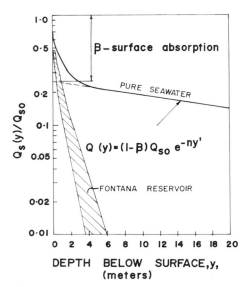

Fig. 12.7. Penetration of solar radiation into a water body as measured by Wunderlich and Gras (1967) and Elder and Wunderlich (1968).

conduction, Q_h, minus the heat emitted from the surface by back radiation, Q_b evaporation, Q_e, and reflected atmospheric radiation, Q_{ra}. This gives (Huber and Harleman, 1968)

$$\rho c_p D_y \frac{\partial T}{\partial y}\bigg|_{y=y_s} = Q_{so} + Q_a + Q_h - Q_{ra} - Q_b - Q_e. \qquad (12.13)$$

The magnitudes of the terms in Eq. (12.13) are discussed in Chapter 13. The bottom boundary condition is that there is no heat flux across the boundary or

$$\frac{\partial T}{\partial y}\bigg|_{y=y_b} = 0. \qquad (12.14)$$

VERTICAL FLOW DISTRIBUTION AND SELECTIVE WITHDRAWAL

Under nonstratified conditions, one could assume that the influent and effluent flows from a reservoir would be distributed either uniformly with depth or in accord with some basic velocity distribution formula. In stratified reservoirs and lakes, however, the influent water has a tendency to move in a layer which has a density similar to that of the influent. The water thus moves as a density current.

Three cases of density currents can be distinguished, as shown in Fig. 12.8. The overflow occurs when the incoming water is warmer or less dense than the surface layer water in the reservoir. The underflow occurs if the incoming water is colder or denser (e.g., sewage or seawater) than the water in the hypolimnion. The interflow occurs when incoming water

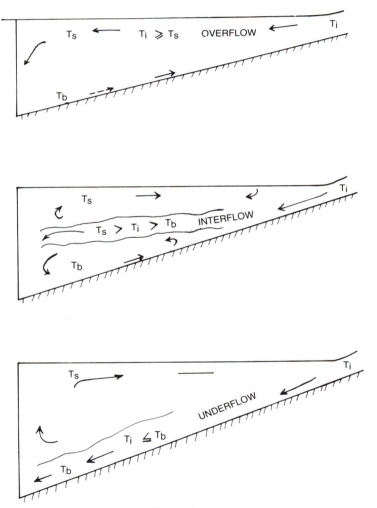

Fig. 12.8. Forms of density currents.

has a temperature or density which is within the range of the reservoir bottom and surface.

The occurrence of overflows, underflows, and interflows explains why, during floods, the high turbidity influent can pass through a reservoir much faster than one would expect in utilizing the theoretical volume to flow ratio.

Density currents, density flows, underflows, overflows, and interflows are all synonomous with stratified flow. As defined by the National Bureau of Standards (1938), "A density current is the movement, without loss of identity by turbulent mixing at the bounding surfaces, of a stream of fluid under, through, or over a body of fluid, with which it is miscible and the density of which varies from that of the current, the density difference being a function of the differences in temperature, salt content, and/or silt content of the two fluids."

Stratified flow has been observed when $\Delta\rho/\rho \geqq 0.005$, where ρ is the fluid density and $\Delta\rho$ is the density differential between the fluid layers (Task Comm. on Sed., 1963). It is interesting to note that this condition is satisfied with water temperatures of 31° and 32.5°C.

Overflows are typical for electric power cooling water discharges or warmer tributary or sewage effluents flowing into colder lakes or into the ocean. Heated discharges may cause stratification due to the overflow density currents in many streams and rivers where, under normal conditions, the cross-sectional temperature distribution would be fairly uniform.

The underflow has been discussed previously in this chapter in the section "Transport and Mixing in Reservoirs." It has been noted at many of the TVA reservoirs and has been observed at all of the mainstream reservoirs on the Tennessee River by Fry *et al.* (1953), who used chloride ion concentrations as a tracer. The concomitant low oxygen concentrations caused by hypolimnetic water releases from one reservoir into another were also confirmed.

Interflows are caused by the discharge of tributary water of an intermediate density into a stratified water body. An example of an interflow is shown in Figs. 12.9 and 12.10. On August 16, 1966, Rhodamine B dye was continuously injected into the flow of the Nantahala River, a tributary to Fontana Reservoir, North Carolina, for 24 hours. At the time of injection, the water temperature was 15°C. During the first phase of the dye's travel through the reservoir, the dye moved along the bottom of the reservoir, as shown in Figs. 12.9(a) and (b). Figures 12.9(c) and (d) show the dye cloud moving toward a level of similar density and Figs. 12.10(a)–(c) show the water traveling at the level selected. The 2.5°C temperature rise can be attributed to mixing and surface heating in the shallow, upper reaches of the reservoir.

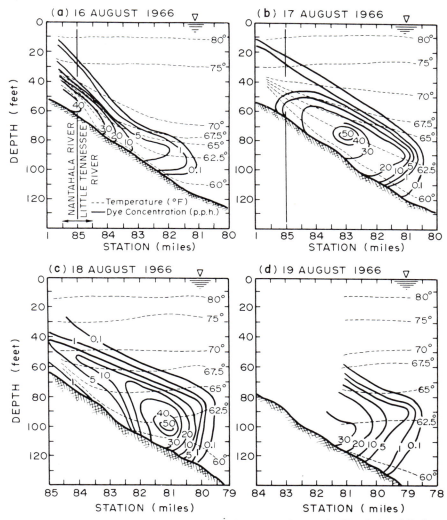

Fig. 12.9. Movement of dye cloud through Fontana Reservoir (Krenkel and Parker, 1969).

One should consider the question of what would happen if the tributary were polluted, or if this was a case of a sewage outfall discharging into a stratified lake. It can be seen clearly that under stratified conditions, the water moves as a relatively narrow density current. To detect such currents, the sampling network must be very detailed and in all depths throughout the vertical profile. The detention time of such a current in the reservoir is shorter than would be estimated from the volume-to-flow

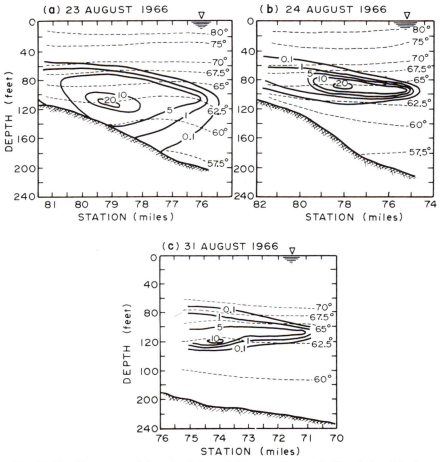

Fig. 12.10. Movement of dye cloud through Fontana Reservoir (Krenkel and Parker, 1969).

ratio. This implies that there must be portions of the reservoir with stagnant water or very little exchange. Obviously, a waste assimilative capacity estimation must consider all of these factors.

At the other end of a reservoir under stratified flow conditions, water is withdrawn from a layer around the reservoir outlet. Some early concepts assumed that the water is withdrawn from the slice which has the same elevation as the outlet and no withdrawal from any other layer. This was a subjective approach, relying on an arbitrarily chosen layer thickness, Δy, as being the thickness zone in which the reservoir water moves toward the sink.

Actually, water is extracted from a stratified reservoir in the form of a generally horizontal bounded layer. The phenomenon is called selective withdrawal and is shown in Fig. 12.11, as contrasted with flow from a non-stratified lake. Selective withdrawal is important to water quality control in reservoirs since zones of poorer water quality can be avoided by withdrawing water from a predetermined and relatively narrow layer. This can be accomplished by multiple intake design on dams.

The thickness of the withdrawal layer, δ, is a function of the distance from the sink and the density and diffusivity characteristics. The value of δ for moderate distances in turbulent flow has been given by Koh (1964), Koh (1966), and Brooks and Koh (1968) as

$$\frac{\delta}{a} = 8.4 \left(k_2 \frac{x}{a} \right)^{1/4}, \tag{12.15}$$

where $a = [q/(g\epsilon)^{1/2}]^{1/2}$, $k_2 = D_y/q \approx 0.001$, $\epsilon = 1/\rho \, d\rho/dy$, and x is the distance from the outlet, q is the discharge per unit width of the layer, g is gravity acceleration, and ρ is the density measured at the outlet layer.

For larger distances ($\delta/a > 13.7$),

$$\frac{\delta}{a} = 7.14 \left(k_2 \frac{x}{a} \right)^{1/4}. \tag{12.16}$$

Not all of the flow in the withdrawal layer will pass through the sink. The-

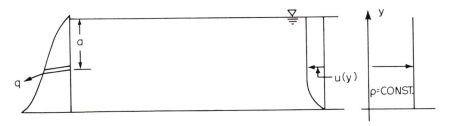

Withdrawal from full depth in a homogeneous fluid.

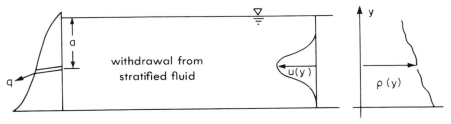

Fig. 12.11. The process of selective withdrawal (Elder, 1964).

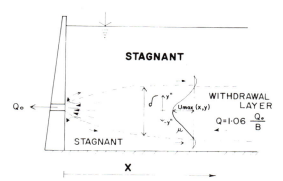

Fig. 12.12.   Selective withdrawal by a line source.

oretically, about 6% of the forward flow is reversed near the outlet to form
a slight reverse flow. Hence,

$$q \approx 1.06 \frac{Q_{out}}{B}.$$ (12.17)

Figure 12.12 is a schematic representation of the selective withdrawal
process by a line source.

To determine how the total withdrawal flow will be distributed to each
computational horizontal slice, it is necessary to estimate the velocity distribution within the withdrawal layer. Since the line sink representation
provides adequate results in most reservoir cases, the Koh (1964, 1966)
solution of the flow equations for a line sink in two-dimensional turbulent
flow may provide the best estimate. However, the available solutions are
only numerical; i.e., no single exact equation is available. An approximate formula of the type

$$u(x,y'') = u(x,0) \cos \left(0.9 \frac{y''}{\delta}\right) e(-0.32 \, y''/\delta)$$ (12.18)

can provide a fair estimate of the velocity distribution which is within
$\pm 3\%$ of the theoretical curve. y'' is a measure from the centerline of the
withdrawal layer, as shown in Figure 12.12. The maximum velocity was
estimated by Koh (1966) as

$$u(x,0) = u_{max}(x) \, 0.284 \frac{\left(\frac{\epsilon g}{D_y \nu}\right)^{1/6} q}{x^{1/3}}.$$ (12.19)

The withdrawal flow in each slice at a distance x from the outlet then becomes

$$q_{out}(y) = B(y) \int_{y-\Delta y/2}^{y+\Delta y/2} u(x,y) \, dy \qquad \text{for } y_0 - \delta/2 < y < y + \delta/2, \quad (12.20)$$

$$q_{out}(y) = 0 \qquad \qquad \text{for } y \geq y_0 + (\delta/2 + \Delta y)$$
$$y \leq y_0 - (\delta/2 + \Delta y),$$

and

$$q_{out} = -1.03 \, Q_{out} \qquad \text{for } y_0 + \delta/2 < y < y_0 + \delta/2 + \Delta y$$
$$y_0 - \delta/2 > y > y_0 - \delta/2 - \Delta y,$$

where $B(y)$ is the average width of the slice.

The distance x can be taken as a half-length of the horizontal centerline of the withdrawal layer between the outlet and the intersection with the reservoir bottom.

There is no scientific investigation available on the flow hydraulics and velocity distribution at the reservoir inlet. However, one can assume that a density current will be formed with the centerline elevation at the reservoir water density closest to that of the incoming water plus or minus the change in the inlet zone. The formulas describing the thickness of the density current probably will be similar to Eqs. (12.15) and (12.16), and, for the velocity distribution, Eq. (12.18) may provide a rough estimate. Then the influent flux into each slice also can be estimated, as shown by Eq. (12.20).

EFFECTS OF STRATIFIED FLOW ON WATER QUALITY

As previously stated, the dynamics of a reservoir under stratified conditions is due to unbalanced heat inputs and reservoir (lake) water mixing.

The effects of a reservoir on downstream water quality are profound and, in many instances, irreversible; i.e., downstream water quality will never be similar to that of the reservoir inflows. Many water quality parameters are affected by the distribution, diffusion, and transformation taking place within the reservoir. The classical concept assuming complete vertical mixing of substances is not applicable to stratified reservoirs or lakes. Any mass balance should include vertical variations as influenced by internal currents as well as the vertical distribution of sinks and sources of the substance (Markowsky and Harleman, 1973).

Dissolved Oxygen

The dissolved oxygen distribution in a reservoir is of primary importance since the oxygen levels determine the character of the biological and biochemical processes taking place at various depths and, in general, determine the overall ecological balance of the reservoir.

Water entering reservoirs contains organic materials. In addition, organic matter may be produced during the warm summer months by the photosynthetic activity of phytoplankton. The organic materials represent an oxygen demand as a result of biological synthesis and respiration. Most of the organic matter is produced in the epilimnion, which is usually fairly stabilized and aerated, receiving solar energy and being mixed by the wind. In the epilimnion, oxygen lost because of the biochemical processes is replaced by surface aeration and photosynthesis.

The major part of the dead microorganisms of the excess of organic particulate matter from the epilimnion sinks into the deeper hypolimnion. Oxygen removed by biological processes in the hypolimnion cannot be replaced because of the low mass transfer rates through the thermocline. In addition, the light penetration into the hypolimnion is minimal. Therefore, the algae which settle into the hypolimnion cannot produce photosynthetic oxygen and, instead, exert a BOD. The net result is a significant depletion of the oxygen resources in the hypolimnion. Figure 12.13 shows a typical dissolved oxygen distribution in a southeastern reservoir.

Because of the lack of mixing, organic wastes discharged into the lower layer are not allowed to benefit from that portion of flow in the upper layer which contains higher concentrations of dissolved oxygen. This results in less oxygen, less dilution, and a more concentrated organic load exertion on the natural waste assimilation and self-purification processes. Also, the rate of oxygen depletion in the lower layer may be increased as a result of the increased concentration of organic materials.

The reduced oxygen concentration and temperature in water withdrawn from the hypolimnion may have detrimental effects on downstream

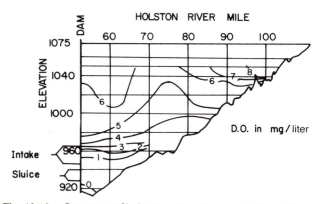

Fig. 12.13.   Oxygen profile in Cherokee Reservoir (Churchill, 1957).

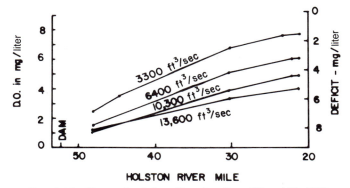

Fig. 12.14. Reaeration below Cherokee Dam (Churchill, 1957).

uses of water. The lower temperatures cause a reduction of BOD removal and decrease the reaeration rate. Consequently, reduced oxygen levels will result in an overall reduction of the downstream waste assimilative capacity (Krenkel *et al.*, 1969). Furthermore, such water may not be suitable for game fishing. Figure 12.14 shows the potential effects of a stratified reservoir on downstream oxygen concentrations. Note that with the existing standard of 5 mg/liter of DO to support a warm water fishery, over 30 miles of the downstream river do not satisfy the standard. Thus, the dam could be considered a large BOD source.

Iron and Manganese

If there is a low oxidation–reduction potential at the mud–water interface, conditions amenable to the dissolution of iron and manganese into the hypolimnetic waters will occur. The mechanism is not clear; however, if these metals are present in the bottom muds, troublesome concentrations may appear in the water under reduced environmental conditions.

It is interesting to note that 50 mg/liter of iron and 15 mg/liter of manganese have been found in the lower layers of the South Holston Reservoir. Since these concentrations do not significantly change with time, the lake can be classified as meromictic because of chemical concentrations and corresponding density increases.

It should be noted that many impounded waters circulate completely, but some, designated as holomictic lakes, circulate only partially. This stable, lower layer can be caused by an accumulation of dissolved or suspended solids in the water and may render this lower portion of the lake unsuitable as a water supply.

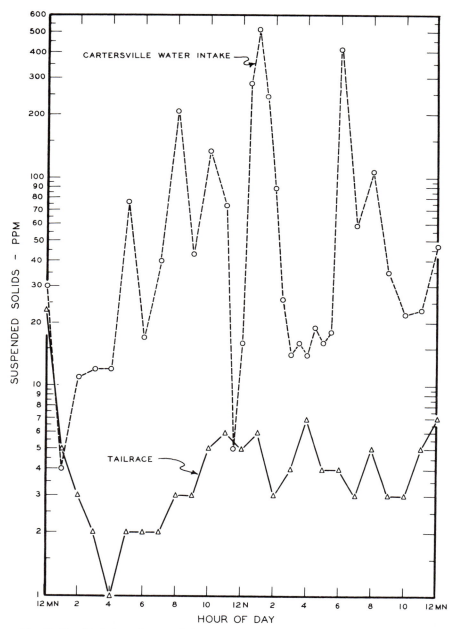

Fig. 12.15. Variation of suspended solids with time on August 20, 1964, at Allatoona tailrace and Cartersville intake (Krenkel, 1965).

Suspended Solids

Reservoirs act as large sedimentation basins and will cause a reduction in suspended solids concentrations in a previously free-flowing river, as shown in Fig. 12.15. The upper curve shows what might be expected in a free-flowing river, while the lower curve shows a relatively constant and low suspended solids concentration emanating from the dam. It should be noted, however, that the solids concentrations labeled as Cartersville Water Intake were subject to discharges from mining operations (Krenkel, 1965).

The sedimentation capabilities are of concern with regard to the life of a reservoir since reservoir siltation can be closely correlated to its efficiency and functioning. The lifetime or efficiency period of some reservoirs built on streams carrying large amounts of suspended and bed load sediments, typical for arid streams, can be reduced to a period of fifty years or less. Upstream settling basins have sometimes been constructed in order to relieve the sediment load and prolong the lifetime of the main basin. However, in several cases, these upstream sediment catch basins were filled in a few years and had to be cleaned or bypassed.

As previously indicated, density currents may significantly reduce the retention of water in the reservoir. By engineering control of the outlets, the amount of sediments deposited in the reservoir can often be significantly reduced.

Temperature

The temperature differential in stratified lakes can be quite significant. Since a major water use in the United States is for cooling purposes, it is obvious that this cooler water is highly desirable for condenser cooling for steam-electrical generation purposes. In fact, the use of this colder water by construction of an underwater dam at the TVA Kingston Steam Plant was reported to have saved the TVA $155,000 in operating costs for the year 1956 alone, which was one-third the cost of the dam (Elder and Dougherty, 1956).

This then raises questions regarding the desirability of artificially over-turning a reservoir. While the stratification may produce a poorer quality water, it may be less expensive to treat this water than to lose the advantage of the colder cooling water available to the steam electrical generating plants. Obviously, what is needed is a thorough systems analysis of these alternatives.

The question posed is "Do we want high dissolved oxygen and high temperature or low dissolved oxygen and low temperature?"

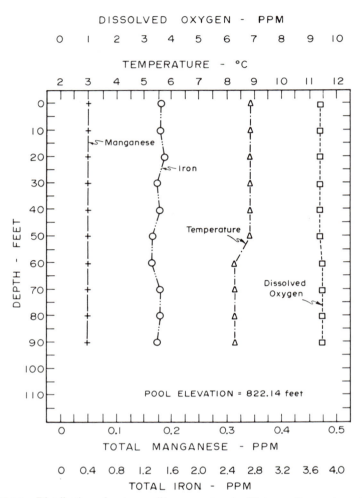

Fig. 12.16.   Distribution of water quality parameters in Allatoona Reservoir on January 15, 1965 (Krenkel, 1965).

Figures 12.16 and 12.17 demonstrate the conditions existing in Allatoona Reservoir, near Atlanta, Georgia, during completely mixed conditions and under a stratified flow regime. Note that in the summer, the temperature varies from over 27° C at the surface to approximately 17° C at the reservoir bottom. Difficulties encountered from the low oxygen and high iron and manganese concentrations are obvious, inasmuch as the water intake is from the lower levels.

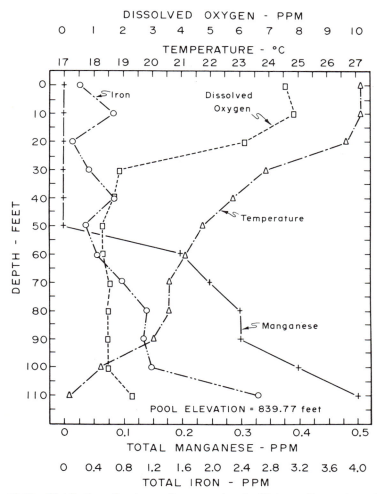

Fig. 12.17. Distribution of water quality parameters in Allatoona Reservoir on August 28, 1964 (Krenkel, 1965).

Note also the variation in dissolved oxygen at the tailrace and four miles downstream from the dam, as shown in Fig. 12.18. The variation at 1000 hours is the result of a peaking power operation. It is instructive to note the rapid rate of oxygen absorbtion as compared to that shown in Fig. 12.14. This comparison lucidly demonstrates the effects of channel geometry and conditions on the reaeration rate.

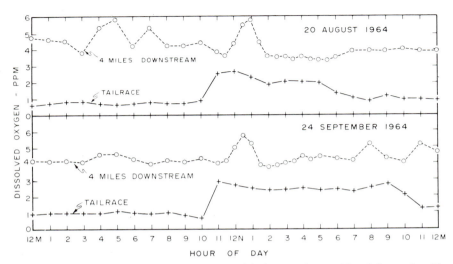

Fig. 12.18. Variation of dissolved oxygen with time on August 20 and September 20, 1964, at Allatoona Dam tailrace and Cartersville intake (four miles downstream) (Krenkel, 1965).

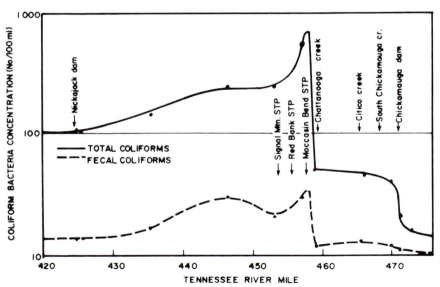

Fig. 12.19. Observed logarithmic mean coliform concentration in Nickajack Reservoir, January to December 1973 (TVA 1976).

Bacteria

Because of its sedimentation basin characteristics, it is obvious that a reservoir will cause a reduction in the bacterial count. This has been studied extensively in the TVA system, the results showing significant declines in coliform bacterial subsequent to their introduction into a reservoir. Figure 12.19 shows the reduction occurring in both fecal and total coliform bacteria for the portion of the Tennessee River between Chickamauga and Nickajack Dams. Note that Chickamauga Dam discharges into the backwaters of Nickajack Dam and that the waters are affected by both industrial and municipal wastewater discharges.

WATER QUALITY MODELING IN LAKES

Mass Conservation Equation

As with the temperature model, the mass balance equation for a one-dimensional horizontal slice can be written as

$$\frac{\partial C}{\partial t} \; v_i \frac{\partial}{\partial y} (Q_v C) \, \Delta y$$

$$= \frac{\partial}{\partial y} \left[A D_y \frac{\partial C}{\partial y} \right] \Delta y + q_{in} C_{in} - q_{out} C + \Sigma S \; V_i, \quad (12.21)$$

where C is the concentration of the substance, C_{in} is the influent concentration to the ith slice, and ΣS represents the sum of sinks and sources of the substance. For a conservative substance, $\Sigma S \to 0$.

Dissolved Oxygen Model

In the dissolved oxygen model, C in Eq. (12.21) represents the dissolved oxygen concentration. The DO sources and sinks are taken as

$$\Sigma S = (P - R) - K_d L - K_N L_N \qquad (12.22)$$

for all except boundary slices.

Both the oxygen production by photosythesis, P, and the respiration oxygen demand, R, are functions of y, time, and temperature.

It is evident that Eq. (12.22) must be coupled with similar equations for the carbonaceous BOD, L, and the nitrogenous oxygen demand, L_N. If

Eq. (12.22) is written for BOD, then the term ΣS would simply be

$$\Sigma S = -K_r L$$

and similarly for NOD

$$\Sigma S = -K_N L_N$$

or

$$\Sigma S = -K'_K,$$

where all the coefficients are temperature dependent. K_N is a coefficient for first-order nitrification, while K'_N is that for a zero-order reaction.

The magnitudes of K_r, K_d, K_N, and K'_N have been discussed in Chapter 10. It should be noted that because of the large volume, the effect of external factors on the magnitude of the coefficients, K_r, K_d, K_N, or K'_N is negligible. Therefore, these coefficients can be approximately of the same order as those measured in a long-term laboratory BOD test with reservoir water. The model does not include the organic matter produced by photosynthetic processes of the phytoplankton nor the effect of settling of the particulate matter from the epilimnion into the hypolimnion. These functions are a part of the process called eutrophication and their general magnitude is discussed in Chapter 13.

Similarly, the function $(P-R)$, which represents the net gain of oxygen from the photosynthetic processes is very difficult to estimate. Some possible ways of representing the $(P-R)$ term have been shown in Chapter 11. Since solar radiation is one of the major inputs to a reservoir model, the photosynthetic oxygen term, P, can be related to the incoming solar radiation at various reservoir levels and the variation of photosynthesis with depth can be established. As shown by Bella (1970), the hypolimnion dissolved oxygen concentration is not excessively sensitive, even to large variations of the P term. Therefore, an approximate estimation of P may be quite sufficient. The hypolimnion values of DO are much more sensitive to the respiration term, R. Thus increased productivity and BOD input to stratified reservoirs may have a profound impact on the DO concentration in the hypolimnion and the lower reservoir outlet.

Boundary Conditions

At the reservoir surface some oxygen is supplied by the atmospheric reaeration which should be included in ΣS. Hence,

$$\frac{\partial}{\partial y}\left[AD_y \frac{\partial C}{\partial y}\right]\Delta y \bigg|_{y=y_s} = AK_L (C_s - C) \qquad (12.23)$$

for the dissolved oxygen balance, and

$$D_y \frac{\partial L}{\partial y} = D_y \frac{\partial L_N}{\partial y} = 0 \qquad (12.24)$$

for the BOD and NOD balances.

At the reservoir bottom, the boundary conditions may be more complicated. The benthic sediments, which consist of organic residues plus deposited suspended solids from tributaries, may be active and exert a significant oxygen demand. In addition, ammonia may be released from decomposing organic matter, thus increasing the nitrogenous oxygen demand. Therefore, the bottom boundary conditions may be written as

$$\frac{\partial}{\partial y}\left[AD_y \frac{\partial C}{\partial y}\right]\Bigg|_{y=y_b} = -S_B \qquad (12.25)$$

and

$$\frac{\partial}{\partial y}\left[AD_y \frac{\partial L_N}{\partial y}\right]\Bigg|_{y=y_b} = S_N, \qquad (12.26)$$

assuming that

$$\frac{\partial L}{\partial y}\Bigg|_{y=y_b} = 0.$$

The magnitudes of the benthal oxygen demand, S_b, or the benthal nitrogenous oxygen demand contribution, S_N, can be determined by calibrating the model against actual measured data.

The value of the reaeration coefficient, K_L, at the water surface should not be determined from traditional formulas developed for channel flow. Because of large depth and irregular flow patterns in two separate layers, use of the common hydraulic formulas is, at least, suspect.

It should be emphasized that for gases with low solubility, such as oxygen, the rate of diffusion through the surface water boundary layer is the controlling factor. Therefore, under normal turbulent open channel flow conditions, the effect of wind is negligible. In quiescent water bodies, the rate of oxygen supply is still controlled by diffusion through the surface water boundary layer and not through the air boundary layer as is the case with heat transfer. However, under quiescent conditions with interflow or underflow density currents, atmospheric reaeration by turbulent motion may drop practically to zero.

The wind effect and subsequent formation of surface waves may substantially increase mixing and diffusion through the surface water boundary layer. The wind action causes the shear stress at the water surface, τ_s, to have a definite value rather than zero as would be assumed from open channel flow theory. Furthermore, when wind blows over the water surface, a drift current is induced in addition to wave formation. The waves also increase the surface roughness, which, in turn, gives rise to a stronger shear stress.

Compared to the amount of research dealing with stream reaeration, there are only a few studies on the increase in aeration by wind. There is little knowledge of the effects of wind on the reaeration of stratified reservoirs, since almost all investigations have been conducted with vertically mixed laboratory basins. From available research data, it can be deduced that wind action increases the rate of diffusion through the water surface layer rather than the rate of surface layer renewal, which is affected by the turbulence in the main water body (Eloubaidy and Plate, 1972).

The wind effect on the reaeration coefficient can be expressed as (Mattingly, 1977)

$$\frac{K_L{}^w}{K_L{}^0} = 1 + 0.2395 \ U_w^{1.646}, \tag{12.27}$$

where $K_L{}^w$ is the surface aeration coefficient under windy conditions, $K_L{}^0$ is the surface coefficient by channel turbulence under quiescent conditions, and U_w is the wind velocity in meters/second. Although the preceding equation yielded very high correlation coefficients under laboratory conditions, its practical use may be limited because of uncertainties associated with determining the channel flow reaeration coefficients, $K_L{}^0$, for reservoirs. Mattingly (1977) used a channel reaeration formula proposed by Bennett and Rathbun (see Chapter 10) which was reported as

$$K_L{}^0 = 0.101 \ \frac{U^{0.607}}{H^{0.684}}, \tag{12.28}$$

where U is the average flow velocity in meters/second and H is the average depth in meters. As previously stated, the use of such equations is suspect for stratified flow conditions. A more useful equation would be the one which correlates turbulent atmospheric reaeration to the hydraulic conditions in the epilimnion. (As a better approximation, H may be considered as the thermocline depth below the surface and U as the average velocity in the epilimnion caused by both wind and back currents rather than the average flow conditions in the entire reservoir.) The phenomenon of wind effects on the reaeration of reservoirs needs further research. The reader is referred to Chapter 10 for additional information on wind-induced reaeration.

RESERVOIR WATER QUALITY PREDICTIONS

Water quality prediction for a reservoir cannot be equated to a simple mass balance. In most cases, the resulting water quality and its distribution are the result of biological and biochemical processes taking place in the reservoir itself rather than a response to external inputs. The dissolved oxygen levels, nutrient and organic matter concentrations, and algal photosynthesis are coupled together, and all are related to the productivity of the reservoir. After reaching a certain state of eutrophication, nutrient levels in reservoirs appear to be less dependent on external nutrient loading. This is probably because nutrients can be released from bottom sediments and/or the nitrogen fixing capabilities of certain algae.

Water quality of reservoirs is both a temporal and spatial phenomenon. The development of planktonic microorganisms, which depends to a certain degree on the nutrient levels in the epilimnion, can be reflected in chemical analyses such as BOD or COD. This phenomenon can cause a substantial apparent increase of BOD and COD levels in the upper layers during the summer months. BOD levels then can reach values exceeding 5–10 mg/liter, even when there is no significant wastewater input.

As we will discuss in Chapter 13, the water quality of a reservoir depends on its trophic state, which can be related to the net rate of production of organic matter in the reservoir. As one would assume, more organic matter is produced in reservoirs and lakes located in warm climates where the waters are fed by nutrient rich lowland tributaries than in lakes in cold climates and higher elevations where tributary waters originate from melted snow and glaciers.

The following factors should be considered as affecting the overall water quality of lakes:

1. *Latitude.* Latitude affects the average temperature in the reservoir and thus the overall rate of biochemical reactions. Lakes in higher and colder latitudes will generally have better water quality than those located in the warmer, lower latitudes. Obviously, the process of degradation in the colder climates will be inhibited because of temperature effects.

2. *Climatic conditions.* Similarly, climatic conditions can affect the average temperature and, in addition, the amount of solar radiation input, reservoir cooling, and nutrient input. Reservoirs with an abundance of sunlight will have poorer quality than those with more cloudy days.

3. *Elevation.* An increase in elevation has the same effect as an increase in latitude. Furthermore, mountain lakes are primarily fed by nutrient poor tributaries and, as a result, usually have lower nutrient and tro-

phic levels. The amount of sunlight also may be limited by surrounding mountains.

4. *Shape and exposure of the inundated volume.* Reservoirs with a lower surface area–volume ratio may demonstrate lower rates of net organic matter production. On the other hand, lower oxygen inputs through the surface aeration of such reservoirs may cause poorer oxygen conditions in the hypolimnion. Shallow, well-mixed reservoirs may have higher oxygen levels but their overall water quality as expressed by other indicators such as turbidity, organic matter content, and taste and odor may be much worse. Thus, shallow lakes may be less suitable for most beneficial uses than deep reservoirs. Also, weed problems are common in shallow lakes, even in colder climatic conditions.

Many reservoirs constructed on larger rivers may have a cross-section consisting of a main, deep navigable channel and shallow, inundated flood plains. The water quality may then differ in the two sections. Most of the flow will be concentrated in the main channel area, leaving the flood plains stagnant and exposed to the sun. Severe water quality problems may be encountered in the shallow zones, such problems may include excessive weed growths and lower oxygen levels during the early morning hours, often resulting in fish kills and deteriorated water quality.

5. *Amount of nutrients in tributary flow.* With the exception of small and/or short detention time reservoirs, the amount of organic matter in the reservoir, which is the primary driving force for the DO levels, depends on the nutrient budget of the reservoir. Thus, in many instances, the amount of nutrients in the tributary flow may be more important than the amount of organics. As previously noted, nutrient contributions are generally higher in lowland rivers than in mountain snow or glacier fed streams. Nutrient levels can be increased by various cultural activities, including agriculture, sewage and industrial wastewater discharges, and erosion.

6. *Retention time of the reservoir.* The theoretical detention time expressed as the volume to flow ratio has no true relevance in stratified reservoirs because of complex flow and current conditions. However, detention time may be correlated roughly to the overall reservoir quality and to the recovery time if some water quality control measures are implemented. Generally, reservoirs with retention times in days or a few weeks do not provide enough time for the full development of the troublesome autotrophic planktonic microorganisms. A longer period of quiescent conditions on the order of 10 to 14 days is necessary before algal blooms can develop. Reservoirs with shorter detention times are more sensitive to a sudden water quality change inasmuch as their water quality is primarily a function of the pollution levels in their tributaries.

If the hydraulic detention time can be roughly related to the residence time of pollutants in the lake or reservoir, a time required to reach 50% of the expected change due to a water quality control measure is 0.69 DT, where DT is the detention time. Also, 95% of the expected change will be reached in a period equaling three residence times.

7. *Reservoir use and management.* Since the currents and the density flows in a reservoir can be affected by the withdrawal elevation, many water quality parameters can be partially controlled. The withdrawal of water from the epilimnion during summer stagnation can prevent downstream septic conditions and low oxygen levels. In addition, epilimnetic withdrawals will also result in lower concentrations of iron and manganese and the absence of odor problems caused by hydrogen sulfide. It should be noted, however, that most dams do not have provisions for selective withdrawal.

Other factors which can locally affect water quality include navigation and boat traffic, residential shoreline density, peaking power generation, and recreation intensity.

8. *Preimpoundment conditions.* Organic soil and forested lands contain large amounts of nutrients and, possibly, iron and manganese oxides. Under the reduced conditions existing during summer stratification, these can be released from the bottom layer into solution and reach the productive epilimnion (euphotic) zones. Although a complete stripping of vegetation and top soil may be economically prohibitive, some engineering measures must be undertaken to remove excessive amounts of nutrients from the inundated areas. The nutrients and organics in place just before the flooding should be included in the overall nutrient balance of the reservoir. It is interesting to note that the TVA Fisheries Division, against the advice of water quality and safety specialists, has insisted on leaving trees in portions of newly constructed reservoirs because the trees allegedly act as "fish attractors."

In spite of all of the modeling techniques available to engineers and scientists involved in reservoir and lake management, reliable water quality prediction in lakes is still an art requiring experience and practical knowledge rather than the simple use of a model. Therefore, it is imperative that the foregoing factors be evaluated and judged prior to making any conclusions from a model.

CONTROL OF HYPOLIMNETIC WATER QUALITY

Cold water stored in the hypolimnion of deep reservoirs and lakes is preferable, for many uses, to water from the epilimnion. In contrast to epilimnetic water, the cold water is not subjected to diurnal variations of temperature and quality, its lower temperatures provide more efficient cooling water, and it contains lower concentrations of planktonic microorganisms and may have, under proper management, fewer odor and turbidity problems. On the other hand, the lower oxygen concentrations and consequent reducing processes occurring in the hypolimnion may often be harmful to many uses, including water supply and aquatic life propagation.

Because most reservoirs do not have provisions for the withdrawal of water from different lake levels, investigations as to the feasibility of "treating the lake" have been studied for several years. According to Symons *et al.* (1968), "Control of the quality of water within a reservoir, rather than attempting to treat and improve the quality of the water as it is withdrawn from the reservoir, prevents downstream water quality problems and assures the usefulness of the entire water volume in the reservoir at all seasons of the year."

The concept of destratification has been controversial, however. For example, the advantage of the colder temperature for cooling water purposes would be lost, changes in the water temperature could affect downstream fisheries, and the effects on the reservoir ecology are not well elucidated. What is really needed is a comprehensive economic study on the costs and benefits of mixing a reservoir as opposed to leaving it stratified.

A summary of the extensive investigations on destratification made by the precursors to the EPA was complied by Symons (1969). Although investigations were limited to small lakes, the information presented on field experiences and water quality changes induced by destratification is instructive.

Control of hypolimnetic water quality could have the following purposes:

1. Improve the oxygen levels to prevent reducing conditions in the lower layers of the reservoir. Aerobic hypolimnion conditions reduce iron and manganese levels, prevent or reduce the release of nutrients from benthic layers, prevent the formation of hydrogen sulfide, and contribute to maintaining the top benthic layer aerobic.

2. Supply oxygen to tailrace water or a water supply intake.

3. Reduce taste and odor problems of withdrawn hypolimnetic water.

4. Equalize some water quality parameters throughout the depth of the reservoir or throughout the hypolimnion.

5. Maintain cold water fisheries in lakes by means of hypolimnic aeration of lakes and reservoirs located on cold water streams.

6. Control downstream water quality and temperature, especially where water from one reservoir is discharged into another.

7. Control eutrophication and excessive algal growths.

Many lakes require winter aeration as well. During the winter months the oxygen penetration through the ice-covered water surface is virtually zero, and despite greater saturation levels of oxygen and reduced deoxygenation rates, DO depletion occurs in many ice-covered lakes. Winter fish kills are common in shallow productive lakes because of high BOD, active mud layers, and reduced photosynthesis.

Fish are known to respond to DO levels. Although some fish prefer deeper, colder water during warm summer periods, the depth distribution of fish can be restricted by low DO concentrations or anoxia in the hypolimnion (Gebhart and Summerfelt, 1976). Most fish species will be limited to the oxygenated epilimnion and thermocline when the lake is stratified. Destratification or hypolimnion aeration can increase the amount of suitable space available for fish. After destratification and the subsequent increase of DO in the hypolimnion, most fish will move to the deeper and colder depths. Destratification can also increase the area and volume of the habitat for warm-water fish (Gebhart and Summerfelt, 1976).

In reservoir water quality control, it is necessary to distinguish between reservoir destratification and hypolimnetic aeration. Both processes are similar, but in destratification, emphasis is put on a local or even complete break of the thermocline and on equalizing the vertical gradient of water quality parameters. Hypolimnetic aeration refers to the process of adding dissolved oxygen to the hypolimnion only, while maintaining thermal stratification. The latter process requires less energy.

Destratification is also effective in controlling eutrophication since colder hypolimnetic water is mixed with water in the epilimnion and the effect of the colder water in combination with increased turbulent mixing can slow down or prevent excessive algal blooms. Both hypolimnetic aeration and destratification can be accomplished by pumping water or by injecting air (pure oxygen) or by a combination of both.

From a water quality standpoint, limited hypolimnetic aeration is preferred. It requires less energy, and the nutrients from the sedimentation zone are not transported to the productive epilimnion. If sufficient dissolved oxygen levels are maintained, better quality and colder hypolimnetic water can be used during the warm summer months.

Reservoir Destratification

If complete destratification is required, the energy input to the destratification device (mechanical or air injection) must be greater than the so-called *stability* of the reservoir, as given by the weight of the water between the centers of gravity of the reservoir water volumes under isothermal (uniform temperature) conditions and under stratified summer or winter conditions (Symons *et al.*, 1967). The energy input, E, in kilowatt hours necessary for complete destratification is then

$$E(\text{kW h}) = 2.777 \times 10^{-7} \, g\rho V_{y_1-y_2} = 2.724 \times 10^{-6} M_{y_1-y_2}, \quad (12.29)$$

where ρ is the reference density, $V_{y_1-y_2}$ and $M_{y_1-y_2}$ are, respectively, the volume and mass of water between the isothermal and stratified centers of gravity of the reservoir in cubic meters and kilograms.

The solution shown in Fig. 12.20 utilizes the stage-volume and stage-density relationships. Under nonstratified conditions, the center of gravity of the reservoir water volume, C.G.$_1$, is located at an elevation, y_1, which divides the stage-volume curve in half. Under stratified conditions the center of gravity of the water mass can be obtained from a stage-mass curve which is a graphical or numerical integration of the mass expressed as

$$M(y) = \int_{y_b}^{y} \frac{dV}{dy} \, \rho g \, dy = \Sigma \, A(y) \, \Delta y \, \rho \, g. \quad (12.30)$$

Then the center of gravity under stratified conditions can be located at the elevation of one-half of the total mass of water in the reservoir. It should

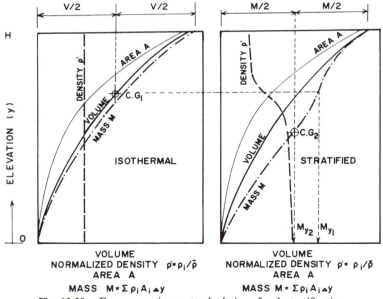

Fig. 12.20.   Energy requirement calculations for destratification.

be noted that because of the stratification, the center of gravity moves down as the lighter water remains near the surface and the heavier water stays near the bottom.

Because the calculated stability of a reservoir computed from the stage-density and stage-mass curves is the theoretical minimum amount of energy required to mix the lake, a destratification efficiency parameter, *DE*, relating the efficiency of various equipment is defined as (Symons, 1969)

$$DE = \frac{\text{New change of stability from } t_1 \text{ to } t_2}{\text{Total energy input from } t_1 \text{ to } t_2} \times 100, \qquad (12.31)$$

where t_1 and t_2 are the times of starting and stopping the mixing. The power capacity of a destratification device in kilowatts then should be

$$P \text{ (kW)} \sim \frac{\max E}{t_{max} - t_0} \frac{100}{DE}, \qquad (12.32)$$

where t_{max} and t_0 are in hours. t_0 is the time of spring (fall) turnover and t_{max} is the time of the maximum stability at the end of summer, as shown in Fig. 12.21.

Similarly, if the primary purpose of destratification is to supply or restore oxygen to the hypolimnion, the oxygenation capacity, *OC*, of a device is defined as

$$OC = \frac{\text{Net change in oxygen balance from } t_1 \text{ to } t_2}{\text{Total energy input from } t_1 \text{ to } t_2}. \qquad (12.33)$$

The units of this parameter are kilograms of O_2/kW h.

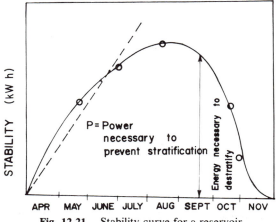

Fig. 12.21. Stability curve for a reservoir.

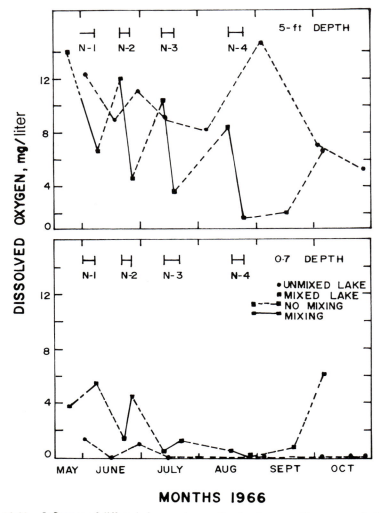

Fig. 12.22.   Influence of diffused-air pumping on dissolved oxygen (Symons *et al.*, 1968).

The equipment used for destratification has included pumps, air injection compressors, pure oxygen injection systems, and combinations thereof.

Localized destratification, which achieves vertical mixing of water only in the vicinity of an outfall or a water intake, is used primarily for the warming of tailrace water or for increasing DO levels in downstream portions of a stream. It does not alleviate the odor or taste problem, since the

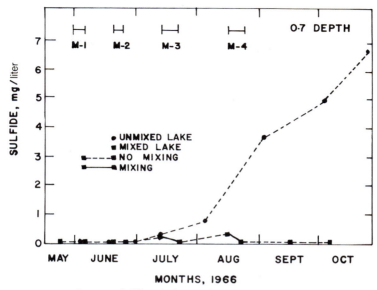

Fig. 12.23. Influence of diffused-air pumping on sulfide (Symons *et al.*, 1968).

resulting water quality may be simply a mixture of poor, oxygen devoid hypolimnetic water with organic rich epilimnion water.

Figures 12.22 and 12.23 show the effects of intermittent diffused air pumping on DO and sulfide content of a stratified lake with a volume of about 3000 acre-ft, as presented by Symons *et al.* (1968). A lake with similar characteristics was used for comparison and is labeled "unmixed."

Hypolimnion Aeration

The advantages of hypolimnetic aeration are obvious. Besides the energy savings, other benefits make it preferable to complete or local destratification. It maintains a cold hypolimnion while reducing the deleterious effects of low oxygen concentrations. Technically and economically feasible hypolimnion aeration systems were developed in the early 1960s in Germany on the Wahnbach Reservoir (Bernhardt, 1974). The basic principle of one such device is shown in Fig. 12.24. It consists of two vertical pipes fixed to an anchored buoy and connected by a gutter pipe which is only partly submerged in the water surface. The air is blown into the longer of the two vertical pipes, and, since the density of the aerated water is lower, it rises to the gutter pipe by means of buoyancy. The water

air

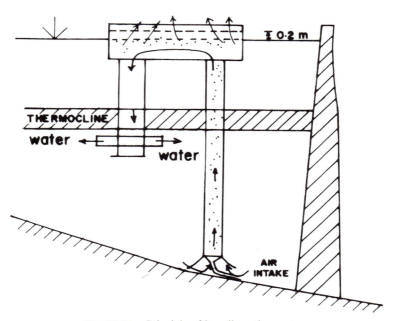

Fig. 12.24.   Principle of hypolimnetic aerator.

degases in the gutter and reeneters the lake below the thermocline through three horizontal distribution pipes. The reported oxygenation efficiency is 1.1 kg of O_2/kW h.

Such systems in which water rises all the way to the water surface are called full air lift aeration. The system generally reduces dissolved iron, manganese, phosphate, and sulfide concentrations while maintaining cold aerated water in the hypolimnion.

Similar full air lift hypolimnetic aerators were used to aerate lakes Tullingesjön Jäilasjön in Sweden (Fast *et al.*, 1976).

Partial air lift systems are similar; i.e., hypolimnetic water is aerated and circulated by an air injection device but the air–water mixture does not upwell to the lake surface, instead, the water and air separate below the thermocline. A most popular system of hypolimetic aeration of this type is shown in Fig. 12.25. The reported oxygenation efficiency of this system is 1.2 kg of O_2/kW h. Figure 12.26 shows the effect of aeration on the water and temperature profile during hypolimnetic aeration.

Some systems use pure oxygen instead of air to prevent saturation of water by gaseous nitrogen which might be harmful to fish and aquatic biota (Rayyan and Speece, 1976).

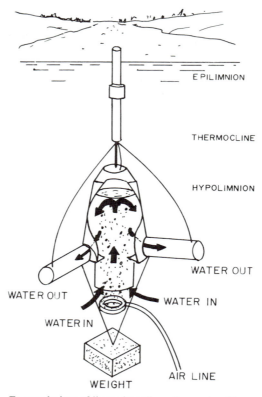

Fig. 12.25. Exposed view of limno hypolimnetic aerator (Fast *et al.*, 1976).

The length of the aeration period depends on the thermal input, the eutrophic conditions of the reservoir, and the reservoir geometry. Lakes and reservoirs in southern, warm climates may require up to nine months of continuous aeration during the warm part of the year, while northern lakes and reservoirs may require only several months of aeration during the summer and another aeration period during winter icy conditions.

An excellent discussion of the various methods of aeration is presented by King (1970). A summary of his comparisons are given in Table 12.1. Table 12.2 is presented in order to allow an estimation of the costs of the systems proposed (Bernhardt, 1969). (Note that the costs must be converted to current dollars.)

Finally, the work at Fort Patrick Henry Dam should be mentioned. The dam impounds a reservoir about 75 feet deep and, on the average, has DO concentrations below 5 mg/liter 103 days per year and below 3 mg/liter 12 days per year. The average unregulated flow at the dam site is 2522

TABLE 12.1 Overall comparison of reaeration methods[a]

Device	Efficiency	Advantages	Disadvantages
Diffusers (including reservoir mixing)	2.8% (est. 1–3 lb/kw h)	Many types available	Clogging. Requires filtering of air. Relatively low transfer efficiency. High first cost
Mechanical aerators	2–5 lb/kw h	Many types available. Natural or forced-air admission. Valves, piping, blowers, etc., are not required	Swimming hazard. Obstructs channel. High first cost
Turbine injection	2–6 lb/kw h (with admission by natural draft)	Natural or forced-air admission. Large discharges. Low first cost, low maintenance cost	Possible power loss and adverse effects on turbine performance. Restricted to sites with hydroplants
Venturi tubes	2–3 lb/kw h	Natural or forced-air admission	Head loss. Small discharges
Cascades	Determined from empirical formula	Maintenance free. Often associated with energy dissipation. Large discharges. No auxiliary equipment required	Head loss may not be allowable. High first cost. Low transfer efficiency

Method	Rate/Efficiency	Characteristics	Remarks
U-tubes	1–5 lb/kw h	Natural or forced air admission. Low maintenance cost. High surface transfer, longer contact time, increased deficit due to pressurization. Can be used in channel	Not yet proven for large discharges. Possible plugging with debris or ice
Hydraulic guns	≈ 2 lb/kw h	Efficient mixing, high surface currents, surface boil	Primarily for mixing. Large bubble size, low transfer efficiency. Small volumes
Pressure injection	No data available		
Fixed-cone valves	No power use or loss is involved	Associated with required energy dissipation. Large volumes. No auxiliary equipment required	Requires reservoir releases, which may not be possible
Mechanical pump mixing	< 1 lb/kw h	Simple equipment. Minimal maintenance	Primarily for mixing. Low transfer efficiency. Verified only for relatively small volumes
Hypolimnion reaeration and mixing	> 2 lb/kw h	Allows stratification to remain undisturbed	Relatively low transfer efficiency
Molecular oxygen	14–55%	Can be used with various types of contact devices. High transfer efficiency. No dissolved nitrogen	High cost of molecular oxygen

[a]From King (1970).

503

TABLE 12.2 Costs of applied aeration systems[a]

	Cascade (weir)	Air hoses (diffusers) Baldeney	Air pipes (diffusers) Lippe	Reservoir circulation	U-tube aerator	Turbine injection system Baldeney	Floating aerator	Turbo oxyder
Oxygenation capacity[b] (kg/hr)	0.88	28	22	16	4.1	120	41	13
Capital costs (DM)[c] (yr of construction)	5,000 (1965)	66,000 (1965)	110,000 (1965)	15,000 (1964)	16,000 (1958)	34,000 (1967)	95,000 (1965)	44,000 (1967)
Relative capital costs[b] (DM/kg O$_2$/hr)	5,700	2,360	5,000	940	4,000	283	2,300	3,280
Energy consumption[b] (kw h/kg O$_2$)	0.85 (gravity flow)	6.60[e]	3.45	2.40	0.67 (gravity flow)	0.98	1.52	2.0
Total costs[d] (DM/kg O$_2$)	0.87 (gravity flow)	0.99	1.05	0.37	0.62 (gravity flow)	0.14	0.51	0.65

[a]From Bernhardt (1969).
[b]With 50% DO deficit.
[c]4DM ≈ $1.00.
[d]With 50% DO deficit, operating 30 days per year, 10-year life, and 0.10 DM/kw h.
[e]Due to local conditions compressed air with 4.5 kg/cm^2 is used.

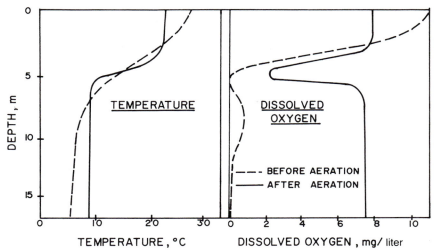

Fig. 12.26. Effect of hypolimnetic aeration on a Lake (Fast *et al.*, 1976).

ft³/sec and, during the stratified period, the minimum daily flow is 750 ft³/sec, which is attained by releasing 3000 ft³/sec for 1 hour during each flow hour period.

Extensive laboratory and field investigations resulted in the recommendation that the most efficient means of raising the DO levels was to use a pure oxygen diffuser system. Table 12.3 gives the details of the recommended system (TVA, 1978).

TABLE 12.3 Characteristics of a typical system

Diffuser type	Marox, aluminum alloy in fiberglass casing
Pore size	1.5–2 μm
Diffuser submergence depth	70 ft
Average distance from dam	150 ft
Number of diffusers	1000
Total diffuser surface area	665 ft²
Average system oxygenation efficiency	98%
Lowest expected weekly average DO concentration in tailrace	4.0 mg/liter
Annual capital cost (1976)	$120,000
Annual oxygen cost (1976)	$230,000
Total annual cost (1976)	$350,000

REFERENCES

Bella, D. A. (1970). Dissolved oxygen variations in stratified lakes. *J. Sanit. Eng. Div., Am. Soc. Civ. Eng.* **96,** 1129–1146.

Bernhardt, H. (1969). *In* "Personal Communication." World Heath Organization, Wroclaw, Poland.

Bernhardt, H. (1975). Ten years experience of reservoir aeration. Proc., Int. Assoc. on Water Pollut. Res., *Prog. Water Technol.* **7** (3/4), 483–495.

Brooks, N. H., and Koh, R. C. Y. (1968). Selective withdrawal from density-stratified reservoirs. Proc. Spec. Conf. Curr. Res. Effects of Reservoirs on Water Quality (R. A. Elder, P. A. Krenkel, and E. L. Thackston, eds.). Rep. No. 17, Dept. of Environ. Water Res. Eng., Vanderbilt Univ., Nashville, Tennessee.

Buliček, J. (1960–1965). Aeration of Moldava River reservoirs. Rep. by Hydraul. Res. Inst., Prague, Czechoslovakia.

Cherkauer, D. S. (1977). Effects of urban lakes on surface runoff and water quality. *Water Res. Bull.* **13,** 1057–1068.

Churchill, M. (1957). Effects of storage impoundments on water quality. *J. Sanit. Eng. Div., Am. Soc. Civ. Eng.* **83,** 1171-1–1171-48.

Elder, R. A. (1964). The causes and persistence of density currents. *Proc.* 3rd Annu. Sanit. Water Resour. Eng. Conf., Vanderbilt Univ., Nashville, Tennessee, (P. A. Krenkel, ed.)

Elder, R. A., and Dougherty, G. V. (1956). Thermal density underflow diversion works for Kingston steam plant. Prepr., Am. Soc. Civ. Eng. Conf., Knoxville, Tennessee (June).

Elder, R. A., and Wunderlich, W. O. (1968). Evaluation of Fontana Field Measurements. Proc. Spec. Conf. Curr. Res. Effects of Reservoirs on Water Quality, (R. A. Elder, P. A. Krenkel, and E. L. Thackston, eds.). Tech. Rept. No. 17, Dept. of Environ. Water Res. Eng., Vanderbilt Univ., Nashville, Tennessee.

Eloubaidy, A. F., and Plate, E. J. (1972). Wind shear turbulence and reaeration coefficient. *J. Hydraul. Div. Am. Soc. Civ. Eng.* **98**, 153–170.

Fast, A. W., Lorenzen, M. W., and Glenn, J. H. (1976). Comparative study with cost of hypolimnetic aeration. *J. Environ. Eng. Div., Am. Soc. Civ. Eng.* **102**, 1175–1188.

Fry, A. S., Churchill, M. A., and Elder, R. A. (1953). Significant effects of density currents in TVA's integrated reservoir system. Proc. Minnesota Int. Hydraulics Conf. Minneapolis, Minnesota.

Gebhart, G. E., and Summerfelt, R. C. (1976). Effect of destratification on depth distribution of fish. *J. Environ. Eng. Div., Am. Soc. Civ. Eng.* **102**, 1215–1228.

Greeson, P. E. (1969). Lake eutrophication, a natural process. *Water Res. Bull.* **5**(4), 16–30.

Harleman, D. R. F., and Huber, W. C. (1968). Laboratory studies on thermal stratification in reservoirs. Proc. Spec. Conf. Curr. Res. Effects of Reservoirs on Water Quality (R. A. Elder, P. A. Krenkel, and E. L. Thackston, eds.). Tech Rep. No. 17, Dept. of Environ. and Water Res. Eng., Vanderbilt Univ., Nashville, Tennessee.

Huber, W. C., and Harleman, D. R. F. (1968). Laboratory and analytical studies of the thermal stratification of reservoirs. Rep. No. 112, Dept. of Civ. Eng., Massachusetts Inst. Tech., Cambridge.

Huber, W. C., Harleman, D. R. F., and Ryan, P. J. (1972). Temperature prediction in stratified reservoirs. *J. Hydraul. Div., Am. Soc. Civ. Eng.* **98**, 645–666.

Hutchinson, G. E. (1957). "A Treatise on Limnology, Vol. 1: Geography, Physics and Chemistry." Wiley, New York.

Jones, S. L. (1964). *In* "Personal Communication." Tenn. Stream Pollution Control Board, Nashville, Tennessee.

King, D. L. (1970). Reaeration of streams and reservoirs: Analysis and bibliography. Eng. Res. Cent., U.S. Bureau of Reclamation, No. REC-OCE-70-55, Denver, Colorado (December).

Koh, R. C. Y. (1964). Viscous stratified flow towards a line sink. Rep. No. KH-R-6, W. M. Keck Lab. Hydraul. and Water Res., California Inst. Technol., Pasadena.

Koh, R. C. Y. (1966). Unsteady flow into a sink. *J. Hydraul. Res.,* Int. Assoc. Hydraul. Res. **4**, 21–35.

Krenkel, P. A. (1965). Investigation of manganese concentration in Allatoona Reservoir, Georgia. Rep. to U.S. Army Corps Eng., Mobile District, Alabama.

Krenkel, P. A., and Parker, F. L. (1969). Engineering aspects of thermal pollution. *In* "Biological Aspects of Thermal Pollution" (P. A. Krenkel and F. L. Parker, eds.). Vanderbilt Univ. Press, Nashville, Tennessee.

Krenkel, P. A., Thackston, E. L., and Parker, F. L. (1969). Impoundment and temperature effects on waste assimilation *J. Sanit. Eng. Div., Am. Soc. Civ. Eng.* **95**, 37–64.

Long, R. R. (1962). Velocity concentrations in stratified fluids. *J. Hydraul. Div., Am. Soc. Civ. Eng.* **88**, 15.

Markowsky, M., and Harleman, R. F. (1973). Prediction of water quality in stratified reservoirs. *J. Hydraul. Div., Am. Soc. Civ. Eng.* **99**, 729–745.

Mattingly, G. E. (1977). Experimental study on wind effects on reaeration, *J. Hydraul. Div., Am. Soc. Civ. Eng.* **103**, 311–323.

Orlob, G. T., and Selma, L. G. (1968). Mathematical simulation of thermal stratification in deep impoundments. Proc. Spec. Conf. Curr. Res. Effects of Reservoirs on Water Quality, Tech. Rep. No. 17, Dept. Environ. Water Res. Eng., Vanderbilt Univ., Nashville, Tennessee.

Rayyan, F., and Speece, R. E. (1976). Hydrodynamics of bubble plumes and oxygen absorption in stratified impoundments, Proc. 8th Int. Conf. on Water Pollut. Res., Sydney, Australia, *Prog. Water Technol.* **9**, 129–142.

Symons, J. M. (1969). Water Quality Behavior in reservoirs. U.S. Dept. of Health, Educ. and Welfare, Public Heath Serv., Cincinnati, Ohio.

Symons, J. M., Irwin, W. H., and Robeck, G. (1968). Control of reservoir water quality by engineering methods. Proc. Spec. Conf. Curr. Res. Effects of Reservoirs on Water Quality (R. A. Elder, P. A. Krenkel, and E. L. Thackston, eds.). Rep. No. 17, Dept. of Env. Water Res. Eng., Vanderbilt Univ., Nashville, Tennessee.

Wunderlich, W. O., and Gras, R. (1967). Heat and mass transfer between a water surface and the atmosphere. Preliminary Rep., Tenn. Valley Auth., Norris, Tennessee.

Wunderlich, W. O. (1968). Discussion to Laboratory studies on thermal stratification in reservoirs. Proc. Spec. Conf. Curr. Res. Effects of Reservoir s on Water Quality, Tech. Rep. No. 17, Dept. of Environ. Water Res. Eng., Vanderbilt Univ., Nashville, Tennessee.

Task Committee on Sedimentation (1963). Sedimentation transportation mechanics: Density currents. progress Rep., *J. Hydraul. Div., Am. Soc. Civ. Eng.* (September).

Tennessee Valley Authority (1976). Quality of water in Nickajack Reservoir. Div. Environ. Planning, Chattanooga. (February).

Tennessee Valley Authority (1978). Evaluation of small-pore diffuser technique for reoxygenation of turbine releases at Ft. Patrick Henry Dam. Chattanooga, Tennessee.

U.S. Bureau of Standards (1938). Report on investigation of density currents. Multilith Rep. (May).

Water Resources Engineers, Inc. (1966). Progress report on development of a mathematical model for prediction of temperature in deep reservoirs. Phase I: Low discharge-volume ratio reservoirs. Walnut Creek, California.

13 | *Eutrophication*

DEFINITIONS

Eutrophication is the natural process of the aging of lakes which can affect all surface water bodies. The process evolves progressively from a lake's geological origin to its extinction. The process proceeds irrespective of man's activities. However, cultural activites such as farming, construction, and wastewater disposal can significantly accelerate the process and shorten the life span of the water bodies.

The process of eutrophication and the relevant trophic states of surface water bodies have recently received much public attention. For example, Lake Erie and Lake Ontario are said to be deteriorating rapidly because of the accelerated production of organic material mass with consequent limitation of their beneficial uses. Many other lakes and reservoirs are also thought to have become highly eutrophic in the last few decades as a result of increased nutrient loadings from intensive farming operations and urban development. Most reservoirs anticipate eutrophication problems subsequent to closure.

The term eutrophication is usually associated with the excessive development of autotrophic aquatic organisms, primarily planktonic algae (phytoplankton) and aquatic weeds (macrophytes). It should be noted that eutrophication is not synonymous with pollution. A body of water may become eutrophic if it receives certain types of contaminants; however, not all pollution results in eutrophication. The term eutrophication refers to natural or artificial addition of nutrients to bodies of water and their effects. Some pollutants such as heavy metals and other toxic components may even inhibit the eutrophication process. It is interesting to note, in this regard, that copper sulfate, which is considered to be a pollutant, has often been used to control excessive algal growths—one of the symptoms of eutrophication.

The word *eutrophication* comes from the Greek *eutrophia* or *eutrophos*, meaning well nourished. Several technical and/or scientific definitions of the term should be considered. In most definitions, eutrophication refers to the natural or artificial addition of nutrients to water bodies and the resultant effects of those added nutrients (Rohlich, 1969).

The problem of eutrophication has been primarily associated with the lowering of the water quality in lakes and reservoirs. Since eutrophication is usually related to the aging of lakes, some disagreement exists as to the applicability of the term to other nonstagnant surface waters. Streams and estuaries do not age in the same sense as lakes do, although added nutrients may also cause water quality problems. For example, excessive growth of aquatic weeds (macrophytes) can develop in shallow reaches of streams; however, their source of nutrients is probably from the sediments (Krenkel and Novotny, 1973).

As previously mentioned, eutrophication is a process in which a water body progresses from its origin to its extinction. The rate at which this process occurs depends on the level of nutrient, organic matter and solids accumulation, and many environmental factors, some of them uncontrollable, as shown in Fig. 13.1. It is a dynamic process with highly variable rates, which differ from year to year, season to season, and even hour to hour.

The youngest stage of a lake is characterized by water with a very low mineral content. Most of the lakes of North America and Northern Europe were formed after glaciers retreated approximately 10,000 years ago. These "young" lakes exhibit low productivity and are called *oligotrophic* (from the Greek *oligo*, meaning "few"). Examples of oligotrophic lakes include Lake Superior and many of the high elevation lakes such as Lake Tahoe in California, Lake Titicaca in Peru, and Lake Baikal in Siberia.

As the nutrient content is increased by runoff and/or wastewater, the photosynthetic (autotrophic) organisms increase in number. These organisms are called *producers,* and they initiate the entire cycle of production of organic matter in the lake.

At a later stage in the development process, the lake becomes mesotro-

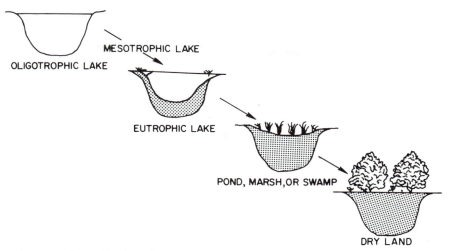

Fig. 13.1. Eutrophication, the process of aging by ecological succession (Greeson, 1969).

phic (from the Greek *meso*, meaning "intermediate"), and when the nutrients and organic matter productivity are high the water body becomes eutrophic. The final stages of lake development prior to extinction are pond, swamp, marsh, and wetland.

The National Technical Advisory Committee on Water Quality Criteria (USEPA, 1972) stated several conditions, analyses, or measures that are considered indicators of eutrophication. Since these parameters are not infallible, experience and judgment must be used in their application. The conditions indicative of organic enrichment are

1. A slow overall decrease year after year in the dissolved oxygen in the hypolimnion as indicated by determinations made a short time before the fall overturn.

2. An increase in dissolved solids—especially nutrient materials such as nitrogen, phosphorus, and simple carbohydrates.

3. An increase in suspended solids—especially organic materials.

4. A shift from a diatom-dominated plankton population to one dominated by blue-green and/or green algae, associated with increases in amounts and changes in relative abundance of nutrients.

5. A steady, though slow, decrease in light penetration.

6. An increase in organic materials and nutrients, in particular an increase in the phosphorus concentrations in bottom deposits.

The photosynthetic process resulting from algae and macrophytes in eutrophic lakes and other surface water bodies can be recognized by the

cyclic fluctuations of the dissolved oxygen concentrations in the euphotic zone. Oxygen is produced during the daytime hours and consumed by respiration during the night. This phenomenon many times results in supersaturation of the water by oxygen on bright sunny days
during the afternoon hours of the productive season and a significant drop
of DO concentrations during the late night and early morning hours as
shown in Fig. 13.2.

The related physical and chemical changes caused by advanced eutrophication (pH variations, oxygen fluctuations, organic substances) may
interfere with recreational and aesthetic uses of water and may also cause
a shift of the fish population from game to rough fish. Excessive aquatic
weed growth may interfere with navigation, recreation, and other uses. In
addition, the taste and odor problems caused by algae can make water
less suitable or desirable for water supply and human consumption.

When the concentrations of phytoplankton algae during the late
summer period exceed certain threshold nuisance values, the term "algal
bloom" is used for describing the situation. An arbitrary number of 500
phytoplankton organisms per milliliter of water was introduced by
Lackey (1949) and accepted by most technical limnologists as the nuisance threshold concentration denoting algal blooms. Generally, the term

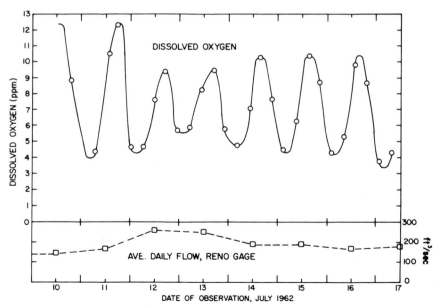

Fig. 13.2.   Dissolved oxygen and flow in the Truckee River (O'Connell *et al.*, 1962).

algal bloom or water bloom (Wasserblute, Fleur d'eau) denotes a prolific growth of plankton (Steward and Rohlich, 1967). A bloom of algae may be so dense that it imparts a greenish, yellowish, or brownish color to the water. In addition, the algae can form large floating chunks of organic matter or cover the water surface in a distinct layer that may be several millimeters or centimeters thick. These growths are not only aesthetically displeasing but may cause a disagreeable odor. Some species of algae may cause fish mortality, and others can be poisonous to cattle and waterfowl and a menace to drinking water supplies.

TROPHIC STATE

Surface waters can be classified and characterized by their trophic state. The best water quality is commonly associated with oligotrophic conditions, while eutrophic conditions often imply a poorer water quality.

Oligotrophic refers to a state of nutrition, specifically, one poor in nutrients. Eutrophic implies one rich in nutrients, and mesotrophic denotes an intermediate state in the nutrient content. Dystrophic refers to bodies of water judged primarily on color and humic matter rather than on nutrient content.

Hutchinson (1969) pointed out that the terms eutrophic and oligotrophic (and also mesotrophic) should be used not only to describe water quality but also should be related to the water bodies and their drainage basins and sediments in forming oligotrophic and eutrophic systems. In this context the term "available nutrients" should be used to describe eutrophication potential. Many nutrients such as phosphates and ammonia may be adsorbed on sediments and/or become incorporated into benthic deposits and, thus, become permanently or temporarily unavailable for phytoplankton development. However, benthic nutrients can be available for growth of macrophytes in the euphotic littoral zones.

As pointed out by Lee *et al.* (1978), the quantity of available phosphorus lies somewhere between the soluble orthophosphate and the total phosphate loads. They proposed that the biologically available phosphorus is approximately equal to the soluble orthophosphate plus 0.2 times the difference between the total phosphorus and the soluble orthophosphate. It is important to note that in most of the models proposed to describe eutrophication and trophic state, differentiation between total and biologically available phosphorus is not made. This could explain much of the variation noted.

The necessity to include watersheds in eutrophication systems was recognized by categorizing lakes (and reservoirs) into autotrophic, i.e., lakes which receive a major portion of the nutrients from internal sources (sediments, atmosphere), and allotrophic, those which receive a major part of their nutrients from external sources. The nutrients and organics present in the lake are autochthonous (those produced in the lake) and allochthonous (those produced or originating from outside the lake).

Eutrophication is closely related to the net production or the increase of organic matter in the surface water body. This process is often called "carbon fixation" inasmuch as the inorganic carbon from the atmospheric or dissolved carbon dioxide or from bicarbonates is "fixed" and converted by the photosynthetic process into new organic matter. The organic matter production is slow during the oligotrophic state and accelerates during the mesotrophic and eutrophic states. In post eutrophication or hypertrophic states, the rate increase diminishes and approaches zero during the final extinction phase as shown in Fig. 13.3.

The rate of organic matter production may be accelerated by the addition of nutrients from runoff and/or wastewater. The effect is more apparent in the early stages than in more advanced stages of eutrophication.

The term "productivity" is widely used by limnologists to describe the rate of organic carbon fixation by photosynthesis. By definition (Steward and Rohlich, 1967), the gross primary production is the quantity of carbon

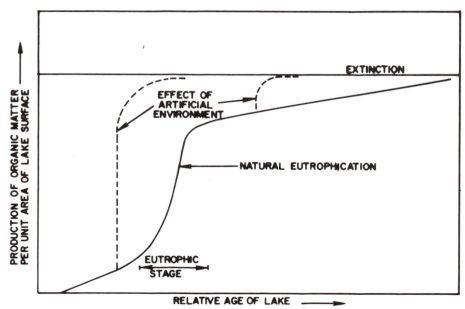

Fig. 13.3. Hypothetical curve of eutrophication (modified by Hasler, 1947).

TABLE 13.1 Classification of productivity in lakes[a]

| | Phytoplankton Production mg C/m²/day | | |
| | Oligotrophic Lakes | Eutrophic Lakes | |
Rate		Natural	Polluted
Mean rates in growing season	30–100	300–1000	1500–3000
Annual rates	7–25	75–250	350–700

[a] After Rodhe (1969).

taken up into organic combination. The net primary production can then be obtained by subtracting the carbon loss by respiration. The gross production should be determined for a period from sunrise to sunset and the loss by respiration for a 24-hour period. Rodhe (1969) suggested the ranges of phytoplankton production in oligotrophic and eutrophic lakes, which are given in Table 13.1.

It should be pointed out that trophic state can be determined by limnologists according to the distribution of the planktonic population. Some organisms are indicative of oligotrophic conditions (e.g., some diatoms), while large populations of blue-green and green algae indicate algal blooms and eutrophication.

TROPHIC INDEX

Determination of the trophic status of lakes based on limnological observations, taxonomy or distribution of organisms and their productivity, and chemical water quality is a tedious process requiring experience and judgment. In engineering water quality management practice, there is a tendency to use a "single number" index to describe a process instead of extensive descriptive observations needed to describe a complex phenomenon. Such an index should be an objective evaluation of the status of the process and should also be predictive, i.e., enable the forecasting of future conditions if inputs or outputs to the system are modified or eliminated. There is presently no such index for the classification of the trophic state of surface water bodies. However, during the past few years, several attempts have been made to establish an approximate estimation of trophic status based on a few controlling parameters and symptoms. A caveat should be stated for all prospective users of such indices inasmuch as a lake or reservoir can be classified as oligotrophic using one parameter and eutrophic using another index.

Before such indices are used, the controlling factor or factors should be established and evaluated. The lack of a precise definition of "trophic status" makes it difficult to develop an accurate or sometimes even acceptable approximate engineering technique to classify water bodies according to their trophic state. To predict the effect of various remedial measures on the trophic status is still more or less guesswork rather than prediction.

Some methods of estimating trophic status based on selected indicators of the process have evolved and have been published and/or used for classifying lakes. Many of the techniques (USEPA, 1973; Lueschow *et al.*, 1970; Shannon and Brezonik, 1972) are all relative systems in which lakes are classified only with respect to each other or to an average quality of a given group of lakes (Shannon and Brezonik, 1972b) and not according to some objective independent scale. These systems often give different weight to various parameters characterizing the trophic status.

The parameters used most frequently are hypolimnetic dissolved oxygen, total phosphorus, transparency by Secchi disk, inorganic nitrogen, and chlorophyll a concentration. Some of the classification systems simply rank the lakes according to the magnitude of these parameters, and the sum of the ranking then gives the trophic state index (TSI) (USEPA, 1973; Lueschow *et al.*, 1970). Shannon and Brezonik's index assigns weights to the standardized values of various parameters (standardized with respect to the group mean and standard deviation), but the results still depend on the sample mean and standard deviation and different TSI values can be obtained if a different group of lakes is analyzed.

Trophic Status Index

This index (Carlson, 1977). was developed for lakes where phosphorus is the limiting nutrient. Carlson based his index (indices) on the fact that there are intercorrelations between the transparency expressed by the Secchi disk depth (the depth at which visibility of the disk disappears), algal concentration expressed as chlorophyll a, and vernal (during spring overturn) or average phosphorus concentrations. This trophic status index was defined as

$$\text{TSI(SD)} = 10 \left(6 - \frac{\ln \text{SD}}{\ln 2} \right), \tag{13.1}$$

where SD is the Secchi disk depth.

The value TSI = 0 was assigned to SD = 64 m, which is the approximate integer magnitude of the highest SD value ever reported. The practical highest limit is TSI = 100, which corresponds to SD = 6.4 cm.

Using the correlations between the chlorophyll a concentrations (Chl) total phosphorus and the Secchi disk value, the other two expressions for the TSI become

$$\text{TSI(Chl)} = 10 \left(6 - \frac{2.04 - 0.68 \ln \text{Chl}}{\ln 2} \right) \qquad (13.2)$$

and

$$\text{TSI(TP)} = 10 \left[6 - \left(\ln\frac{48}{\text{TP}} \middle/ \ln 2 \right) \right], \qquad (13.3)$$

where Chl is the concentration of chlorophyll a and TP is the concentration of the total phosphorus.

This method offers three indices instead of one single number. The best indicator of trophic status varies from lake to lake and from season to season. It should be noted that Secchi disk values may be erroneous in lakes where turbidity is caused by factors other than algae, and priority should be given to chlorophyll a as the best indicator.

The TSI index was not related to the trophic status, and accurate ranges of this TSI index to previously defined trophic status could not be established. Figure 13.4 shows the ranking of some Wisconsin lakes according to the "classical" nomenclatural definitions of the trophic status along with their TSI values. As one would expect, difficulties have been encountered in the use of the TSI index for lakes which are not phosphorus limited (Sloey and Spangler, 1978).

Trophic Characterization System for Wisconsin Lakes

The lake condition index (LCI) was developed by Uttormark and Wall (1975). This system is based on evaluation of eutrophication symptoms (transparency, use impairments, fish kills) rather than on causes (nutrient content). In this method "penalty points" were assigned to lakes depending on the degree to which they exhibited undesirable symptoms of eutrophication. Four parameters were selected for analysis, and ranges of values for each parameter were specified to depict lake conditions ranging from desirable to undesirable. The parameters and the ranges are listed in Tables 13.2 to 13.5. The parameters are treated separately and composite lake ratings were determined by summing the number of points assigned

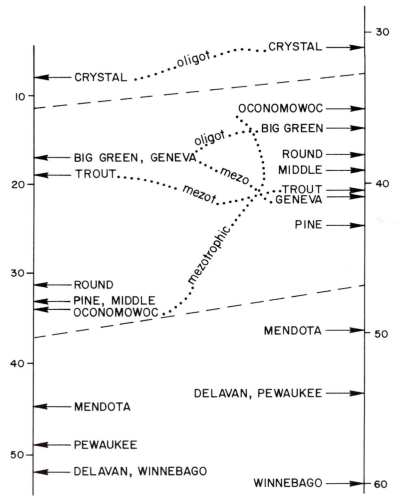

Fig. 13.4. Rank ordering of 12 Wisconsin lakes by the multiparameter method and the trophic status index method. Data from Lueschow *et al.* (1970) and Sloey and Spangler (1978).

in each of the four categories. The sum is termed an LCI. The index is not synonymous to ''trophic status'' or to ''productivity.''

Several other parameters were considered for the LCI system (alkalinity, conductivity), but they were omitted when it was determined that they had little, if any, effect on the results.

TABLE 13.2 Transparency penalty points for lake classification index[a]

	Transparency conditions[b]	
Max. Range, m	Min. Range, m	Penalty Points
0.0–0.45	0.0–0.45	4
0.0–0.45	0.45–3	3
0.0–0.45	3–7	2
0.0–0.45	>7	2
0.45–3	0.45–3	2
0.45–3	3–7	1
0.45–3	>7	0
3–7	3–7	0
3–7	>7	0

[a]After Uttormark and Wall (1975).
[b]Annual maxima and minima.

TABLE 13.3 History of fish kills

Fish Kills	Penalty Points
None	0
Positive, max depth <10 m	3
Positive, max depth >10 m	4

TABLE 13.4 Dissolved oxygen points for lake classification index

	Penalty Points	
Dissolved Oxygen Conditions[a]	max depth 10 m	max depth 10 m
Dissolved oxygen in hypolimnion greater than 5 mg/liter virtually all times	0	0
Concentration in hypolimnion less than 5 mg/liter but greater than 0 mg/liter	1	2
Portion of hypolimnion void of oxygen at times	3	4
Entire hypolimnion void of oxygen at times	5	6

[a]Lakes which do not stratify receive few or no penalty points.

TABLE 13.5   Recreational use impairment for lake classification index

	Penalty Points		
Use impairment	Weeds Only	Algae Only	Weeds + Algae
No impairment of use			
Very few algae present, no "bloom" conditions, and/or very few weeds in the littoral zone	0	0	0
Slight impairment of use			
Occasional "blooms," primarily green species of algae, and/or moderate weed growth in the littoral zone	2	2	2
Periodic impairment of use			
Occasional "blooms," predominantly blue-green species, and/or heavy weed growth in the littoral zone	3	4	5
Severe impairment of use			
Heavy "blooms" and mats occur frequently, blue-green species dominate, and/or excessive weed growths over entire littoral zone	6	7	9

The authors suggested the assignment of trophic characterization based on the following ranges of the lake classification index.

LCI	Trophic classification
0–1	Very oligotrophic
2–4	Oligotrophic
5–9	Mesotrophic
10–12	Eutrophic
13–	Hypertrophic

The use of symptomatic lake classification indices, such as LCI, instead of causative ones has a disadvantage if one intends to use such indices for evaluating future water quality conditions. Estimation of the results of future water quality abatement measures is impossible, and, therefore, the symptomatic indices have little value for such purposes. However, the authors did suggest that correlation may exist between the LCI index and nutrient levels in the lake, a possible propect for future research.

FACTORS AFFECTING EUTROPHICATION

There is no single prevailing factor which can be depicted as the primary driving force of eutrophication. Even though the nutrient content may appear to be the most important parameter, there are other factors which must be considered. The factors pertinent to the analysis include nutrients, light intensity and penetration, temperature, mixing, CO_2 content, grazing by predators, average depth, availability of nutrients, chemistry of water, and concentration and types of organisms. The factors affecting eutrophication may be divided into three groups—physical, chemical (nutrients), and biological (limnological).

Physical Factors

Most oligotrophic lakes and reservoirs are found at higher elevations and colder climates. Therefore, it is obvious that water temperature and amount of solar radiation affects the rate of the eutrophication process.

These factors primarily affect the growth rates of photosynthetic organisms and, in addition, the thermal regimes of the water bodies (stratification). All of the primary production takes place in the euphotic zone (which often coincides with the epilimnion).

Many authors stress the importance of morphology, especially mean depth, as an index of the productivity and classification of surface water bodies. Shallow water bodies tend to be more productive and, also, in a more advanced stage of eutrophication than deeper lakes and reservoirs.

Other important factors affecting eutrophication include the shape of the surface area and the detention time of water in the lake. It should be noted that the theoretical detention time (volume divided by flow) cannot be substituted for the actual residence time of water or a nutrient in the productive zone. Since algae require relatively long time periods for full development, the residence time necessary for optimal production should be long, possibly weeks or months.

The same factors that affect lakes will also affect man-made reservoirs. The most common differences between lakes and reservoirs are water fluctuations and better mixing in reservoirs, which favors lower productivity.

Although the eutrophication process is not commonly associated with streams and estuaries, the longer detention times in estuaries favor phytoplankton development, with the same resulting effects that one would find in lakes. Shallow rivers and streams rich in nutrients may develop prolific dense growth of aquatic weeds (macrophytes) which are not common in deep rivers. Stream velocity, in addition to depth, also appears to be an important factor adversely affecting the growth of photosynthetic organisms in streams. Water bodies with higher turbulence and mixing do not provide an optimal environment for the development of planktonic algae.

Chemical Factors

Many elements and chemicals are necessary for the growth of aquatic organisms. Common chemical nutrients required for the promotion of growth and proliferation of algae include carbon, hydrogen, oxygen, sulfur, potassium, calcium, magnesium, nitrogen, phosphorus, molybdenum, and other elements and chemicals. Theoretically, according to Liebig's law of the minimum, any of these chemicals may become limiting to the algal communities, but usually, nitrogen and phosphorus control the development of aquatic blooms and luxurious growths of aquatic macrophytes.

Nitrogen is an essential component of proteins, nucleic acids, and other biologically important materials. Sources of nitrogen are primarily inorganic nitrates and nitrites and, to a lesser degree, ammonia. Fixation of atmospheric nitrogen by blue-green algae can also supply large quantities of nitrogen to aquatic environments. In addition, inorganic nitrogen can be produced by nitrifying bacteria and/or by the decomposition of organic matter containing proteins.

Phosphorus is used by living cells for energy transfer. The sole original source of phosphorus is the weathering of phosphate-containing minerals, but it is intensively recycled by microorganisms and all living forms. Phosphorus is available to algae and macrophytes only in the phosphate form.

During investigations of the eutrophication problems of Madison (Wisconsin) lakes, Sawyer (1974) noted that algal blooms occurred when concentrations of inorganic nitrogen (NH_3, NO_2^-, NO_3^-) and inorganic phosphorus exceeded respective values of 0.3 mg N/liter and 0.01 mg P/liter. It should be noted that very low concentrations of inorganic N and P may be measured during the productive period because of the uptake by algae. Therefore, the critical nutrient concentrations should be evaluated during the spring overturn or before the beginning of the productive season.

Vollenweider (1968) used the critical concentrations suggested by Sawyer (1947) and estimated the following critical values in terms of annual loadings (quoted in Porcella, et al, 1974): P, 0.2–0.5 g/m²/yr and N, 5–10 g/m²/yr. Another way to estimate critical loadings to a lake based on the

Vollenweider work was developed by Schindler *et al.* (quoted in Porcella *et al*, 1974). The loading functions are related to mean depths as

Permissible loadings: Below those levels, probably oligotrophic

$$\log_{10} P_A = 0.6 \log_{10} H - 1.6, \tag{13.4}$$
$$\log_{10} N_A = 0.6 \log_{10} H - 1.3. \tag{13.5}$$

Dangerous loadings: Above these levels, probably eutrophic.

$$\log_{10} P_D = 0.6 \log_{10} H - 0.43, \tag{13.6}$$
$$\log_{10} N_D = 0.6 \log_{10} H - 0.13. \tag{13.7}$$

The phosphorus loadings, P_A and P_D, and the nitrogen loadings, N_A and N_D, are in g/m²/yr and the mean depth, H, is in meters. Oligotrophic lakes are supposed to occur at loadings below the permissable levels, while eutrophic lakes occur above the dangerous levels and mesotrophic lakes lie in between the permissible and dangerous levels. Table 13.6 summarizes the results.

Estimating the Limiting Nutrient.

Although many nutrients contribute to algal growths, according to Liebig's law of the minimum, only one limits the growth rate. Most of the life-supporting elements and chemicals are required only in trace quantities and exist in nature in amounts exceeding the requirements for optimal growth of algae. Only carbon, hydrogen, oxygen, nitrogen, and phosphorus are required in appreciable quantities. Because of the relative abundance of carbon (from alkalinity or atmospheric CO_2) and water, nitrogen and phosphorus are thought to be the primary nutrients controlling

TABLE 13.6 Specific loading levels for lakes expressed as total nitrogen and total phosphorus in g/m⁻²/year⁻¹ [a]

Mean Depth up to	Permissible Loading, up to		Dangerous Loading in Excess of	
	N	P	N	P
5 m	1.0	0.07	2.0	0.13
10 m	1.5	0.10	3.0	0.20
50 m	4.0	0.25	8.0	0.50
100 m	6.0	0.40	12.0	0.80
150 m	7.5	0.50	15.0	1.00
200 m	9.0	0.60	18.0	1.20

[a] After Vollenweider (1968).

the rate of eutrophication. It should be noted, however, that some controversy exists over the role of carbon.

From analyzing algal growths, it was found that when the nitrogen/phosphorus ratios were greater than 15:1, phosphorus was the possible limiting nutrient, and when the ratios were less than 15:1, nitrogen was limiting. However, these ratios may vary from 5:1 to 20:1.

A laboratory procedure for determining the limiting nutrient and eutrophication potential of surface waters was published by Maloney *et al.* (1972). In this technique, test algae are added to a bottle containing a water sample and the increase in the organic biomass under standard laboratory conditions is then measured. Nutrients can be added (nitrogen, phosphorus, and carbon) both separately and in combination, and the growth then compared to a blank (no nutrient additions) sample. The ratio of added nutrients should be $C:N:P = 200:21:1$ in order to approximate the proportional nutrient demand of the test algae. If a significant increase in the algal growth rates is noticed after a nutrient is added, it can be deduced that the particular nutrient may be limiting. The test is now a standard EPA procedure for evaluating the algal growth potential of surface waters.

By plotting phosphorus versus nitrogen concentrations during various seasons, a straight line relationship can usually be established. It is assumed that the limiting nutrient will be exhausted first, which will show as a negative intercept of the line on the axis of the limiting nutrient (Fig. 13.5). The one which is not limiting will remain in solution yielding a positive intercept of the line of best fit. In Fig. 13.5(a), the phosphorus concentrations are limiting, while Fig. 13.5(b) indicates that nitrogen can be limiting.

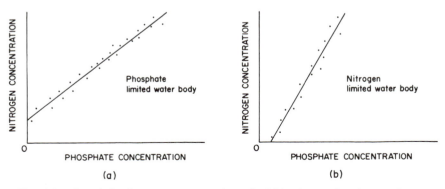

Fig. 13.5. Correlation between concentrations of soluble nitrate, phosphate, and oxygen in eutrophic lakes.

Biological Factors and Considerations

Eutrophication is a biological process. Although the physical and chemical factors can affect the rate of eutrophication, it is the photosynthetic process by chlorophyll containing organisms which produces new organic matter from inorganic nutrients. The chemistry of water and physiological factors represent the constraints of the process rather than the process itself.

As the eutrophication process progresses from the original oligotrophic to more advanced stages, the organisms' quantity and quality vary according to the level of nutrients and organics present in the water body. In Nauman's original work on eutrophication (Nauman, 1919, quoted by Steward and Rohlich, 1967), it was suggested that increases in nutrients might be noted in the plankton as soon as or sooner than they were measured chemically. Although this statement may reflect the chemical technology known at that time, there is no doubt that a skilled limnologist can evaluate the trophic states using indicator organisms and their distribution, composition, and tolerance to various levels of pollution and nutrient content.

The process of eutrophication affects not only algae but also the composition and distribution of the entire aquatic biota. The rate of eutrophication depends on the growth of algae and benthic macrophytes, their grazing by predators, decay and decomposition of organic matter by bacteria, inflow–outflow balance of organic matter, etc. The net productivity is a balance between the organic matter gains and losses.

Biological activity does not depend passively on the nutrient level but can actively affect it. The living processes occurring in water bodies represent, in fact, a continuous transformation of nutrients from inorganic to organic forms and vice versa.

The eutrophication process is a chain in the formation of organic matter and its transformation from one trophic level to another. The chain starts with fixation of inorganic carbon and nutrients by algae and macrophytes, is followed by their grazing by herbivorous zooplankton, and continues to higher trophic levels as shown in Fig. 13.6. These processes take place both in the water and in the benthos, and there is usually a balance among the phytoplankton, the zooplankton, and higher trophic species. Algal blooms occur when the balance is distorted; i.e., when growth of a certain algal species is not followed by grazing by zooplankton.

A general consequence of the enrichment of water ecosystems by nutrients is an increase in the standing crop of the phytoplankton and the planktonic herbivores. If phytoplankton were the only group which would benefit from the nutrient enrichment, the phytoplankton biomass would

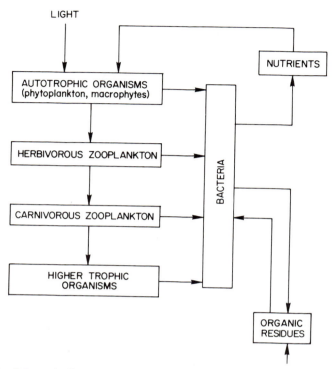

Fig. 13.6. Schematic diagram of the transformation of organic matter by various trophic levels.

be in close relationship to the concentration of the limiting nutrient. Predation on the phytoplankton and also on zooplankton by higher trophic organisms may account for the fact that often the concentrations and distribution of organisms do not behave according to the available supply (Brooks, 1969).

The rate of eutrophication, since it depends on the concentration of photosynthetic organisms, can be affected by the concentration and quality of herbivorous zooplankton. Brooks (1969) quoted a case reported by Hrbáček (1962) in which the rate of eutrophication in a pond was reduced significantly by altering the fish stock in the pond. When the fish stock was reduced, *Daphnia* developed as a dominant larger size planktonic herbivore. The pond transparency increased significantly resulting in a classical appearance of oligotrophic conditions. But when small-bodied planktonic herbivores predominated (fish stock high), the same pond had the classic characteristic of eutrophy. However, the control of standing crop of algae by planktonic herbivores and fish population has not yet been demonstrated in a large lake.

The increase of phytoplanktonic organisms, especially blue-green algae, also has an almost immediate effect on fish. The algae not only reduce oxygen levels because of their endogenous respiration demand, but may also have a toxic effect on the fish (Larkin and Northcote, 1969). An increase of algal concentration may, therefore, cause a shift in fish population from game fish (less tolerant) to more tolerant species. In North America, outstanding examples are the changes in fish population of the lower Great Lakes. Catches of whitefish, sauger, walleye, and blue pike dropped drastically in the last few decades and were replaced by warm-water more tolerant species such as fresh-water drum, carp, yellow perch, and smelt.

MODELING EUTROPHICATION

The classical oxygen balance concept may not work for lakes and other surface water bodies that are in an advanced stage of eutrophication. In these water bodies, the production of organic matter by phytoplankton, attached algae, and macrophytes may greatly exceed the BOD contributions from runoff and wastewaters. There are also other problems associated with the use of the traditional dissolved oxygen balance schemes. Oxygen levels are influenced by photosynthesis and respiration, and BOD concentrations are affected by the concentration of planktonic organisms.

The state of the art of water quality modeling has reached levels where even such complex systems as the eutrophication process can be modeled. The modeling of eutrophication actually involves the simulation of a significant part of or possibly the entire ecological aquatic system. Figure 13.7 is a schematic picture of the components of the system. The primary driving force (input) is the sunlight energy subject to nutrient constraints. Lakes and reservoirs are usually stratified; therefore, it is necessary to break them into the upper (epilimnion) and lower (hypolimnion) zones. Photosynthesis takes place only in the euphotic part of the lake, while decomposition takes place in the lower part of the water body.

The system can be characterized as a composite of three subsystems:

1. Hydraulic and thermal subsystem.
2. Chemical subsystem including nitrification, denitrification, sedimentation, adsorption–desorption, exchange of chemicals between the lower water zone and bottom sediments, and atmospheric exchange of chemicals and volatilization.

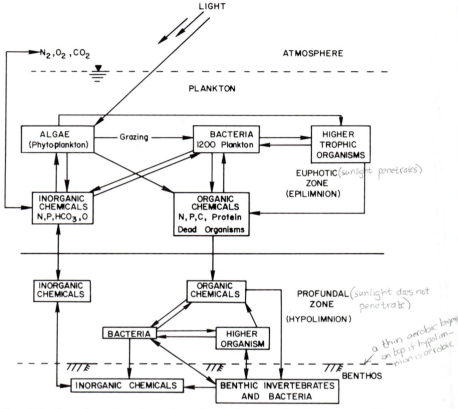

Fig. 13.7. Transformation of organic matter during eutrophication in a deep lake.

3. Biological subsystem including photosynthesis, zooplankton grazing, and higher trophic organisms' growth and decomposition.

Aquatic rooted plants (macrophytes) are important in shallow water
bodies. The conceptual representation of interactions and processes occurring in lakes is shown in Fig. 13.8. The macrophytes may derive their
nutrients from sediments and may not be affected by the nutrient concentration in water. They may, however, strongly affect the dissolved oxygen
and carbonaceous system of the water body.

Because of the general difficulty of measuring or mathematically
describing biological processes, all equations are rough approximations.
and the coeffients should be accepted as a range of values rather than

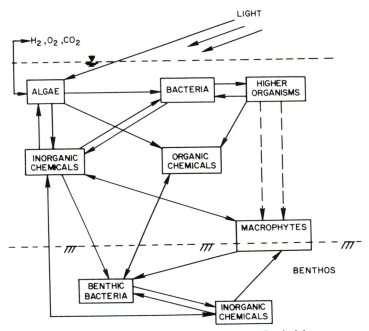

Fig. 13.8. Interactions and processes occurring in lakes.

accurate estimates. The process of calibration and verification is, therefore, very important.

The basic modeling techniques for lakes and reservoirs as well as for shallow water bodies have been discussed in the preceding chapters. The modeling process involves segmentation of the water body into horizontal slices (deep lakes and reservoirs) or longitudinal segments (shallow water bodies).

The basic structure of the model consists of three parts (Dahl-Madsen and Gargas, 1974; Thomann *et al.*, 1975; Chen and Orlob, 1972): transport and dispersion subsystem, biological subsystem, and chemical subsystem.

Since transport, dispersion, and most of the chemical models have been discussed elsewhere, only the biological component will be briefly introduced here. It should be emphasized that the state of the art of transport and chemical component modeling is more advanced than that of the biological subsystem.

The basic equations for conservation of mass of a substance (biological or chemical) in a three-dimensional representation can be written as (Thomann *et al.*, 1975)

$$\frac{\partial s_k}{\partial t} = \frac{\partial}{\partial x}\,(us_k) - \frac{\partial}{\partial y}\,(vs_k) - \frac{\partial}{\partial z}\,(ws_k)$$

$$+ \frac{\partial}{\partial x}\left(D_x\,\frac{\partial s_k}{\partial x}\right) + \frac{\partial}{\partial y}\left(D_y\,\frac{\partial s_k}{\partial y}\right) + \frac{\partial}{\partial z}\left(D_z\,\frac{\partial s_k}{\partial z}\right)$$

$$\pm \overset{\circ}{S}_k(x,y,z,s_k,s_1) + W_k(x,y,z,t),$$

$$k = 1\ \ldots\ m,\, l = 1\ \ldots\ m,\, k \neq 1, \quad (13.8)$$

where s_k is the kth dependent variable (biological or chemical), u, v, and w are the velocity components in the x-, y-, and z-directions, D_x, D_y, D_z are the dispersion coefficients in each respective spatial direction, $\overset{\circ}{S}_k$ represents the kinetic interactions between the variables, and W is the direct input of the substance, s_k.

In most cases, the three-dimensional concept can be simplified by a one- or two-dimensional approximation and/or by neglecting some terms of smaller order of magnitude (e.g., dispersion in streams, convection in some estuaries).

The use of Eq. (13.8) presumes that the advection and dispersion terms are known either from measurements or a hydrodynamic model. The possible techniques and methods have been discussed elsewhere.

The number of variables, s_k, may vary; however, the following variables should be considered as a minimum.

Biological subsystem:
1. Phytoplankton chlorophyll (as a measure of biomass), s_1
2. Herbivorous zooplankton (as biomass or carbon), s_2
3. Carnivorous zooplankton (as biomass or carbon), s_3
4. Higher organisms (as biomass or carbon), s_4
5. Macrophytes (for shallow water bodies—as biomass or carbon), s_5

Chemical sybsystem:
6. Organic carbon—dissolved or total, s_6
7. Organic nitrogen, s_7
8. Ammonia, s_8
9. Nitrates, s_9
10. Nitrites, s_{10}
11. Organic (total) phosphorus, s_{11}
12. Available phosphorus, s_{12}

Other variables may be included if a better representation (which may not necessarily be more accurate) is desired. These include dissolved nitrogen gas (N_2), inorganic carbon system (CO_2, HCO_3^-, pH), and benthic bacteria.

Phytoplankton Chlorophyll

Variation of the phytoplankton chlorophyll can be described as

$$\mathring{S}_1 = (\mu - \rho_1)\, s_1 - \delta_2\, s_2/\alpha_0 - v_s\, s_1/H, \qquad (13.9)$$

where s_1 and s_2 are resepctively the concentrations of the phytoplankton chlorophyll and the herbivorous zooplankton, v_s is the settling velocity of the phytoplankton, and H is the depth of the segment or the thickness of the computational layer. A description of the coefficients and their approximate magnitudes are given in Table 13.7.

The growth rate expression is based on the basic law of microbial growth by photosynthesis (Thomann *et al.*, 1975; Porcella, *et al*, 1974), which is

$$\mu = \mu_{max}\, \phi_T\, \phi_I\, \phi_N, \qquad (13.10)$$

where μ_{max} is the maximum growth rate, ϕ_T is a function expressing the effect of temperature on the growth rate, ϕ_I is a function reflecting the effect of light intensity, and ϕ_N denotes the effect of nutrients.

The maximum growth rate, μ_{max}, represents the photosynthetic chlorophyll production when all environmental factors are optimal. This is rarely the case, furthermore, the value of μ_{max} is only a rough estimate because of the variability of phytoplankton composition as subjected to various factors. The reported literature values of μ_{max} range from $\mu_{max} = 1.0\ \text{day}^{-1}$ (Eppley, quoted by Thomann *et al.*, 1975) to $\mu_{max} = 3.0\ \text{day}^{-1}$ (Roesner *et al.*, 1973) at 20°C. Obviously μ_{max} is one of the variables which should be investigated during calibration.

The temperature function assumes the familiar form

$$\phi_T = \Theta^{(T-20)}, \qquad (13.11)$$

where T is the temperature of water in °C and Θ is the thermal factor, ranging from 1.047 to 1.066.

The light extinction function, which also incorporates the self-shading effect, can be written as (Thomann *et al.*, 1975)

$$\phi_I = \frac{2.718\, f}{K_e\, H}\, (e^{-\xi_1} - e^{-\xi_0}) \qquad (13.12)$$

TABLE 13.7 Some coefficients and their ranges in the model of eutrophicated water bodies

Coefficient	Units	Range	References
α_0	chlorophyll content of algae	mg Chl/g algae	20–100	Roesner et al., 1973
α_1	nitrogen content of algae	mg N/mg algae	0.08–0.09	Roesner et al., 1973
α_k	nitrogen content of higher trophic forms	mg N/mg s_k		Roesner et al., 1973
β_1	phosphate content of algae	mg P/mg algae	0.012–0.015	Roesner et al., 1973
β_k	phosphate content of higher trophic forms	mg P/mg s_k		
μ_{max}	max growth rate of phytoplankton	day^{-1}	1.0–6.0	Thoman et al., 1975 Roesner et al., 1973
ρ_1	endogenous respiration of algae	day^{-1}	0.02–0.5	Roesner et al., 1973
ρ_2	respiration of herbivorous zooplankton	day^{-1}	0.5	Dahl-Madsen and Gargas, 1974

Symbol	Description	Units	Value/Range	Reference
ρ_k	endogenous respiration of higher trophic forms	day^{-1}		
δ_2	phytoplankton grazing by herbivorous zooplankton	day^{-1}		
δ_k	grazing by higher trophic forms	day^{-1}		
K_{mn}	half-saturation constant for nitrogen	μg/liter	lakes and estuaries 5–50	Thomann et al., 1975; Dahl-Madsen and Gargas, 1974
			streams 200–400	Roesner et al., 1973
K_{mp}	half-saturation constant for phosphorus	μg/liter	lakes and estuaries 1–10	Thomann et al., 1975; Dahl-Madsen and Gargas, 1974
			streams 30–50	Roesner et al., 1973

where $K_e = K_e' + 0.0088\, s_1 + 0.054\, s_1^{0.66}$

$$\xi_1 = \frac{I_a}{I_s}\, e^{-K_e H},$$

$$\xi_0 = \frac{I_a}{I_s},$$

where K_e' is the light extinction coefficient at zero chlorophyll, f is the photoperiod, I_s is the light intensity at μ_{max}, I_a is incoming solar radiation, and H is the depth of the segment. All variables are functions of x, y, and z.

The nutrient effect function uses the Michaelis–Menton representation of microbiological growth. Hence,

$$\phi_N = \frac{s_9 + s_{10}}{K_{mn} + (s_9 + s_{10})} \frac{s_{12}}{K_{mp} + s_{12}}, \tag{13.13}$$

where K_{mn} and K_{mp} are the half-saturation constants for nitrogen and phosphorus, respectively, and s_9, s_{10}, and s_{12} are the respective concentrations of nitrite, nitrate, and available phosphorus.

Phytoplankton Settling Rate

Settling of phytoplankton represents a significant loss of the biomass from the computational segment or from the entire water body. Estimating the settling rate is a very difficult problem. Assuming that the computational segment behaves like an ideal settling tank, the settling velocity becomes

$$v_s = w_s - v,$$

where w_s is a settling velocity of the phytoplankton under quiescent conditions and v is the vertical flow velocity component or the surface overflow rate, i.e., the flow-through rate divided by the surface area. It has been observed (Thomann *et al.*, 1975) that the settling rate varies with the lake depth and the state of the phytoplankton biomass. Recommended values of v_s for QUAL II are within 0 and 1.8 m/sec.

Herbivorous Zooplankton Biomass (Carbon)

The concentration of the herbivorous zooplankton depends on the amount of their food supply (zooplankton), the death rate, and grazing by higher trophic level organisms. The mass balance then becomes

$$\overset{\circ}{S}_2 = (\delta_2 - \rho_2)\, s_2 - \delta_3\, s_3, \tag{13.14}$$

where δ_2 can again be approximated as

$$\delta_2 = \frac{\mu_{2\,max}\, s_1}{K_g + s_1}, \tag{13.15}$$

where K_g is the half-saturation constant for s_1.

Mass balance equations for all trophic groups higher than herbivorous zooplankton are similar and can be written as

$$\overset{\circ}{S}_k = (\delta_k - \rho_k)\, s_k - \delta_{k+1} s_{k+1}. \tag{13.16}$$

The equations for the biological portion of the ecological model are gross oversimplifications of the real phenomena. The major assumption on which the model is based is that the processes are linear, i.e., that the growth of a group of organisms is based on the concentration of organisms in a trophic level immediately below. This may not be true, especially in the case of the carnivorous zooplankton, which may also obtain their food supply from other dissolved and suspended organic matter besides herbivorous zooplankton. Thus, an increase of carnivorous zooplankton may not result in a decrease in phytoplankton concentration as would be indicated by this conceptual model. A feedback connected to the BOD model may partially alleviate the problem, however.

Modeling of higher trophic levels such as fish should assume that they will be free swimmers (Chen and Orlob, 1972), will be distributed according to the available food supply, and will not be affected by water currents.

Chemical Subsystem

Many chemical components can affect or be affected by phytoplankton and macrophyte growths. The major chemicals are the nutrients— inorganic carbon, nitrogen, and phosphorus—but many other chemicals may also become potentially limiting. In most practical models, only nitrogen and phosphorus are considered.

Nitrogen Cycle

As has been discussed previously, nitrogen can assume various forms, but only five basic species of nitrogen are important in modeling ecological aquatic systems. Nitrification, which is a semichemical process depending on nitrifying and denitrifying organisms, is the primary bridge connecting two inorganic forms of nitrogen, i.e., ammonia and nitrate nitrogen. The overall process occurring during eutrophication is shown in Fig. 13.9. It can be seen that the nitrogen (and also phosphorus) subsystem is a constant transformation from organic forms to inorganic forms and vice versa. It can be assumed that chlorophyll containing organisms utilize only inorganic forms, nitrates, and, to a lesser degree, ammonia and dissolved nitrogen gas, while bacteria derive their nitrogen needs mainly from organic forms and ammonia. Nitrifying bacteria represent a special case of autotrophic organisms since they use the process of oxidation of ammonia to nitrates as their energy source.

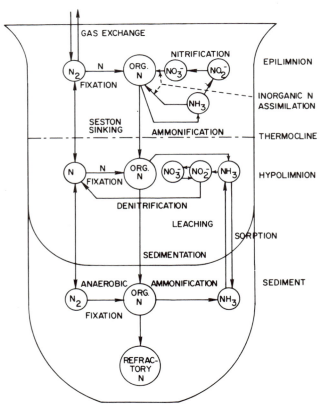

Fig. 13.9.   Nitrogen cycle reactions in an idealized stratified lake (Shannon and Brezonik, 1972).

The benthos may be a significant medium in which a large portion of the nitrogen (and phosphorus) transformation process takes place. Significant quantities of nitrifying and denitrifying bacteria have been found in benthic layers of lakes or reservoirs (Keeney, 1972, 1973; Lee, 1970). The benthos also represents a nutrient pool or storage. When the concentration of nutrients in the overlying water drops, nutrients are released from this storage source.

Ammonia nitrogen can be represented by the following mass balance equation:

$$\mathring{S}_{11} = \frac{\beta_1}{\alpha_0} (\rho_1 - \kappa_p \mu) \, s_1 + \sum_{2}^{4} \beta_k \rho_k s_k, \tag{13.17}$$

where s_8 is the concentration of ammonia and K_N is the rate coefficient for the conversion of ammonia to nitrite. For layers adjacent to the bottom, it is necessary to include the ammonia release from the benthic layer in the mass balance equation. Nitrite nitrogen is an intermediate step of the ammonia oxidation to nitrate, and it rarely occurs in appreciable concentrations.

Nitrate nitrogen represents the primary nitrogenous nutrient source for algae, with its mass balance as

$$\mathring{S}_{10} = K_N s_8 - \alpha_1 \mu \, s_1/\alpha_0. \tag{13.18}$$

Under anaerobic conditions, nitrate may be lost by denitrification, i.e., conversion to nitrogen gas.

Phosphorus Cycle

The phosphorus model appears to be less complicated than that for nitrogen since the model considers only the interactions of phosphorus and algae. However, as previously noted, only a portion of the phosphorus is available for algal and macrophyte growth. Phosphorus adsorbed on suspended particles and organic phosphates are generally not available for biological use. Some functional relationship which would correlate the available phosphate to the total phosphate levels should be developed.

The phosphorus model can be divided into two parts, total phosphorus and available phosphorus. The total phosphorus mass balance becomes

$$\mathring{S}_{11} = \frac{\beta_1}{\alpha_0} (\rho_1 - \kappa_p \mu) \, s_1 + \sum_{2}^{4} \beta_k \rho_k s_k, \tag{13.19}$$

where κ_p represents a factor relating total phosphorus to available phosphorus. Hence,

$$s_{12} = \kappa_p \, s_{11} \qquad \text{or}$$

$$\frac{s_{11} - s_{12}}{SS} = \frac{1}{\kappa'_p} \, s_{12}, \tag{13.20}$$

where κ_p is a function of suspended solids and κ'_p is factored out by the suspended solids concentration.

The phosphorus release rate from the benthic layer must be added to the phosphorus mass balance equations for segments adjacent to active benthic layers (i.e., those containing large amounts of organic materials and/or clay deposits).

Release of Nutrients from Benthic Sediments into Water

Nutrient–water–sediment interchange represents a significant mass balance factor in the nutrient balance of segments adjacent to benthic layers. Nutrients in the particulate phase, e.g., those adsorbed on sediments or incorporated into organic matter, will settle under quiescent conditions into the benthic layer. Thus, the benthos represents a storage pool of nutrients which, in this form, may be available only to rooted macrophytes. If there is a large concentration gradient between the nutrients in the bottom mud and overlying water, the nutrients can diffuse back in the water. Under quiescent conditions, the diffusion coefficients of nutrients are of the order of magnitude of 10^{-6} cm^{2}sec^{-1} (Stumm and Leckie, 1970) for phosphates (because of smaller molecular weight the diffusion coefficient for ammonia will be higher). The release rate can be computed as

$$j \simeq D \frac{\Delta c}{\sigma}, \tag{13.21}$$

where j is the release rate in g/m^2/day, D is the diffusion coefficient in m^2/day, Δc is the concentration difference between the nutrient in the sediment and that in water in mg/liter, and σ is the thickness of the so-called diffusion layer in meters.

Davis (1972), following a theory of diffusive transport in a turbulent environment developed by Levich, estimated the thickness of the diffusive layer to be

$$\frac{\sigma}{H} = 4.5 \, (\mathrm{Re})^{-7/8}, \tag{13.22}$$

TABLE 13.8 Nutrient release rates from sediments

Sediment Character and Site	Release Rate $(mg/m^{-2}day^{-1})$		Author
Phosphate			
Lake Baldeggersee	9–10	aerobic	Vollenweider (1968)
Muddy River	96	anaerobic	Fillos and Swanson (1975)
	9.6	aerobic	
Lake Warner	26	anaerobic	
	1.2	aerobic	
Ammonia			
Lake Warner	120		Fillos and Swanson (1975)
Muddy River	360		

where H is the depth of flow and Re is the Reynolds number (Re = UH/ν).

From these considerations, it may be concluded that the nutrient release rate will be higher in turbulent streams than in lakes and reservoirs. Table 13.8 shows the magnitudes of measured release rates. It should be noted that the experiments by Fillos and Swanson (1975) indicate about an order of magnitude higher release rate for phosphate if the water is devoid of oxygen. Since no such increase was observed for ammonia, one could deduce that the increase is largely due to a chemical change rather than to increased turbulence by escaping gases or increased biological activity.

LAKE REHABILITATION

The problems and signs of eutrophication include prolific weed growths, nuisance algal blooms, deteriorating fisheries, and impaired water quality, all of which reduce the beneficial uses of the affected water bodies. Since the primary beneficial uses of lakes and reservoirs are recreation, water supply, and shoreline developments, there is increasing public pressure for improving the water quality of affected areas.

Lake rehabilitation can be defined as the manipulation of a lake ecosystem to effect an in-lake improvement in degraded or undesirable conditions (Dunst *et al.*, 1974). The terms rehabilitation, renovation, and renewal are often used synonymously. Similar techniques can be also applied to reservoirs; however, reservoirs are more manageable since reservoir management methods may help to alleviate some problems.

Approaches to lake restoration fall into two general categories: (1) methods to limit fertility and/or sedimentation in lakes, and (2) procedures to manage consequences of lake aging.

Limiting fertility and controlling sedimentation can be achieved by reducing nutrient and sediment input into the lake from nonpoint and point sources, dredging of sediment in the lake, precipitation of nutrients in lake waters, biotic harvesting, and sealing the lake bottom.

The second group of remedial measures is only cosmetic since these methods do not remove the problem but make the consequences of eutrophication more tolerable or limited during a certain critical period. These techniques include application of algicides, weed cutting, fish stock reduction or elimination of certain species, hypolimnetic aeration and/or lake destratification, and manipulation of water levels.

Controlling the Eutrophication Process

Since eutrophication is apparently related to nutrients, the effort to control eutrophication must be oriented toward controlling nutrient levels and inputs. The critical nutrient concentrations suggested by Sawyer (1947) or permissible loading values by Vollenweider (1968) or Schindler *et al.* (quoted in Porcella, *et al,* 1974) indicate that excessive algal growths can be reduced if concentrations of inorganic nitrogen and phosphorus are kept below 0.3 mg/liter and 0.01 mg/liter, respectively, prior to stratification. These loading values provide a basis for estimating the degree to which nutrient levels must be reduced in order to minimize algal growth. Although the data base for Eqs. (13.4)–(13.7) included values from 30 lakes (12 from Central Europe, 10 from North America, and 8 from Northern Europe), they still must be considered provisional subject to confirmation and modification. For example, nutrient loading studies conducted on Florida lakes with a mean depth near 2 m indicated that these lakes are able to withstand much higher loadings than those suggested by Eqs. (13.4) to (13.7) (Shannon and Brezonik, 1972b).

Reduction of nutrient input can be achieved by the following measures:
1. *Nitrification–denitrification and/or phosphorus removal from wastewater treatment plant effluents.* The removal of phosphorus is relatively less complex and less expensive than nitrogen removal processes. It should be noted that the majority of the wastewater treatment plants in the United States and throughout the world were primarily designed to remove suspended solids and biodegradable carbonaceous organics (CBOD). Although some nitrogen and phophorus is incorporated into the biological flocs and subsequently removed with the waste sludge, these

removals may not be significant. Typical nutrient removal efficiencies for various treatment processes are given in Table 13.9.

Significant reduction of phosphorus can also be achieved by their elimination at the source, e.g., by eliminating or minimizing the phosphate content of detergents. According to Lee *et al.* (1978), 35% of the phosphorus in domestic wastewaters originates from phosphate detergents. Furthermore, the domestic wastewater phosphorus content has decreased since the early 1970s by several milligrams per liter. The prime reason for this decrease is that detergents of 1970 contained some 12% phosphorus, while today's (1980) content is approximately 5%.

2. *Diversion of wastewater outside or downstream from the lake where it can be less harmful.* Diversion without treatment is not an acceptable solution; however, effluents treated by conventional techniques are less dangerous to flowing waters than to stagnant water bodies.

Appreciable improvements in water quality of Madison, Wisconsin, lakes were reported after the city sewage effluents were diverted (Dunst *et al.,* 1974; Sonzogni *et al.,* 1975). The experience with diversion of wastewaters into Lake Washington was similar (Edmondson, 1977). It should be noted, however, that diversion of wastewaters does not limit the causes of eutrophication acceleration to phosphorus and nitrogen, since many other contaminants are contained in wastewaters.

3. *Land use practices.* A significant portion of nutrients originates

TABLE 13.9 Comparison of nutrient removal processes for domestic waste[a]

	Removal Efficiency (%)	
Process	Nitrogen	Phosphorus
Ammonia stripping	80–98	
Anaerobic denitrification	60–95	
Algae harvesting	50–50	varies
Conventional biological treatment	30–50	10–30
Ion exchange	80–92	86–98
Electrochemical treatment	80–85	80–85
Electrodialysis	30–50	30–50
Reverse osmosis	65–95	65–95
Distillation	90–98	90–98
Land application	varies	60–90
Modified activated sludge	30–50	60–80
Chemical precipitation		88–95
Chemical precipitation with filtration		95–98
Sorption		90–98

[a] After Eliassen and Tchobanoglous (1969).

from nonpoint sources. The best management practices, as defined and required by the Water Pollution Control Amendments of 1972 (PL 92-500), are to be oriented toward the reduction of nutrient loadings from these difficult-to-control sources. Efforts to limit nutrient and sediment inputs from lands within drainage basins generally follow two lines: (a) structural and land treatment means to intercept or reduce nutrients and sediment before they reach water bodies, and (b) regulatory approaches, particularly land use control, to restrict the use of lands which may have direct or indirect pollution potential or effect.

A significant portion of the nutrients is carried by, or associated with, sediments, and efforts to control soil losses in a watershed should also lower nutrient loadings.

Methods to Reduce In-Lake Nutrient Concentrations

Nutrient cycling and transformation processes in lakes are still not well understood. Both water and sediment nutrient levels must be reduced if appreciable improvements in lake water quality are to be achieved. The most critical period for algae development is during summer stratification. However, the critical nutrient levels must be below the limits before the production period, i.e., prior to the onset of stratification.

In-lake removal schemes may have an advantage to methods which would reduce the nutrient input. These methods are mostly instantaneous, while curbing the nutrient levels may not show an effect until after a prolonged period of time. When an input is instantaneously stopped, it takes three retention times to achieve 95% reduction of the concentrations in the lake, assuming that the lake is completely mixed. In using this rough "rule of thumb," it should be remembered that the hydraulic detention time (volume divided by flow) is not the same as the retention time of nutrients in the lake. For example, much shorter residence times were estimated for phosphorus in the Great Lakes when the interaction of sediments and water was taken into account, resulting in the prediction of a more rapid recovery (Sonzogni *et al.*, 1976).

Dredging of Sediments

Since sediments represent a large stored source of nutrients, sediment removal has often been proposed as a means of reducing eutrophication. Before such measures are evaluated, however, the potential impact of bottom sediments on the lake should be considered. Table 13.8 demonstrates that the maximum diffusional release of phosphorus in aerobic lakes is in the order of 1 $mg/m^{-2}/day^{-1}$ or approximately 0.3 to 0.4

$g/m^{-2}/year^{-1}$. These levels may be below dangerous loadings if the average depth is greater than 50 m as estimated by Eq. (13.6). Although some remedial effects of dredging may be acknowledged, the overall impact of the dredging process will probably adversely affect the lake environment. Stumm and Leckie (1970) state the following reasons why dredging should not be used as a means of curbing eutrophication: (1) the capacity of sediment to store nutrients and make them unavailable is reduced; (2) the disruption of well-consolidated portions of bottom sediments by the dredging operation would expose large quantities of stored nutrients and make them immediately available to algae; (3) the removal of sediments and the disturbance of the relationship between benthos and plankton would upset the ecological balance of the lake, obliterate its buffer capacity, and reduce its resistance to external changes. In addition, the cost of dredging is prohibitive in most lake situations.

Nutrient Inactivation

Nutrients are primarily available to algae in their ionic dissolved form. Therefore, nutrient inactivation by chemical additions should accomplish the following: (1) change the form of the nutrients from available forms to less available or unavailable ones; (2) remove the nutrients from the euphotic zone; and (3) prevent the release or recycling of the nutrients back into the euphotic zone (Dunst *et al.*, 1974).

The addition of chemicals is most attractive for phosphorus control and has been used for controlling eutrophication in phosphorus limited lakes. Metal addition has been used for the removal of phosphates from wastewater effluents, and research results indicate that good removals of phosphorus from lake water can be achieved by adding metallic coagulant aids, namely, aluminum sulfate. Consequently, by adding a coagulant, other water quality parameters such as turbidity and color may also be improved. The precipitated flocs which settle to the bottom may provide a protective cover reducing the nutrient release from benthal layers. The application rates of aluminum sulfate have ranged from 700 kg/ha to 4650 kg/ha (Dunst *et al.*, 1974), and its efficiency varied with the coagulant dosage, phosphate concentration and form, and the chemical composition of the water (pH, alkalinity). Dilution and flushing have been attempted in order to reduce excessive algal growth by lowering the nutrient levels within a lake. This can be accomplished by replacing nutrient-rich lake water with nutrient poor water and causing subsequent washout of phytoplankton from the lake. Lowering of water levels below the groundwater levels by pumping water out of the lake will partially replace the lake water with nutrient poor groundwater. Generally, spring-fed lakes show better quality than those fed by runoff. It may also be possible to bring in nutrient poor water by pumping from other sources such as municipal supplies or a nearby watercourse. Biotic harvesting is often used as a maintenance and cosmetic relief for the symptoms of eutrophication; i.e., excessive algal and macrophyte growth are harvested by floating weed-cutters to improve the appearance of a lake. In addition, significant amounts of nutrients can be also removed from the lake in this manner.

Biotic harvesting has also been attempted as a means of removing nu-trients from sewage and wastewater effluents in algal ponds. This method may be effective in small ponds, but the effects of harvesting in larger lakes have so far been only cosmetic and limited to the removal of aquatic weeds.

Aeration and Destratification

Partial or full destratification does not reduce the problem of eutrophi-cation for it is mainly designed to improve the dissolved oxygen condi-tions. Destratification is often accomplished by hypolimnetic aeration by means of high-rate pumps or diffusers. Aeration with lower energy will re-sult only in increased oxygen levels in the hypolimnion since high-energy inputs are necessary to ''break'' the thermocline. Such a scheme has been used successfully on the Weinbach Reservoir in Germany (Bernhardt, 1967).

However, as stated before, increased oxygen levels in the hypolimnion may reduce phosphate releases from the sediments into the hypolimnion and thus reduce the nutrient levels in the lake. In addition, total destratifi-cation reduces the residence time of algae in the euphotic zone, and, as a consequence, a concomitant reduction in the algal concentrations occurs.

Selective Withdrawal

The hypolimnion and thermocline areas of eutrophic lakes are usually low or devoid of dissolved oxygen and, as a consequence, rich in nu-trients because of the increased release from sediments under anaerobic conditions. By increasing the discharge from lower outlets (sluiceways) of reservoirs or by siphoning or pumping water from the hypolimnion of nat-ural lakes, aerobic conditions can be maintained and high nutrient levels can be reduced.

It should be realized that selective withdrawals cause increases of tem-perature in the hypolimnion, and, therefore, the dissolved oxygen in-crease may be negated by increased rates of biochemical oxygen de-manding reactions. Also, the discharge of the anaerobic high nutrient waters may result in some problems downstream from the lake such as increased weed growth, hydrogen sulfide odors, and/or fish kills by the low oxygen containing waters. In-stream or turbine aerations may help to alleviate these problems. Other inlake control methods include sediment exposure by drawdown, thus exposing the sediments to the atmosphere and reducing their oxygen demand and lake bottom sealing, which should prevent the release of nutrients. Sealing can be accomplished by covering the bottom with plastic materials, clay, or deposits of hydrous metal coagulant flocs.

It must be emphasized that with few exceptions all of the discussed methods are still in an early research stage, since their overall benefits or possible damages to the aquatic ecosystems are uncertain, and their costs are quite high.

Control of the Symptoms of Eutrophication

The symptoms of eutrophication include prolific algal and macrophyte growths, water devoid of or low in oxygen during night hours in the epilimnion and/or in the hypolimnion, taste and odor problems, and limited recreational use. Several methods are oriented toward reducing the impact of eutrophication rather than treating the causes.

Weed and algae harvesting is the most common example of cosmetic treatment of symptoms although some nutrients are also removed from the lakes. Taste and odor problems are commonly treated by altering or improving water treatment technology, such as the application of activated carbon, ozonation, etc.

The control of nuisance algal blooms can be also achieved with a variety of chemicals. Chemical treatment has the greatest utility and justification in highly eutrophic lakes in which the nutrient supply cannot be curbed or effectively controlled and where other managment practices are not feasible (Dunst *et al.*, 1974). Chemical controls can be divided into three categories: algicides, herbicides, and piscicides.

Several algicides are registered for use in water by the EPA. Copper sulfate is the most common, but others such as acrolein, dichlone, or sodium pentachlorophenate have been used in the past. The dosage of copper sulfate varies from 0.05 to 4.0 mg/liter depending on the form and stage of eutrophication. The effectiveness of copper sulfate is limited in hard water because of carbonate combination with the copper to form an insoluble precipitate of basic copper carbonate.

Herbicides are used for controlling the growths of macrophytes. For many years prior to the 1960s, sodium arsenite was the only herbicide used effectively to control macrophytes (Dunst *et al.*, 1974). However, presently there are at least 20 chemicals registered with the EPA ranging from simple inorganic chemicals, such as sodium arsenite and sodium hypochlorite, to more complex organic herbicides.

Herbicides are usually used when the bottom of the littoral zone is exposed. Herbicides must be used with caution and the use of persistent chemicals must be avoided. Professional experience and *in situ* testing should be applied to arrive at effective dosages with minimal harmful environmental effects.

Piscicides are toxicants used mainly for fish population control. Although their primary use is to improve unbalanced fish population and fishing, their usage often results in improved water quality. Excessive turbidity and the absence of macrophytes is often associated with an overabundant carp population, and many lake managers believe that limiting the carp population will reduce the symptoms of eutrophication. However, the effect of carp elimination on water quality has rarely been documented. It is interesting to note that European experience in commercial fish management does not share this opinion. Carp have been stocked and commercially grown in lakes and ponds for more than five hundred years with no apparent significant effect on accelerating the eutrophication process.

Any use of chemicals may produce side effects. By definition, adding, for example, copper sulfate to waters represents pollution and should be judged accordingly. Some of the commonly used chemicals may be cumulative toxicants with yet unknown environmental effects.

Biological Methods

The use of biological methods to control excessive aquatic growths has received much attention in recent years. For example, the introduction of the white amur (*Ctenopharyngodon idella Val.*), a native of the Amur River between Manchuria and Russia, has received much attention by the popular press. This fish, an exotic species to the United States, consumes large quantities of aquatic vegetation and results in edible fish flesh. However, even though the introduction of these herbivorous fish has resulted in a significant reduction in aquatic growths, many biologists have strenuously objected to their presence in United States waters. In fact, many states have made the introduction of the amur illegal. The reasons for concern are (1) biological control agents often are nonspecific for the target species, (2) limiting ecological parameters were not properly identified and considered for the successful introduction of the control agent into its new environment, and (3) the lack of knowledge of the possible antagonism between the biological control agent and companion aquatic animals. The senior author observed the introduction of these fish into some Polish lakes with somewhat anomalous results inasmuch as the fish reduced the aquatic weed problem, but also reduced the population of indigenous fish.

Studies by the TVA's Division of Environmental Planning and the Office of Agricultural and Chemical Development confirmed the effectiveness of the Amur and the concerns over its introduction in 1972. However, the Director of the TVA's Division of Forestry, Fisheries, and Wildlife, a forester, was totally unaware of these studies and attempted to promote the instigation of these fish (unsuccessfully) into TVA lakes in 1977.

The Corps of Engineers is presently investigating several other biological agents for controlling aquatic weed infestation. These studies include the use of pathogens and insects for the control of the waterhyacinth, a moth for the control of Eurasian watermilfoil, and a midge for the control of a hydrilla.

The TVA Experience

The problems of weed control are exemplified by the experience of the TVA in controlling weeds in their impoundments. The major problem in the TVA reservoirs is caused by Eurasian watermilfoil (*Myriophyllum spicatum L.*), although other species of aquatic weeds are emerging in the system. The effects of these hydrophytes include increased mosquito production, inhibition of recreation, aesthetic degradation, clogging of water intakes, and inhibition of desirable waterfowl food plants.

The primary means of controlling Eurasian watermilfoil infestations are water level fluctuations and the application of 2,4-D (2,4-dichlorophenoxyacetic acid). In 1976, it was estimated that 16,000 acres of the TVA lakes were infested with this macrophyte. The restrictions placed on the registration of chemicals by the EPA have resulted in only one supplier being registered under the provisions of the Federal Insecticide, Fungicide and Rodenticide Act (FIFRA), with a concomitant cost increase for 2,4-D of $35 per acre in 1974 to over $100 per acre treated in 1977. Thus, if all 16,000 acres were treated each year, the cost would exceed $1.5 million per year. It should be noted that the use of mechanical harvesters in this instance is impractical because of the potential spread of the weed by fragmentation. In addition, the use of new, possibly more efficient, herbicides is inhibited because of the lack of registration by the EPA.

Evaluation of Lake Improvement

A major problem encountered in instigating lake rehabilitation processes is the lack of a predictive mechanism that will demonstrate potential improvement in lake water quality as a result of a specific treatment technique. For example, will a demonstrable improvement result in the Great Lakes as a result of the 1 mg/liter phosphorus discharge limitation imposed by the International Joint Commission (IJC) on phosphate detergent limitations? In Las Vegas, Nevada, $62 million is being invested in a phosphorus removal facility with no assurance that the water quality of Lake Mead will be significantly improved. This does not include operation and maintenance costs, which are substantial.

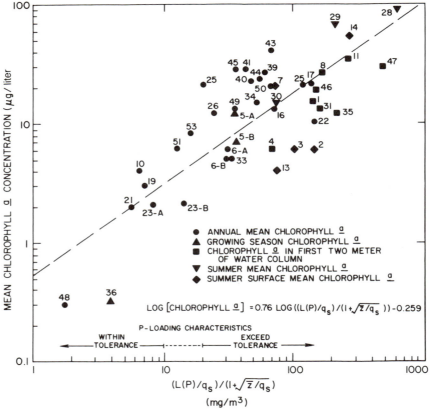

Fig. 13.10. Phosphorus loading characteristics and mean chlorophyll *a* relationship applied to United States OECD water bodies (Lee *et al*, 1978).

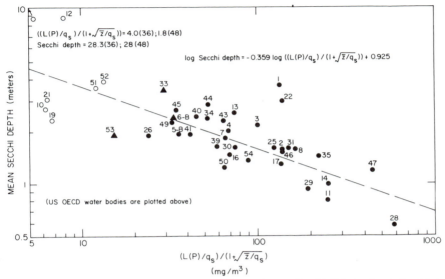

Fig. 13.11. Phosphorus loading characteristics and mean Secchi disk depth relationships as applied to United States OECD water bodies (Lee *et al.*, 1978).

Assuming that phosphorus is the limiting nutrient, Lee *et al.*, (1978) have expanded the Vollenweider relationships to now include chlorophyll a, Secchi disk depth, and hypolimnetic oxygen depletion. These relationships were developed as a result of the North American Project of the Organization for Economic Cooperation and Development (OECD) studies on eutrophication effects and control. The relationships show surprisingly good correlations when their nature is considered. For example, different investigators contributing the data used different analytical techniques and concomitant errors. The lakes used possessed differing characteristics and the measurements are based on total phosphorus. Because of the large amount of data existing on the TVA lakes, it is unfortunate that the agency chose not to participate in the study because the general manager's office thought eutrophication to be unimportant.

Figures 13.10 to 13.12 show the results of these studies where on the abscissa $L(P)$ = Surface area total phosphorus loading (mg P/m²/yr); q_s = hydraulic loading (m/yr) = $\bar{z}/\tau_w$; $\bar{z}$ = mean depth (m) = water body volume (m³) ÷ surface area (m²); and τ_w = hydraulic residence time (yr) = water body volume (m³) ÷ annual inflow volume (m³/yr).

In addition to demonstrating the relationships between these measures of eutrophication and phosphorus loading, the plots show relative insensi-

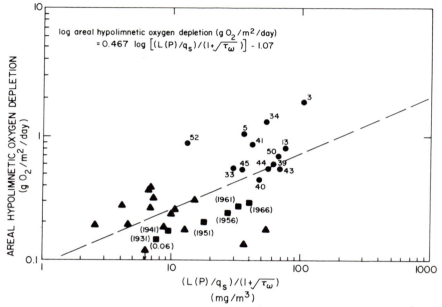

Fig. 13.12. Phosphorus loading characteristics and hypolimnetic oxygen depletion relationship in natural waters (Lee *et al,* 1978). Legend: ■ Central basin of Lake Erie (Year of Data Record); ● U.S. OECD water bodies; ▲ Washington, U.S.A., and Ontario, Canada, water bodies (Data taken from, and water bodies identified in Welch and Perkins, 1977).

tivity of the chlorophyll a levels, water clarity, and hypolimnetic oxygen rates to phosphorus loading inasmuch as log–log plots are utilized. The studies also demonstrated that non–point source control of phosphorus is essential in many lakes if phosphorus is indeed the limiting nutrient controlling eutrophication.

REFERENCES

Bernhardt H. (1967). Aeration of Wahnbach Reservoir without changing the temperature profile. *J. Am. Water Works Assoc.* (August).

Brenzonik, P. L. (1973). Nitrogen sources and cycling in natural waters. USEPA-660/3-73-002, Washington, D.C. (July).

Brooks, J. L. (1969). Eutrophication and changes in the composition of the zooplankton. *In* Eutrophication: Causes, consequences, correctives pp. 236–255 (G. A. Rohlich, ed.). Natl. Acad. Sci., Washington, D.C.

Carlson, R. E. (1977). A trophic state index for lakes. *Limnol. Oceanogr.* **22,** 361–369.

Chen, C. W., and Orlob, G. T. (1972). Ecologic simulation. OWRR C-2044, Water Res. Eng., Walnut Creek, California.

Dahl-Madsen, K. I., and Gargas, E. (1974). A preliminary eutrophication model of shallow fjords. Paper pres. at Int. Conf. Int. Assoc. Water Pollut. Res., 7th, Paris, France.

Davis, J. T. (1972). "Turbulence Phenomenon." Academic Press, New York.

Dunst, R. C., Born, S. M., Uttormark, P. D., Smith, S. A., Nichols, S. A., Peterson, J. O., Knauer, D. R., Serns, S. L., Winter, D. R., and Wirth, T. L. (1974). Survey of lake rehabilitation techniques and experiences. Tech. Bull. No. 75, Wisconsin Dept. Nat. Resour., Madison.

Edmondson, W. T. (1977). Trophic equilibrium of Lake Washington. USEPA-600/3-77-087, Corvallis, Oregon (August).

Eliassen, R., and Tchobanoglous, G. (1969). Removal of nitrogen and phosphorus from wastewater. *Environ. Sci. Technol.* **3,** 536–541.

Fillos, J., and Swanson, W. R. (1975). The release rate of nutrients from river and lake sediments. *J. Water Pollut. Control Fed.* **47,** 1032–1042.

Greeson, P. E. (1969). Lake eutrophication, a natural process. Water Resour. Bull. **5**(4), 16–30.

Hasler, A. D. (1974). Eutrophication of lakes by domestic drainage. *Ecology* **28**, 383–395.

Hrbáček, (1962). Species composition and the amount of the zooplankton in relation to fish stock. *Rozpr. Cesk. Akad. Ved. Mat. Prir. Ved.* **72**, 1–116, Prague, Czechoslovakia.

Hutchinson, G. E. (1969). Eutrophication, past and present. *In* Eutrophication: Causes, consequences, correctives, pp. 17–26 (G. A. Rohlich, ed.). Natl. Acad. Sci., Washington, D.C.

Keeney, D. R. (1972). The fate of nitrogen in aquatic ecosystems. Univ. of Wisconsin Water Resour. Center, Eutrophication Info. Program, *Lit. Rev.*, No. 3, Madison.

Keeney, D. R. (1973). The nitrogen cycle in sediment-water system. *J. Environ. Qual.* **2**, 15–29.

Krenkel, P. A., and Novotny, V. (1973). The assimilative capacity of the South Fork Holston River and Holston River below Kingsport, Tennessee. Rep. to Tennessee Eastman Co., Kingsport.

Lackey, J. B. (1949). Plankton as related to nuisance conditions in surface waters. (F. R. Moulton, and L. Hitzel, eds.) *In* Limnological aspects of water supply and waste disposal. *Am. Assoc. Adv. Sci. Bull.*, Publication No. 28, pp. 56–63, Washington, D.C.

Larkin, P. A., and Northcote, T. G. (1969). Fish as indices of eutrophication. *In* Eutrophication: Causes, consequences, correctives, pp. 256–273 (G. A. Rohlich, ed.), Natl. Acad. Sci., Washington, D.C.

Lee, G. F. (1970). Factors affecting the transfer of materials between water and the sediment. Univ. of Wisconsin Water Resour. Center Eutrophication Info. Program, *Lit. Rev.*, No. 1, Madison.

Lee, G. F., Rast, W., and Jones, R. A. (1978). Eutrophication of water bodies: Insights for an age-old problem. *Environ. Sci. Technol.* (August).

Lueschow, L. A., Helm, J. M., Winter, D. R., and Karl, G. W. (1970). Trophic nature of selected Wisconsin lakes. *Wis. Acad. Sci. Arts Letters* **58**, 237–264.

Maloney, T. E. *et al.* (1972). Use of algal assays in studying eutrophication problems. *Proc. 6th Int. Conf. Int. Assoc. Water Pollut. Res.*, Jerusalem, Israel (June). 205–214.

O'Connell, R. L., Geckler, J. R., Clark, R. M., Cohen, J. B., and Hirth, C. R. (1962). Report of survey of the Truckee River. U.S. Dept. of Health, Educ. and Welfare, Public Health Serv., Cincinnati, Ohio. (July).

Porcella, D. B. Bishop, A. B., Anderson, J. C., Asplund, O. W., Crawford, A. B., Greeney, W. J., Jenkins, D. I., Jurinek, J. J., Lewis, W. D., Middlebrooks, E. J., and Walkinshaw, R. W. (1974). Comprehensive management of phosphorus water pollution. USEPA-600/5-74-01, Washington, D.C.

Rodhe, W. (1969). Crystallization of eutrophication concepts in Northern Europe. *In* Eutrophication: Causes, consequences, correctives, pp. 50–64 (G. A. Rohlich, ed.). Natl. Acad. Sci., Washington, D.C.

Roesner, L. A., Monser, J. R., and Evenson, D. E. (1973). The stream quality model QUA-II. EPA Contract No. 68-01-0739, Water Res. Eng., Walnut Creek, California.

Rohlich, G. A., ed. (1969). Eutrophication: Causes, consequences, correctives, pp. 3–7. Natl. Acad. Sci., Washington, D.C.

Sawyer, C. N. (1947). Fertilization of lakes by agricultural and urban drainages. *J. N. Eng. Water Works Assoc.* **51**, 109–127.

Shannon, E. E., and Brezonik, P. L. (1972a). Eutrophication analysis: A multivariate approach. *J. Sanit. Eng. Div., Am. Soc. Civ. Eng.* **98**, 37–58.

Shannon, E. E., and Brezonik, P. L. (1972b). Relationship between lake trophic state and nitrogen and phosphorus loading rates. *Environ. Sci. Technol.* **6**, 719–725.

Sloey, W. E., and Spangler, F. L. (1978). Trophic status of the Winnebago Pool Lakes. Proc. Ann. Am. Water Resour. Assoc. (Wisconsin), 2nd, Water Res. Center, Univ. of Wisconsin, Madison.

Sonzogni, W. C., Fitzgerald,G. P., and Lee, G. F. (1975). Effect of wastewater diversion on the Lower Madison Lakes. *J. Water Pollut. Control Fed.* **47,** 532–542.

Sonzogni, W. C., Uttormark, P. C., and Lee, G. F. (1976). A phosphorus residence time model: theory and application. *Water Res.* **10,** 429–436.

Steward, K. M., and Rohlich, G. A. (1967). Eutrophication—a review. Rep. to State Water Qual. Board of California, Water Res. Cent., Univ. of Wisconsin, Madison.

Stumm, W., and Leckie, J. O. (1970). Phosphate exchange with sediments: Its role in the productivity of surface waters. Pap. Pres. Int. Conf. Int. Assoc. Water Pollut. Res., 5th, San Francisco, California (July).

Thomann, R. V., DiToro, D. M., Windfield, R. P., and O'Connor, D. J., (1975). Mathematical modeling of phytoplankton in Lake Ontario. 1: Model development and verification USEPA-660-3-75-005, Washington, D.C.

U.S. Environmental Protection Agency (1972). Water quality criteria. Natl. Advisory Comm. Rep., pp. 55–56, Washington, D.C.

U.S. Environmental Protection Agency (1973). An approach to a relative trophic index system for classifying lakes and reservoirs. Natl. Eutrophication Survey, Work Paper No. 24, Pacific Northwest Lab. Corvallis, Oregon.

Uttormark, P. D., and Wall, J. P. (1975). Lake classification—a trophic characterization of Wisconsin lakes. USEPA-660-3-75-033, Office Res. and Devel., Washington, D.C.

Vollenweider, R. A., (1968). Scientific fundamentals of the eutrophication of lakes and flowing waters, with particular reference to nitrogen and phosphorus as factors in eutrophication. Organ. for Econ. Cooperation and Devel., Directorate of Scientific Affairs, Rep. No. DAS/CSI/68.27, Paris, France.

Welch, E. B., and Perkins, M. A. (1977). Oxygen deficit rate as a trophic state index. Draft manuscript, Dept. of Civ. Eng., Univ. of Washington, Seattle.

14 | *Thermal Pollution*

INTRODUCTION

Considerable time has elapsed since the scientific community and regulatory agencies officially recognized that the addition of large quantities of heat to a receiving water possesses the potential for causing ecological harm. It is interesting to note that a "comprehensive" water pollution survey conducted by the United States Public Health Service in 1962 (USDHEW, 1963) did not even include a significant power plant in the study, while today the Environmental Protection Agency (EPA) has demonstrated inclinations toward precluding the use of once-through cooling.

It is obvious that the thermal pollution problem must be placed in its proper perspective in order to prevent the imposition of unnecessary and pointless costs to the American consumer/taxpayer. For example, it has been shown by the Utility Water Act Group (UWAG), which is composed of the majority of electrical power generating industry in the United States, that should the EPA be successful in enforcing its requirements for cooling devices, the average cost to each American household would be on the order of $250 per year—$125 through increases in the residential electrical bill and $125 through increased costs of goods and services provided by industrial and commercial consumers of electricity (Edison Electric Institute, 1974).

The temperature of receiving waters is considered and measured as a physical quantity. However, temperature levels are quite significant in water quality management. Probably the single most important water quality parameter affecting aquatic life is temperature. All biochemical and most physical–chemical reactions are temperature dependent, inasmuch as most reaction rates, whether physical, chemical, or biological in nature, are approximately doubled for each 10°C increase in temperature. Thus, the modeling of temperature in receiving waters is of great import.

The problem is quite complex and involves many disciplines. In order to understand the plethora of claims stated by various investigators, it is necessary to explore both natural and technological additions of heat to our waterways. Furthermore, the effect(s) of these heat additions on biological, physical, and chemical environments must be elucidated. It is to this end that this chapter is directed.

OVERFLOW VERSUS COMPLETELY MIXED HEATED DISCHARGES

If the discharge from a power plant is in the form of an overflow, as described in Chapter 12, mixing between the upper and lower layers is inhibited, thus minimizing oxygen replacement and self-purification in the lower layer. Because of lack of mixing, organic wastes discharged into the lower layer do not have access to the oxygen in that portion of the stream flowing in the upper layer. Thus, there is less dissolved oxygen, less dilution water, and a more concentrated organic load in the lower layer leading to an acceleration of the dissolved oxygen depletion. The net result may be a considerable reduction in the waste assimilative capacity of the receiving water.

If the heated discharge is completely mixed with the receiving water, some of the aforementioned effects are eliminated. However, the rise in temperature still causes a decrease in the ability of water to hold dissolved oxygen, an increase in the metabolic activity of organisms, an increased rate of biochemical oxygen demand exertion, and a possible reduction in waste assimilative capacity.

A discussion of whether the heated water should be discharged as a layer or completely mixed with the receiving water must be based on conditions existing in the receiving water and the amount of heat to be dissipated. Obviously, the heat will be dissipated more quickly if it is concentrated in a lesser volume of water. The effects of the resulting stratified flow on the biota are not so obvious, however. While it is true that fish will

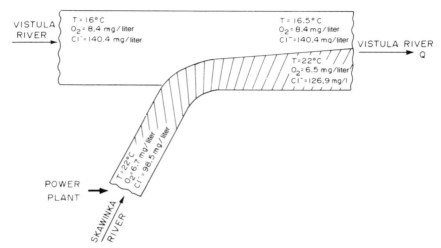

Fig. 14.1.   Power plant discharge into Vistula River (Krenkel, 1967).

have passage through the cool, lower layers of water, they cannot utilize this route if the water is devoid of oxygen.

Another possible route for heated water effluents is to "hug" the shoreline and not gain entry into the major portion of the river flow for many miles downstream. Such a situation was noted in Poland as illustrated in Fig. 14.1 (Krenkel, 1967). This separation of flow was reported to have existed for over 10 km downstream from the confluence of the two rivers. Note also the division of water quality characteristics indicating the two flow entities. These conditions delight biologists, since fish will have a portion of the river unaffected by the heated water discharge in which to pursue their natural activities.

The question of the effects of layered flow in the ecological system must be subjected to study as the biological changes induced by the stratified flow conditions are relatively unknown.

CULTURAL HEAT CONTRIBUTIONS

Industrial Heat Contributions

Though the most significant contribution to thermal pollution is from the steam electrical generating industry, it should be noted that many wet process industries do contribute waste heat to our environment. It is more difficult to quantitize these effects. However, it has been estimated that over 70% of the process water withdrawn for industrial uses is for cooling purposes.

In terms of the total thermal pollution problem, it is estimated that the contribution of industry to thermal pollution is in the order of 20 to 30% of the total heat rejected to our receiving waters. It should also be noted that these contributions are usually less concentrated heat loads and therefore

TABLE 14.1 Heat absorbed by cooling water for various processes[a]

Item	Heat Load Absorbed by Cooling Water	Btu's per
Alcohol	20,000	gal
Aluminum	31,000	lb
Beer	91,000	bbl
Butadiene	31,000	lb
Cement	150,000	ton
Refined Oil	150,000	bbl
Soap	97,000	ton
Sugar	200,000	ton
Sulfuric Acid	650,000	ton
Cooling power equipment		
Air compressors		
Single stage	380	BHP-h[b]
Single stage with aftercooler	2,540	BHP-h
Two stage with intercooler	1,530	BHP-h
Two stage with intercooler and aftercooler	2,550	BHP-h
Diesel engine jacket water and lube oil (incl. dual fuel)		
Four cycle, supercharged	2,600	BHP-h
Four cycle, nonsupercharged	3,000	BHP-h
Two cycle, crankcase compr.	2,000	BHP-h
Two cycle, pump scavenge (large)	2,300	BHP-h
Two cycle, pump scavenge (high speed)	2,100	BHP-h
Natural gas engines		
Four cycle (250 psi compr.)	4,500	BHP-h
Two cycle (250 psi compr.)	3,000	BHP-h
Refrigeration		
Compression	250	min-ton
Adsorption	500	min-ton
Steam jet refrigerator condenser (100 psi dry steam supply, 2 in. Hg cond.)	1,100	lb of steam
Steam turbine condenser	1,000	lb of steam

[a] After McKelvey and Brooke (1959).
[b] BHP-h = brake horsepower-hour.

do not cause the problems resulting from a single, large point source of heat such as that emanating from a steam electrical generating plant. Even sewage contributes some temperature rise to the receiving water, although its effect is negligible. The heat absorbed by cooling water in some industrial processes is shown in Table 14.1, which was taken from McKelvey and Brooke (1959).

The really significant heat loads result from the discharge of condenser cooling water from the ever-increasing number of steam electrical generating plants. Electrical power generation in the United States has doubled every ten years since 1945, and all indications are that the rate of increase will be even greater during the next few decades, although the current energy crisis may help to mitigate the problem. Furthermore, the potential problems associated with these concentrated heat loads are compounded by the increasing size of individual power plants and the greater quantities of heat discharge by equivalent-sized nuclear power reactors. As will be shown subsequently, large nuclear power plants cur-

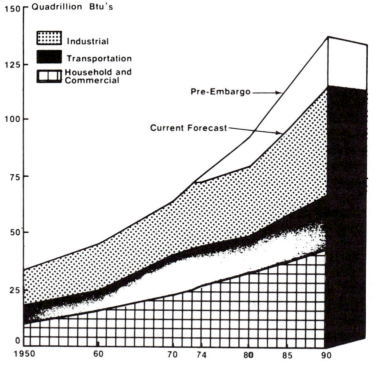

Fig. 14.2. Projected energy demand in the United States (FEA, 1976).

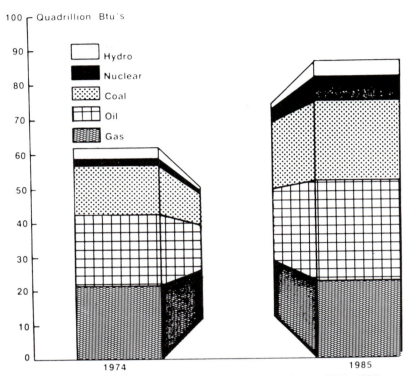

Fig. 14.3. Means of supplying energy in the United States (FEA, 1976).

rently require approximately 50% more cooling water for a given temperature rise than do fossil-fueled plants of equal size. If the breeder reactor were allowed to be developed in the United States, the amount of heat dissipated to our waterways would approach that of fossil-fueled plants (Water Resources Council, 1968).

As shown in Fig. 14.2 (Federal Energy Administration, 1976), the *rate* of increase of energy demand in the United States is continuing to rise, even after a short decrease, because of the 1974 embargo on oil. The projected means of supply of this energy is shown in Fig. 14.3, which indicates even more rejected heat to our waterways.

Table 14.2 indicates the anticipated growth of the power industry as projected by the Federal Power Commission (1971). Even a cursory examination of these data demonstrates a dramatic increase in energy requirements with time. It is interesting to note the percentage of power production that is and will be contributed by nuclear power.

In 1970 hydroelectric plants provided 15.2% of the capacity and energy

TABLE 14.2   Projection of generating capacity[a]

Generator	1970 Megawatts	Percent	1980 Megawatts	Percent	1990 Megawatts	Percent
Conventional hydro	51,700	15.2	68,000	10.4	82,000	6.5
Pumped storage hydro	3,600	1.1	27,000	4.0	71,000	5.6
Fossil steam	260,300	76.5	393,000	59.0	557,000	44.6
Internal combustion and gas turbine	18,300	5.4	30,000	4.5	50,000	3.9
Nuclear	6,100	1.8	147,000	22.1	500,000	39.4
Total	340,000	100.0	665,000	100.0	1,260,000	100.0

[a] After Federal Power Commission (1971).

produced; indications are that this will continue to decline as shown in Table 14.2. There are very few undeveloped hydroelectric power sites remaining, although potential sites for pumped storage installations still exist in most areas of the United States (Water Resources Council, 1968). The optimal use of pumped storage is for peaking power and reserve capacity since the large steam electric plants now being constructed operate best at high plant factors which are complemented by the low plant factors of the pumped storage plants.

Population Growth and Per Capita Consumption

It is instructive to note the relative effects of population growth and per capita consumption on electrical energy demands. Contrary to popular belief, increasing per capita consumption of electrical energy accounts for the majority of increasing energy requirements. According to Resources for the Future (1970), "Ninety percent of the growth in power generation in the last thirty years has been caused by higher per capita consumption and only ten percent by population growth." Also, the Northwest Public Power Association has predicted that "Only 1½ percent of the estimated 6 percent load growth per year in the next 20 years is attributable to population growth. The rest will come from increase in per capita consumption" (Northwest Public Power Assoc., 1970).

Environmentalists have suggested that so-called "luxury" uses of electricity should be curtailed and thus alleviate the energy problem. As dem-

TABLE 14.3 Estimated electrical energy requirements for the South Central region for 1990[a]

Use Category	Projected Electrical Energy Requirements for 1990 (Million kW h)
Farm (excluding I and D pumping)	8,613
Irrigation and drainage pumping	984
Nonfarm residential	242,760
Commercial	134,235
Industrial	418,963
Street and highway lighting	4,265
All other	18,355
Losses and energy unaccounted for	72,205
Total	900,380

[a] After Federal Power Commission (1971).

onstrated by Koelzer and Tucker (1971), however, the savings in electrical energy demand incurred by such a restriction would not be great. In order to place the proposal in its proper perspective, they used detailed energy consumption data for the South Central Region which was prepared by the Federal Power Commission (Federal Power Commission, 1971), as shown in Table 14.3.

If the use of freezers, dishwashers, disposals, clothes dryers, etc., were prohibited, the total energy projection for the South Central Region would be reduced by less than 2% or 885,170 million kW h. Adding air conditioning to the list of prohibited uses of electricity would lower the projection to 834,420 million kW h, for a total reduction of 7%. In either case, the savings in energy would not be greatly effective.

While arguments would be presented that the bans proposed would also decrease industrial electrical energy demands because of decreased production, the overall conclusions would not change significantly. The problem posed then is concerned with "trade-offs" between economic growth, increased standard of living, and total protection of the environment. The current energy crisis has made it quite clear that all of these objectives cannot be attained. Thus, it may be concluded that the addition of heat to the environment will continue to increase.

Cooling Water Requirements

The Water Resources Council (1978) estimates that steam electric generation will decrease from 88.9 million gallons per day in 1975 to 79.5 mil-

TABLE 14.4 Average condenser water requirement and consumptive use for fossil-fueled, steam-electric power plants, 1965–2020[a,b]

Year	Condenser Requirements	Consumptive Use		
		Once Through	Cooling Ponds	Cooling Towers
1965–1979	40	0.3	0.4	0.5
1980–1999	35	0.2	0.3	0.4
2000–2019	30	0.15	0.25	0.35
2020	25	0.1	0.2	0.3

[a] After Water Resources Council (1968).
[b] In gallons per kW h.

lion gallons per day in the year 2000. Actually, the major portion of this use is of a nonconsumptive nature, thus not precluding that portion of water for other beneficial utilization. Examination of Table 14.4, which is the Water Resources Council's estimate of cooling water requirements and consumptive use based on a temperature rise of 15°F (8.3°C.), demonstrates the consumptive use in fossil-fueled steam electric power plants (Water Resources Council, 1968). The predicted decrease in unit water requirements is based on improved technology. It is interesting to note that while the latest Water Resources Council (1978) report estimates that the fresh-water withdrawals for steam electric generation will decrease by 11% by the year 2000, the consumptive use is projected to increase from 2% of the total fresh-water withdrawals for steam electric generation in 1975 to 13% by the year 2000. This significant increase is expected because of the increase in the use of cooling devices.

The degree of thermal pollution depends on thermal efficiency which is determined by the amount of heat rejected to the cooling water. Thermodynamically, heat should be added at the highest possible temperature and rejected at the lowest possible temperature if the greatest amount of work is to be gained and the highest thermal efficiency realized.

Current generally accepted maximum operating conditions for conventional thermal stations are 1000°F and 3500 psi, with a corresponding heat rate of 8700 Btu's kW h, 3413 Btu's resulting in power production and 5287 Btu's being wasted. Plants have been designed for 1250°F and 5000 psi; however, metallurgical problems have held operating conditions to lower levels.

Nuclear plants operate at temperatures of from 500° to 600°F (260° to 315.6°C) and pressures of up to 1000 psi, resulting in a heat rate of approximately 10,500 Btu/kW h. Thus, for nuclear plants, 3413 Btu's may be used for useful production and 7087 Btu's may be wasted.

Table 14.5 is presented in order that a comparison of various steam

TABLE 14.5 Heat characteristics of typical steam electric plants[a,b]

Plant Type	Thermal Efficiency (percent)	Required Input (Heat Rate)	Required Input Minus Heat Equivalent[c]	Lost to Boiler Stacks (etc.)[d]	Heat Discharged to the Condenser	Cooling Water Requirement (ft³/sec per MW of capacity)[e]
Fossil fuel	33	10,500	7,100	1,600	5,500	1.6
Fossil fuel	40	8,600	5,200	1,300	3,900	1.15
Light water reactor	33	10,500	7,100	500	6,600	1.9
Breeder reactor	42	8,200	4,800	300	4,500	1.35

[a]This table is based, with minor modification, on information contained in an FPC staff study entitled "Problems in Disposal of Waste Heat from Steam-Electric Plants, June 1969."

[b]Heat values in Btu per kW h.

[c]The heat equivalent of 1 kw h of electricity is 3413 Btu's.

[d]Approximately 10 to 15% × required input for fossil fuel and approximately 3 to 5% × required input for nuclear.

[e]Based on a temperature rise of 15°F (8.3°C).

TABLE 14.6   Typical energy balance[a]

Assumed overall efficiency	40%
Assumed generator efficiency	97.5%
Heat equivalent of 1 kW h	3413 Btu
Fuel energy required, 3413/0.40	8533 Btu
Heat losses from boiler furnace, at 10% of fuel use	853 Btu
Energy in steam delivered to turbine 8533 − 853	7680 Btu
Heat loss from electric generator at 2.5% of generator input	87 Btu
Electric generator output	3413 Btu
Energy required for generator equals energy output from turbine	3500 Btu
Energy remaining in steam leaving turbine, removed in condenser 7680 − 3500	4180 Btu
Total cooling water required, 10° rise $\dfrac{4180 + 87}{10 \times 8.33}$	51 gal
Total cooling water required, 15° rise $\dfrac{4180 + 87}{15 \times 8.33}$	34 gal

[a] After Cootner and Lof (1965).

electric plants of different design may be made. It should be noted that even though the development of the breeder reactor will increase the efficiency of nuclear plants, it will not significantly change heat discharges as compared to the best designed fossil-fueled plant.

Table 14.6 shows a typical energy balance for a steam electric power plant producing 1 kW h net electrical output (Cootner and Lof, 1965).

For comparative purposes, a nuclear power plant requires approximately 10,500 Btu's to produce 1 kW h of electricity. Thus 7087 Btu's are wasted into the cooling water, or 2907 more Btu's than the fossil-fueled plant described in Table 14.5. It should be noted, however, that the plant efficiencies assumed are not average and the plants may not be operating at peak loads, thus somewhat modifying the comparison.

Increasing efficiencies have resulted from significant improvements in the conventional steam cycle; however, future heat rate gains will be minimal. Present-day central station average thermal efficiency is said to be 33% and the best present performance near 42%.

EFFECTS OF HEAT ON THE ENVIRONMENT

Fluctuating Discharges of Heated Water

Most steam-powered electrical generating plants are operated at varying load factors, and, consequently, the heated discharges demonstrate wide

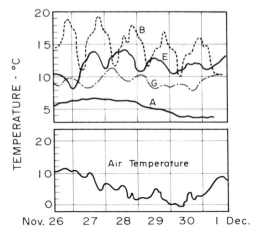

Fig. 14.4. Variation in temperature of River Lea (Gameson *et al.*, 1959).

variation with time. Thus, the biota is not only subjected to increased or decreased temperature, but also to a sudden or "shock" temperature change. An attempt to quantify the effects of the rate of temperature change on aquatic biota was made by Speakman and Krenkel (1972). It was found that bluegill were much more susceptible to decreasing temperatures than increasing temperatures.

This type of power plant operation is exemplified by Fig. 14.4, which was taken from an investigation by Gameson *et al.* (1959). At station B, a "short distance below the outfall," the diurnal variation is as great as 8°C. Even at station E, which was about two miles downstream, the effects are significant, approximately one-half those at station E. Obviously, the biota would have a most difficult time adjusting to these temperature changes. It is important to note that "shock" loadings may be more harmful to fish than continual exposure, as observed by Cairns (1955). The relationship between flow and temperature increase should be taken into account, however.

Effects of Heated Discharges on Waste Assimilation

Former Senator Edmund Muskie stated that "It is the opinion of the Senate Subcommittee on air and water pollution that excessive heat is as much a pollutant as municipal wastes or industrial discharges" (Anon., 1967). The reasons for this statement have been alluded to in a previous section. It is instructive to note the following case history which demonstrates the validity of Muskie's statement.

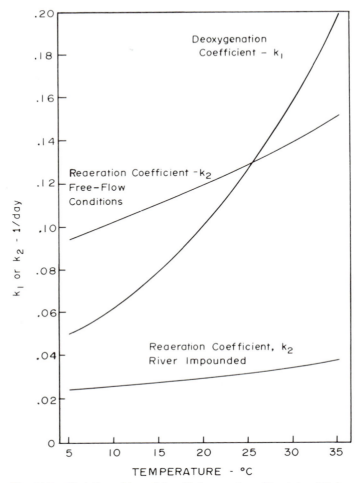

Fig. 14.5.   Variation of k_1 and k_2 with temperature (Krenkel and Parker, 1969).

A classic example of the change in river waste assimilation capacity, caused by both temperature and impoundment, was presented by Krenkel *et al.* (1965). A paper mill of the Georgia Kraft Company, which had previously satisfactorily discharged its waste effluent into the free-flowing Coosa River, must now discharge into the backwaters of a downstream impoundment and receive its flow as regulated by a peaking hydroelectric power plant upstream and the releases from a pumped storage operation. The daily fluctuation in discharge may range from 200 to 7800 ft³/sec. In addition, a steam electric generating plant, with a capacity of 300,000 kW, has located adjacent to the mill, and the condenser water raises the temperature of the stream several degrees.

The effects of temperature on the stream self-purification process are demonstrated in Fig. 14.5, which shows the variation of the rate constants k_1 (deoxygenation) and k_2 (reaeration) with respect to temperature. Examination of this relationship demonstrates that an increase in temperature causes a considerable increase in k_1. While k_2 also increases with increasing temperature, it is negated by the combination of a lesser dissolved oxygen content and a greater rate of change of k_1 with temperature.

The overall effect of the impoundment on the rate of oxygen recovery is demonstrated by the lower curve, which depicts the reaeration rate constant under existing, impounded conditions. Note that, while k_1 at a given temperature is unchanged, the value of k_2 at any temperature if significantly reduced.

In order to illustrate the effect of temperature on the waste assimilative capacity of the Coosa River, observed data were used to obtain Fig. 14.6, which depicts the oxygen balance in the Coosa River prior to the previously mentioned water resources developments. Note that under these conditions, the Coosa River easily assimilated 28,000 lb/day of BOD at the existing river temperature of less than 25°C. Even under the free-flowing conditions, however, a temperature of 30°C would cause the dis-

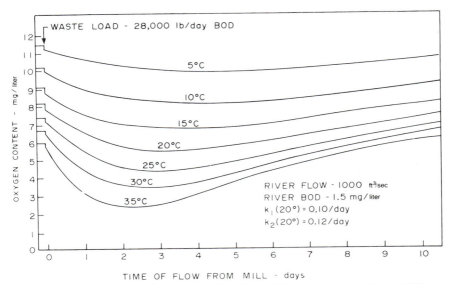

Fig. 14.6. Oxygen sag curve, free-flowing condition (Krenkel and Parker, 1969).

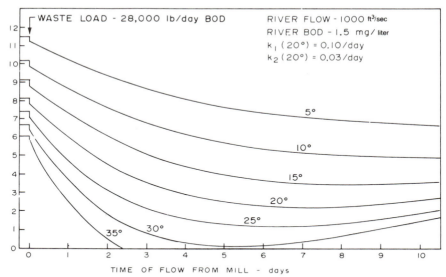

Fig. 14.7. Oxygen sag curves, river impounded (Krenkel and Parker, 1969).

solved oxygen level to fall below 4.0 mg/liter, the minimum required to
satisfy the existing stream standards in 1964.

Figure 14.7 is presented to show that, because of the combined effects
of the water resources developments and the observed temperature increases, the Coosa River can no longer satisfactorily assimilate the
28,000-lb BOD load, as under the previously existing free-flowing conditions.

The quantitative effect of temperature on the waste assimilative capacity of the Coosa River can be shown as in Fig. 14.8, where the possible
waste load that would not deplete the dissolved oxygen content of the
river below 4.0 mg/liter is plotted versus the river water temperature
(note that EPA standards now require 5 mg/liter of dissolved oxygen).

It may be concluded from Fig. 14.8 that if the river temperature were
25°C, a 5°C increase in temperature under free-flow conditions is equivalent to 1000 lb/day of BOD, and a 5°C increase in temperature under the
existing, impounded conditions is equivalent to 5200 lb/day of BOD.

Note the significant reduction in assimilative capacity already caused
by the impoundment and the observed increase in temperature, as demonstrated by a comparison of the curve for free-flow conditions and the
curve for impounded conditions. It is obvious from these curves, which
were computed from observed conditions, that the addition of heat and
the impoundment of the Coosa River have had the same end result as if an

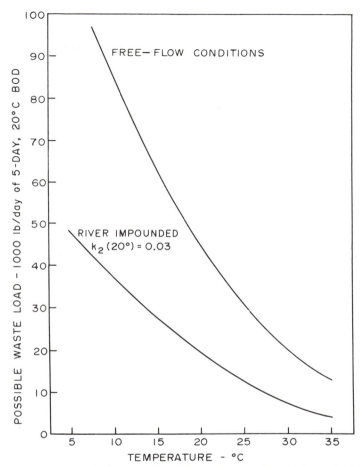

Fig. 14.8. Waste assimilative capacity of Coosa River at Georgia Kraft Co. Mill, river flow = 940 ft³/sec (Krenkel and Parker, 1969).

equivalent amount of sewage or other organic waste material were added to the river.

Because of the previously described reduction in the waste assimilative capacity of the Coosa River, the mill reduced its waste load to the river to less than 20,000 lb/day of BOD, which was the allowable load permitted by the Georgia Water Quality Control Board in order to maintain an adequate dissolved oxygen content. Of course, under the new law (PL-92-500) further reduction in BOD will be required.

Bohnke (1961, 1966) states that an increase in temperature of the river water of 6°C will cause a 6% reduction in self-purification capacity at a minimum oxygen content of 3 mg/liter. The loss of assimilative capacity due to a specific power plant on the Lippe River was said to be 10,000 population equivalents at low flow.

It may be concluded from these studies that the addition of heated water to a receiving water can be considered equivalent to the addition of sewage or other organic waste material, since both pollutants may cause a reduction in the oxygen resources of the receiving water.

EFFECTS OF HEAT ADDITIONS ON AQUATIC LIFE

There is no doubt that temperature is the single most important water quality parameter affecting the biota. However, temperature changes affected by cooling water must be placed in their proper perspective. In most instances, the increases in temperature are not large and probably do not cause biological harm outside the mixing zone. In fact, little data exist to support the claims of extensive heat damage from power plants on the biota. Furthermore, outside of entrainment problems, few substantiated fish kills have been reported as a result of power plant operations.

The possible effects of heat on fish may be summarized here.

(a) Direct death from excessive temperature rise beyond the thermal death point
(b) Indirect death due to
    (1) less oxygen available
    (2) disruption of the food supply
    (3) decreased resistance to toxic materials
    (4) decreased resistance to disease
    (5) predation from more tolerant species
    (6) synergism with toxic substances
(c) Increase in respiration and growth
(d) Competitive replacement by more tolerant species
(e) Sublethal effects

While each of these factors could be important at a specific location, the temperature rises typical of most power plants are not usually sufficient to be of concern. It is interesting to note that field research studies on the effects of heat on the environment have been hindered, both in the United States and Europe, because of the lack of sufficient temperature elevations below existing power plants.

A brief discussion of these factors is in order. Obviously, if the temperature becomes high enough, an organism can succumb. However, temperatures rarely reach such levels. It should be noted in this regard that fish are poikilotherms; i.e., they have no internal mechanism to compensate for temperature changes in their environment. (Man does possess the ability to adjust to temperature change and is called a homiotherm.) Organisms are further classified into groups known as stenotherms and eurytherms. Eurytherms possess a wide range of temperature tolerance while stenotherms are rather narrow in their ability to resist temperature change. This becomes quite important when considering the application of uniform temperature standards no matter what the geographic location; i.e., fish in the extreme south may already be near their temperature tolerance. The relationship is shown in Fig. 14.9 as taken from Hawkes (1969).

Obviously, as the temperature increases, an organism increases its rate of metabolism and thus its oxygen consumption. Inasmuch as the solubility of oxygen decreases with increasing temperature, the combination of lower oxygen content and increased oxygen requirements may cause stress on the organism. Increased temperatures could also cause a disruption or change in the food supply for the fish. These factors may cause the fish to be more susceptible to toxic materials, less resistant to disease, and less able to escape from predation by other organisms. Thus, increasing temperatures may exert some rather subtle effects.

It is important to note the synergistic action of temperature with toxic materials. It is generally accepted, for example, that heavy metals have a much greater toxicity to fish when the temperature is raised. Cyanide toxicity demonstrates a two to threefold increase in the rate of lethal action according to McKee and Wolf (1963). Hawkes (1969) indicates some

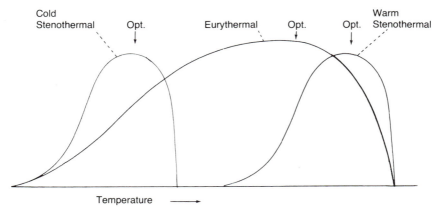

Fig. 14.9. Eurytherms and stenotherms (Hawkes, 1969).

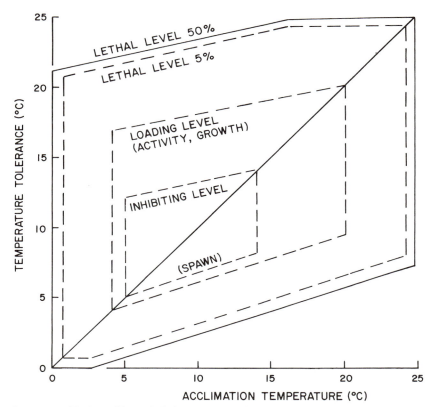

Fig. 14.10.   Upper and lower lethal temperatures for young sockeye salmon (Brett, 1960).

questionability of the generalization of the synergistic effects of heat, however.

The effects of temperatures on fish are determined not only by the temperature level, but by the history of temperature exposure by the organisms. Thus, the acclimation time, the acclimation temperature, and the rate of temperature change all affect the significance of a particular temperature on fish (Speakman and Krenkel, 1972).

Little is known concerning the sublethal effects of temperature on organisms. For example, does an increase in temperature cause damage to a fish's reproductive capacity or change its longevity? Some insight into this problem may be gained from previous paragraphs.

Figure 14.10 shows the complex nature of temperature regimes on the life cycle of a fish, and Fig. 14.11 demonstrates changes in metabolic activity with temperatures. Figure 14.12 shows that increased tempera-

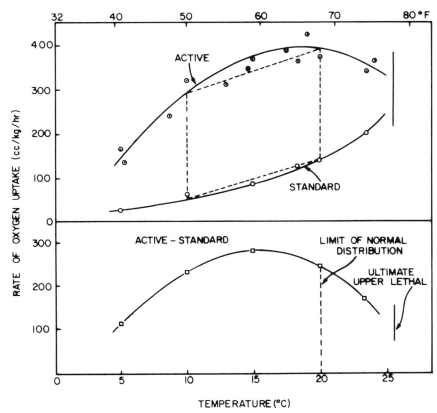

Fig. 14.11. Relation of oxygen uptake to temperature in acclimated speckled trout (Fry *et al.*, 1946).

tures cause the growth of undesirable algae, and Fig. 14.13 illustrates that *Escherichia coli* multiply much more rapidly at elevated temperatures. Finally, Fig. 14.14 shows the type of information that is available, which gives median tolerance limits for fish acclimated to 20°C.

It may easily be concluded from this brief discussion that a serious paucity of data exists on the effects of temperatures on fish. As stated by Jones (1964) in his treatise on fish and river pollution: "Whether fish are killed in significant numbers by thermal pollution is doubtful, but they may disappear from heated regions of rivers, at least during the warm months of the year."

The results of the Section 316(a) studies described in Chapter 2 also indicate that the thermal problem is not as serious as once thought.

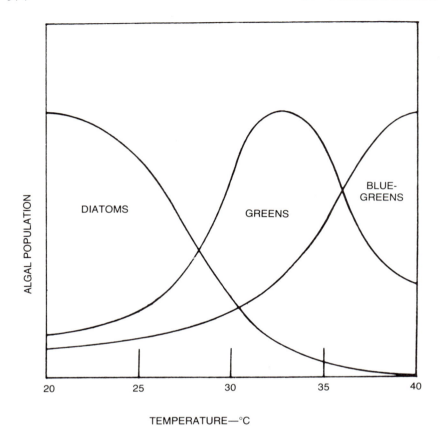

TEMPERATURE—°C

Fig. 14.12.   Algae population shifts with change in temperature (Cairns, 1955).

MECHANISMS FOR THE NATURAL DISSIPATION OF HEAT

The need for adequate temperature criteria that will permit the sustenance and propagation of aquatic life is obvious. After the biologists have delineated the temperature levels that must be maintained in our receiving waters, control measures to limit the amount of heat discharges must be utilized. The concepts of heat dissipation and the methodology available will now be examined.

The Energy Balance

Obviously, the simplest method of disposing of waste heat is to discharge it directly to the receiving water and then allow natural forces to

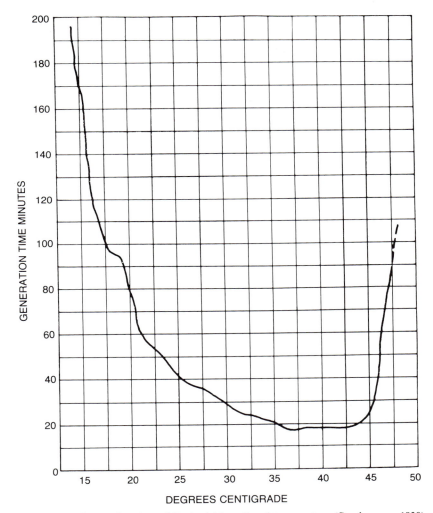

Fig. 14.13. Generation time of *Escherichia coli* and temperature (Stephenson, 1930).

bring the water back to an equilibrium temperature. This process is known as once-through cooling and is shown schematically in Fig. 14.15. In order to predict the behavior of these heated effluents, it is necessary to resort to an energy balance.

The energy or heat budget was used by Schmidt (1915) to approximate ocean evaporation and has since been applied to compute evaporation from water bodies of all sizes. The relatively recent development of more sophisticated instrumentation has allowed the energy budget to be utilized with a fair degree of reliance.

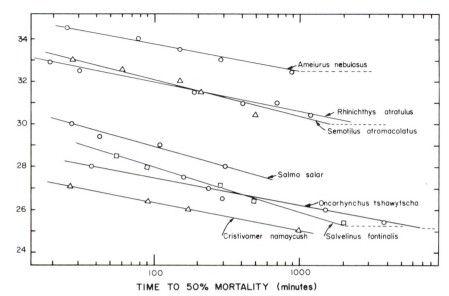

Fig. 14.14.   Median resistance times for fish acclimated to 20°C (Brett, 1952).

The energy budget was first tested against a water budget control at Lake Hefner in 1950–1951 (Anderson, 1954). It was concluded that satisfactory results were obtained for periods of ten days or more. A second check against a water budget was made at Lake Colorado City, Texas, in 1954–1955. Here, inflow was extremely small and outflow was zero (Harbeck *et al.*, 1959). The components of the energy budget per unit surface area of a reservoir per unit time may be written as (World Meterorological Organization, 1966)

$$Q_s - Q_r + Q_a - Q_{ar} - Q_{bs} + Q_v - Q_e - Q_h - Q_w = Q, \quad (14.1)$$

where Q_s = short-wave radiation incident to the water surface; Q_r = reflected short-wave radiation; Q_a = incoming long-wave radiation from the atmosphere; Q_{ar} = reflected long-wave radiation; Q_{bs} = long-wave radiation emitted by the body of water; Q_v = net energy brought into the body of water in inflow, including precipitation and accounting for outflow; Q_e = energy utilized by evaporation; Q_h = energy conducted from the body of water as sensible heat; Q_w = energy carried away by the evaporated water; and Q = increase in energy stored in the body of water.

Edinger and Geyer (1965) have depicted the heat transfer terms across a water surface, as shown in Fig. 14.16, noting temperature dependent terms and typical values obtained in English units. The addition of heated

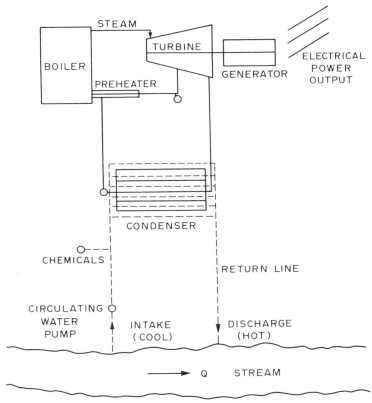

Fig. 14.15. Schematic diagram of once-through cooling and steam electrical generation (Krenkel and Parker, 1969).

water discharges simply superimposes the heat addition upon the other dissipations and additions of energy.

Figure 14.17 (U.S. Congress, Senate, 1968) shows the results of calculations by Bergstrom (1968) for a water surface in central Illinois. The data demonstrate the relationship of rate of heat dissipation to elevation of the water surface temperature over natural temperature and the mechanisms by which this dissipation is achieved. It is significant to note that the rate of heat dissipation for a given rise in temperature is greater in summer than in winter. Also, the heat dissipation by evaporation is much greater in summer than in winter. These calculations would appear to support the contention that heated effluents should be discharged in the most concentrated form possible (neglecting any biological or other effects) in order to dissipate the heat most rapidly.

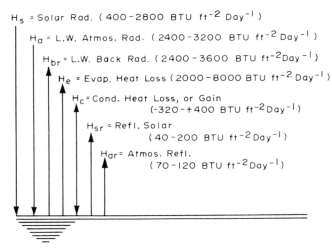

H_s = Solar Rad. (400 - 2800 BTU ft^{-2} Day^{-1})

H_a = L.W. Atmos. Rad. (2400 - 3200 BTU ft^{-2} Day^{-1})

H_{br} = L.W. Back Rad. (2400 - 3600 BTU ft^{-2} Day^{-1})

H_e = Evap. Heat Loss (2000 - 8000 BTU ft^{-2} Day^{-1})

H_c = Cond. Heat Loss, or Gain
(-320 - +400 BTU ft^{-2} Day^{-1})

H_{sr} = Refl. Solar
(40 - 200 BTU ft^{-2} Day^{-1})

H_{ar} = Atmos. Refl.
(70 - 120 BTU ft^{-2} Day^{-1})

NET RATE AT WHICH HEAT CROSSES WATER SURFACE

$$\Delta H = (H_s + H_a - H_{sr} - H_{ar}) - (H_{br} \pm H_c + H_e) \text{ BTU ft}^{-2} \text{ Day}^{-1}$$

H_R Temp. Dependent Terms

Absorbed Radiation
Independent of Temp.

$$H_{br} \sim (T_s + 460)^4$$

$$H_c \sim (T_s - T_a)$$

$$H_e \sim W (e_s - e_a)$$

Fig. 14.16. Mechanisms of heat transfer across a water surface (Edinger and Geyer, 1965).

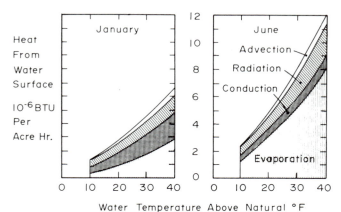

Fig. 14.17. Various modes of heat dissipation from water surface (U.S. Congress, Senate, 1968). Heat dissipation from water surface by evaporation, radiation, conduction, and advection during January and June.

578

While the terms in the energy budget are discussed in detail by Anderson (1954) and Edinger and Geyer (1965), brief comments pertaining to their determination are in order.

Short-Wave Radiation, Q_s

Short-wave radiation originates directly from the sun, although the energy is depleted by absorption by ozone, scattering by dry air, absorption scattering by particulates, and absorption and scattering by water vapor. It varies with latitude, time of day, season, and cloud cover. Thus, while this quantity can be empirically calculated, it is much better to measure it with a pyrheliometer, which will give the accuracy required for the energy budget.

Long-Wave Atmospheric Radiation, Q_a

Long-wave atmospheric radiation depends primarily on air temperature and humidity, and increases as the air moisture content increases. It may be a major input on warm cloudy days when direct solar radiation approaches zero. It is actually a function of many variables, including carbon dioxide and ozone, although it can be fairly accurately calculated by means of empirical formulation as demonstrated in the Hefner studies. It can be measured with the Gier–Dunkle flat plate radiometer, although it is more convenient to calculate than measure.

Reflected Short-Wave and Long-Wave Radiation, Q_r and Q_{ar}

Solar reflectivity, (R_{sr}), is more variable than atmospheric reflectivity, (R_{ar}), inasmuch as the solar reflectivity is a function of sun altitude and cloud cover, while atmospheric reflectivity is relatively constant. The Lake Hefner studies demonstrated the atmospheric reflectivity to be approximately 0.03, while on an annual basis the solar reflectivity was 0.06. The Hefner studies used the equation

$$R_{sr} = a \, S_a^{\,b} \tag{14.2}$$

to determine solar reflectivity, where S_a is the sun altitude in degrees and a and b are constants depending on cloud cover. Note that $R_{sr} = Q_r/Q_s$ and $R_{ar} = Q_{ar}/Q_a$.

Long-Wave or Back Radiation, Q_{bs}

Water sends energy back to the atmosphere in the form of long-wave radiation and radiates almost as a perfect blackbody. Thus, the Stefan–

Boltzmann fourth-power radiation law can be utilized, or

$$Q_{bs} = 0.97\sigma(T_0 + 273)^4, \tag{14.3}$$

where Q_{bs} = long-wave radiation in calorie/cm²/day, σ = Stefan–Boltzmann constant = 1.171×10^{-7} calorie/cm²/deg⁴/day, and T_0 = water surface temperature in °C.

All that is required to compute Q_{bs} is the water surface temperature, and a table giving the value of Q_{bs} for any temperature, T_0, is usually available.

Energy Utilized by Evaporation, Q_e

Each pound of water evaporated carries its latent heat of vaporization of 970 Btu's; thus Q_e is a significant term in the energy budget. The Lake Hefner study was explicitly promulgated for determining correct evaporation relationships and resulted in this equation:

$$Q_e = 11.4W\,(e_s - e_a) = \text{Btu/ft}^2/\text{day}, \tag{14.4}$$

which is of the general type of evaporation formula

$$E = (a + bW_x)\,(e_s - e_a), \tag{14.5}$$

where a, b = empirical coefficients; W_x = wind speed at some elevation, x; e_a = air vapor pressure; e_s = saturation vapor pressure of water determined from water surface temperature; and E = evaporation = $Q_e/\rho L$ (ρ = density of evaporated water, L = latent heat of vaporization).

Many expressions have been developed for estimating the evaporation rate. The coefficients differ because of variation in the reference height for measurement of wind speed and vapor pressure, the time period over which measurements are averaged, and local topography and conditions. As stated by Edinger and Geyer (1965), "It would also be expected that the coefficients would be much different for rivers and streams than for lakes and might well be dependent on water velocity and turbulence, particularly in the case of smaller rivers."

Energy Conducted as Sensible Heat, Q_h

Heat enters or leaves water by conduction if the air temperature is greater or less than water temperature. The rate of this conductive heat transfer is equal to the product of a heat transfer coefficient and the temperature differential.

A single direct measurement of this quantity is not available, and recourse to an indirect method is necessary. The method involves using average figures of air temperature, water surface temperature, and humidity for the period in question, and computing the ratio of Q_h to Q_e, which is known as the Bowen ratio and expressed as

$$R_B = \frac{Q_h}{Q_e} = \frac{0.61\,P\,(T_0 - T_a)}{1000\,(e_0 - e_a)}, \tag{14.6}$$

where P = atmospheric pressure in millibars, T_a = temperature of air in °C, T_0 = temperature of water surface in °C, e_0 = saturation vapor pressure corresponding to temperature of water surface in millibars, and e_a = vapor pressure of air at height at which T_a is measured in millibars.

Energy Carried Away by Evaporated Water, Q_w

Water being evaporated from the surface is at a higher temperature than the lake water, and thus energy is being removed. Though some believe this term is included in the conductive energy term (World Meteorological Organization, (1966), it is relatively small and can be readily computed from

$$Q_w = \rho_e\,c\,E\,(T_e - T_b) = \frac{\text{cal}}{\text{cm}^2/\text{day}}, \tag{14.7}$$

where ρ_e = density of evaporated water, g/cm³, c = specific heat of water, cal/g, E = volume of evaporated water, g/cm²/day, T_e = temperature of evaporated water, °C, and T_b = base or reference temperature, °C.

Advected Energy, Q_v

The net energy contained in water entering and leaving the lake may be computed from this expression:

$$\begin{aligned}
Q_v = [\,&c_{si}\,V_{si}\,\rho_{si}\,(T_{si} - T_b) + c_{gi}\,V_{gi}\,\rho_{gi}\,(T_{gi} - T_b) \\
&- c_{so}\,V_{so}\,\rho_{so}\,(T_{so} - T_b) - c_{go}\,V_{go}\,\rho_{go}\,(T_{go} - T_b) \\
&+ c_p\,V_p\,\rho_p\,(T_p - T_b)]/A, \tag{14.8}
\end{aligned}$$

in which Q_v = advected energy in cal cm⁻²/day⁻¹, c = specific heat of water (≈ 1 cal g⁻¹ deg⁻¹), V = volume of inflowing or outflowing water in cm³/day⁻¹, ρ = density of water (≈ 1 cal g⁻¹ deg⁻¹), T = temperature of water in °C, and A = average surface area of reservoir in cm². The subscripts are: si = surface inflow, gi = groundwater inflow, so = surface outflow, go = groundwater outflow, p = precipitation, and b = base or reference temperature, usually taken as 0°C.

Since some of the terms in Eq. (14.8) may not be measurable, a water budget is performed for the same period, evaporation is estimated, and the unknown terms are found by trial and error.

Increase in Energy Stored, Q

The change in storage in the energy-budget equation may be either positive or negative, and it is found from properly averaged field measurements of temperature and Eq. (14.9).

$$Q = [c\rho_1 V_1 (T_1 - T_0) - c\rho_2 V_2 (T_2 - T_0)]/At, \qquad (14.9)$$

in which Q = increase in energy stored in the body of water in cal cm^{-2}/day^{-1}, c = specific heat of water ($\simeq 1$ cal g^{-1}); ρ_1 = density of water at T_1 ($\simeq 1$ g cm^{-3}), V_1 = volume of water in the lake at the beginning of the period in cm^3), T_1 = average temperature of the body of water at the beginning of the period in °C, ρ_2 = density of water at T_2 ($\simeq 1$ g cm^{-3}), V_2 = volume of water in the lake at the end of the period, in cm^3, T_2 = average temperature of the body of water at the end of the period in °C, T_0 = base temperature in °C, A = average surface area in cm^2 during the period, and t = length of period in days.

From this necessarily brief discussion of the various parameters comprising the energy balance, it may be concluded that it is possible to predict heat dissipation using these concepts. Obviously, the reliability of the results will depend on the degree of sophistication used in the theoretical approach and the frequency and accuracy of the measurements taken.

Temperature Modeling in Flowing Waters

Because the temperature of the water in streams and estuaries is a response to several qualitatively different inputs, the solution and mathematical descriptions are usually quite complex. Heat transfer in flowing waters is a combination of convective motion and radiation transfer together with emission, scattering, and absorption of radiation energy. Three basic equations govern or describe heat transfer in streams and estuaries. The first two are the equation of continuity and the equation of motion which have previously been discussed. The third equation is the energy or heat balance equation, which for one-dimensional stream flow can be simplified to

$$\frac{\partial T}{\partial t} + U \frac{\partial T}{\partial x} + \frac{\partial \langle v'T' \rangle_{av}}{\partial y} = D_t \frac{\partial^2 T}{\partial x^2} + \frac{Q_s}{H \rho_w C_p}, \qquad (14.10)$$

where T is the water temperature, U is the mean stream velocity, ρ_w is the density of water, C_p is the heat capacity of water, v' is the vertical velocity fluctuation, H is the depth of flow, Q_s is the short-wave radiation input, D_t is the longitudinal turbulent heat transfer mixing coefficient, x is the direction of flow, y is the vertical direction, and t is time.

From dimensional analysis it can be demonstrated that the differential change of the correlation moment, $\partial \langle v'T' \rangle_{av}/\partial y$, can be assumed to be a differential change of the heat transfer over y. Because $\bar{v} \simeq 0$, where $\bar{v}$ is the average vertical flow velocity, this may be interpreted as meaning that the heat transfer over the depth is a result of only the turbulent velocity pulsations. According to Minskij (1952), the vertical pulsation velocity is independent of the relative depth, and, therefore, the differential change of the correlation moment will depend only on the charge of temperature as given by the difference of the convection heat flux between the upper and lower layers. If $Q + \Delta Q$ is the amount of heat per unit time originating from the upper layer of thickness $(H - y)$ from the direction of a heat source, and Q is the heat extracted into the lower layer of thickness y, then the differential change of the correlation moment, which is actually a change of heat flux in the layer Δy, is

$$\frac{\partial (v'T')}{\partial y} \simeq \frac{\Delta Q/\rho_w\, C_p}{\Delta y} \simeq \frac{Q_w/\rho_w\, C_p}{H}, \qquad (14.11)$$

where Q_w is an energy input to the boundary of the system as shown in Fig. 14.16. This assumption is possible because other energy inputs related to long-wave radiation, evaporation, and condensation can only affect the very thin upper surface layer and the temperature in the bulk of the water is equilibrated by turbulence.

The value of the energy input, Q_w, to the surface of the water body can be assumed to be heat transfer through the surface layer. This can be approximated as

$$Q_w \simeq \alpha\, \rho_w\, C_p (T_s - T), \qquad (14.12)$$

where T_s is the surface temperature and α is the surface layer renewal coefficient.

The final form of the heat transfer equation in rivers and unstratified estuaries can be written as

$$\frac{\partial T}{\partial t} - D_L \frac{\partial^2 T}{\partial x^2} + U \frac{\partial T}{\partial x} - \frac{\alpha}{H}(T_s - T) - \frac{Q_s}{H\, \rho_w\, C_p} = 0, \qquad (14.13)$$

where the longitudinal dispersion coefficient, D_L, has been substituted for D_T.

Surface Temperature Computation

The surface temperature differs from that of the main bulk of the water, with the difference possibly as much as several tenths degrees Centigrade. Using an energy budget approach, the surface temperature can be calculated by assuming that $\Delta Q = 0$ at the boundary interface. Then,

$$Q_a - Q_{ra} - Q_b - Q_e - Q_h = Q_w + \frac{\partial Q_s}{\partial y_{wat}} \delta_{wat}, \qquad (14.14)$$

where Q_s = the rate of heat flow into a control volume from solar radiation, Q_a = the rate of heat flow into the surface layer from atmospheric radiation, Q_{ra} = the rate of heat reflected from the water surface, or the reflected atmospheric radiation, Q_b = the rate of heat flow from the water surface by back radiation, Q_e = the evaporative heat loss from the water surface, Q_h = the rate of heat loss by conduction at the air-boundary layer, Q_w = the rate of heat flow by surface layer renewal, and δ_{wat} = boundary layer of water at air–water interface.

In addition, it is assumed that only short-wave radiation can penetrate into the water body. Other heat inputs contribute only to surface heating, the heat of which is then transferred into the main body of the water by turbulent surface renewal phenomena.

In the air boundary, $\partial Q_s / \partial y \to 0$, and in the water boundary layer, $\partial Q_s / \partial y = Q_{s(H)} (1 - e^{-\eta\delta}w)$, where $Q_{s(H)}$ is the solar radiation reaching the water surface, η is the extinction coefficient of short-wave solar radiation in water, and δ_w is the thickness of the water surface boundary layer.

It has been found that with a high degree of accuracy, the long-wave radiation input, evaporative energy input, and head conductivity input can be approximated as in the next section (Novotny and Krenkel, 1971).

Total Long-Wave Radiation Flux

All equations describing the long-wave radiation input at the boundary are based on the Stefan–Boltzmann fourth-power radiation law and are quite similar. It is possible to combine equations and approximate the net long-wave radiation input by

$$\Delta Q_a = Q_a - Q_{ra} - Q_b = A_a + B_a (T_A - T_S) + C_a T_A, \qquad (14.15)$$

where $A_a = 695.04 (\beta - 0.874)$ cal cm^{-2}/day^{-1}, $B_a = 11.42$ cal cm^{-2}/day^{-1} °C^{-1}, and $C_a = 10.18 (\beta - 1.123)$ cal cm^{-2}/day^{-1} °C^{-1}. β is a coefficient accounting for cloud cover and humidity as defined by Raphael (1962) and shown in Fig. 14.18, and T_A is the air temperature.

Evaporation

Knowing the relative humidity, f, and the air temperature, T_A, the evaporative heat flux can be approximated as

$$Q_e = A_e + K_e(T_s - T_A), \qquad (14.16)$$

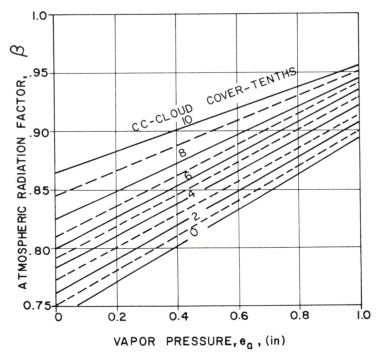

Fig. 14.18. Atmospheric radiation factor β (Raphael, 1962).

where $A_e = 0.625\ h_r e^{0.0625 T_A}\ (1 - f)$ cal cm^{-2}/day^{-1}, and $K_e = 0.0166$ $h_r e^{0.0625 T_A}$ cal cm^{-2}/day^{-1} °C^{-1}. Applying the Harbeck formula (1962), the vapor transfer coefficient, h_r, can be computed from

$$h_r = \frac{E\phi}{e_w - e_a} = 392\ X^{0.1}\ U, \qquad (14.17)$$

where X is a characteristic length expressed in meters and U is a relative water surface velocity expressed in meters per second. For the velocity, U, the vector subtraction, $U = |\mathbf{U}_A - \mathbf{U}_S|$ is understood. In the preceding discussion, $\mathbf{U}_A$ is the air (wind) velocity, $\mathbf{U}_S$ is the water surface velocity, e_w is the saturation vapor pressure, e_a is the vapor pressure of air, ϕ is a conversion factor, and E is the evaporation rate.

Heat Conduction in Air Boundary Layer

Since both evaporation and conduction are similar and follow the same laws, both processes have approximately the same transfer coefficients, or

$$h_{\text{vapor}} \approx h_{\text{heat}}.$$

In this case, the heat conductivity can be computed using the same transfer coefficient, or

$$Q_h = h_{\text{heat}} \rho_A c_{pA} (T_S - T_A) = K_h (T_S - T_A), \qquad (14.18)$$

where ρ_A is the density of air, and c_{pA} is the heat capacity of air.

The Surface Renewal Heat Transfer

This factor can be approximated as

$$Q_w = \frac{\alpha \, \rho_w c_{pw}}{H} (T_S - T) = K_w (T_S - T). \qquad (14.19)$$

The surface temperature then becomes

$$T_s = \frac{Q_s(1 - e^{-\eta \delta w}) + A_a - A_e + T_A(B_a + C_a + K_e + K_h) + K_w T}{B_a + K_e + K_h + K_w}. \qquad (14.20)$$

Inserting Eq. (14.20) into Eq. (14.13) and defining a tentative base temperature, T_0, as

$$T_0 = T_A + \frac{A_a - A_e + C_a T_A}{K_A} + \frac{Q_s(K_w + K_A)}{K_w + K_A}, \qquad (14.21)$$

where the overall air heat transfer coefficient is defined as

$$K_A = B_a + K_h + K_e.$$

One can obtain an equation describing the heat transfer as

$$\frac{\partial T}{\partial t} - D_L \frac{\partial^2 T}{\partial x^2} + U \frac{\partial T}{\partial x} + \frac{K_w K_a}{H c_p \rho_w (K_w + K_A)} (T - T_0) = 0. \qquad (14.22)$$

Since in turbulent streams, the magnitude of K_w is more than two orders of magnitude greater than K_A, the value of K_w has little influence on the process of turbulent heat transfer in streams, and the heat transfer coefficient can be simplified as

$$K = \frac{K_w K_A}{H c_p \rho_w (K_w + K_A)} \approx \frac{K_A}{H c_p \rho_w}. \qquad (14.23)$$

Application of the Model to a River Temperature Computation

It is obvious that the general type of temperature equation, Eq. (14.13), is also valid for describing the temperature occurring in streams under

natural meteorological conditions. This enables one to relate the temperature increase due to a thermal pollution load to the natural temperature
which would occur in the stream if no thermal addition was present.
Mathematically,

$$\frac{\partial(T - T_N)}{\partial t} - D_L \frac{\partial^2(T - T_N)}{\partial x^2} + U \frac{\partial(T - T_N)}{\partial x} + K(T - T_N) = 0, \quad (14.24)$$

where T_N is a natural temperature with no thermal pollution.

In most engineering considerations, the heat load (e.g., from thermal
power plants) is constant and $T_i - T_N =$ constant, where T_i is the initial
temperature of the reach under consideration. It should be noted that this
is a steady-state condition: if the natural temperature is considered as
being a base temperature and the only time varying input is the natural
temperature. Thus, the time varying solution can be divided into two
parts: (a) the time varying natural temperature, and (b) the steady-state
decay of $\Delta T = T - T_N$, which is superimposed on the naturally occurring

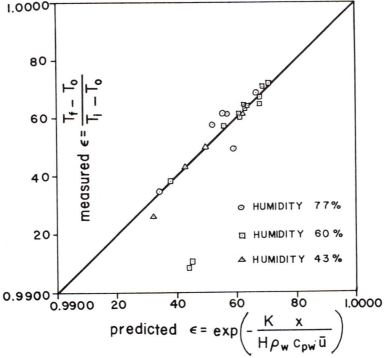

Fig. 14.19. Results of the cooling experiments in a laboratory flume (Novotny and
Krenkel, 1971).

temperature. Therefore,

$$T(t) = \Delta T_{ss} + T_N(t). \tag{14.25}$$

This approach differs from generally accepted practice, which uses the so-called equilibrium temperature as defined by Edinger and Geyer (1965). When the course of the natural temperature is known, computation of the heat load decay is quite simple, and a computation of the equilibrium temperature is not necessary. Since Eq. (14.25) is generally valid, the methodology developed herein may also be applied to cooling ponds, etc.

Also, in most streams, the effects of longitudinal mixing can be neglected. Under these conditions, Eq. (14.24) becomes an exponential first-order equation, or

$$\Delta T_{ss}(t) = \Delta T_i e^{-Kt}, \tag{14.26}$$

where ΔT_i represents an initial temperature increase in the stream.

This model was tested both in the laboratory and in prototype rivers. The results obtained in a laboratory flume are presented in Fig. 14.19 and the field data, which were obtained on the Cumberland River below Wolf Creek Dam, Kentucky, are shown in Fig. 14.20. In both cases, agreement of the measured and computed data is satisfactory.

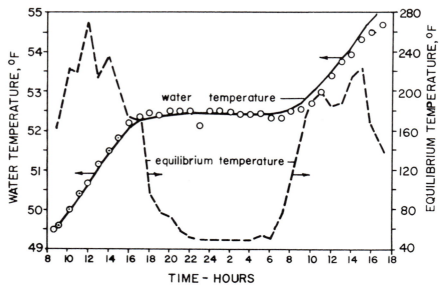

Fig. 14.20. Comparison of computed and predicted river temperatures, Cumberland River below Wolf Creek Dam, Kentucky, July 29–30, 1971, ○ = measured values; -- = computed temperatures (Novotny and Krenkel, 1971).

Alternative Methods for Cooling Water

If predictive techniques demonstrate that the heated waters must be cooled prior to their introduction into the receiving water, recourse to cooling devices must be made. A spectrum of sophistication is available for this process, the most sophisticated of which is the dry cooling tower, which does not receive any cooling water, and the least sophisticated is a simple cooling pond, which may be constructed solely for the purpose of heat dissipation. These devices have been classified by McKelvey and Brooke (1959) as shown in Table 14.7. Relative ground areas required for the same heat load are included. Schematic diagrams of various types of cooling devices are shown in Fig. 14.21.

When examining the various types of cooling devices available, it should be kept in mind that the rate of heat transfer is dependent on the area of water surface in contact with air, the relative velocity of air and water during contact, the time of contact between air and water, and the difference between wet-bulb temperature of air and inlet temperature of water.

Cooling Pond

The cooling pond is the simplest and most economical method of water cooling (assuming land is inexpensive and available); however, it is also the most inefficient. It may be constructed simply by erecting an earth dyke 6 to 8 feet high and may operate for extended periods with no makeup water.

TABLE 14.7 Cooling alternatives[a]

Device	Relative Ground Area
Ponds	
Cooling	1000
Spray	50
Atmospheric (natural draught)	
Spray filled	15
Wood filled	4
Chimney towers (natural draught)	
Mechanical draught	
Forced draught	
Induced draught	
Counter flow	1.5
Cross flow	1–2
Dry cooling	

[a] After McKelvey and Brooke (1959).

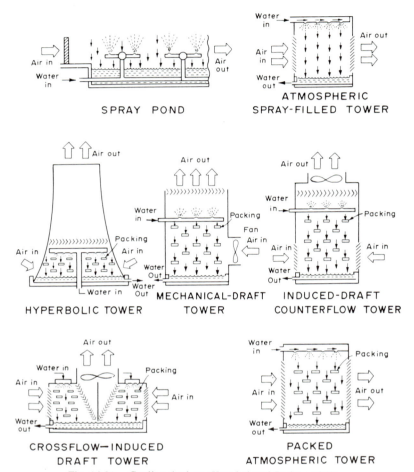

Fig. 14.21.   Cooling devices (Krenkel and Parker, 1969).

Its main disadvantages are the low heat transfer rate and the large areas required. For a still pond, the heat dissipated averages 3.5 Btu/hr/ft² surface /degree temperature difference between pond surface and air.

Spray Pond

Spray ponds may handle as many as 120,000 gpm (gallons/minute) of water and their low head requirements result in lower pumping costs than for cooling towers. Water is sprayed into the atmosphere some 6 to 8 feet above the pond, and the water is cooled as it mixes with the air and a portion evaporates.

Performance is limited, however, by the relatively short contact time of air and water spray. Also, water loss is high and impurities may easily enter the system. Properly designed spray ponds may produce overall cooling efficiencies up to 60% (Marks, 1963).

Atmospheric Tower

An atmospheric tower implies that air movement through the tower is only dependent on atmospheric conditions. A spray-filled tower depends solely on spray nozzles for increasing the air–water interface while the packed tower sprays the water over filling or packing. The packed tower is no longer common, however.

Atmospheric spray towers are of the simplest design and may cool up to 1.5 gpm of water per square foot of active horizontal area with the wind blowing at 5 mph (Marks, 1963). Their advantages include no mechanical parts, low maintenance costs, no subjection to recirculation of used air, and long trouble-free life.

Disadvantages include high initial cost (approximately identical to a mechanical draft tower), high pumping head, location limited to an unobstructed area, require great length because of rather narrow construction, high wind losses, and nozzle clogging. This design is well suited for small operation, however.

Hyperbolic Tower

The hyperbolic tower operates similarly to a huge chimney. The heavier outside air enters at the tower base, displaces the lighter saturated air in the tower, and forces it out the top. The initial cost is higher, but this is balanced against savings in power, longer life, and less maintenance. Its operation is countercurrent; it can cope with large water loads, and it requires a relatively small area.

This tower will probably become common in the United States as cooling tower requirements expand.

Mechanical Draft Tower

As implied, the mechanical draft tower utilizes fans to move the air through the tower. Thus, no dependence is placed on natural draft or wind velocity. The arrangement of the fans dictates the method in which the air is moved through the system. Each arrangement has certain advantages and disadvantages.

As elucidated by McKelvey and Brooke (1959) the advantages of a mechanical draft tower are

1. close control of cold water temperature;

2. the small ground area in which it can be maintained;
3. the generally low pumping head;
4. the location of the tower is not restricted;
5. more packing per unit volume of tower;
6. a closer approach and longer cooling range are possible;
7. capital cost is less than for a natural draught chimney.

Likewise, the purported disadvantages are:

1. A considerable expenditure of horsepower is required to operate the fans.
2. It is subject to mechanical failure.
3. It is subject to recirculation of the hot humid exhaust air vapors into the air intakes.
4. Maintenance costs are high.
5. Operating costs are high;
6. Performance will vary with wind intensity. Unlike the atmospheric tower, the performance decreases with increase in wind strength, until a certain critical velocity is reached, after which the performance improves, because of a falloff in recirculation.
7. Exhaust heat loading and climatic conditions can be very prejudicial to the economic use of a mechanical draught tower.

The Dry Cooling Tower

The dry cooling tower is not an evaporative cooling device but, instead, cools fluids by forcing or inducing atmospheric air across a coiled cross section. It eliminates water problems such as availability, chemical treatment, water pollution, and spray nuisance, and there is no upper limit to which air can be heated.

However, the dry cooling tower is much less economical than an evaporative cooling device. The specific heat of air is only one-fourth that of water, and maintenance costs are high. In addition, the application of these devices to large central station generating plants has not been proved.

Thus, the cost of dry cooling towers is presently thought to be prohibitive for most installations, even though many environmentalists believe that these are the only answer to the alleged thermal pollution problem.

Problems with Cooling Towers

In addition to the cost of installing and operating evaporative type cooling towers, several other considerations should be studied.

The consumptive loss of water amounts to approximately 2 to 3% of the circulated water, including blow-down and drift loss. The consumptive loss of water for various cooling processes varies considerably, being approximately 1% of condenser flow for once through cooling and some 2 to 3% for cooling towers. Jaske (1971) has shown that the increased loss of water incurred by using cooling towers instead of once through cooling for a 1000 MW plant with 32% efficiency is sufficient to supply the water requirements for a city of 100,000.

It has also been postulated that the increased moisture reaching the atmosphere from the cooling towers could cause significant changes in the climate. The drift could certainly cause ice and fog problems in the immediate vicinity. In addition, the combination of drift and SO_2 from the stacks may form a highly corrosive environment. Quite possibly, the emissions from cooling towers may violate the equivalent opacity standards imposed by air pollution control regulations.

Other environmental impacts of cooling towers are associated with their size (the three TVA Paradise cooling towers are each 437 feet high and 320 feet in diameter), the high bearing strength required for their foundation (structural design requirements), and excessive noise.

If salt water is used as a coolant, the effect of salt emissions on the environment must be considered. For example, it has been estimated that the five cooling towers for the Calvert Cliffs plant will emit 210 tons of salt per day resulting in a groundwater salinity of 840 ppm for 10 square miles surrounding the plant (USAEC, 1971).

Finally, it should be noted that the use of cooling towers requires from 5 to 8% of the energy produced by the plant. Thus, more power is required, resulting in additional air pollution and increased fuel consumption. Also, no mention of the blow-down characteristics, which may be highly pollutional in nature, has been made. It is obvious that as water is continuously evaporated from the cooling towers, the solids content of the recirculated waters will increase. When these waters are periodically released and replaced with ''new'' water, the water discharged, known as ''blow-down,'' may have detrimental effects on the aquatic environment. It is interesting to note, in this regard, that the TVA Phipps Bend nuclear plant, using high concentration factors, will violate many of Tennessee's water quality standards because of the existence of otherwise innocuous materials from upstream wastewater discharges.

The capital and operating costs of cooling towers are quite high. While the costs of TVA construction and operation are probably high because of increases due to the government's inherent inefficiency, Table 14.8 is indicative of the costs of constructing and operating closed-cycle cooling systems (Krenkel, 1978). The coal-fired plants have been given variances because of some 316 successful demonstrations and the Browns Ferry

TABLE 14.8   Summary of capital costs and associated operating costs for closed cycle cooling systems

| | Thermal | | |
| | Capital | Operating | |
Plant	Cost ($ × 10⁶)	Cost ($ × 10⁶)	Total ($ × 10⁶)
Coal-Fired			
Bull Run	61.5ᵃ	17.6	57.7
Cumberland 1 and 2	43.8	48.7	92.5
Gallatin 1–4	34.5	22.6	57.1
John Sevier 1–4	71.7ᵃ	13.4	40.4
Total	201.6	102.3	303.9
Nuclear Plants			
Browns Ferry 1–3	57.0	37.0	94.0
Sequoyah 1 and 2	56.0	24.0	80.0
Watts Bar 1 and 2	46.0	ᵇ	46.0
Bellefonte 1 and 2	77.0	ᵇ	77.0
Hartsville 1–4	189.0	ᵇ	189.0
Phipps Bend	105.0	ᵇ	105.0
Yellow Creek	120.0	ᵇ	120.0
Total	650.0	61.0	711.0
System Totals	851.6	163.3	1014.9

ᵃ1976 dollars, all other cost are in terms of 1974 dollars.
ᵇOperating penalties not shown since plants were originally designed for closed cycle cooling.

and Sequoyah cooling towers were constructed because lawyers indicated that the costs of the cooling towers would be less than those incurred by litigation caused by interveners. Thus, the cooling towers at Browns Ferry and Sequoyah are the results of lawyers' decisions rather than rational engineering technology.

BENEFICIAL USE OF WASTE HEAT

If a feasible use of the heat rejected by power generation could be found, the problem of thermal pollution would become academic. However, the heat is low grade, and thus far a practical use for this heat has not been found, although several research projects are still being pursued. Two prime requisites must be met for waste heat utilization. One is that is must be capable of using the majority of the rejected heat on a continuous basis, and the other is that it must be economical. Thus far, no proposal has succeeded in meeting both of these requirements.

Space Heating

It would appear logical that waste heat could be used for space heating. However, the question immediately arises as to what to do with the heated waters during the warm seasons. Warsaw, Poland, is heated by heat from a central plant, but contrary to popular belief, the heat used is not waste heat but heat generated specifically for space heating.

Wastewater Treatment Plants

Inasmuch as biological treatment of wastewaters is accelerated with increasing temperatures, the addition of heated waters from a power plant to a wastewater treatment plant appears to be a reasonable alternative. However, a few fundamental calculations demonstrate that the hydraulic load on the treatment plant from the power discharges precludes this idea, from the standpoints of both size and dilution (Krenkel and Englande, 1971).

Irrigation

The use of heated waters for irrigation is being investigated in several locations. It is true that the additional heat will extend the growing season and increase productivity. However, it has been estimated that a 1000 MW plant would require 100 to 200 square miles of farmland. Furthermore, the water cannot continuously be applied.

Aquaculture

Inasmuch as fish and shellfish will demonstrate increased productivity at their optimal temperatures, the maintenance of these temperatures by the use of power plant discharges is an interesting possibility. Several endeavors of this type have been successful; however, it should be kept in mind that this use of heated waters is location specific. Thus, it is not universally applicable.

Greenhouses

It has been proposed that greenhouse heating by waste heat is economical if space is available. Hutant (1969) states that a 1000 MW nuclear plant could heat 4.4 square miles of greenhouse, although substantiating data are not presented. The TVA was also successful in using simulated waste heat for greenhouse propagation; however, the area requirements were prohibitive.

Miscellaneous Applications

Waste heat has been shown to be beneficial in the maintenance of ice-free channels and the inhibition of frost formation for crops. It has also been proposed for use in absorption type air conditioning and in the de-icing of airport runways (Hutant, 1969). However, a reasonable and economically feasible method of beneficial waste heat utilization at all plant sites has yet to be realized.

Conclusions

While the beneficial use of waste heat appears to be attractive, the relatively low heat content in relation to the volume of water makes waste heat utilization impractical except for certain site selective operations. While for the most part unreferenced, a review of the potential for waste heat utilization by Rimberg (1974) reaches the same conclusion.

REFERENCES

Anderson, E. R. (1954). Water loss investigations: Lake Hefner studies. Tech. Rep., *U.S. Geol. Surv. Prof. Pap.* 269.

Anon. (1967). Thermal pollution: Senator Muskie tells AEC to cool it. *Science Mag.* **158,** 755–756.

Bergstrom, R. N. (1968). Hydrothermal effects of power stations. Pap. pres. at Am. Soc. Civ. Eng. Water Resour. Conf., Chattanooga, Tennessee.

Bohnke, N. (1961). Effects of organic wastewater and cooling water on self-purification of waters. Proc. 22nd Purdue Ind. Waste Conf., Lafayette, Indiana.

Bohnke, N. (1966). New method of calculation for ascertaining the oxygen conditions in waterways and the influence of the forces of natural purification. *Proc. Int. Conf. Int. Assoc. Water Pollut. Res.,* 3rd, Munich.

Brett, J. R. (1952). Temperature tolerance in young Pacific Salmon, genus *Oncorhynchus.* J. Fish. Res. Board Can. **9,** 265.

Brett, J. R. (1960). Thermal requirements of Fish—Three Decades of Study, 1940–1970. *In* Biological problems in water pollution. Tech. Rep. W60-3, U.S. Public Health Serv., Cincinnati, Ohio.

Cairns, J. (1955). The effects of increased temperatures upon aquatic organisms. Proc. Purdue Ind. Wastes Conf., 10th, Lafayette, Indiana.

Cootner, P. H., and Lof, G. O. (1965). "Water Demand for Steam Electric Generation." For Resources for the Future, John Hopkins Press, Baltimore, Maryland.

Edinger, J. E., and Geyer, J. C. (1965). Heat exchange in the environment. Cooling Water Studies for Edison Electric Inst., Johns Hopkins Univ., Baltimore, Maryland.

Edison Electric Institute (1974). Comments on EPA's proposed #304 guidelines and #306 standards of performance for steam electric power plants. Utility Water Act Group Johns Hopkins Univ., Baltimore, Maryland.

Federal Energy Administration (1976). National energy outlook, Report No. FEA-N-75/713 (Feb.), Washington, D.C.

Federal Power Commission (1971). "The 1970 National Power Survey," Part III. U.S. Govt. Printing Office, Washington, D.C.

Fry, F. E. J., Hart, J. S., and Walker, K. F. (1946). Lethal temperature relations for a sample of young Speckled Trout, *Salvelinus fontinalis.* Univ. of Toronto Studies in Biology, No. 54., Toronto, Ontario, Canada.

Gameson, A. L. H., Gibbs, J. W., and Barrett, M. J. (1959). A preliminary temperature survey of a heated river. *Water Water Eng.* (Jan), 13–17.

Harbeck, G. E., Jr. (1962). A practical field technique for measuring reservoir evaporation utilizing mass-transfer theory. *U.S. Geol. Surv. Prof. Pap.* 272-E, Washington, D.C.

Harbeck, G. E., Koberg, G. E., and Hughes, G. H. (1959). The effect of the addition of heat from a power plant on the thermal structure and evaporation of Lake Colorado City, Texas. *U.S. Geol. Surv. Prof. Pap.* 272-B, Washington, D. C.

Hawkes, H. A. (1969). Ecological changes of applied significance induced by the discharge of heated waters. *In* "Engineering Aspects of Thermal Pollution" (F. L. Parker and P. A. Krenkel, eds.). Vanderbilt Univ. Press, Nashville, Tennessee.

Hutant, J. A. (1969). Utilizing waste heat for urban systems. *In* "Electric Power and Thermal Discharges" (M. Eisenbud and G. Gleason, eds.). Gordon & Breach, New York.

Jaske, R. T. (1972). A future for once through cooling *Power Eng.* **76,**(1), 44–47.

Jones, J. R. E. (1964). "Fish and River Pollution." Butterworth, London.

Koelzer, V. A., and Tucker, R. C. (1971). Water use and management aspects of steam electric generation. Prepr., Energy, Environ. and Educ. Symp., Univ. of Arizona, Tucson.

Krenkel, P. A. (1967). Rep. to World Health Organization on Poland 0026 Project. Copenhagen, Denmark.

Krenkel, P. A. (1978). Energy, Resource Planning, and Environmental Legislation. *Progr. Water Technol.* **10**(3/4), 81–94.

Krenkel, P. A., and Englande, A. J. (1971). Feasibility of utilizing power plant waste heat for biological waste treatment. Rep. to URS Co., San Mateo, California.

Krenkel, P. A., and Parker, F. L. (1969). Engineering aspects, sources and magnitude of thermal pollution. *In* "Biological Aspects of Thermal Pollution," (P. A. Krenkel and F. L. Parker, eds.). Vanderbilt Univ. Press, Nashville, Tennessee.

Krenkel, P. A., Cawley, W. A., and Minch, V. A. (1965). The effects of impoundments on river waste assimilative capacity. *J. Water Pollut. Control Fed. Fed.* **37,** 9.

Marks, P. H. (1963). Cooling towers. *Power.,* **107**(3), S1–S16.

McKee, J. E., and Wolf, H. W. (1963). Water quality criteria, 2nd ed. State Water Qual. Control Bd., Sacramento, California.

McKelvey, K. K., and Brooke, M. (1959). "The Industrial Cooling Tower." Elsevier, Amsterdam.

Minskij, E. M. (1952). Turbulence of channel flow. *Gidrometeoizdat* Leningrad, USSR (in Russian).

Northwest Public Power Association (1970). *Bull.* **24,**(June-July), Vancouver, Washington.

Novotny, V., and Krenkel, P. A. (1971). Heat transfer in turbulent streams. Res. Rep. No. 7, Dept. of Environ. and Water Resour. Eng., Vanderbilt Univ., Nashville, Tennessee.

Parker, F. L., and Krenkel, P. A. (1969). Thermal pollution, state of the art. Rep. No. 3, Dept. of Environ. and Water Resour. Eng., Vanderbilt Univ., Nashville, Tennessee.

Raphael, J. M. (1962). Prediction of temperature in rivers and reservoirs. *J. Power Div., Am. Soc. Civ. Eng.* **88** (P02), 157–181.

Resources for the Future (1970). Resour. Newsletter No. 34, Washington, D.C.

Rimberg, D. (1974). Utilization of waste heat from power plants. Noyes Data Corp., Park Ridge, New Jersey.

Schmidt, W. (1915a) *Annalen der Hydrographie* **43**, 111–124.

Speakman, J. N., and Krenkel, P. A. (1972). Qualification of the effects of rate of temperature change on aquatic biota. *Water Res.* **6**, 1283–1290.

Stephenson, M. (1930). "Bacterial Metabolism," p. 101. Longmans, Green, New York.

U. S. Atomic Energy Commision (1971). Thermal effects and U.S. nuclear power stations. Report No. WASH 1169, Washington, D.C.

U.S. Congress, Senate, Subcommittee on Air and Water Pollution, *Thermal Pollution,* Public Works Comm., (1968). 90th Cong., Part I, 22–64.

U.S. Department of Health, Education and Welfare (1963). Report on Coosa River system, Georgia–Alabama. Public Health Serv., Taft Sanit. Eng. Cent., Cincinnati, Ohio.

Water Resources Council (1968). "The Nation's Water Resources." U.S. Govt. Printing Office, Washington, D.C.

Water Resources Council (1978). "The Nation's Water Resources, 1975–2000." Vol. 1, Summary. Washington, D.C.

World Meteorological Organization (1966). Measurement and estimation of evaporation and evapotranspiration. Tech. Note No. 83, Geneva, Switzerland.

15 | Mixing Zones and Outfall Design

INTRODUCTION

While the concept of mixing zones was discussed in Chapter 9 in the context of mixing and dispersion, further discussion is warranted here because of the relationship between outfall design and initial mixing. As previously demonstrated, the mixing zone is that reach of receiving water bodies where the most profound spatial water quality changes take place. Previous definitions have equated mixing zones to a reach of the receiving water body between the outfall and a cross-section where the receiving flow is fully mixed with the waste discharge. However, after reviewing the concepts of stratified flow, the reader should immediately realize that this definition would be limited to vertically and laterally mixed channels. A more general definition of a mixing zone may be based on the assumption that it is the zone with the most significant spatial changes of a substance discharged from an outfall (outfalls) with an upper boundary as a point where no particle from the outfall can reach, and the lower boundary as a point where the first derivative of the longitudinal mass flow of a conservative substance discharged from the outfall would be zero.

It is evident that water quality problems of mixing zones may be more severe than those in reaches where the waste is fully mixed. Higher concentrations of pollutants because of less dilution can be expected with more serious environmental and ecological consequences to the use of water and aquatic life.

Receiving water quality standards are applied to waters excluding mixing zones. For all practical purposes, their application to mixing zones would mean that all effluents would have to comply with the stream standards. This would impose severe and, for the most part, unjustified water quality limitations on most wastewater discharges.

As pointed out in Chapter 9, initial EPA guidelines allowed for a limited mixing zone as a zone of initial dilution in the immediate area of a point or nonpoint source of pollution (USEPA, 1976). Careful consideration must be given to the appropriateness of a mixing zone in which a substance discharged into the receiving water is bioaccumulative and persistent or carcinogenic, mutagenic or teratogenic. In such cases, the states, which are responsible for managing water quality, must consider the ecological and human health effects of assigning a mixing zone including bioconcentrations in sediments and aquatic biota, bioaccumulation in the food chain, and the known or predicted safe exposure levels for the substance. In some instances, the ecological and human health effects may be so adverse that a mixing zone is not appropriate.

Any mixing zone should be free of point or nonpoint sources pollution related to (USEPA, 1976):

(a) Materials in concentrations that exceed the 96-hour LC_{50} (50% lethal concentration-$TL_m{}^{96}$) for biota significant to the indigenous aquatic community;

(b) Materials in concentrations that settle to form objectionable deposits;

(c) Floating debris, oil, scum, and other matter in concentrations that form nuisances;

(d) Substances in concentrations that produce objectionable color, odor, taste, or turbidity; and

(e) Substances in concentrations which produce undesirable aquatic life or result in a dominance of nuisance species.

The basic hydraulics of diffusion and its mechanisms as it relates to mixing zones have been discussed in Chapter 9. In this chapter, attention will be focused on the engineering management and design of mixing zones so that the pollution impact of higher concentrations of undiluted or insufficiently diluted sewage and wastewaters will be minimal and the zone itself will be as small as possible.

The evaluation of mixing zones and proper outfall design is important for large rivers, stratified reservoirs, lake and ocean outfalls. Poorly eval-

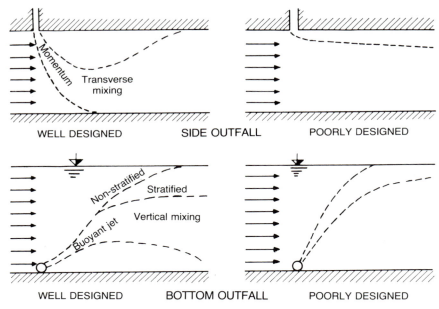

Fig. 15.1.   Well and poorly designed outfall and mixing zones.

uated and designed mixing zones may result in excessive pollution near river and reservoir banks or lake and ocean beaches reducing or even negating their use for recreation. Higher pollution may also result in lower oxygen concentrations. Although fish can escape to less polluted sections, shellfish and benthic biota cannot escape, and permanent damage can be done to shallow near-shore growing and spawning grounds of less mobile aquatic inhabitants.

The problem of mixing zones and their engineering evaluation can be roughly divided into two categories: Design of the location and configuration of waste outfalls so that high initial dilution and spread of the waste are achieved; and evaluation of transverse and vertical mixing so that a proper location and configuration of the outfall will result in the shortest mixing zone.

Figure 15.1 shows cases of poorly designed and well-designed waste outfalls and adjacent mixing zones.

OUTFALL DESIGN

The function of a waste or sewage outfall is to convey wastewater to a point of suitable disposal where the effect of the waste on the receiving

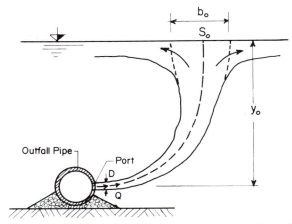

Fig. 15.2. Definition sketch for the rising plume of wastewater discharged from a diffuser port.

water body is minimal, even at the point of initial mixing. As stated before, it may be desirable to avoid mixing zone problems by treating the waste to a level which would comply with the receiving water standards (stream standards). However, this approach is uneconomical and often unnecessary, in spite of all public and environmentalist pressure. In the design of waste outfalls, it is important to locate the point of discharge so that excessive concentrations are avoided and mixing is encouraged.

The mixing zone can be broken into two areas:

1. A zone in which jet momentum and buoyancy of the wastewater discharge is prevailing in determining the initial shape and mixing of the discharge with the ambient flow;

2. A zone in which lateral and vertical dispersion is prevailing in mixing the plume with the ambient water.

The state of the art of the mathematical modeling of jet and plume discharges and the experiments on which the models are based have reached a stage which provides a basis for adequate design of outfalls for wastewater disposal, whether single or multiple jet type.

Figure 15.2 is a definition sketch of a plume discharge into a uniform nonflowing ambient fluid. In stratified environments, the plume may sometimes not reach the surface because of the density differential barrier as shown in Fig. 15.3.

An outfall can be designed either as a single port or multiple port (line or slot) source. Multiple port outfalls obviously increase the initial dilution and decrease the initial momentum.

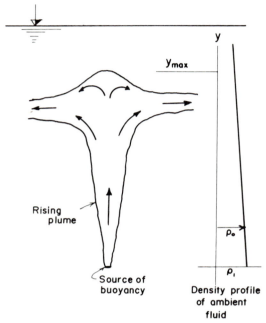

Fig. 15.3.   Rising plume in a stratified environment.

In order to design wastewater outfalls, the following considerations should be taken into account and investigated during the preliminary design surveys:

(a) *Water quality requirements.* Criteria for use of water and shoreline must be defined. The primary concern in the design of sewage outfalls is the dilution of coliform bacteria and their concentration in water, near the shore, and at the nearest water supply intake. In addition to bacterial pollution, other parameters of interest include carcinogenic, toxic, and biogenic cummulative components, turbidity, oil and grease, floatables, and taste and odor-causing materials.

(b) *Hydraulic (oceanographic) survey.* The convective and diffusive transport of pollutants and their spread depends effectively on the currents near the outfall. The survey data should include the general direction of currents, their magnitude and frequency, their variation with depth, and their relationship to tides. In addition, it is necessary to evaluate the lateral and vertical dispersion intensity. Stratification near the outfall may determine whether the outfall plume will remain submerged or reach the surface. The density distribution can be obtained from the measured total dissolved solids or the vertical distribution of conductivity in addition to a temperature profile. Furthermore, the surveys should provide information on submarine topography, marine life distribution, and eutrophication potential.

(c) *Site selection.* Based on the survey data, the outfall should be placed in a location with favorable currents, i.e., currents with prevailing directions away from the shore or near water intakes. The currents represent the convective component of the pollutant transport. The magnitude of dispersion, both lateral and vertical, determine how effectively the plume will spread. The site selection for the outfall should not interfere with the marine or aquatic biota and should be a safe distance from shellfish and fin fish grazing and growing grounds.

Types of Outfalls

1. *A bank or shoreline outfall* is probably the most common type of outfall but also the least favorable one. Side outfalls are simple pipe or channel outflows into the receiving water. In many cases, the sewage, with a little dilution, remains near the bank or shoreline for long distances. Often, no hydraulic engineering is involved in its design. The functioning of side outfalls can be improved if surface jets have sufficient momentum to carry the sewage water farther from the shore.

2. *Submerged diffusers* are transverse pipes with openings (ports) located on the bottom of the receiving water body. The diffusers can be either a single port or multiple port (line) type and are more effective in dispersing pollutants with ambient water than bank outfalls.

Hydraulic Models for Outfall Design

Jet and Plume Mixing

A buoyant jet is a turbulent-free shear flow which has both momentum and buoyancy at the source. If the momentum is negligible, the jet becomes a rising plume. A typical buoyant jet from an outfall may be jetlike near the outlet but becomes plumelike at some distance from the outlet (Brooks, 1973).

There are several models describing buoyant jets and plumes. The reader is referred to the reference section of this chapter for additional information. These models are usually solved only for certain simplified situations. However, in most cases, as in many engineering design problems, a complete answer and detailed analysis is not required and the design may be based on simplified models applied to critical situations.

In the design of a wastewater outfall, one must know the dilution of the wastewater at the surface, at a shoreline, and at a certain distance from the outlet.

Buoyant Rising Jets

Figures 15.2 and 15.4 represent the problem. The outlet can be either a single port or multiple port (line) source. The single port solution can be extended to more than one outlet provided that the spacing of the outlets is far enough apart to prevent significant interference. If the spacing is closer, a row of jets will produce a flow pattern similar to that of a line source (slot jet) viewed from a moderate distance, as shown in Fig. 15.4. The effective width, B, of a multiple port outfall is

$$B = \frac{\pi D^2}{4S},\tag{15.1}$$

where D is the diameter of a port and S is the port spacing.

The rising jet is assumed to be fully turbulent. Because of the turbulent mixing at the jet boundary, it entrains the ambient fluid, resulting in mixing of the effluent with the ambient water. The coefficient of entrainment, α, has been defined as (Brooks, 1973)

$$\alpha = \frac{dQ/ds}{Pu},\tag{15.2}$$

where Q is the volume flux across the jet cross-section, s is the jet centerline trajectory, P is the perimeter of the jet, and u is the characteristic velocity along the s-axis. The values of the entrainment coefficient have been found to be approximately $\alpha = 0.082$ for round (single) port jets and $\alpha = 0.14$ for slot (line) jets (Fan and Brooks, 1969).

All of the mathematical modeling applies only to the zone of established flow, that is, the zone beyond the distance of $6D$ for round jets and $5B$ for slot jets.

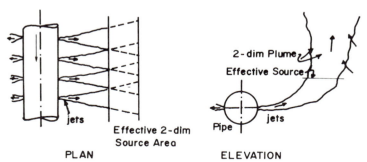

PLAN   ELEVATION

Fig. 15.4. ·Effective line source formed by a row of jets.

The models dealing with buoyant jets can only be solved numerically. In order to simplify the solution, the basic equations can be integrated across the plume, which reduces the plume system to only one variable, s, the distance along the axis of the plume (Brooks, 1973; Fan and Brooks, 1969; Motz and Benedict, 1970). The cross-sectional velocity, momentum, and mass distribution is assumed to be of the Gaussian profile, described by the equation

$$u = U \exp(-\xi^2/b^2) \quad \text{and} \quad b = \sqrt{2}\,\sigma \qquad (15.3)$$

where u is the velocity at any point in the cross-section, U is the centerline velocity, σ is the standard deviation of the velocity profile, and ξ is the distance along the axis perpendicular to the s-axis. Similar equations can also be written for the other variables.

The integrated mass and momentum equations give a system of five equations with five unknowns (Brooks, 1973): (1) Entrainment (mass conservation), (2) horizontal momentum flux, (3) vertical momentum flux, (4) buoyancy (density flux), and (5) pollutant mass or heat flux. In addition, the geometry equations will add two more unknowns, x and y: (6) Horizontal displacement and (7) vertical displacement.

The seven unknowns are all functions of s. The numerical solution provides the centerline trajectory of the buoyant plume as a function of x and y (x and z for laterally spreading plumes and jets), the plume width, average velocity, concentration, and temperature distribution. A large number of normalized solutions were developed and standard computer programs are available (Fan and Brooks, 1969; Sotil, 1971; Baumgartner et al., 1971; Ditmars, 1969).

Solution and Example for Buoyant Jets in Stagnant Environments

The numerical solutions yield the centerline (minimum) dilution in the buoyant jet discharged from a single pipe into a uniform stagnant environment. The centerline dilution, S_0, which is the same as the centerline concentration divided by the average concentration at the end of the zone of establishment, can be related to the dimensionless height above the outlet, y/D, and the outlet densimetric Froude number as

$$F = \frac{Q}{\frac{\pi}{4} D^2 \left(\dfrac{\rho_0 - \rho_1}{\rho_0} g D \right)^{1/2}} 1.07 = 1.07 \frac{U_0}{\left(\dfrac{\Delta\rho}{\rho_0} g D \right)^{1/2}}, \qquad (15.4)$$

where Q is the outlet discharge, D is the outlet diameter, ρ_0 is the density of the ambient fluid, and ρ_1 is the density of the discharged waste. The multiplier 1.07 accounts for the approximative difference between the

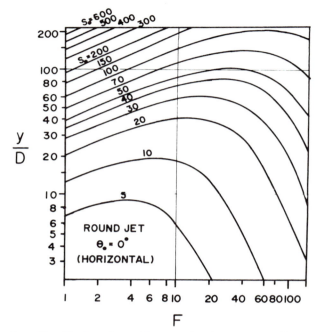

Fig. 15.5. Centerline dilution of round bouyant jets in stagnant uniform environments (Fan and Brooks, 1969) $O_0 = 0°$ (horizontal). To get centerline dilution relative to the nozzle, multiply S_0 by 1.15 to adjust for the zone of flow establishment; for average multiply by 2.

outlet Froude number and that at the end of the zone of establishment, and U_0 is the discharge velocity.

Figures 15.5–15.8 show the results of the numerical solutions for a single pipe discharge into a stagnant environment for both the horizontal and vertical pipe directions.

For $(y/D)/F > 30$, the plume solution for the centerline dilution can be approximated as

$$S_0 = 0.089 \frac{g'^{1/3} y^{5/3}}{Q^{2/3}}, \tag{15.5}$$

where

$$g' = g \frac{\rho_0 - \rho_1}{\rho_0}.$$

The half-width of the jets shown in Figs. 15.7 and 15.8 is defined as two standard deviations of the velocity distribution according to Eq. (15.3).

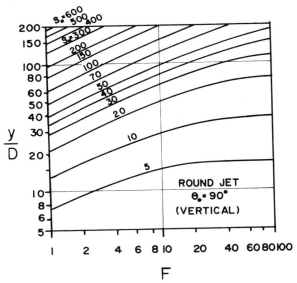

Fig. 15.6. Dilution of round buoyant jets in stagnant uniform environments (Fan and Brooks, 1969). $O_0 = 90°$ (vertical). To get dilution relative to the nozzle, multiply S_0 by 1.15 to adjust for the zone of flow establishment; for average, multiply by 2.

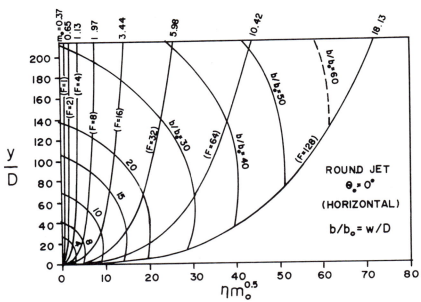

Fig. 15.7. Trajectory and half-width b/b_0 of round buoyant jets in stagnant uniform ambient fluids (Fan and Brooks, 1969). $O_0 = 0°$ (horizontal). (F and Y/D scales based on $\alpha = 0.082$ and $\lambda = 1.16$.)

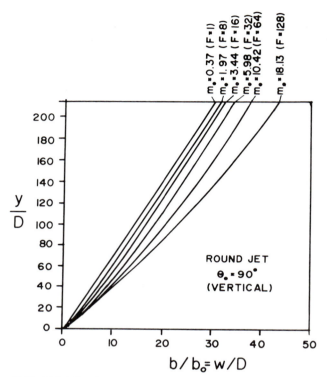

Fig. 15.8. Half-width b/b_0 of round buoyant jets in stagnant uniform ambient fluids (Fan and Brooks, 1969). $O_0 = 90°$ (vertical). Trajectories are all vertical lines. (F and Y/D scales based on $\alpha = 0.82$ and $\lambda = 1.16$.)

Hence,

$$w = \sqrt{2}\, b$$

Example. An effluent outfall located 20 m below the surface of a stagnant nonstratified water body is discharging a waste flow of $Q = 2.25$ m³/sec. There is sufficient head to discharge the wastewater at a velocity of 4.5 m/sec. The density of the ambient (sea) water is 1.025 g/cm³ and that of the wastewater is 0.999 g/cm³. Compare the wastewater dilution at the water surface if the waste is discharged by 50 or five horizontal ports. Assume no contraction and no interference of the jets.

Required total area of the jets, A

$$A = \frac{2.25 \text{ m}^3/\text{sec}}{4.5 \text{ m}/\text{sec}} = 0.5 \text{ m}^2.$$

For 50 ports

$$D = \left(\frac{4\,A}{\pi\,50}\right)^{1/2} = \left(\frac{4*0.5}{\pi*50}\right)^{1/2} = 0.118 \text{ m} = 118 \text{ mm}.$$

$$y/D = \frac{20}{0.118} = 169.$$

$$F = 1.07 \frac{U_0}{\left(\frac{\Delta\rho}{\rho_0}\,g\,D\right)^{1/2}} = \frac{4.5*1.07}{\left(\frac{0.026}{1.025}\,9.81*0.118\right)^{1/2}} = 28.$$

From Fig. 15.5, $S_0 = 100$.

The dilution related to the wastewater discharge concentration then becomes

Centerline dilution at the top of the plume:

$$S_{od} = 1.15\,S_0 = 115.$$

Average dilution at the top of the plume:

$$S_{ad} = 2\,S_0 = 200.$$

For five ports

$$D = \left(\frac{4\times0.5}{\pi\times5}\right)^{1/2} = 0.357 \text{ m} = 357 \text{ mm}.$$

$$y/D = \frac{20}{0.357} = 56.$$

$$F = 1.07 \frac{4.5}{(0.026\ 9.81\ 0.357)^{1/2}} = 16.$$

From Fig. 15.5,

$$S_o = 28,$$
$$S_{od} = 32,$$
$$S_{ad} = 56.$$

Since no interference between the rising plumes was assumed, the ports must be spaced accordingly. Figure 15.7 gives the dimensionless centerline trajectories of the plumes for various Froude numbers overlaid with the dimensionless plume half-widths, w/D. Then for 50 ports ($F = 28$, $y/D = 169$) $w/D = 33$; therefore, the diameter of the plume at the head of the plume is approximately: $2*w = 2*33*0.118 = 7.8$ m. Similarly

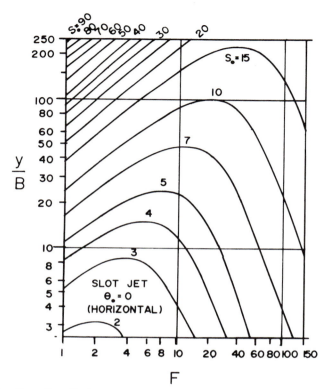

Fig. 15.9. Centerline dilution of slot buoyant jets in stagnant uniform environments (Fan and Brooks, 1969). $O_0 = 0°$ (horizontal). (For average dilution, multiply by $\sqrt{2}$.)

for five ports ($F = 16, y/D = 56$) $w/D = 13$ and the full width of the plume at the water surface is $2 * w = 2 * 13 * 0.373 = 9.7$ m.

The dilution ratio can be slightly improved if the ports are spaced closer together so that the outfall would resemble a slot discharge as shown in Fig. 15.9.

Round or Line Sources in Stratified Environment with No Current

Under stratified conditions, the plume may rise only to a certain depth and stay submerged. The important characteristics of the outfall are then the maximum height of the rise, y_{max}, and the dilution of the top of the rising plume. Figure 15.3 shows the basic flow pattern for this case.

The solution can again be obtained by numerical computer programs or with the aid of design graphs based on dimensionless characteristics of the plume trajectory. All the variables necessary for the design can be related to the densimetric Froude number

$$F = 1.07 \frac{U_0}{\left(\frac{\rho_0 - \rho_1}{\rho_0} g D\right)^{1/2}} \tag{15.6}$$

and a stratification number

$$T = 1.07 \frac{\rho_0 - \rho_1}{D\left(-\frac{d\rho_a}{dy}\right)}, \tag{15.7}$$

where ρ_1 is the density of the discharged waste, ρ_0 is the reference density (ambient fluid at the source level), and ρ_a is the ambient density.

Then, define the following variables (Brooks, 1973):

$$m_0 = 0.324 \; F^2 T^{-1} \tag{15.8}$$
$$\mu_0 = 2.38 \; F^{1/4} \; T^{-5/8} \tag{15.9}$$
$$S_{td} = 1.15 \; \mu_t/\mu_0 \quad \text{(centerline dilution related to the discharge)} \tag{15.10}$$
$$S_{ad} = 2 \; \mu_t/\mu_0 \quad \text{(average dilution related to the discharge)} \tag{15.11}$$
$$\frac{y_{max}}{D} = 1.37 \; \xi_t \; F^{1/4} \; T^{3/8} \tag{15.12}$$

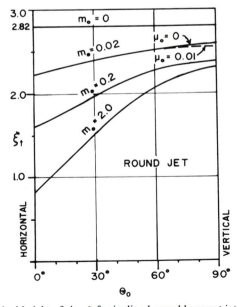

Fig. 15.10. Terminal height of rise ξ_t for inclined round buoyant jets with $m_0 = 0$ to 0.01 (Fan and Brooks, 1969).

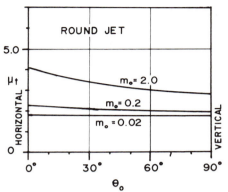

Fig. 15.11. Terminal volume flux parameter, μ_t, for inclined round buoyant jets with $\mu_0 = 0$ to 0.01 (Fan and Brooks, 1969).

The coefficients ξ_t and μ_t can be read from Figs. 15.10 and 15.11 (round jets) and 15.12 and 15.13 (slot jets).

For most practical cases, it is sufficient to use the plume solution; i.e., the discharge momentum is assumed to be negligible. Then $m_0 = \mu_0 = 0$

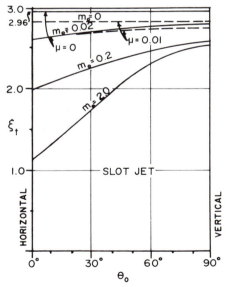

Fig. 15.12. Terminal height of rise ξ_t for inclined slot buoyant jets with $\mu_0 = 0$ to 0.01 (Fan and Brooks, 1969).

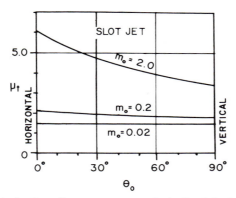

Fig. 15.13. Terminal volume flux parameter, μ_t, for inclined slot buoyant jets with $\mu_0 = 0$ to 0.01 (Fan and Brooks, 1969).

and the plume rise and dilution can be computed as

$$y_{max} = 3.75 \, (Q \, g'_a)^{1/4} \left(- \frac{g}{\rho_0} \frac{d\rho_a}{dy} \right)^{3/8} \tag{15.13}$$

and

$$y_{max} \, S_{td} = 3.19 \frac{\rho_0 - \rho_1}{- \dfrac{d\rho_a}{dy}}, \tag{15.14}$$

where $g'_a = g(\rho_0 - \rho_1/\rho_0)$.

Example. Estimate the dilution and a plume rise for a plume rise from a diffuser discharging sewage at a flow rate, $Q = 2.25$ m³/sec. The diffuser is located at a depth $H = 30$ m below the sea surface. There is an approximate linear density distribution with the bottom seawater density of $\rho_a = \rho_0 \, 1.0258$ g/cm³ and that at the surface of $\rho_a = 1.0246$ g/cm³. The density of sewage is $\rho_1 = 0.9995$ g/cm³. Determine whether the sewage field will be submerged if the diffuser has (a) 50 ports or (b) 5 ports. Assume no interference between the jets.

$$- \frac{d\rho_a}{dy} = \frac{-(\rho_a - \rho_0)}{H} = \frac{-(1.0246 - 1.0258)}{30} = 0.00004 \, \frac{\text{g/cm}^3}{\text{m}}.$$

For 50 ports, from the previous example, $F = 28$ and $D = 0.118$ m,

$$T = 1.07 \frac{\rho_0 - \rho_1}{D \left(- \dfrac{d\rho_a}{dy} \right)} = 1.07 \frac{1.0258 - 0.9995}{0.118 * 0.00004} = 5962.$$

Then

$$m_0 = 0.324\ F^2\ T^{-1} = 0.324 * 28^2/5962 = 0.42,$$
$$\mu_0 = 2.38\ F^{1/4}\ T^{-5/8} = 0.0239.$$

From Figs. 15.10 and 15.11, $\xi_t = 2.0$ and $\mu_t = 3.00$. Then the plume top centerline dilution

$$S_{td} = 1.15\ \mu_t/\mu_0 = 1.15\ \frac{3.00}{0.0239} = 144,$$

$$S_{ad} = 2.0\ \mu_t/\mu_0 = 251,$$

and the maximum plume rise

$$y_{max} = 1.37\ \xi_t\ D\ F^{1/4}\ T^{3/8} = 19.37\ m < 30\ m.$$

The plume rise is less than the depth of the outfall below the surface and, therefore, the plume will be submerged.

For 5 ports, in a similar way,

$$S_{td} = 82,$$
$$S_{ad} = 143,$$

and

$$y_{max} = 34\ m > 30\ m,$$

and the plume will not be submerged.

It can be seen that the increased number of ports greatly enhances the chance of generating a submerged sewage field.

Spread of Pollutants in a Current

The jets and buoyant plumes represent the initial stage of the process of wastewater mixing with the receiving flow. Only very rarely is the width of the diffuser the same as the width of the receiving stream, river, or estuary. In many cases, it is important to know how rapidly the pollutants will spread laterally across the flow and how soon a reasonably uniform distribution of the pollution discharge across a cross-section will be achieved.

The initial mixing of jets produces a broad field of wastewater. Depending on a particular situation, diffusers, both single and multiple port types, can be considered as either a point source or a line source. The individual plumes from a multiple source diffuser blend together and lose their identity. The length of a line source is then the length of the diffuser plus one single jet diameter to account for the spread of the sewage field. If the diffuser is located at an angle other than normal to the current, the effective length of the diffuser is then its projection perpendicular to the current.

The jets and plumes extend from the diffuser to a cross-section where no further rise of the plume trajectory can be observed. The second part of the mixing zone then follows this initial stage of the sewage field formation and extends to the downstream boundary of the mixing zone.

In general, the spread of a substance in the second part of the mixing zone is three-dimensional, and no accurate analytical solution can be obtained. For most cases in practice, two-dimensional simplified solutions are adequate.

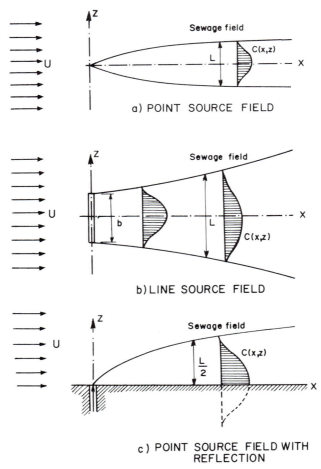

Fig. 15.14. Sewage field in a current.

The two-dimensional model analyzes the lateral spread of the substance along with the longitudinal movement by a wide uniform current. Both vertical mixing (after the plume reaches the surface or its final elevation) and the longitudinal mixing of the substance are assumed to be negligible. It should be noted that these assumptions are conservative and should provide some degree of safety in the design.

The basic differential equation for the spread of a substance is then

$$\frac{\partial}{\partial z}\left(-D_z \frac{\partial C}{\partial z}\right) + U \frac{\partial C}{\partial x} + KC = 0, \tag{15.15}$$

where D_z is the lateral dispersion coefficient, U is the flow (current) velocity, K is the decay coefficient, and C is the concentration of the substance.

For a point source in a uniform infinitesimally wide current as shown in Fig. 15.14(a) and a conservative substance, the solution can be expressed analytically by the normal distribution function as (Edinger and Polk, 1969)

$$C(x,z) = \frac{C_1}{2\sqrt{\xi}} e^{-z^2/4\xi}, \tag{15.16}$$

where C_1 is a constant of integration to be determined from the boundary conditions and ξ is a longtudinal coordinate given by

$$\xi = \frac{xD_z}{U}.$$

The constant of integration, C_1, can be evaluated by considering the source discharge to be located at a point $\xi = \xi_s$, $z = 0$ and having a concentration C_0. This gives

$$\frac{C(x,z)}{C_0} = \left(\frac{\xi_s}{\xi}\right)^{1/2} e^{-z^2/4\xi}. \tag{15.17}$$

The source location, ξ_s, can be obtained from the equation

$$\xi_s = \frac{1}{\pi}\left(\frac{Q_p}{UH}\right)^2, \tag{15.18}$$

where Q_p is the pollutant (heat) discharge flow rate from the outfall, U is the current (stream) velocity, and H is the average depth of the receiving water body.

For a nonconservative substance, Eq. (15.17) must include a decay term. The analytical solution then gives

$$\frac{C(x,z)}{C_0} = \left(\frac{\xi_s}{\xi}\right)^{1/2} e^{-z^2/4\xi}\, e^{-\alpha(\xi-\xi_s)}, \tag{15.19}$$

where $\alpha = K/D_z$.

For a sewage field discharging into an ocean or lake, the lateral dispersion coefficient, D_z, is affected by the scale of turbulence which can be related to the geometrical dimensions of the sewage field. Using the four-thirds law, the magnitude of the dispersion coefficient can be defined as

$$\frac{D_z(x)}{D_{z0}} = \left(\frac{L}{b}\right)^{4/3}, \tag{15.20}$$

where D_{z0} is the initial value of the lateral dispersion coefficient.

Brooks (1960) included the preceding relationship and developed a model for a line source located in a wide current as shown in Fig. 15.14(b). Brooks' solution for the centerline concentration ($z = 0$) yielded

$$C_{max}(x) = C_0\, e^{-K\,x/U}\, \mathrm{erf}\left(\frac{\frac{3}{2}}{\left(1 + \frac{2}{3}\,\beta\,\frac{x}{b}\right)^3 - 1}\right)^{1/2}, \tag{15.21}$$

where b is the width of the line source and erf denotes the standard error function, and

$$\beta = \frac{12\,D_{z0}}{Ub}.$$

The width of the sewage field is

$$\frac{L}{b} = \left(1 + \frac{2}{3}\,\beta\,\frac{x}{b}\right)^{3/2}. \tag{15.22}$$

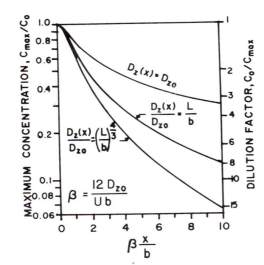

Fig. 15.15. Dilution along the centerline of a sewage field in an ocean current according to various diffusion laws (Brooks, 1960).

Brooks' equation is plotted in Fig. 15.15 for a conservative substance ($K = 0$). For a nonconservative substance, multiply $C_{max}(x)/C_0$ from Fig. 15.15 by the term $e^{-Kx/U}$.

It should be noted that the preceding equations assume no reflection at stream boundary. For a point source located at a shore-or bank line [Fig. 15.14(c)], the width of the sewage field would be equal to the half-width of that for an infinitesimally wide current. However, since the half-width now becomes the full width of the field, the concentrations within the sewage field must be doubled to account for the pollutant reflection.

Example. A sewage outfall is located 5 km upstream from a shoreline recreational area. The main current of the receiving water body is in the direction from the outfall to the area. The effluent discharge is $Q_p = 10$ m³/sec and the concentration of fecal coliforms in the effluent must comply with the effluent standards of a maximum of 400/100 ml. The average velocity of the current is 0.1 m/sec and the average depth of the receiving water body is 10 m. Estimate the sewage dilution and the concentration of coliform bacteria at the recreational site if (a) the outfall is located at the bank, and (b) the outfall is located 250 m from the bank in the current. Assume a point source and no effect of the sewage field on dispersion. The lateral dispersion coefficient was established by field measurements as being $D_z = 0.05$ m²/sec, and the coliforms die-off coefficient was found to be $K = 1.0$ day$^{-1} = 0.00001157$ sec^{-1}. Then

$$\xi = \frac{xD_z}{U} = \frac{5000 \times 0.05}{0.1} = 2500,$$

$$\xi_s = \frac{1}{\pi}\left(\frac{Q_p}{UH}\right)^2 = \left(\frac{10}{0.1 \times 10}\right)^2 = 31.83,$$

$$\frac{K}{D_z} = \frac{0.00001157}{0.05} = 0.000231.$$

Outfall located at shoreline ($z = 0$):

The dilution of sewage, S, can be computed from Eq. (15.17), assuming reflection of sewage on the shoreline as

$$\frac{1}{S} = 2\frac{C_{max}(x)}{C_0} = 2\left(\frac{\xi_s}{\xi}\right)^{1/2} = 2\left(\frac{31.83}{2500}\right)^{1/2} = 0.226$$

and

$$S = 4.43.$$

The concentration of fecal coliforms then becomes from Eq. (15.19)

$$C'(x) = \frac{1}{S}e^{-\alpha(\xi-\xi_s)}C_0' = 0.226 * e^{(-0.000231(2500-31.83))} * 400$$

$$= 51/100 \text{ ml}.$$

Outfall located 250 m from the shore (z = 250):

$$\frac{1}{S} = \frac{C(x,z)}{C_0} = \left(\frac{\xi_s}{\xi}\right)^{1/2} e^{-z^2/4\xi} = \left(\frac{31.83}{2500}\right)^{1/2} e^{-250^2/4*2500} = 0.000218.$$

The dilution of the sewage at the recreational site is then

$$S = 4591.$$

The fecal coliform concentration is

$$C'(x,z) = \frac{1}{S} e^{-\alpha(\xi-\xi_s)} C'_0 = 0.000218 * 0.565 * 400 = 0.04/100 \text{ ml.}$$

It can be seen that the location of the outfall further from the shore greatly reduces the danger of shoreline pollution.

REFERENCES

Baumgartner, D. J., Trent, D. S., and Byranm, K. V. (1971). Users guide and documentation for outfall plume model. Working Pap. No. 80, U.S. Environ. Prot. Agency, Pacific Northwest Lab., Corvallis, Oregon.

Brooks, N. H. (1960). ''Diffusion of Sewage Effluent in an Ocean Current. Waste Disposal in the Marine Environment.'' Pergamon, Oxford.

Brooks, N. H. (1973). Dispersion in hydrologic and coastal environments. USEPA-/600/3-73-010, Washington, D.C.

Ditmars, J. D. (1969). Computer programs for round buoyant jets into stratified ambient environments. W. N. Keck Lab. of Hydraul. and Water Resour., Tech. Memo 69-1, Calif. Inst. of Technol., Pasadena, California.

Edinger, J. E., and Polk, E. M. (1969). Initial mixing of thermal discharges into a uniform current. Rep. No. 1, Dept. of Environ. and Water Res. Eng., Vanderbilt Univ., Nashville, Tennessee.

Fan, L. N., and Brooks, N. H. (1969). Numerical solutions of turbulent buoyant jet problems. W. N. Keck Lab. Hydraul. and Water Resour. Rep. No. KN-R-18, Calif. Inst. of Technol., Pasadena, California.

Motz, L. H., and Benedict, B. A. (1970). Heated surface jet discharged into a flowing ambient stream. Rep. No. 4, Dept. of Environ. and Water Res. Eng., Vanderbilt Univ., Nashville, Tennessee.

Sotil, C. A. (1971). Computer program for slot buoyant jets into stratified environment. W. M. Keck Lab. of Hydraul. and Water Resour., Tech. Memo 71-2, Calif. Inst. of Technol., Pasadena, California.

U.S. Environmental Protection Agency (1976). Guidelines for state and areawide water quality management program development. Washington, D.C.

16 | *Pollutant Movement and Transformation in Soils*

INTRODUCTION

The groundwater contamination problem as exemplified by the "Love Canal" incident which occurred in the 1970s, caused intense interest in the pollutant movement in groundwater sources. An EPA–HEW panel report confirmed the health hazards to people living in the area surrounding the former chemical dumping site and justified the evacuation of some 239 families from the area (Anon., 1979). It has been estimated that approximately 51,000 chemical dumps exist in the United States with the potential for other "Love Canals." Thus, an understanding of the behavior of contaminants in groundwater systems is essential to the water quality engineer.

The chemical composition of many surface waters during low-flow periods is primarily the result of various processes taking place in soils and underlying mineral and bedrock layers through which water must percolate before it reaches an outlet to the surface waters. The most significant changes in the composition of infiltrated water occur in the upper soil layers, namely in the A and B horizons. Figure 16.1 shows a typical distribution of soil layers and their relationship to water quality changes of infiltrated water.

Soils can retain, modify, decompose, or adsorb pollutants. Every year

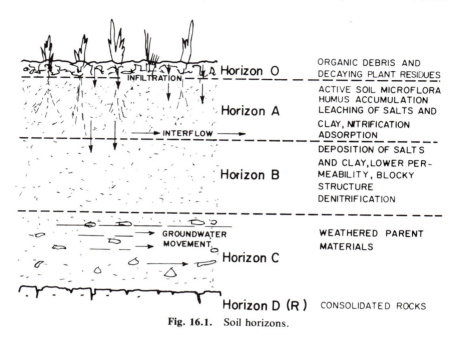

Fig. 16.1.   Soil horizons.

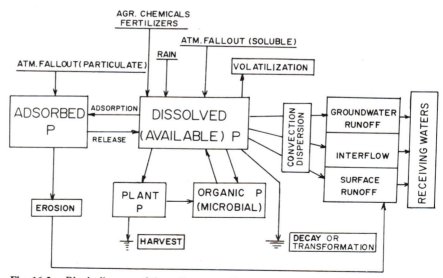

Fig. 16.2.   Block diagram of the soil-water-pollutant interaction (Novotny, *et al.*, 1978).

enormous amounts of organic materials, atmospheric pollution, and other liquid and solid wastes are deposited and incorporated into soils and safely decomposed into their basic constituents, carbon dioxide, nitrogen, phosphorus, and other residues of biological decomposition. Because of a high concentration of soil bacteria, the decomposition processes are quite intensive and effective, and represent one of the best natural recycling processes. The processes which participate in the soil decomposition and removal of pollutants include adsorption, filtration, ion exchange, biochemical action of soil microorganisms, particle charge interaction, pH effect, and precipitation. On account of the adsorption and retention processes of the soil, almost all of the phosphorus, heavy metals, many pesticides, and organic chemicals remain near the point of application and move only with eroded soil. The exception is in sandy or peat soils, which have little tendency to adsorb pollutants. The residual pollution which passes the soil layers plus that elutriated from the soils enters the groundwater aquifer. Figure 16.2 shows a block diagram of possible pathways and processes participating in pollutant movement and transformation through the upper soil layers.

POLLUTANT MOVEMENT AND TRANSFORMATION MODEL

The transformation of pollutants in soil is a complex system which includes the processes of convection due to the downward movement of soil water, diffusion due to the concentration gradient in the soil water, adsorption and desorption on soil particles, uptake by plants' roots., transformation from or into another form, sublimation and volatilization, uptake by soil microorganisms, and decay.

A model which describes the aforementioned system should include the following three components: hydraulic model including the equation of continuity and motion, dissolved component model, and adsorbed component model.

The inputs to the system include the pollutant contribution from rainfall, dust and dirt fallout, fertilizers, and other agricultural chemicals. Most of the inputs are related to land use. The output from the system is the pollutant distribution between the soil solution and the soil particles (adsorbed phase). The pollutants adsorbed on soil particles in the upper soil layer may be transported by erosion processes and the dissolved pollutants may be transported at the lower boundary of the system to groundwater. The lower boundary of soil adsorption may be related to the depth of the root zone or to the tillage depth of the crop land. In many cases the soil zone depth coincides with the depth of the A horizon.

SOIL WATER MOVEMENT

Water is the primary vector in pollutant movement and transport through soil and geological subsoil layers. The infiltrated water carries dissolved pollutants previously deposited on the surface or from atmospheric pollution. The infiltration rate also determines how much of the precipitation will result in surface runoff, thus affecting the amount of erosion and pollutant washout.

Hydrologic Characteristics of Soil Types

Soils have been classified into four hydrologic groups (Chow, 1964):

Group A. Soils of low total runoff potential have high infiltration rates, even when thoroughly wetted. These consist chiefly of deep, well-to-excessively drained sand or gravel, and, therefore, possess a high rate of water transmission.

Group B. Soils of low–moderate total runoff potential have moderate infiltration rates when thoroughly wetted, and range from moderately deep, moderately well to well-drained soils having moderately fine to moderately coarse texture. Consequently, these soils have a moderate rate of water transmission.

Group C. Soils of high–moderate total runoff potential have slow infiltration rates when thoroughly wetted, and consist primarily of soils with a layer that impedes the downward movement of water, or soils with moderately fine to fine texture. These soils have a resultant slow rate of water transmission.

Group D. Soils of high total runoff potential have very slow infiltration rates when thoroughly wetted, and mainly consist of clay soils with a high swelling potential, soils with a permanently high water table, soils with a clay pan or clay layer at or near the surface, and shallow soils over nearly impervious material. These have the expected very slow rate of water transmission.

The potential storage of soil moisture can be partitioned into two moisture classes: (1) gravitational water, i.e., that held between saturation and 0.33 bar tension, and (2) plant available water, or that held between 0.33 and 15 bar tension. The moisture content at 0.33 bars is assumed to repre-

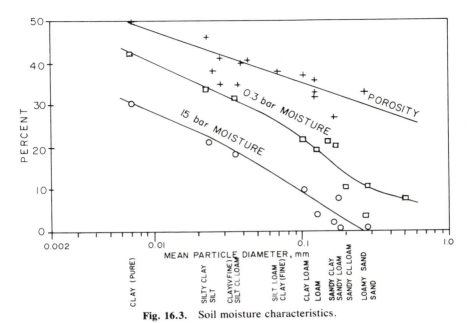

Fig. 16.3. Soil moisture characteristics.

sent the field moisture capacity or the lower limit of gravitational water, and 15 bar moisture represents the permanent wetting percentage in medium textured soils. The amount of water that can be drained by gravity from a soil column is derived by subtracting the 0.33 bar moisture from the actual soil moisture. The plant available moisture may be obtained by subtracting the 15 bar moisture from the actual soil moisture. Porosity, 0.3 bar moisture, and 15 bar moisture tend to be dependent on the soil texture characteristics. Figure 16.3 shows an approximate correlation relationship of these characteristics to the mean soil particle diameter, which can also be roughly related to soil texture.

Soil Permeability Rates

Soil permeability is that property of a soil that allows it to transmit water, Soil permeability rates may be determined by tests on representative samples of soil and expressed as the coefficient of permeability (or hydraulic conductivity). This term equals the apparent velocity of water flow under a hydraulic gradient of unity which exists when the pressure head on a soil specimen, divided by the depth of the specimen, is equal to 1. A slope of unity is typical for saturated vertical seepage into soils. Permeability classes according to the Soil Conservation Service (SCS) classi-

the inconsistencies that arise in the definition and identification of soil textural classes, and the difference in soil permeability because of other factors besides texture. To avoid such difficulties, Horn (1971) proposed a method for estimating permeability rates based on the mean particle size rather than a general textural description. Figure 16.4 is based on Horn's model and represents a relationship of permeability rates to the mean particle size with approximate textural classes and the corresponding mean particle sizes given in Table 16.2. Approximate curves that establish graphically the general limits of permeability rates for natural soils are indicated in Fig. 16.4 by the heavy dashed lines. For example, the curve on the far right in Fig. 16.4 represents the approximate upper limits of permeability for well-structured soils; at the top left is a curve representative of poorly structured, high-swelling soils. These curves describe the extremes of permeability.

Depending on the intended application of the curves, permeability in a vertical direction for an entire soil section may be rated on the basis of the least permeable layer, recognizing that for some soils, the presence of

TABLE 16.2 Soil texture, representative particle size contents, and mean diameters[a]

Textural Class (USDA)	Sand, as a Percentage[b]	Silt, as a Percentage	Clay, as a Percentage	Mean Diameter in Millimeters
Sand	95	3	2	0.285
Loamy sand	83	10	7	0.250
Sandy clay loam	58	15	27	0.176
Sandy loam	55	25	10	0.167
Sandy clay	52	6	42	0.157
Loam	40	40	20	0.124
Clay loam	33	33	34	0.103
Clay (fine)	(a)40	10	50	0.122
	(b)25	25	50	0.0785
	(c)10	40	50	0.035
Silt loam	(a)34	53	13	0.107
	(b)32	65	13	0.0726
	(c) 7	80	13	0.029
Silt clay loam	10	55	35	0.0362
Clay (very fine)	(a)22	1	77	0.067
	(b)10	13	77	0.0328
	(c) 1	22	77	0.007
Silt	5	90	5	0.0240
Silty clay	6	47	47	0.0236

[a] After Horn (1971).
[b] Note: b describes central point; a and c describe the range.

very thin continuous clay laminas or other abruptly contrasting layers may exert a profound influence on the permeability behavior of the entire soil body. Lateral permeabilities for soil sections may be rated according to the layer of greatest permeability. Alternatively, average permeability rates may be taken for the entire section.

UNSATURATED FLOW IN POROUS MEDIA

The process of water movement in soils occurs primarily when the soil is not saturated, i.e., water does not occupy all the available voids. Assuming an element of soil to be partially saturated with water and the instantaneous water content to be Θ, the equation of continutity becomes

$$\frac{\partial \Theta}{\partial t} = -\left(\frac{\partial q_1}{\partial x_1} + \frac{\partial q_2}{\partial x_2} + \frac{\partial q_3}{\partial x_3}\right), \tag{16.1}$$

where q_j is the flow in the direction, j. If Darcy's law is applied to an unsaturated soil medium, it becomes

$$q_j = K(\Theta)\frac{\partial h}{\partial x_j}, \tag{16.2}$$

where $K(\Theta)$ is the hydraulic conductivity of the soil and h is the water potential or head. Equation (16.1) then becomes

$$\frac{\partial \Theta}{\partial t} = \sum_{1}^{3}\frac{\partial}{\partial x_j}\left[K(\Theta)\frac{\partial h}{\partial x_j}\right], \tag{16.3}$$

which was developed by Richards (1931). In most soil cases, the total head is equal to the sum of the pressure and position heads, or

$$h = -\psi + z, \tag{16.4}$$

where ψ is the pressure head and z is the position head. Define

$$D(\Theta) = K(\Theta)\frac{\partial x}{\partial \Theta} \tag{16.5}$$

as the water diffusivity. Substituting Eqs. (16.4) and (16.5) into Eq. (16.3) and assuming only vertical movement, one obtains

$$\frac{\partial \Theta}{\partial t} = \frac{\partial}{\partial z}\left[D(\Theta)\frac{\partial \Theta}{\partial x}\right] - \frac{\partial K(\Theta)}{\partial z}. \tag{16.6}$$

This model was further developed by Philip (1969) and Parlange (1971).

Infiltration

Water enters the soil surface because of the combined influence of gravity and capillary forces. As the process proceeds, the capillary pore spaces become filled with water to greater depths as percolation continues. Normally, the gravitational water encounters increased resistance to flow due to a reduction in the extent or dimension of the flow channels, an increase in the length of the channels, or the presence of an impermeable barrier such as rock or clay. The result is a rapid reduction in infiltration rate in the first few hours of a storm, after which the rate remains essentially constant for the remainder of the period of rainfall excess.

The infiltration rate of a given soil may be governed by any of the following three separate processes (Gray, 1973): (1) entry of water into the surface layer of the soil, (2) the downward movement or percolation of water through the soil profile, or (3) flow through deep cracks in the profile.

Several empirical and semiempirical formulas are commonly used for the estimation of infiltration rates.

Horton (1940):

$$f = f_c + (f_0 - f_c)e^{-kt}, \tag{16.7}$$

where f is the infiltration rate in cm/hr (in./hr), f_c is the infiltration rate which represents a reasonably steady-state rate of water adsorption reached after water has been applied continuously for a long period of time, f_0 is the initial rate of infiltration, k is a constant, and t is time in hours from the beginning of the storm.

Philip (1969):

$$f = \tfrac{1}{2}St^{-0.5} + A(t), \tag{16.8}$$

where S is the sorptivity of the soil and A is taken as the hydraulic conductivity at the wetting front (approximately 40–60% of the saturated permeability). Use of Philip's equation requires knowledge of the moisture distribution throughout the depth of the unsaturated soil medium, and, generally, the solution can be accomplished only by numerical integration (Philip, 1957; Rogowski, 1971).

Recognizing the difficulties of relating infiltration rates to soil characteristics and soil moisture distribution, Holtan (1961) proposed a formula which would relate infiltration rate to the exhaustion of storage.

Holtan (1961):

$$f = a(S - F)^n + f_c = aF_p{}^n + f_c, \tag{16.9}$$

TABLE 16.3 Holtan equation vegetation factor[a]

Cover	Value of Vegetative Factor, K
Bluegrass	1.0
Crabgrass and alfalfa	0.70
Lasfedesa and timothy	0.45
Alfalfa	0.35
Weeds	0.30

[a] After Holtan (1961).

where S is the volume of storage above the control horizon, F is the cumulative infiltration, and a and n are coefficients with proposed average values of $a = 0.17$ cm/hr and $n = 1.387$. $F_P = S - F$ is a measure of the soil moisture remaining in the soil column at any time. While n is approximately constant for most soils, a can vary in range from 0.1 to 0.7 depending upon the type of vegetative cover.

The Holtan equation was proposed for periods when there is rainfall excess, i.e., during ponding. He found from experimental data that $[F_p]_0$ can be expressed as

$$[F_p]_0 = K\, S_0,$$

where the zero subscript indicates evaluation at time zero, K is a vegetative factor (Table 16.3), and S_0 is the available pore space between 0 and 54 cm depth.

SOIL ADSORPTION

The process of the fixation of pollutants by soil and dust particles can be accomplished by either precipitation or adsorption. Precipitation refers to a process in which pollutants form difficultly soluble compounds (e.g., phosphates at high pH values and some metals). Adsorption is a physicochemical process by which molecules or ions are immobilized by soil particles. Adsorbed pollutants may also undergo chemical interaction with ionic sites of soil particles, which is referred to as "chemical" adsorption, "activated" adsorption, or "chemosorption." In the case of chemical precipitation reactions, the amount of pollutants in the particulate fraction depends on their solubility in the soil environment. If adsorption is the dominant process, the removal of pollutants from the soil solution is governed by the concentration of pollutants in the solution, which, in turn, is in dynamic equilibrium with the soil adsorbed component. The preferred form of describing this equilibrium distribution between the dissolved and adsorbed components is to express the quantity S_e as a function of C_e at a fixed temperature, where S_e is the amount of a pollutant adsorbed on soil particles and C_e is the concentration of the pollutant in the soil solution in equilibrium.

Several mathematical descriptions of adsorption isotherms have evolved in the literature. The Langmuir and Freundlich ones are the most common and widely used.

The Langmuir Adsorption Isotherm

The Langmuir isotherm model was developed theoretically for single layer adsorption. However, it has been found to closely describe soil ad-

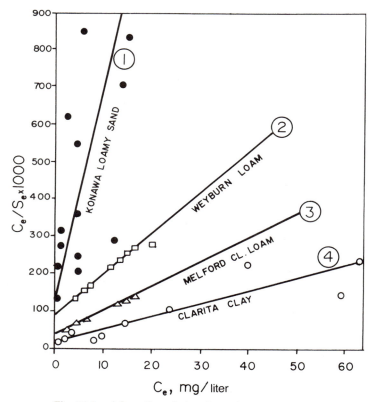

Fig. 16.5. Adsorption of phosphates by various soils.

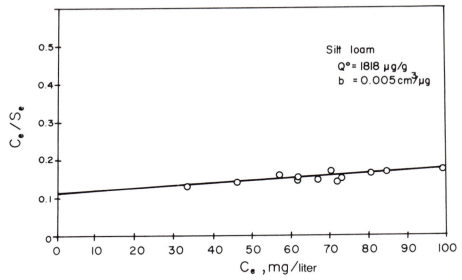

Fig. 16.6. Adsorption of ammonia by various soils (Misra *et al.*, 1974).

sorption phenomenon. It is based on the assumptions that maximum adsorption corresponds to a saturated monolayer of solute molecules on the adsorbent surface, that the energy of adsorption is constant, and that there is no transmigration of adsorbate in the surface phase. The Langmuir isotherm is (Weber, 1972)

$$S_e = \frac{Q^0 \, b \, C_e}{1 + C_e}, \qquad (16.10)$$

where Q^0 is the number of moles (or mass) of solute adsorbed per unit weight of adsorbent (soil) during the maximum saturation of soil and b is a constant related to the net energy of enthalpy of adsorption. The accepted units for Q^0 and b are $\mu g/g$ and liters/mg ($\mu g/ml$), respectively.

Figures 16.5 and 16.6 show linearized Langmuir isotherms for the adsorption of phosphates and ammonia by various soils.

The Freundlich Isotherm

The Freundlich isotherm is useful if the energy term, b, in the Langmuir isotherm varies as a function of surface coverage, S. The Freundlich equation has the general form

$$S_e = K \, C_e^{1/n}, \qquad (16.11)$$

where K and N are constants.

The Freundlich equation is basically empirical, but it is often useful as a means for data fitting.

Estimation of the Isotherm Parameters from Soil Characteristics

Two parameters describe the sorption process identified by the Langmuir isotherm, i.e., Q^0, which can be related to the mass of a pollutant adsorbed per unit weight of soil, and b, which is related to the energy of the net enthalpy of adsorption. In a heterogenous medium, such as soil, both parameters should be considered as statistical quantities with certain probabilities or ranges of occurrence. In some special cases, laboratory adsorption studies may be available which can provide the values of Q^0 and b. However, for most modeling and pollution transport studies these variables can be estimated only roughly from a few routinely measured soil parameters.

The sorption characteristics of soils are primarily related to the surface active components of the soil. Clay particles, because of their large specific area, active surface adsorption sites (mostly Al and Fe), and a negative surface charge, are quite active in retaining phosphates, and, to a lesser extent, pesticides, ammonia, and some organics. Organic matter in soils is also known to have a high affinity for adsorbing various components. The clay adsoprtion activity, which is mainly chemical in nature, depends on the pH of the soil.

The adsorption process is best described for phosphorus, but similar considerations can be applied to other pollutants. The soil sorptivity for phosphorus is controlled by several factors. Aluminum and iron oxides and hydroxides are largely responsible for phosphate retention in acid soil (Hsu, 1965; Vijayachandran and Harter, 1975; Tandon, 1970), calcium compounds may fix phosphate in calcareous soils (Hsu, 1965), and organic matter may also significantly contribute to phosphate adsorption (Tandon, 1970; Syers *et al.*, 1973).

Several authors have attempted to correlate phosphate sorptivity to various soil parameters. Representative data for 102 soils gathered from the literature (Tandon, 1970; Syers *et al.*, 1973; Gunary, 1970; Ballaux and Praske, 1975) were selected for statistical multiregressional analysis. In summarizing the literature findings, it was shown that the parameters controlling phosphate adsorption by soils are pH, aluminum, iron, clay, and organic matter content (Novotny *et al.*, 1978).

Soil adsorption characteristics for many commonly used organic chemicals have been reported extensively in a publication by Goring and Hamaker (1972), in which it was suggested that the single most important factor affecting the adsorption of organic chemicals by soils is the soil's organic carbon content. The extent of the adsorption of pesticides by soils varies over a wide range, with dicamba being an example of a weakly adsorbed material and DDT a strongly adsorbed material.

Two empirical isotherms were developed by Chesters (1967) using multiple regression analysis of lake sediment adsorption data. The parameters considered X_1-lindane concentration, X_2-sediment concentration, X_3-ratio of lindane:sediment, X_4-organic matter, and X_5-clay content as independent variables and lindane adsorption as the dependent variable. The best fit equations were

$$y = -0.063 + 0.093\ X_1 - 0.013\ X_2 + 0.034\ X_3 + 0.0034\ X_4 + 0.0058\ X_5 \qquad \text{Correlation coefficient, } r = 0.92$$

and

$$y = 0.203 + 0.095\ X_1 - 0.018\ X_2 + 0.042\ X_3,$$
$$\text{Correlation coefficient, } r = 0.94,$$

where the adsorption, Y, is expressed in $\mu g/mg$ of sediment.

The predicted and observed values of the adsorption of lindane are compared in Fig. 16.7.

The adsorption of Zn, Cd, Pb, and Hg by clay and clayey soils follows the same pattern. The distribution of lead between the dissolved and adsorbed phase is mostly determined by its solubility in the soil solution.

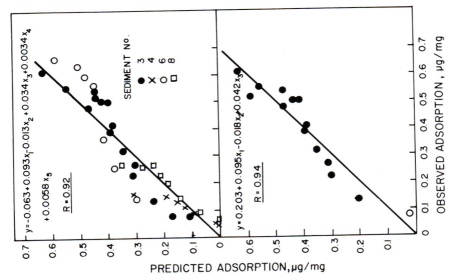

Fig. 16.7. Scatter diagram of observed lindane adsorption predicted from regression equations (Chesters, 1967).

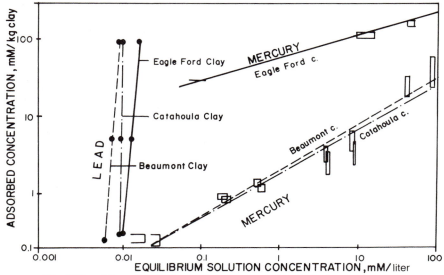

Fig. 16.8.   Adsorption isotherm for mercury and lead (Sanks *et al.*, 1976).

During soil adsorption experiments conducted on some clayey Texas soils (Sanks *et al.*, 1976), the equilibrium solution concentration of Pb never exceeded 0.03 *mM*/liter. In addition, a higher percentage of the Pb was precipitated. Other more mobile toxic metals exhibit typical adsorption characteristics on clay and organic particles of soils. Typical adsorption isotherms for mercury and lead are shown in Fig. 16.8.

Kinetics of Adsorption

The kinetics of the soil adsorption process can be expressed as

$$\text{Soluble pollutant} \underset{1-K}{\overset{K}{\rightleftharpoons}} \text{Adsorbed pollutant.}$$

where K is the adsorption coefficient. Very few data are available which allow the quantification of the adsorption kinetics. Most of the available information relates to phosphorus. From the limited amount of available data (Chesters, 1967; Ryden *et al.*, 1972; Enfield, 1974), it is evident that phosphate and pesticide sorption are not instantaneous processes. Adsorption studies with a duration of several days revealed that there is an initial adsorption stage lasting for minutes or hours with a relatively fast adsorption rate followed by a slow adsorption process lasting for days or weeks. For some metals and pesticides, the process is essentially completed within several hours.

A first-order adsorption model was assumed to represent a reasonable approximation of the process, i.e.,

$$\frac{dS}{dt} = K_\alpha (S_e - S), \qquad (16.12)$$

where K is the adsorption kinetics coefficient and S and S_e are, respectively, the amount of pollutant adsorbed, and the adorption equlibrium determined by an isotherm based on a solution concentration, C. Ryden *et al.* (1972) estimated the adsorption kinetics coefficient for phosphorus to be 0.12 h^{-1}. Enfield (1974) discussed two simplified kinetic models as described by the equations

$$\frac{dS}{dt} = \alpha (KC - S), \qquad (16.13)$$

where C is the equilibrium solution concentration, and

$$\frac{dS}{dt} = a \, C^b \, S^c. \qquad (16.14)$$

In the preceding equations, α, K, a, b, and c are statistical constants. It should be noted that Eq. (16.12) is almost identical to Eq. (16.13), assuming that the extent of adsorption is linearly proportional to the pollutant concentration in the soil solution. The experimental data by Enfield (1974) confirms the approximate magnitude of the adsorption coefficient as mentioned previously.

Decay, Sublimation, and Transformation

Although not important for some materials, the process of decay, sublimation, and tranformation must be included in a model if it is to describe the behavior of such pollutants as pesticides and ammonia. These processes are usually described by a first-order rate reaction:

$$\frac{dC}{dt} = -K_t \, C - \frac{K_s}{D_x} \, C, \qquad (16.15)$$

where K_t is the decay or transformation rate, K_s is the sublimation or stripping rate, and D_x is the depth of the upper soil zone.

Rates of decay for organic chemicals are largely related to the organic matter content of the soil (Goring and Hamaker, 1972). For the transformation of ammonia, the temperature and pH control the rate of decay

(Starr et al., 1974; Justice and Smith, 1962; Stojanovic and Broadbent, 1956).

Pollutants Movement with Adsorption

In addition to the equations of continutity and motion of soil water, we have the governing equations for movement of pollutants.

For the free phase:

$$\Theta \frac{\partial C}{\partial t} = D_L \frac{\partial^2 C}{\partial z^2} - \frac{q}{A} \frac{\partial C}{\partial z} - \rho \frac{\partial s}{\partial t} \pm \Sigma N. \tag{16.16}$$

For the sorbed phase:

$$\frac{\partial S}{\partial t} = K_\alpha (S_e - S), \quad \text{where } S_e = \frac{Q^0 b \, C}{1 + bC} \text{ (or } S_e = K \, C^{1/n}). \tag{16.17}$$

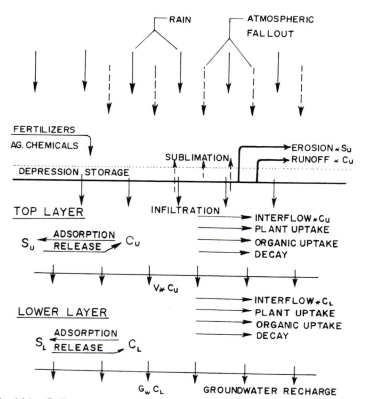

Fig. 16.9. Pollutant transport and transformation processes in soil columns.

Equations (16.16) and (16.17) constitute a general kinetic model of chemical movement with sorption described by the Langmuir (Freundlich) isotherm. For this model, C is the concentration of dissolved pollutant (μg/cm^3), S is the amount of pollutant sorbed on soil particles (μg/g), ρ is the bulk density of the soil (g/cm^3), D_L is apparent dispersion coefficient (cm^2/hr), q is downward water flow (cm^3/hr), A is the horizontal cross-sectional area (cm^2), ΣN is the sum of sinks and sources of the pollutant within the soil volume (μg/cm^3/hr), b is the partition coefficient (ml/μg), K and n are coefficients for the Freundlich isotherm, Q^0 is the adsorption maximum of the soil for the particular pollutant (μg/g), K_α is the adsorption or release rate coefficient for packed bed sorption (hr^{-1}), S_e is the equilibrium of the sorbed phase with the free phase (μg/g), t is the time (hr), z is the depth (cm), and Θ is the soil moisture (cm^3/cm^3).

The aforementioned model is nonlinear and can only be solved numerically.

The schematic representation of the solution of the model is shown in Fig. 16.9. To simplify this solution, the soil zone is divided into an upper (boundary) layer exposed to the atmosphere and one or several lower layers extending to the B horizon. For the numerical solution, the following relationships replace the analytical form of Eqs. (16.16) and (16.17) (Novotny *et al.*, 1978).

For the free phase of the upper layer model, U,

$$\theta \frac{C_U^{j+1} - C_U^j}{DT} \times \text{VOL}_U = (\text{RAIN} \times \text{CR} \times A) + (\text{ATMFL}_S + \text{FERTIL}/\text{DT})A$$

$$+ (\text{ORGREL} - \text{PLANTU})(\text{VOL}_U) - \rho \frac{\partial S_U}{\partial t}(\text{VOL}_U)$$

$$- \frac{C_U^{j+1} + C_U^j}{2}(V \times A + K_d \times \text{VOL}_U \times \theta + K_{\text{SUB}} \times A$$

$$+ \text{ANRAIN} \times A). \qquad (16.18)$$

For the sorbed phase of the upper layer model:

$$\rho \frac{S_U^{j+1} - S_u^j}{DT} \text{VOL}_U = K_\alpha \text{VOL}_U \left[S_e - \frac{S_U^{j+1} + S_U^j}{2} \right]$$

$$- \left[\text{SOILLS} \times \frac{S_U^{j+1} + S_U^j}{2} \right] + \text{ATMFL}_P. \qquad (16.19)$$

For the free phase of the lower zone model:

$$\theta \frac{C_L^{j+1} - C_L^j}{2DT} \text{VOL}_L = \left[V \frac{C_U^{j+1} + C_U^j}{2} - \text{GINFIL} \frac{C_L^{j+1} + C_L^j}{2} \right] A$$

$$+ (\text{ORGREL} - \text{PLANTU})(\text{VOL}_L) - \rho \frac{\partial S_L}{\partial t}(\text{VOL}_L). \qquad (16.20)$$

For the sorbed phase of the lower zone model:

$$\frac{\partial S_U}{\partial t} \text{VOL}_L = K_\alpha \left[S_e - \frac{S_L^{j+1} + S_L^j}{2} \right] \text{VOL}_L \qquad (16.21)$$

where, in addition to the variables described previously, VOL is the volume of the soil layer ($= A \times Dx$), A is the surface area, RAIN is rainfall intensity, CR is the concentration of pollutants in the rain, FERTIL is the pollutant contribution from fertilizers, ATMFL is atmospheric fallout (P= particulate, S= soluble), ORGREL is the release of pollutant from soil organic matter, PLANTU is the uptake of pollutant into crop tissues, GINFIL is the groundwater recharge, SOILLS is soil loss by erosion, j is the time superscript, DT is the time step, and Dx is the depth of soil layer.

Using the Langmuir isotherm, the adsorbed equilibrium concentration becomes

$$S_e = \frac{bQ^0 (C^j + C^{j+1})}{2 + b(C^j + C^{j+1})}, \qquad (16.22)$$

and the Freundlich model would yield

$$S_e = \frac{K}{2} (C^j + C^{j+1})^{1/n}. \qquad 16.23$$

NITROGEN ACCUMULATION AND LEACHING BY SOILS

Nitrogen is one of the four essential elements (carbon, oxygen, hydrogen, and nitrogen) which form the basic structure of organic proteins. Nitrogen is also the most abundant in the atmosphere since nitrogen gas accounts for about 80% of its gaseous content. In addition, nitrogen is an important nutrient for algal and plant aquatic growths, contributing significantly to accelerate eutrophication.

Nitrogen transformation in nature can be described by the nitrogen cycle schemes shown in Figs. 3.1 and 3.2. The following processes are part of the overall nitrogen cycle in soils:

Nitrogen fixation by which algae and soil microorganisms utilize atmospheric nitrogen and change it to an organic form.

Nitrogen assimilation by algae or heterotrophic microorganisms which utilize ammoniacal or nitrate nitrogen to form protein, thus converting it to an organic form.

Deamination, a process by which the organic protein nitrogen is decomposed into ammonia.

Hydrolysis of urea, $(CO(NH_2)_2)$, which is a waste product of life and is readily hydrolyzed to ammonia in the presence of the enzyme urease. Urease can be supplied by many aquatic and soil microorganisms and the process is relatively rapid.

Nitrification, a process by which ammonia is oxidized by *Nitrosomonas* to nitrite and further by *Nitrobacter* to the nitrate form. The reaction is strictly aerobic requiring approximately 4.33 parts of oxygen per one part of ammonia.

Denitrification, a reducing process occurring in an anaerobic environment where nitrate becomes the electron acceptor and is reduced to nitrogen gas and its oxides and/or ammonia and subsequently lost, mostly to the atmosphere.

Ammonia and nitrite immobilization in which ammonia and nitrite can be adsorbed on soil adsorption sites or by organic matter. In this form, both components are unavailable for plant growth. An adsorption equilibrium exists between the dissolved and adsorbed phases.

Ammonia stripping, in which calcareous soils at high pH values can cause ammonia ion to be converted to ammonia gas which can then be stripped to the atmosphere.

Unlike phosphorus and some organic chemicals which are adsorbed on soil and can be controlled by erosion control practices, nitrogen can be highly mobile and may be lost by leaching. The loss by leaching depends on the form in which nitrogen exists in the soil. Ammonia may be fixed by soil particles and/or soil organisms and may move to surface waters with the sediment. Nitrate nitrogen is dissolved and moves primarily with the groundwater. Therefore, leaching and runoff contamination by nitrogen cannot be evaluated without a detailed understanding of the nitrogen transformation processes taking place in the soil zone and knowledge of the water balance and water movement in the soil.

Nitrogen Transformation Model

There has been a tendency to associate the suspected increased nitrogen content of surface waters with increased fertilizer use. However, the behavior of nitrogen in soils is highly complex, and such a simple correlation is not possible. In addition to the nitrogen added to soils in the form of fertilizers, one must consider the organic matter in the soil and the rate at which it is mineralized, the atmospheric nitrogen which is fixed, the nitrogen involved in crop utilization and leaching, the nitrogen assimilation by microorganisms, nitrogen addition from septic tank effluents, and nitrogen returned to the atmosphere.

and 1.1, respectively, if the rate changes are expressed in mg/liter per g of soil per day at 20°C.

If oxygen is absent, or if the oxygen supply rate is not sufficient to satisfy the oxygen demand, nitrates may be reduced to nitrogen gas (or its oxides) and ammonia, i.e.,

$$NO_3^- \longrightarrow N_2$$
$$\downarrow$$
$$NH_3$$

The process of denitrification usually occurs in subsoils with lower permeability or when soils are saturated with water for extended periods of time (during a higher supply of infiltrated water). Denitrification accounts for about 30% of the average nitrogen losses in Illinois (Stanford *et al.*, 1970). Another study (Pratt *et al.*, 1972) reported nitrogen losses by denitrification up to 52% for soils that had clayey horizons overlying sands.

Carter and Allison (1960) concluded that very little, if any, nitrogen loss occurred under aerobic conditions except when higher dosages of dextrose were added in the presence of nitrates. They also stated that denitrification is of minor importance in soils that are kept strictly aerobic. But it was also emphasized by the authors that in heavy soils and in medium textured soils following heavy rains there were often periods of several hours or even days when normal field soils were not well aerated.

The denitrification reaction in soils has been described as a first-order reaction (Starr *et al.*, 1974; *Misra et al.*, 1974):

$$-\frac{d(NO_3^-)}{dt} = K(NO_3^-). \tag{16.25}$$

The values of the denitrification coefficient were measured (Misra *et al.*, 1974) for sandy and silt loams. No apparent effect of texture could be detected from the measured data. Figure 16.12 summarizes the results of the denitrification experiments.

The denitrification rate is a function of easily decomposable carbon and is expected to decrease with the decrease in the available energy source.

Nitrogen Fixation

Legumes (soybean, peas, beans) fix nitrogen from the atmosphere by symbiotic microorganisms living on their roots. It has been assumed that the rate of nitrogen fixation can be related to the root growth (Duffy *et al.*, 1975) as

$$N_f = K_f r_g, \tag{16.26}$$

where N_f is the rate of N fixation (mg N/day^{-1}/cm^{-2}), K_f is a constant ($\approx$ 0.011 mg N/cm^3), and r_g is the rate of root growth (cm/day).

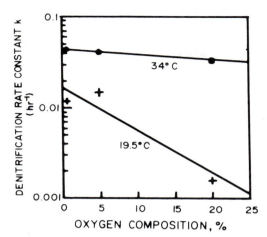

Fig. 16.12. Nitrogen reduction rate as a function of oxygen concentration at 19.5 and 34.5°C (Misra *et al.*, 1974).

Ammonification of Soil Organic Nitrogen

Three ways of producing ammonia from organically bound nitrogen can be distinguished (Painter, 1970): from extracellular organic nitrogen compounds (e.g., urea), chemically or biochemically; from living bacterial cells during endogenous respiration, when cells are becoming smaller; and from dead and lysed cells.

Very little data can be found in the literature for the breakdown of various nitrogen-containing components to ammonia. Complex organic compounds are first deaminated to ammonia by appropriate exoenzymes or enzymes on the cells, and a permease transports the ammonia into the cell where it is used for synthesis. McLaren (1969) reported that in a number of soils, urea decomposes at a maximum rate from 0.02 to 0.15 moles/g of soil/hr, which is equivalent to 34 mg/g of soil/day. This would lead to this equation for the decomposition of urea:

$$\frac{d(U)}{dt} = \frac{34\,(U)}{500 + (U)}. \tag{16.27}$$

The rate of hydrolysis of urea varies between soils and is temperature dependent. It proceeds at a much faster rate than nitrification, and, because of the large saturation constant, K_s, it usually follows a first-order reaction.

The process of urea hydrolysis is important if manurial fertilizer is applied to a farm field or if a land is irrigated by a wastewater-containing urea.

Soil organic nitrogen other than urea breaks down at a much slower rate than urea. Most of the organic nitrogen is not directly available for plant growth inasmuch as it first must be converted to ammonia or nitrate. Stanford *et al.* (1973) concluded that in the mineralization process of soil organic nitrogen, the rate limiting factor is the process of ammonification. A first-order reaction approximates the process as

$$\frac{dN}{dt} = KN, \tag{16.28}$$

where N is the concentration of the organic nitrogen remaining in the soil, and the K is the rate coefficient.

The coefficient, K, was found statistically uniform for all soils investigated by the authors, having a value of (0.29 ± 0.007) weeks^{-1} at 25°C. As with all biochemical reaction rate coefficients, it is also temperature dependent. The Arrhenius plot yielded this equation for the overall organic nitrogen mineralization reaction:

$$\log K = 6.16 - \frac{2299}{T}, \tag{16.29}$$

where the temperature, T, is in °K. Figure 16.13 shows the Arrhenius plot

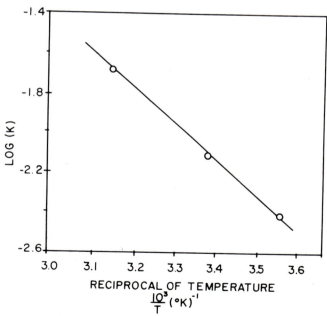

Fig. 16.13.   Arrhenius plot for determination of the activation energy for urea hydrolysis (Larsen *et al.*, 1972).

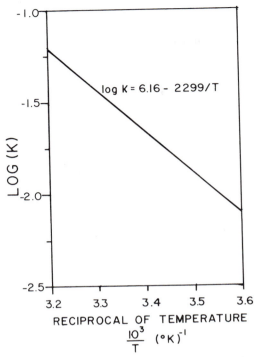

Fig. 16.14. Arrhenius plot for the overall mineralization of organic nitrogen (Stanford *et al.*, 1973).

for urea hydrolysis. The Arrhenius plot for the overall mineralization of organic nitrogen is shown in Fig. 16.14.

Nitrogen Immobilization

Part of the soil nitrogen is dissolved and can readily move with soil water and groundwater. The dissolved fractions involve nitrates and soluble ammonia. However, a significant fraction of the nitrogen components, especially ammonia and almost all of the organic nitrogen, can become immobilized. Nitrogen immobilization in soils is a result of physical–chemical attractive forces, chemical precipitation, and biochemical reactions. The known nitrogen fixation processes include (Stanford *et al.*, 1970) (a) ammonium fixation by clay minerals, (b) ammonium fixation by lignin-derived substances in soil organic matter, (c) reactions of amino acids derived from plant materials and microbial synthesis with quinones and subsequent polymerization, (d) biological immobilization which involves ammonia uptake by heterotrophic bacteria participating in the decay processes or organic matter in the soil.

The immobilized nitrogen is not readily available for mineralization, and it appears that the deamination, i.e., liberation of ammonia from the fixed nitrogen is much slower than the rate of nitrification.

The tieup (immobilization) or release (deamination) of soil nitrogen depends on the chemical composition of the material undergoing decomposition, primarily on its carbon/nitrogen ratio. Plant residues having large percentages of readily available carbon will stimulate the growth of microbial cells when incorporated into the soil and decomposed under aerobic conditions.

Soil itself has an affinity for the adsorption of positively charged ammonia ions. It has been assumed that the nitrate nitrogen, since it is an anion and forms soluble salts, does not react with the soil.

Thomas (1972), on the other hand, states that for most soils, with the exception of very sandy soils low in organic matter, there are both positive and negative charged sites. An example is a soil in which iron oxides thickly coat the negatively charged clay particles. The positively charged sites will react with nitrate and retard its progress through the soil with the water movement. In most soils that possess bright red subsoil colors, the movement of nitrates is substantially retarded. Soils with a very dense negative charge on their surfaces do not only fail to attract nitrate, they actively repel it.

Shaffer *et al.* (1969) statistically analyzed the rate of ammonia immobilization by a multiple regression technique. The proposed equation was quoted as

$$R = \frac{d(\mathrm{NH_4}^+)}{dt} = 0.892 - 0.00216\, T - 0.027\, (\mathrm{org.\ N})$$
$$+\ 0.392\, \log_{10}(\mathrm{NH_4}^+), \quad (16.30)$$

where T is temperature.

Preul and Schroepfer (1968) investigated nigrogen adsorption by soils in great detail. The adsorption isotherms (Freundlich) for three soils are shown in Fig. 16.15.

Since the immobilization of ammonia is for all practical purposes an adsorption process, an adsorption isotherm kinetic model can be used for its description:

$$\frac{d(\mathrm{NH_4})_s}{dt} = K_{sa}\, (S_e - [\mathrm{NH_4}^+]_s), \quad (16.31)$$

where the sorbed equilibrium concentration, S_e, may be described by an adsorption isotherm, such as the Freundlich or Langmuir model. The data

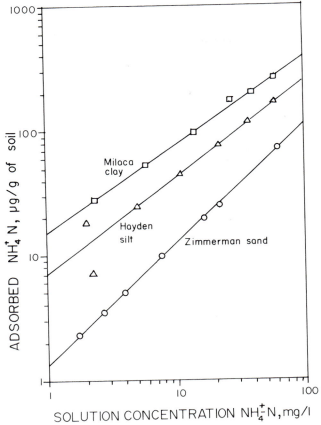

Fig. 16.15. Ammonia adsorption isotherms for three soils (Preul and Schroepfer, 1968).

by Preul and Schroepfer (1968) also enable the estimation of the kinetic coefficient of adsorption, K_{sa}, which for the three investigated soils in Fig. 14.15 was 0.42–0.69 hr^{-1}.

Formulation of the Nitrogen Transformation Model

A model of the mobile nitrogen movement and transformation in soils was formulated by Cho (1971) as

$$\frac{\partial C_i}{\partial t} = D \frac{\partial^2 C_i}{\partial z^2} - v \frac{\partial C_i}{\partial z} \pm \sum_{j=1}^{n} \phi_{ij}, \qquad (16.32)$$

where C_i is the concentration of the ith component of the mobile nitrogen

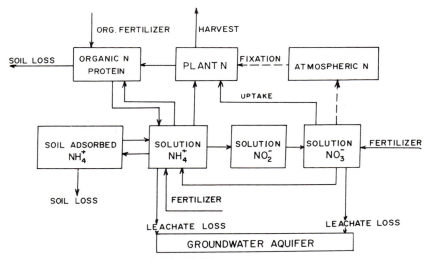

Fig. 16.16.   Block diagram of nitrogen transformation in soils.

(mg/liter); D is the apparent diffusion coefficient (cm^2/hr); z is the distance within the soil column (cm); v is the apparent water velocity (cm/hr); t is the time; and ϕ_{ij} is the rate production (+ sign) or consumption (− sign) of the ith nitrogen component from the jth source or sink.

Referring to Fig. 16.16 and Eq. (16.32), the sources and sinks in the mobile nitrogen balance model become those shown here.

Solution for NH_4^+ (i = 1):

$$\Sigma\phi_{1j} = -(K_2 - K_{-5})\,C_1 + K_5 C_0 + K_4 C_3 - \rho\,\frac{dS_N}{dt}$$
$$- (C_1 \text{ plant uptake}) + (\text{Fertilizer } C_1). \qquad (16.33)$$

Solution for NO_2^- (i = 2):

$$\Sigma\phi_{2j} = K_2\,C_1 - K_3\,C_2. \qquad (16.34)$$

Solution for NO_3^-:

$$\Sigma\phi_{3j} = K_3\,C_2 - (K_4 + K_6)C_3 - (C_3 \text{ plant uptake})$$
$$+ (\text{Fertilizer } C_3). \quad (16.35)$$

In addition to the mobile nitrogen compound model, the fixed nitrogen must be balanced, too.

Adsorbed NH_4^+ (S_N):

$$\frac{dS_N}{dt} = K_1\,(S_e - S_N), \qquad (16.36)$$

where the equilibrium concentration of sorbed ammonia is

$$S_e = \frac{bQ°C_1}{1 + bC_1} \quad \text{(Langmuir model)}$$

or

$$S_e = \hat{K}C_1^{1/n} \quad \text{(Freundlich model)}.$$

Crop N (C_p):

$$\frac{dC_p}{dt} = (C_1 \text{ plant uptake}) + (C_3 \text{ plant uptake})$$
$$+ (N_2 \text{ fixation}) - K_7 C_p - (\text{harvest}). \quad (16.37)$$

Organic N (C_0):

$$\frac{dC_0}{dt} = K_{-5}C_1 - K_5 C_0 + K_7 C_p + (\text{Org. fertilizer } C_0). \quad (16.38)$$

The fixed nitrogen in the upper soil layer can be subjected to the erosion process and move with soil and surface runoff during a large storm event or whenever the surface runoff takes place.

REFERENCES

Anon. (1979). *Environment Reporter* **10**(14), 905.

Ballaux, V. C., and Praske, D. E. (1975). Relationship between sorption and resorption of phosphorus by soils. *Soil Sci. Soc. Am., Proc.,* **39,** 275–280.

Carter, J. N., and Allison, F. E. (1960). Investigation of denitrification in well-aerated soils. *Soil Sci. Soc. Am., Proc.* **90,** 173–177.

Chesters, G. (1967). Terminal report on phase I of insecticide adsorption by lake sediments as a factor controlling insecticide accumulation in lakes. Office of Water Resour. Res. Grant No. B-008-Wis., Univ. of Wisconsin, Madison.

Cho, C. M. (1971). Convective transport of ammonium with nitrification in soil. *Can. J. Soil Sci.* **51,** 339–350.

Chow, V. T., ed. (1964). Runoff. *In* "Handbook on Applied Hydrology." McGraw–Hill, New York.

Duffy, J., Chung, C., Boast, C., and Franklin, M. (1975). A simulation model of bio-physicochemical transformation of nitrogen in tile-drained corn belt soil. *J. Environ. Qual.* **4,** 477–485.

Enfield, C. G. (1974). Rate of phosphorus sorption by five Oklahoma soils. *Soil Sci. Soc. Am., Proc.* **38,** 404–407.

Goring, C. A. I., and Hamaker, J. W. (1972). "Organic Chemicals in the Soil Environment." Dekker, New York.

Gray, D. M. (1973). Handbook on the principles of hydrology. Water Info. Center, Port Washington, New York.

Gunary, D. (1970). A new adsorption isotherm for phosphate in soil, *J. Soil Sci.* **21,** 72–77.

Holtan, H. N. (1961). A concept for infiltration estimates in watershed engineering. USDA Agric. Res. Serv., ARS 41-51, Washington, D.C.

Holtan. H. N., England, C. B., and Whelan, D. E. (1967). Hydrologic characteristics of soil types. *J. Irrig. Drain. Div., Am. Soc. Civ. Eng.* **93,** 33–41.

654

Horn, M. E. (1971). Estimating soil permeability rates. *J. Irrig. Drain. Div., Am. Soc. Civ. Eng.* **97**, 263–274.

Horton, R. E. (1940). An approach towards a physical interpretation of infiltration capacity. *Soil Sci. Soc. Am., Proc.* **5**, 399–417.

Hsu, P. H. (1965). Fixation of phosphate by aluminum and iron in acidic soils. *Soil Sci. Soc. Am., Proc.* **99**, 398–402.

Justice, J. K., and Smith, R. L. (1962). Nitrification of ammonium sulfate in calcareous soil as influenced by combination of moisture, temperature, and levels of added nitrogen. *Soil Sci. Soc. Am., Proc.* **26**, 246–250.

Larsen, V., Axley, J. H., and Miller, G. L. (1972). Agriculatural waste water accomodation and utilization of various forages. Tech. Rep. No. 19, Water Resour. Res. Cent., Univ. of Maryland, College Park.

McLaren, A. D. (1969). Steady state studies of nitrification in soil: Theoretical considerations. *Soil Sci. Soc. Am., Proc.* **33**, 273–276.

McLaren, A. D. (1971). Kinetics of Nitrification in Soil: Growth of the Nitrifiers. *Soil Sci. Soc. Am., Proc.* **35**, 91–95.

Misra, C., Nielsen, D. R., and Biggar, J. W. (1974). Nitrogen transformation in soil during leaching: III. nitrate reduction in soil. *Soil Sci. Soc. Am., Proc.* **38**, 300–304.

Misra, C., Nielsen, D. R., and Biggar, J. W. (1974). Nitrogen transformation in soils during leaching: I–III. *Soil Sci. Soc. Am., Proc.* **38**, 289–304.

Novotny, V., Tran, H., Simsiman, G. V., and Chesters, G. (1978). Mathematical modeling of land runoff contaminated by phosphorus. *J. Water Pollut. Control Fed.* **50**(1), 101–112.

Painter, H. A. (1970). A review of literature on inorganic nitrogen metabolism in microorganisms. *Ware Res.* **4**, 393–450.

Parlange, J. Y. (1971). Theory of water movement in soils. *Soil Sci. Soc. Am., Proc.* **3**, 170–174.

Philip, J. R. (1957). The theory of infiltration: I. The infiltration equation and its solution. *Soil Sci. Soc. Am., Proc.* **83**, 345–357.

Philip, J. R. (1969). Theory of infiltration. *In* ''Advances in Hydroscience (W. T. Chow, ed.). Academic Press, New York.

Pratt, P. F., Jones, W. W., and Hunsaker, V. E. (1972). Nitrate in deep soil profiles in relation to fertilizer rates and leaching volume. *J. Environ. Qual.* **1**, 97–102.

Preul, H. C., and Schroepfer, G. J. (1968). Travel of nitrogen in soils. *J. Water Pollut. Control Fed.* **40**, 30–48.

Richards, L. S. (1931). Capillary conduction through porous media. *Physics* **1**, 318–333.

Rogowski, A. S. (1971). Watershed physics: Model of the soil moisture characteristics. *Water Resour. Res.* **7**, 1575–1582.

Ryden, J. C., Syers, J. K., and Harris, R. F. (1972). Potential of an eroding urban soil for the phosphorus enrichment of streams. *J. Environ. Qual.* **1**, 430–438.

Sanks, R. L., LaPlante, J. M., and Gloyna, E. F. (1976). Survey—suitability of clay beds for storage. Tech. Rep. EHE-76-04-CRWR-128, Cent. for Res. in Water Resour., Univ. of Texas, Austin.

Shaffer, M. J., Dutt, G. R., and Moore, W. J. (1969). Predicting changes in nitrogen compounds in soil–water systems, collected papers: Nitrates in agricultural waste waters. Fed. Water Qual. Admin., Water Pollut. Control, Dept. of the Interior, Res. Series 13030 ELY 12/69, 15–28.

Stanford, G., England, C. B., and Taylor, A. W. (1970). Fertilizer use and water quality. USDA, Soil and Water Conser. Res. Div., ARS-41-168, Beltsville, Maryland.

Stanford, G., Free, M. H., and Schwaninger, D. H. (1973). Temperature coefficient of soil nitrogen mineralization. *Soil Sci. Soc. Am., Proc.* **115**, 321–328.

Starr, J. L., Broadbent, F. E., and Nielsen, D. R. (1974). Nitrogen transformation during continuous leaching. *Soil Sci. Soc. Am., Proc.* **38**, 283–289.

Stojanovic, B. J., and Broadbent, F. E. (1956). Immobilization and mineralization rates of nitrogen during decomposition of plant residues. *Soil Sci. Soc. Am., Proc.* **20**, 213–218.

Syers, J. K., Browman, M. G., Smillie, G. W., and Corey, R. B. (1973). Phosphate sorption by soils evaluated by the Langmuir adsorption equation. *Soil Sci. Soc. Am., Proc.* **37**, 358–368.

Tandon, H. L. S. (1970). Fluoride extractable aluminum in soils: 2. As an index of phosphate retention by soils. *Soil Sci. Soc. Am., Proc.* **109**, 13–18.

Thomas, G. W. (1972). The relation between soil characteristics, water movement, and nitrate contamination of ground water. Tech. Rep. No. 52. Univ. of Kentucky, Water Resour. Instit., Lexington.

Vijayachandran, P. K., and Harter, R. D. (1975). Evaluation of phosphorus adsorption by a cross section of soil types. *Soil Sci. Soc. Am., Proc.* **119**, 119–126.

Weber, W. J. (1972). "Physicochemical Processes for Water Quality Control." Wiley (Interscience), New York.

Wilde, H. E., Sawyer, C. N., and McMahon, T. C. (1971). Factors affecting nitrification kinetics. *J. Water Pollut. Control Fed.* **43**, 1845.

Zanoni, A. (1969). Secondary effluent deoxygenation at different temperatures. *J. Water Pollut. Control Fed.* **41**, 640.

Index